Plant Life and Environment Dynamics

This book series aims to publish volumes focusing on plant responses to changing environmental conditions. The series is cenrted on the theme of plant-environmnet interaction, ecosystem changes, plant physiology, various physio-biochemical and molecular mechanism of adpatation wokring within the plants. The book series also provides knowledge and understanding of plant's stess tolerance strategies such as ionomic transcriptomic, proteomic, phytohaormoes, cellular redox regulation, energy mentabolism, nanomateiral in plant protection and genomics of plants under environmnetal stress metabolomic approaches which are currently being used in order to protect plant life against adverse effects of changing environment. This series will cover book ideas focusing on all aspects of plant's responses to the change in the respective environment. The books published in this series are a useful source of information for academicians and researchers working in the field of agriculture, botany, agronomy, biotechnology, plant science, crop physiology, plant stress physiology, plant and environment interactions related courses.

Series plans to cover dynamics of plant-environment interaction via following broad themes:

(a) Abiotic stress and plant development

(b) Photosynthesis and energy metabolism

(c) Bio-stimulants and plant life

(d) Phytohormones in stress tolerance

(e) Cellular redox regulation (ROS, RNS and RSS), oxidative stress and antioxidant defense system

(f) Transcriptomics, Proteomics and Functional Genomics

(g) Ionomics: Uptake and Assimilation

(h) Nanomaterials and plant life

Ajay Kumar • Pardeep Singh • Suruchi Singh •
Bhupinder Singh
Editors

Wild Food Plants for Zero Hunger and Resilient Agriculture

Springer

Editors
Ajay Kumar
Department of Plant Science
Central University of Kerala
Kasaragod, Kerala, India

Pardeep Singh
Department of Environmental Studies
University of Delhi, PGDAV College
New Delhi, India

Suruchi Singh
Sunbeam College for Women
Mahatma Gandhi Kashi Vidyapith
Varanasi, Uttar Pradesh, India

Bhupinder Singh
Indian Agricultural Research Institute
Centre for Environment Science and Climate
Resilient Agriculture (CESCRA)
New Delhi, India

ISSN 2730-6755 ISSN 2730-6763 (electronic)
Plant Life and Environment Dynamics
ISBN 978-981-19-6501-2 ISBN 978-981-19-6502-9 (eBook)
https://doi.org/10.1007/978-981-19-6502-9

This Springer imprint is published by the registered company Springer Nature Singapore Pte Ltd.
The registered company address is: 152 Beach Road, #21-01/04 Gateway East, Singapore 189721, Singapore

*I dedicate this book to my beloved brother **Anil Kumar** who took his last breath on 26th January 2023 at a very young age of 36 while we were editing this book. I have no words to express the pain of his absence in our family. May his soul attain Sadgati*

Ajay Kumar

Preface

Wild food plants (WFPs) are an essential part of many ethnic communities across the globe, and they are relevant even today. Indigenous communities collect and consume diverse WFPs, and we know little about their nutritional and health benefits. It is only recently that we have realized their huge benefits. Wild food plants are an important part of diverse cultures and ethnic diets across the globe. This book *Wild Food Plants for Zero Hunger and Resilient Agriculture* has been my dream project. Hailing from the Western Himalayas, our diets in the village have been highly diversified, and I remember my mother used to collect and cook wild plants such as *Fagopyrum esculentum*, *Diplazium esculentum*, *Amaranthus viridis*, *Phytolacca acinosa*, *Capsella bursa-pastoris*, and *Lepidium sativum*. Then, there were wild fruits such as *Berberis asiatica*, *Rubus ellipticus*, *Fragaria vesca*, and *Rubus niveus* which she used to collect and keep in the steel jars, and when we visited villages, she offered us. Sometimes, we also used to climb up mountains on a family trip for the collection of the wild fruits of *Viburnum grandiflorum*. While going to school, we used to hunt *Elaeagnus umbellata* fruits. As I am placed in another corner of India, I rarely get a chance to visit my village, and the consumption of wild foods has also become rare. With these fond memories and the recent decline of wild foods from our diets, this has been my dream to work and write on these wild food plants. This book provides an interesting update on the wild food plants, their diversity, traditional systems of food based on wild food plants and their roles in food security and achieving zero hunger. Further, considering the vulnerabilities associated with mainstream foods, the wild foods have shorter supply chains and are more resilient. Therefore, they are most relevant in the current context.

This book has 14 chapters, all authored by eminent scholars in the field. The book opens with an introductory Chap. 1 contributed by Anju and coauthors. Anju et al. introduces the concept of wild food plants, the diversity of wild food plants and their contributions to achieving food security, particularly zero hunger and resilience in agriculture. This chapter also discusses wild food plants' potential medicinal, socio-cultural, and economic roles. Chapter 2 is contributed by Avik and Rajasri who are eminent scholars of ethnobotany. This chapter poetically explains the relevance of wild food plants and traces the history of the WFPs to prehistoric times. This chapter finally outlines the implications of WFPs on food security policies in improving dietary diversity and creating a sustainable food system. Chapter 3 is another exciting topic contributed by Harsha and coauthors and discusses the history of WFPs and the impacts of globalization on the loss of food diversity. This chapter further provides a conceptual framework linking traditional conservation practices and the promotion of sustainable use of WFPs and policy interventions needed to promote and conserve the use of WFPs. Chapter 4 is authored by Sajana and her colleagues. In this article, the authors provide extensive details about the traditional food systems that are based on WFPs. Comprehensive survey of the global traditional food systems based on wild food plants provides an important insight into the huge diversity of alternative food systems and their roles in nutritional security.

In Chap. 5, Nilay et al. provide an interesting account on the usage of wild edible plants among upland indigenous communities of Northeast India. This chapter explores the beautiful states of Northeast India for their rich wild food plants. This chapter provides examples of wild food plants from each state of Northeast India. Chapter 6 contributed by Stanzin and coworkers discusses an important high-altitude wild fruit plant, sea buckthorn. It is a high-altitude plant and is highly nutritious. Many ethnic communities consume this plant in various forms and have health benefits. Stanzin et al. discusses all these medicinal and nutritional benefits of sea buckthorn. Nutritional potential of wild edible rose hips for food security is discussed in Chap. 7 by Kanwal and Sumeet. Yet another interesting article on the high-altitude wild roses, it covers various aspects of wild roses in food security.

Chapter 8 provides a brief account of eight wild food plants that are consumed in the Indian Himalayas. With detailed and unique aspects about their preparation, this chapter is contributed by Malini. Her personal experiences in the Himalayas have motivated her to write this chapter. This chapter also discusses the phytochemical constitutes, nutritional value, and medicinal importance of the eight selected wild food plants. Chapter 9 contributed is by Adya, Anju and Ajay. In this chapter, they discussed the reemergence of the pseudocereals in recent times. They provide unique aspects of pseudocereals and their nutritional and medicinal importance. Chapter 10, contributed by Abha et al., provides a detailed account of nutraceutical potential of tropical wild edible plants of India with selected examples.

Since some WFPs are also relatives of the cultivated crops, they can be used for the improvement of domesticated crops. In Chap. 11, Anju and Ajay discusses the utilization of the CWRs in crop improvement programs using modern tools such as gene editing. In Chap. 12, Anupama et al. discuss conservation of wild food plants and crop wild relatives, planning, strategies, priorities, and legal frameworks. This chapter outlines the need for the conservation of the WFPs and the strategies, as to how to conserve them. Databases have become increasingly important in any research. Therefore, Sushil and coworkers in Chap. 13 provide a short but worth reading account of the databases relevant to WFPs and their importance. In the last chapter (Chap. 14), Avinash et al. comprehensively discuss the traditional agricultural knowledge of farmers in India. They give examples of traditional knowledge of various plants that are used in crop production and management, crop protection, farm machinery and tools, soil and water management, medicinal and aromatic plants for disease diagnosis, animal husbandry, stored grain pests' management, weed management, and value-added food products. The farmers preserve and transfer the information to the rural community.

This book is written in simple language and is helpful for the students and common citizens who have an interest in wild food plants. This is also an important book for the researchers who wish to work on various aspects of wild food plants such as their economic, nutraceutical, and dietary potential.

On behalf of myself and all the co-editors, I firstly thank all the contributors for their valuable contributions. I also acknowledge the support of my Ph.D. (Anju, Malini, and Ashifa) and M.Sc. (Saritha, Nidhi, Zehra, Anagha, and Chithra and Rituparna) students. Last but not least, I thank my wife Komal for her support and encouragement. On behalf of myself and co-editors, I also acknowledge the support and help of the whole staff of Springer Nature particularly Aakanksha Tyagi, Priyanga Kabali, Selevakumar Rajendran, and Ashok Kumar at all the stages of the publication of this book.

Kasaragod, Kerala — Ajay Kumar
New Delhi, Delhi — Pardeep Singh
Varanasi, Uttar Pradesh — Suruchi Singh
New Delhi, Delhi — Bhupinder Singh

Contents

Editors and Contributors

About the Editors

Ajay Kumar is working as Assistant Professor of Plant Science at Central University of Kerala, India. He obtained his PhD from School of Life Sciences, Jawaharlal Nehru University, New Delhi, under the guidance of Dr. Nirala Ramchiary who is a renowned crop scientist. Dr. Ajay has more than seven years of teaching and research experience and has published several papers, review articles and book chapters in reputed journals and books, respectively. His areas of research include phytochemical genomics, wild food plants, traditional food systems, food security and abiotic stresses. He has been a member of Academic Council, Central University of Kerala in the past, and currently he is the member of University Court. He has successfully guided more than 30 post-graduate students for their dissertations.

Pardeep Singh is presently working as an Assistant Professor (Department of Environmental Studies, PGDAV College University of Delhi, New Delhi, India). He has obtained his master's degree from the Department of Environmental Science, Banaras Hindu University, Varanasi, India and his doctorate degree from the Indian Institute of Technology (Banaras Hindu University), Varanasi. He has published more than 75 research articles in international journals in the field of Environmental Management and edited more than 45 books.

Suruchi Singh is presently working as Faculty in the Department of Botany, Sunbeam College for Women, India. She obtained her PhD from Banaras Hindu University, India, in Botany with Environmental Science as major field. Dr. Singh received her postdoctoral experiences in the Department of Botany, Banaras Hindu University under Council of Scientific and Industrial Research and Department of Science and Technology. She has published articles in various international journals and books series (>50 peer-reviewed articles). In her research, she has identified processes and activities where environmental efficiencies of selected crop plants can be increased against UV-B and tropospheric ozone.

Bhupinder Singh joined the Indian Council of Agricultural Research (ICAR) as a Scientist (Plant Physiology) in 1991 and was posted at the Nuclear Research Laboratory, Indian Agricultural Research Institute (IARI), New Delhi in 1992. Presently, he is working as a Principal Scientist and Head at the Centre for Environment Science and Climate Resilient Agriculture (CESCRA), ICAR-IARI. He is also discharging the additional responsibility of the Radiological Safety Officer (RSO), at the ICAR-IARI, New Delhi since 2006. He has more than 25 years of research and teaching experience and has focused his research pursuits in the key area of nutriophysiology and radiophysiology. His scientific interests have helped in gaining insight into the role of phytosiderophore biosynthesis in improving Fe-Zn efficiency under limited nutrient availability and in phytoremediation of heavy metals. He has more than 90 research publications in reputed international and national journals. He has guided more than 13 PhD and MSc students and has several awards and recognitions to his credit such as ICAR-Lal Bahadur Shastri Young Scientist Award, DST BOYSCAST Fellowship, INSA-DAAD International Scientific Exchange Fellowship, RD Asana Gold Medal and R.H. Dastur (AAAS) awards. His research pursuits have been supported by several funding agencies such as DST, ICAR, BARC, Tata Steel, Ltd., Ministry of Steel and DBT. He is also a fellow of the Indian Society for Plant Physiology (ISPP) and was its elected Vice President for the years 2014–2015. He was a member of the scientific panel on biological hazard (2009–2015); member, Scientific Panel on the genetically modified organisms and foods (2015–2018) and member of both Panels at FSSAI, Govt. of India (2019–2022).

Contributors

K. Abha Manohar Department of Agronomy and Agroforestry, Centurion University of Technology and Management, R. Sitapur, Odisha, India

Vaishnav Suresh Babu Department of Plant Science, School of Biological Sciences, Central University of Kerala, Kasaragod, Kerala, India

Yani Baja Department of Agriculture, Kamle District, Arunachal Pradesh, India

Malini Bhattacharyya Department of Plant Science, School of Biological Sciences, Central University of Kerala, Kasaragod, Kerala, India

Sumit Chakravarty Department of Forestry, Uttar Banga Krishi Viswavidyalaya, Pundibari, West Bengal, India

Sunandani Chandel ASPEE College of Horticulture and Forestry, Navsari Agricultural University, Navsari, Gujarat, India

Kiran Vyshnav Eliyan Department of Plant Science, School of Biological Sciences, Central University of Kerala, Kasaragod, Kerala, India

Sumeet Gairola Plant Sciences and Agrotechnology Division, CSIR-Indian Institute of Integrative Medicine, Jammu, Jammu and Kashmir, India

Chandan Kumar Gupta Division of Plant Physiology and Biochemistry, Indian Institute of Sugarcane Research, Lucknow, India

R. P. Harisha Centre for Environment and Development, Bengaluru, India

Ajay Kumar Department of Plant Science, Central University of Kerala, Kasaragod, Kerala, India

Nilay Kumar College of Horticulture and Forestry, Central Agricultural University, Pasighat, Arunachal Pradesh, India

Sushil Kumar Department of Botany, Government Degree College, Kishtwar, Jammu and Kashmir, India

Stanzin Ladol Department of Zoology, Central University of Jammu, Jammu and Kashmir, India

Chowlani Manpoong Faculty of Agricultural Sciences, Arunachal University of Studies, Namsai, India

Kalkame Cheran Momin College of Horticulture and Forestry, Central Agricultural University, Pasighat, Arunachal Pradesh, India

Chau Chiktiya Mounglang Faculty of Agricultural Sciences, Arunachal University of Studies, Namsai, India

Faisal Mushtaq Department of Botany, Government Degree College, Kishtwar, Jammu and Kashmir, India

Dileep Kumar Pandey College of Horticulture and Forestry, Central Agricultural University, Pasighat, Arunachal Pradesh, India

Himanshu Pandey Division of Plant Physiology and Biochemistry, Indian Institute of Sugarcane Research, Lucknow, India

Adya Ram Department of Plant Science, School of Biological Sciences, Central University of Kerala, Kasaragod, Kerala, India

G. Ravikanth SMS Foundation Centre for Biodiversity and Conservation, Ashoka Trust for Research in Ecology and the Environment (ATREE), Royal Enclave, Bengaluru, India

Avik Ray Center for Studies in Ethnobiology, Biodiversity, and Sustainability (CEiBa), Malda, West Bengal, India

Rajasri Ray Center for Studies in Ethnobiology, Biodiversity, and Sustainability (CEiBa), Malda, West Bengal, India

Anupama Sajeev Department of Plant Science, School of Biological Sciences, Central University of Kerala, Kasaragod, Kerala, India

Golla Nagaraju Gari Saritha Department of Plant Science, School of Biological Sciences, Central University of Kerala, Kasaragod, Kerala, India

Avinash Sharma Faculty of Agricultural Sciences, Arunachal University of Studies, Namsai, India

Deepak Sharma School of Life Sciences, Jawaharlal Nehru University, New Delhi, India

Nidhi Sharma Department of Plant Science, School of Biological Sciences, Central University of Kerala, Kasaragod, Kerala, India

Sarang Viswanathan Sherly Department of Plant Science, School of Biological Sciences, Central University of Kerala, Kasaragod, Kerala, India

Gopal Shukla Department of Forestry, Uttar Banga Krishi Viswavidyalaya, Pundibari, West Bengal, India

R. Siddappa Setty Centre for Environment and Development, Bengaluru, India

Kanwaljeet Singh Plant Sciences and Agrotechnology Division, CSIR-Indian Institute of Integrative Medicine, Jammu, Jammu and Kashmir, India

Mayanglambam Sanjit Singh Faculty of Agricultural Sciences, Arunachal University of Studies, Namsai, India

Praveen Kumar Singh Center for Protected Cultivation Technology, ICAR - Indian Agricultural Research Institute, Pusa, New Delhi, India

Swarn Singh Department of Botany, Padma Shri Padma Sachdev Government PG College for Women, Gandhi Nagar, Jammu, Jammu and Kashmir, India

Sajana Sreedharan Department of Plant Science, School of Biological Sciences, Central University of Kerala, Kasaragod, Kerala, India

Abhirami Surendran Department of Plant Science, School of Biological Sciences, Central University of Kerala, Kasaragod, Kerala, India

Anju Thattantavide Department of Plant Science, School of Biological Sciences, Central University of Kerala, Kasaragod, Kerala, India

Vangapandu Thriveni Department of Agronomy and Agroforestry, Centurion University of Technology and Management, R. Sitapur, Odisha, India

Indhukumari Uthirchakkavu Department of Plant Science, School of Biological Sciences, Central University of Kerala, Kasaragod, Kerala, India

Chapter 1
An Introduction to Wild Food Plants for Zero Hunger and Resilient Agriculture

Anju Thattantavide, Sajana Sreedharan, Nidhi Sharma, Indhukumari Uthirchakkavu, Abhirami Surendran, and Ajay Kumar

Abstract The noncultivated or nondomesticated food plants that are collected from the wild natural habitats for consumption are called wild food plants (WFPs). They enrich the dietary diversity and significantly contribute to the micro and macronutrients of the body. WFPs are popular as a potential source of income for the local communities, and they ensure food supply during famine periods. The capacity to survive in harsh environmental conditions contributes to the stress adaptation potential of the these plants. Therefore, WFPs can be explored as a climate change adaptation strategy or resilient plants for agriculture. In the scenario of climate change, nutritional and yield reduction of the staple crops, wide diversity of WFPs with regional adaptations can be utilized for crop improvement programs and can significantly improve food security. The popularity of WFPs among ethnic communities is a link between traditional knowledge and modern scientific systems. But the traditional knowledge related to WFPs is being lost. Loss of traditional knowledge related to the WFPs has also threatened their existence. Therefore, it is essential to document and conserve the WFPs and promote their consumption globally. The popularization of WFPs can help to eradicate the hidden hunger and malnutrition among the population since they are a cheap source of locally available nutrient-rich food. The genetic diversity of the WFPs should be protected, and several crop improvement programs can be applied to improve the traits of the plant. The effective exploration of the diversity of WFPs can directly contribute to achieving zero hunger by 2030.

Anju Thattantavide and Sajana Sreedharan contributed equally to this work.

A. Thattantavide · S. Sreedharan · N. Sharma · I. Uthirchakkavu · A. Surendran
Department of Plant Science, School of Biological Sciences, Central University of Kerala, Kasaragod, Kerala, India

A. Kumar (✉)
Department of Plant Science, Central University of Kerala, Kasaragod, Kerala, India

A. Kumar et al. (eds.), *Wild Food Plants for Zero Hunger and Resilient Agriculture*, Plant Life and Environment Dynamics,
https://doi.org/10.1007/978-981-19-6502-9_1

Keywords Wild food plants · Food security · Climate resilience · Food security · Nutritional deficiency

1.1 Introduction

Attaining zero hunger by 2030 is the second goal in the 2030 Agenda for Sustainable Development (Morton et al. 2017). Elimination of hunger and attaining a state of food security are interlinked (FAO 2018a, b). According to FAO's definition, food security is not merely the access to food in sufficient quantities but also fulfilling nutritious and dietary requirements needed for maintaining a healthy and active life (FAO 2002). Availability and access to food and its effective utilization are crucial components of food security, which is why eradicating hunger and malnutrition is challenging as well as important in attaining food security (Mc Carthy et al. 2018; OECD 2013). According to FAO's statistics, in a time period of 5 years between 2014 and 2019, more than 60 million suffered from hunger. As per the data from 2019 based on a study among children aged below five, about 144 million were found to be stunted, 38.3 million were overweight, and 340 million or more faced micronutrient deficiencies. Furthermore, in 2020, COVID-19 pandemic also negatively affected the global food system slowing down the efforts toward sustainability (FAO et al. 2020). Alternative sources of foods should be explored to solve these challenges. Wild food plants (WFPs) are underutilized resources that have the potential to become a crucial part in attaining global food security (Borelli et al. 2020). Their utilization can be dated back to hunter-gatherer societies that used wild edible plants as a source of food (Ong and Kim 2017). Various communities have depended on WFPs, which became part of their diet until the advancement of agriculture (Akinola et al. 2020). Some WFPs are diverse and widely distributed and have similar or higher nutritional components compared to cultivated crops (Motti 2022). To fulfill the needs of the increasing population, the current food production capacity needs to scale up. At the same time, there are environmental and health concerns (Godfray and Garnett 2014). Thus, WFPs are a healthier and more nutritious alternatives to the conventional diet, providing an avenue for food security with the help of sustainable agriculture (Shava et al. 2009; Duguma 2020). WFPs are distributed in harsh environmental conditions and adapted to changing climatic conditions (Jordanovska et al. 2020). Current agricultural activities give priority to productivity over resilience (Musvoto et al. 2015) and environmental issues (Kerr 2012; Pingali 2012). An alternative food system of crops such as WFPs that are capable of withstanding climate change and ensuring nutritional diversity need to be developed (Shaheen et al. 2017). It is the need of the era to conserve knowledge regarding wild food plants and integrate it into current food practices to overcome present food security issues and for a sustainable food system.

1.2 Different Categories of WFPs

The edible plants that are not domesticated or cultivated are considered WFPs, which are capable of eradicating hunger, diversifying the food basket, reducing malnutrition in rural communities, generation of income for villagers, and ensuring food security for the consumers (Bhatia et al. 2018). WFPs include cereals, legumes, leafy vegetables, plants bearing edible seeds and fruit, as well as root crops (Asfaw and Tadesse 2001).

1.2.1 Grain legumes

Legumes are a collective name for members of the family Leguminosae (Fabaceae), whereas food legumes or grain legumes are prominent food plants of the family Fabaceae (Maphosa and Jideani 2017) and are considered to be important group of food crops after cereals (Graham and Vance 2003). Grain legumes support food security, income, and nutritional improvement and also help to maintain soil fertility in the tropics (Abate 2012). Grain legumes with still unexplored nutritional, ecological, and economic potentials and hold future promising potential are generally called minor food grains (Chibarabada et al. 2017). *Abrus precatorius* L. is a wild legume with amino acid richness that is widely distributed in low land and extremely dry conditions (Bhat and Karim 2009). *Sphenostylis stenocarpa* Hochst. ex A. Rich. is another wild legume that is popularly consumed in western, eastern, and central Africa (Aletor and Aladetimi 1989). The capacity to grow in sandy, humid, leached soils makes them highly adaptable plants to harsh conditions. *A. precatorius* L. (crab eye), *Acacia leucophloea* (Roxb.) Willd (white-bark acacia), *Bauhinia purpurea* L. (butterfly tree), *Canavalia* spp. (sword bean), *Clitoria fairchildiana* Howard. (butterfly pea tree), *Entada phaseoloides* (L.) Merr. St. (thomas bean), *Galactia longifolia* (Jacq.) Benth. ex (hoehne milk peas), *Lupinus* spp. (lupins), *Mucuna monosperma* DC ex (velvet bean), *Neonotonia wightii* (Wright and Arn.) Lackey (perennial soybean), *Pachyrhizus erosus* (L.) Urb (Mexican yam bean), *Parkia filicoidea* L. (locust beans), *Prosopis glandulosa* Torr. (honey mesquite), *Saponaria vaccaria* L. (cow cockle), *Teramnus labialis* (L.) Spreng. (pulivaalkodi), and *Vigna trilobata* (L.) Verdc. (African gram) are highly valuable nutritionally rich and stress-tolerant wild legumes (Bhat and Karim 2009). The nutritional value of wild *Cordyla, Mucuna*, and *Canavalia* legumes is highly significant and can be utilized for programs to eradicate protein-energy malnutrition. The antinutrients present in some of the wild legumes (eg. *Sesbania* and *Pachyrhizus erosus*) is drawback associated with these plants; therefore, proper processing methods should be applied to detoxify the legumes before consumption.

1.2.2 Root and Tuber Crops

Food plants with consumable subterranean starchy roots, rhizomes, tubers, corms, or stems are designated as root and tuber food crops (Ugwu 2009; Liu et al. 2014). They are rich sources of carbohydrates (Saranraj et al. 2019), one of the very important food groups in the human diet (Chandrasekara and Josheph Kumar 2016). Globally, in the list of the most important groups of food crops, root and tuber crops come after cereals and grain legumes (Ray and Sivakumar 2009). *Manihot esculenta* Crantz (cassava), *Ipomoea batatas* (L.) Lam. (sweet potato), *Solanum tuberosum* L. (potato/Irish potato), *Dioscorea* spp., *Colocasia esculenta* (L.) Schott (taro) are some of the widely cultivated root or tuber crops in tropical regions (Ray and Sivakumar 2009). There are many other wild plants that are less popular, for example, *Maranta arundinacea* L. (arrowroot), *Solenostemon rotundifolius* (Poir.) J. K. Morton (country potato) (Saranraj et al. 2019), *Xanthosoma sagittifolium* (L.) Schott (tannia/yautia), *Alocasia macrorrhizos* (L.) G. Don (giant taro) and *Cyrtosperma merkusii* (Hassk.) Schott (swamp taro) (Taylor et al. 2018), *Amorphophallus paeoniifolius* (Dennst.) Nicolson (elephant foot yam) and *Canna edulis* Ker Gawl. (cannas) were also used in many countries as a part of their food system (Chandrasekara and Josheph Kumar 2016). These tuber crops are a good source of folic acid and vitamin C (FAO 2019). Wild tubers are capable of withstanding unfavorable climatic and edaphic conditions in marginal areas and have the potential to strengthen the food security of nations (Muthoni and Nyamongo 2010; Enyiukwu et al. 2014).

1.2.3 Leafy Vegetables

Leafy vegetables were a very important element of the diet for different communities (Muthoni and Nyamongo 2010; Konsam et al. 2016). Certain leafy vegetables were a significant item in the daily diet of only some communities, but some others were in widespread usage (Maundu 1997). There exists a wide diversity of leafy vegetables used by different communities across the world (Mathieu and Meissa 2008; Orech et al. 2007; Dansi et al. 2008; Turner et al. 2011; Maseko et al. 2017). *Amaranthus* spp., *Basella rubra* L., *Bauhinia purpurea* L., *Boerhaavia diffusa* L., *Cassia tora* L., *Oxalis corniculata* L., *Marsilea vestita* Hook. and Grev., (Chauhan et al. 2014), *Basella alba* L., *Ipomoea aquatica* Forssk., *Chenopodium album* L., *Hibiscus sabdariffa* L., *Moringa oleifera* Lam. (Ravishankar et al. 2015), *Euphorbia hirta* L., *Centella asiatica* (L.) Urb., *Indigofera tinctoria* L., (Gupta et al. 2017), *Marsilea quadrifolia* L., *Mollugo oppositifolia* L., *Portulaca oleracea* L., *Marsilea minuta* L., *Allium cepa* L., *Mentha spicata* L., *Leucas aspera* L., *Alternanthera sessilis* (L.) R. Br. ex Dc., *Polygonum plebeium* R. Br. (Parida and Mahalik 2020), *Achyranthes aspera* L., *Cassia occidentalis* L., *Melochia corchorifolia* L., *Scoparia dulcis* L., (Kumari and Solanki 2019), *Vernonia squarrosa* Less., *Wrightia arborea*

(Dennst.) Mabb., *Ophioglossum reticulatum* L., *Eclipta prostrata* (L.) L. (Mallick et al. 2020), *Allmania nodiflora* (L.) R. Br. ex Wight, *Cocculus hirsutus* (L.) W. Theob., *Cardiospermum halicacabum* L., *Digera muricata* (L.) Mart., *Clerodendrum serratum* (L.) Moon., and *Rivea ornata* Choisy (Reddy et al. 2007) are very few examples of wild leafy vegetables used by various tribal groups in India. In the case of a majority of the wild leafy vegetables, beyond their regional diversity, the nutritional profile data is ill-developed (Afolayan and Jimoh 2009). But the superiority of wild leaves over commercial cabbage and spinach was scientifically proved in different studies (Turan et al. 2003; Gupta et al. 2005). In several areas of the globe, they are considered a weed, and people show reluctance to include them in their diet. At the same time, Modi et al. (2006) reported that, in rural areas, the leaves are found abundantly, and villagers used to consume them regularly; hence the consumption of these leaves is directly associated with the poverty level of the people. Several studies reported the pharmacological and antioxidant activities of wild leafy vegetables, which indicate their health benefits (Bvenura and Sivakumar 2017).

1.2.4 Fruits

Wild plant species bearing edible fruits are a sustainable food source in developing countries (Suwardi et al. 2020), especially during seasons of food shortage (Duguma 2020). Wild fruits had significant roles in tribal households; the people of Ethiopia dominantly consumed fruits of wild plants more than other parts (Duguma 2020). Indigenous groups consumed fruits either raw or made into some dishes; the plants bearing edible fruits include trees (e.g., *Aegle marmelos* (L.) Correa, *Ficus benghalensis* L., *Solanum torvum* Sw., *Limonia acidissima* Groff and *Syzygium cumini* (L.) Skeels), shrubs (e.g., *Flacourtia indica* (Burm. f.) Merr. and *Ziziphus oenoplia* (L.) Mill.) and climbers (*Bauhinia vahlii* Wight and Arn., *Uvaria hamiltonii* Hook. f. and Thomson, *Ziziphus rugosa* Lam., and *Olax psittacorum* (Lam.) Vahl) and people depended on them also for other products such as timber, fuel wood, and medicine (Mahapatra and Panda 2012). Abundant diversity of wild plants bearing edible fruits are counted across the world (Cambie and Ferguson 2003; Musinguzi et al. 2011; Lulekal et al. 2011a, b; Dansi et al. 2012; Cheikhyoussef and Embashu 2013; van der Hoeven et al. 2013; Ju et al. 2013; Mahklouf 2019; Suwardi et al. 2020) and thousands of species remain unexploited (Chakravarty et al. 2016). These species of *Ficus* are very important in Asia, Africa, and Europe (Sardeshpande and Shackleton 2019).

1.2.5 Oilseed Crops

Other than as a source of food, there are various wild plants that are used for several other economic uses. The utilization of seeds for the extraction of edible oil is one among them. Seeds of certain plants are a source of seed oil; generally, they are called oilseed crops (Waseem et al. 2017). Seed oil from underutilized wild plants was used by rural communities; for example, Tibetans in Yunnan, China, used the seed oil from wild plants such as *Prinsepia utilis* Royle, and *Noccaea yunnanensis* (Franch.) Al-Shehbaz, *Toxicodendron succedaneum* (L.) Kuntze and *Toxicodendron vernicifluum* (Stokes) F. A. Barkley, *Cornus ulotricha* C. K. Schneid. and Wangerin, *Cornus schindleri* Wangerin, and *Cornus macrophylla* Wall. (Ju et *al.* 2013). In some Asian countries, seed oil from *Guizotia abyssinica* (L. f.) Cass. (in Nepal and South Asia) (Sthapit and Padulosi 2011), *Perilla frutescens* (L.) Britton (in Central Himalaya) (Maikhuri et al. 1999) and *Brassica juncea* (L.) Czern. (Malkanthi 2017) traditionally used for cooking food.

1.2.6 Other Crops

Not only roots, rhizomes, grains, leaves, seeds and fruits but also other parts of certain plants such as flowers, latex, pith, arils, bark, and pith are edible, and hence they were included in the diet of consumers of various regions of the world (Ray et al. 2020). Flowers of *Cassia fistula* L., *Moringa oleifera* Lam. are consumed by tribes of Central India as emergency food (Jain and Tiwari 2012). Flowers of *Agave americana* L., *Artemisia scoparia* Waldst. and Kitam., *Bauhinia variegata* L., and *Indigofera cassioides* DC., are also edible and utilized as vegetables (Bhatia et al. 2018). The villagers of Mizoram consumed flowers of 59 wild species, some of them are consumed raw (e.g., *Aeschynanthus parviflorus* (D. Don) Spreng., *Bauhinia variegata* L., *Bauhinia purpurea* L., *Persicaria chinensis* (L.) H. Gross, *Rhododendron arboreum* Sm., *Smilax perfoliata* Lour., *Vaccinium sprengelii* (G. Don) Sleumer, and *Begonia longifolia* Blume) or eaten boiled, fried, cooked with rice, meat, fish, or egg or made into salads or chutney (Khomdram et al. 2019). The indigenous communities of Udhampur used the inner bark of *Flacourtia indica* (Burm. f.) Merr. to prepare tea (Bhatia et al. 2018). In West Sikkim, the bark of the tree *Cinnamomum tamala* (Buch.-Ham.) T. Nees and Eberm. was used as a condiment (Mahendra et al. 2017). Kadar tribes of Kerala used the bark of *Cinnamomum malabatrum* (Burm. f.) J. Presl as a spice (Chaithanya et al. 2015).

1.3 Diversity of Wild Food Plants

The majority of the WFPs serve as a major food crop for indigenous communities, and it serves as a minor source of food for other populations across the globe (Menendez-Baceta et al. 2012). Therefore, the frequent WFP consumption is usually mostly dominated among indigenous or ethnic communities. Several ethnobotanical studies documented the WFP diversity in the food basket of different communities (Huai and Pei 2000; Alemayehu et al. 2015; Kidane and Kejela 2021). The WFP diversity, richness, and differential pattern between the two Mapuche communities of Argentina were reported by Ladio and Lozada (2003). 64 WFP from 47 genera was recorded by Agarwal and Chandra (2019) from Mandal Chopta forest, Uttarakhand, in which fruit species dominated the list. Singh et al. (2016) documented 111 WFPs from Bandipora district Kashmir Himalaya, and leafy vegetables were predominant in the list followed by fruits. *Angelica archangelica, Amaranthus caudatus, Berberis lycium, Asparagus racemosus, Fragaria vesca, Oxyria digyna, Hippophae rhamnoides,* and *Juglans regia* were highly used by the population in the study. Similarly, WFP diversity in different regions of the globe was reported in various studies, such as Nepal (Uprety et al. 2012), Manipur state, India (Pfoze et al. 2011), Indonesia (Sujarwo et al. 2016), Ethiopia (Berihun and Molla 2017), Turkey (Yeşil et al. 2019), Uganda (Ojelel et al. 2019) and Zimbabwe (Maroyi 2011). Photographs of some WFPs consumed by the tribal communities of Northern Kerala are shown in Figure 1.1. Some WFPs utilized in different part of the globe are listed in Table 1.1.

Fig. 1.1 Photographs of some WFPs used by the tribals of the North Kerala region of India. (**a**) *Solanum aturense* Dunal, (**b**) *Talinum triangulare* (Jacq.) Willd, (**c**) *Amaranthus polygamus* L., (**d**) *A. dubius* Mart. ex Thell. (**e**) *Flacourtia inermis* Roxb, (**f**) *Colacasia* spp., (**g**) *Averrhoa carambola* L., (**h**) *Amaranthus palmeri* S. Watson, (**i**) *Canna indica* Mill., (**j**) *Momordica dioica* Roxb. ex Wild, (**k**) *Canavalia ensiformis* (L.) DC, (**l**) *Euphorbia hirta* L., (**m**) *Amplelocissus arachnoidea* (Hausskn.) Planch, (**n**) *Dioscorea oppositifolia* L., (**o**) *Laportea interrupta* (L.), (**p**) *Colacasia* spp., (**q**) *Boerhavia diffusa* L. and (**r**) *Achyranthes aspera* L.

Table 1.1 Examples of some WFPs utilized in different parts of the globe

The scientific name of the plant	Family	Country/region	Reference
Utilized part—Root and tuber			
Adenia hondala (Gaertn.) W. J. de Wilde	Passifloraceae	Kerala (India), Ethiopia, Sri Lanka	Narayanan et al. (2011); Das et al. (2013): Nazarudeen et al. (1996); Mosissa and Abraha (2018); Ranil et al. (2021)
Amorphophallus commutatus (Schott) Engl.	Araceae	Kerala (India), Indonesia	Chaithanya et al. (2015); Raj et al. (2022); Wahidah et al. (2021)
Amorphophallus paeoniifolius (Dennst.) Nicolson	Araceae	Kerala, Karnataka (India), Indonesia	Yesodharan and Sujana (2007); Madhurima et al. (2012); Haridas and Kunhikannan (2020); Mutaqin (2020)
Asparagus racemosus Willd	Asparagaceae	Kerala (India), Nepal, Ethiopia	Yesodharan and Sujana (2007); Binu (2010); Narayanan et al. (2011); Chaithanya et al. (2015); Haridas and Kunhikannan (2020); Shrestha and Dhillion (2006); Lulekal et al. (2011a, b); Mosissa and Abraha (2018)
Cheilocostus speciosus (J. Koenig) C. D. Specht	Costaceae	Kerala (India), China, Myanmar	Yesodharan and Sujana (2007); Narayanan et al. (2011); Chaithanya et al. (2015); Haridas and Kunhikannan (2020); Xu et al. (2020); Shin et al. (2018)
Colocasia esculenta (L.) Schott	Araceae	Kerala, Chhattisgarh (India), Philippines, Malaysia	Yesodharan and Sujana (2007); Chaithanya et al. (2015), Matthews and Ghanem (2021); Banik (2014); Matthews et al. (2012); Zulhazman et al. (2021)
Dioscorea bulbifera L.	Araceae	Kerala, Andhra Pradesh, Assam, Meghalaya (India)	Yesodharan and Sujana (2007); Chaithanya et al. (2015); Reddy et al. (2007): Dutta (2015), Sheikh (2013)

(continued)

Table 1.1 (continued)

The scientific name of the plant	Family	Country/region	Reference
Dioscorea hamiltonii Hook. f.	Dioscoreaceae	Kerala, Odisha, Assam (India), Nepal	Narayanan et al. (2011); Kumar (2017), Dutta (2015); Ghimeray et al. (2010)
Dioscorea hispida Dennst.	Dioscoreaceae	Kerala, Odisha, Assam (India), Ethiopia	Yesodharan and Sujana (2007); Narayanan et al. (2011); Chaithanya et al. (2015); Kumar (2017); Dutta (2015); Haridas and Kunhikannan (2020); Mosissa and Abraha (2018)
Dioscorea intermedia Thwaites	Dioscoreaceae	Kerala (India)	Narayanan et al. (2011)
Dioscorea oppositifolia L.	Dioscoreaceae	Kerala, Odisha, Assam, Meghalaya (India), Ethiopia, Nepal	Binu (2010); Narayanan et al. (2011); Chaithanya et al. (2015); Kumar (2017), Dutta (2015), Sheikh (2013); Haridas and Kunhikannan (2020); Mosissa and Abraha (2018); Ghimeray et al. (2010)
Dioscorea pentaphylla L.	Dioscoreaceae	Kerala, Odisha, Assam (India), Meghalaya, Ethiopia	Yesodharan and Sujana (2007); Binu (2010); Narayanan et al. (2011); Chaithanya et al. (2015); Kumar (2017); Dutta (2015), Sheikh (2013); Haridas and Kunhikannan (2020); Mosissa and Abraha (2018)
Dioscorea tomentosa J. Koenig ex Spreng.	Dioscoreaceae	Kerala (India)	Narayanan et al. (2011); Chaithanya et al. (2015)
Dioscorea wallichii Hook. f.	Dioscoreaceae	Kerala, Odisha (India)	Narayanan et al. (2011); Chaithanya et al. (2015); Kumar (2017)
Dioscorea deltoidea Wall. ex Griseb. (Bhyakur)	Dioscoreaceae	Nepal, tropical and Sub-tropical regions of world	Shrestha and Dhillion (2006); Semwal et al. (2021)

(continued)

Table 1.1 (continued)

The scientific name of the plant	Family	Country/region	Reference
Ipomoea mauritiana Jacq.	Convolvulaceae	Kerala, Odisha (India), Southern Burkina Faso, Kenya	Narayanan et al. (2011); Haridas and Kunhikannan (2020); Saranraj et al. (2019); Guigma (2014); Mutie et al. (2020)
Maranta arundinacea L.	Marantaceae	Kerala (India), Sri Lanka	Binu (2010); Haridas and Kunhikannan (2020); Saranraj et al. (2019); Mudannayake et al. (2015)
Manihot esculenta Crantz (Sim tarul)	Euphorbiaceae	Nepal	Shrestha and Dhillion (2006)
Pecteilis gigantea (Sm.) Raf.	Orchidaceae	Kerala	Narayanan et al. (2011)
Utilized part: stem and shoot			
Arenga wightii Griff.	Arecaceae	Kerala (India)	Chaithanya et al. (2015)
Bambusa bambos (L.) Voss	Poaceae	Kerala (India), Nigeria, Thailand, Vietnam, Pakistan	Chaithanya et al. (2015); Ogunjinmi et al. (2009); Shirai and Rambo (2014); Nguyen et al. (2021); Khan et al. (2003)
Begonia picta Sm. (Magar kanche)	Begoniaceae	Nepal	Shrestha and Dhillion (2006)
Caryota urens L.	Arecaceae	Kerala (India), Nepal, Sri Lanka	Yesodharan and Sujana (2007); Chaithanya et al. (2015); Ghimeray et al. (2010); Mudannayake et al. (2015)
Centella asiatica Urban (Ghod tapre)	Umbelliferae	Nepal, India (India)	Shrestha and Dhillion (2006)
Cissus quadrangularis L.	Vitaceae	Kerala (India)	Yesodharan and Sujana (2007)
Cleome viscosa L.	Cleomaceae	Kerala (India)	Yesodharan and Sujana (2007)
Cheilocostus speciosus (J. Koenig) C. D. Specht	Costaceae	Kerala (India)	Yesodharan and Sujana (2007)
Cymbopogon citratus Stapf (Kagati ghans)	Poaceae	Nepal	Shrestha and Dhillion (2006)
Dryopteris cochleata	Dryopteridaceae	Nepal, India (India)	Shrestha and Dhillion (2006); Mehra and Mittal (1962)

(continued)

Table 1.1 (continued)

The scientific name of the plant	Family	Country/region	Reference
Dendrocalamus hamiltonii Nees and Arn. ex Munro (Bans)	Poaceae	Nepal, Sikkim (India)	Shrestha and Dhillion (2006); Pradhan et al. (2021)
Drepanostachyum falcatum (Nees) Keng f. (Nigalo)	Poaceae	Nepal	Shrestha and Dhillion (2006)
Ensete superbum (Roxb.) Cheesman	Musaceae	Kerala (India)	Yesodharan and Sujana (2007)
Glycyrrhiza glabra L. (Jethimau)	Fabaceae	Nepal	Shrestha and Dhillion (2006)
Girardinia diversifolia (Link) Friss (Allo)	Urticaceae	Nepal	Shrestha and Dhillion (2006); Subedee et al. (2022)
Mentha arvensis L. (Babari)	Labiatae	Nepal	Shrestha and Dhillion (2006)
Polystichum squarrosum	Dryopteridaceae	Nepal	Shrestha and Dhillion (2006)
Persicaria mollis (D. Don) Gross (Thotne)	Polygonaceae	Nepal	Shrestha and Dhillion (2006)
Rheum emodi Wall. (Padamchal)	Polygonaceae	Nepal	Shrestha and Dhillion (2006)
Saccharum spontaneum L.	Poaceae	Kerala (India)	Yesodharan and Sujana (2007); Mary et al. (2006)
Spinacia oleracea L.	Chenopodiaceae	Nepal, Iran	Shrestha and Dhillion (2006); Sabaghnia et al. (2014)
Ophioglossum reticulatum L. (Jibre sag)	Ophioglossaceae	Nepal	Shrestha and Dhillion (2006)
Oxalis corniculata L. (Chari amilo)	Oxalidaceae	Nepal	Shrestha and Dhillion (2006)
Osyris wightiana Wall. ex Wight (Nundhiki)	Santalaceae	Nepal	Shrestha and Dhillion (2006)
Smilax macrophylla Roxb. (Kukur daino)	Smilacaceae	Nepal	Shrestha and Dhillion (2006)
Urtica dioica L. (Sisnu)	Urticaceae	West Bengal (India), Nepal	Mandal et al. (2009); Shrestha and Dhillion (2006)
Utilized part: leaves			
Alternanthera brasiliana (L.) Kuntze	Amaranthaceae	Kerala (India)	Narayanan et al. (2011)
Alternanthera sessilis (L.) R. Br. ex DC.	Amaranthaceae	Kerala (India)	Yesodharan and Sujana (2007); Binu (2010); Narayanan et al. (2011); Chaithanya et al. (2015)

(continued)

Table 1.1 (continued)

The scientific name of the plant	Family	Country/region	Reference
Amaranthus spinosus L.	Amaranthaceae	Kerala, Odisha, West Bengal, Himachal Pradesh (India), Vietnam	Yesodharan and Sujana (2007); Binu (2010); Chaithanya et al. (2015); Kumar et al. (2014); Biswas and Das (2011); Chand et al. (2017); Nguyen et al. (2021)
Amaranthus viridis L.	Amaranthaceae	Kerala (India)	Yesodharan and Sujana (2007); Binu (2010); Chaithanya et al. (2015)
Allmania nodiflora (L.) R. Br. ex Wight	Amaranthaceae	Kerala (India)	Binu (2010)
Allium wallichii Kunth (Jimbu)	Liliaceae	Uttrakhand (India), Nepal	Pandey et al. (2021); Shrestha and Dhillion (2006)
Bambusa bambos (L.) Voss	Poaceae	Kerala (India), Viet-nam, Pakistan	Yesodharan and Sujana (2007); Chaithanya et al. (2015); Nguyen et al. (2021); Khan et al. (2003)
Boerhavia diffusa L.	Nyctaginaceae	Kerala, Gujarat, Himachal Pradesh (India)	Binu (2010); Chaithanya et al. (2015); Bedi (1978); Chand et al. (2017)
Centella asiatica (L.) Urb.	Apiaceae	Kerala, West Bengal, Himachal Pradesh (India), Vietnam	Yesodharan and Sujana (2007); Narayanan and Kumar. (2007); Biswas and Das (2011); Chand et al. (2017); Nguyen et al. (2021)
Centella asiatica (L.) Urb. *(Ghod tapre)*	Umbelliferae	Nepal	Shrestha and Dhillion (2006)
Colocasia esculenta (L.) Schott	Araceae	Kerala, Jammu and Kashmir, Maharashtra, West Bengal, Himachal Pradesh (India)	Yesodharan and Sujana (2007); Chaithanya (2015); Kumar and Hamal (2009); Lokhande (2020); Narayanan and Kumar (2007); Chand et al. (2017)

(continued)

Table 1.1 (continued)

The scientific name of the plant	Family	Country/region	Reference
Cymbopogon citratus Stapf (Kagati ghans)	Poaceae	Nepal	Shrestha and Dhillion (2006)
Commelina benghalensis L.	Commelinaceae	Kerala, Maharashtra, Gujarat, Chhattisgarh (India)	Yesodharan and Sujana (2007); Lokhande (2020); Chauhan et al. (2014)
Cyathula prostrata (L.) Blume	Amaranthaceae	Kerala (India)	Narayanan et al. (2011)
Euphorbia hirta L.	Euphorbiaceae	Kerala (India)	Yesodharan and Sujana (2007)
Girardinia diversifolia (Link) Friss (Allo)	Urticaceae	Nepal (India)	Subedee et al. (2020); Shrestha and Dhillion (2006)
Oxalis corniculata L.	Oxalidaceae	Kerala, Maharashtra (India) and Nepal	Yesodharan and Sujana (2007); Chaithanya et al. (2015); Lokhande (2020); Ghimeray et al. (2010)
Osyris wightiana Wall. ex Wight (Nundhiki)	Santalaceae	Nepal	Shrestha and Dhillion (2006)
Ophioglossum reticulatum L. (Jibre sag)	Ophioglossaceae	Nepal	Ojha and Devkota (2021); Shrestha and Dhillion (2006)
Portulaca oleracea L.	Portulacaceae	Kerala (India)	Yesodharan and Sujana (2007)
Phytolacca acinosa Roxb. (Jaringo)	Phytolaccaceae	Nepal	Shrestha and Dhillion (2006)
Senna tora (L.) Roxb.	Fabaceae	Kerala (India)	Narayanan et al. (2011); Chaithanya et al. (2015)
Smilax macrophylla Roxb. (Kukur daino)	Smilacaceae	Nepal	Shrestha and Dhillion (2006)
Osyris wightiana Wall. ex Wight (Nundhiki)	Santalaceae	Nepal	Shrestha and Dhillion (2006)
Urtica dioica L. (Sisnu)	Urticaceae	Europe, Asia, North Africa, North America, Nepal	Bhusal et al. (2022); Shrestha and Dhillion (2006)
Vigna radiata (L.) R. Wilczek	Fabaceae	Kerala (India), Vietnam, Pakistan	Yesodharan and Sujana (2007); Nguyen et al. (2021); Khan et al. (2003)
Utilized part: fruits			
Aegle marmelos Corr.	Rutaceae	Kerala (India)	Yesodharan and Sujana (2007); Narayanan et al. (2011)

(continued)

Table 1.1 (continued)

The scientific name of the plant	Family	Country/region	Reference
Aglaia perviridis Hiern	Meliaceae	Kerala (India)	Binu (2010)
Alangium salviifolium (L. f.) Wangerin	Cornaceae	Kerala, Rajasthan, Hyderabad (India), China, Thailand, Philippines, Indonesia, Papua New Guinea, Tropical Australia, Africa, Madagascar, Southern and Eastern Asia, Western Pacific Ocean Islands and New Caledonia	Yesodharan and Sujana (2007); Binu (2010); Shravya et al. (2017)
Ailanthus triphysa (Dennst.) Alston	Simaroubaceae	Kerala (India)	Yesodharan and Sujana (2007)
Antidesma montanum Blume	Euphorbiaceae	Kerala (India), Bangladesh ,China, Indonesia, Myanmar, Vietnam, Thailand, Malaysia	Yesodharan and Sujana (2007); Binu (2010); Zaman et al. (2022)
Aporosa cardiosperma (Gaertn.) Merr.	Phyllanthaceae	Kerala (Part of Agasthyamala Biosphere Reserve), India	Narayanan et al. (2011); Chaithanya et al. (2015); Chand and Azeez (2021)
Artocarpus heterophyllus Lamk	Moraceae	Kerala (India), Sri Lanka	Yesodharan and Sujana (2007); Narayanan et al. (2011); Chaithanya et al. (2015); Ranasinghe et al. (2019)
Artocarpus hirsutus Lamk	Moraceae	Kerala, Western Ghats of India	Yesodharan and Sujana (2007); Narayanan et al. (2011); Chaithanya et al. (2015); Solanki et al. (2020)
Baccaurea courtallensis (Wight) Müll. Arg.	Phyllanthaceae	Kerala (India)	Binu (2010); Narayanan et al. (2011); Chaithanya et al. (2015); Pradheep et al. (2021)
Berberis asiatica Roxb. ex DC. (Chutro)	Berberidaceae	Nepal	Shrestha and Dhillion (2006)
Calamus rotang L.	Arecaceae	Kerala	Yesodharan and Sujana (2007)
Canthium rheedei DC.	Rubiaceae	Kerala (India)	Yesodharan and Sujana (2007)

(continued)

Table 1.1 (continued)

The scientific name of the plant	Family	Country/region	Reference
Coccinia grandis (L.) Voigt	Cucurbitaceae	Kerala (India), North central east Africa, Asia, Australia, Caribbean, Southern United States	Yesodharan and Sujana (2007); Binu (2010); Chaithanya et al. (2015); Muniappan et al. (2009)
Cucumis melo L.	Cucurbitaceae	Kerala (India)	Chaithanya et al. (2015)
Cycas circinalis L.	Cycadaceae	Kerala (India)	Narayanan et al. (2011)
Diospyros malabarica (Desr.) Kostel.	Ebenaceae	Kerala (India), Mid hill regions of western Nepal	Yesodharan and Sujana (2007); Shrestha et al. (2021)
Elaeagnus parvifolia Wall. ex Royle (Gunyeli)	Elaeagnaceae	Nepal	Shrestha and Dhillion (2006)
Ficus racemosa L. (Athi)	Moraceae	Kerala (India), Nepal, Thailand, Vietnam, Pakistan, Sri Lanka	Narayanan et al. (2011); Galil (1984); Nguyen et al. (2021); Khan et al. (2003); Mudannayake et al. (2015)
Fragaria nubicola (Hook. f.) Lindl. ex Lacaita (Bhui ainselu)	Rosaceae	India, Jammu and Kashmir, Uttarakhand (India), Nepal, Bhutan, Pakistan	Shrestha et al. (2021); Shrestha and Dhillion (2006)
Flacourtia montana J. Graham	Salicaceae	Kerala, Semi-evergreen and moist deciduous forests of Western Ghats, India	Binu (2010); Narayanan et al. (2011); Chaithanya et al. (2015); Mundaragi et al. (2019)
Ficus auriculata Lour. (Nimaro)	Moraceae	Nepal	Shrestha and Dhillion (2006)
Ficus neriifolia Sm. (Dudhilo)	Moraceae	Nepal	Walker et al. (1999); Shrestha and Dhillion (2006)
Ficus semicordata Buch.-Ham. ex Sm. (Khanew)	Moraceae	Nepal	Shrestha and Dhillion (2006)
Gaultheria fragrantissima Wall. (Matchhyanu)	Ericaceae	Nepal	Shrestha and Dhillion (2006)
Garcinia gummi-gutta (L.) Roxb.	Clusiaceae	Kerala (India)	Yesodharan and Sujana (2007); Narayanan et al. (2011); Vishnu et al. (2022)

(continued)

Table 1.1 (continued)

The scientific name of the plant	Family	Country/region	Reference
Glycosmis pentaphylla (Retz.) DC.	Rutaceae	Kerala (India)	Binu (2010)
Gmelina arborea Roxb.	Lamiaceae	Kerala (India), Costa Rica	Yesodharan and Sujana (2007); Roque (2004)
Grewia tiliifolia Vahl	Malvaceae	Kerala (India)	Yesodharan and Sujana (2007); Narayanan et al. (2011); Chaithanya et al. (2015)
Hibiscus surattensis L.	Malvaceae	Kerala (India), tropical and subtropical regions worldwide	Yesodharan and Sujana (2007); Akarca (2019)
Holboellia latifolia Wall. (Gofle lahara)	Lardizabalaceae	Nepal	Shrestha and Dhillion (2006)
Ixora brachiata Roxb.	Rubiaceae	Kerala (India)	Yesodharan and Sujana (2007)
Ixora coccinea L.	Rubiaceae	Kerala (India)	Narayanan et al. (2011)
Lantana Camara Linn.	Verbenaceae	Kerala (India)	Yesodharan and Sujana (2007); Narayanan et al. (2011)
Leea indica (Burm. f.) Merr.	Vitaceae	Kerala (India), South-east Asia	Narayanan et al. (2011); Singh et al. (2019)
Lindera pulcherrima (Nees) Hook. f. (Sis)	Lauraceae	Nepal	Shrestha and Dhillion (2006)
Mesua ferrea Linn.	Calophyllaceae	Kerala (India), South-east Asia in tropical evergreen forest	Yesodharan and Sujana (2007); Chahar (2013)
Mimusops elengi L.	Sapotaceae	Kerala (India), Pakistan	Yesodharan and Sujana (2007); Narayanan et al. (2011); Jahan et al. (1995)
Mahonia nepaulensis DC. (Jamunemandro)	Berberidaceae	Nepal	Shrestha and Dhillion (2006)
Momordica charantia L. (Tite karela)	Cucurbitaceae	Tamil Nadu (India), Nepal	Sathishsekar and Subramanian (2005); Shrestha and Dhillion (2006)
Morus serrata Roxb. (Kimbu)	Moraceae	Kenya, Nepal	Peris et al. (2014); Shrestha and Dhillion (2006)

(continued)

Table 1.1 (continued)

The scientific name of the plant	Family	Country/region	Reference
Myrica esculenta Buch.-Ham.ex D. Don (Kaphal)	Myricaceae	Nepal	Shrestha and Dhillion (2006)
Olea dioica Roxb.	Oleaceae	Kerala (India)	Yesodharan and Sujana (2007)
Passiflora foetida Linn.	Passifloraceae	Kerala (India), USA, Mexico, Caribbean, Central and South America	Yesodharan and Sujana (2007); Binu (2010); Narayanan et al. (2011); Patil et al. (2013)
Phyllanthus emblica L.	Phyllanthaceae	Kerala (India), China	Yesodharan and Sujana (2007); Chaithanya et al. (2015); Mirunalini and Krishnaveni (2010)
Physalis minima L.	Solanaceae	Kerala (India), East Malaysia	Binu (2010); Narayanan et al. (2011); Lem et al. (2022)
Piper longum L.	Piperaceae	Kerala (India), Indonesia	Yesodharan and Sujana (2007); Kumar et al. (2011)
Piper nigrum L.	Piperaceae	Kerala (India), South and Southeast Asia, Brazil, Vietnam, Indonesia	Yesodharan and Sujana (2007); Takooree et al. (2019)
Paris polyphylla Sm. (Satuwa)	Liliaceae	China, India, Nepal	Chhetri et al. (1970); Cunningham et al. (2018); Shrestha and Dhillion (2006)
Potentilla indica (Andr.) Wolf (Bhui khaphal)	Rosaceae	Nepal	Shrestha and Dhillion (2006)
Prunus domestica L. (Alucha)	Rosaceae	Southwestern Asia, Europe, Nepal	Lim (2012); Shrestha and Dhillion (2006)
Pyrus pashia Buch.-Ham. ex D. Don (Mayal)	Rosaceae	Nepal	Shrestha and Dhillion (2006)
Prunus cerasoides D. Don (Painyu)	Rosaceae	Nepal	Shrestha and Dhillion (2006)
Pyracantha crenulata (D. Don) M. Roem. (Ghangaru)	Rosaceae	Nepal	Shrestha and Dhillion (2006)
Rhus javanica L. (Bhak amilo)	Anacardiaceae	Nepal	Shrestha and Dhillion (2006)
Rhus wallichii Hook. f. (Chaysi)	Anacardiaceae	Nepal	Shrestha and Dhillion (2006)

(continued)

Table 1.1 (continued)

The scientific name of the plant	Family	Country/region	Reference
Rubus ellipticus Sm. (Ainselu)	Rosaceae	Nepal	Shrestha and Dhillion (2006)
Schleichera oleosa (Lour.) Merr.	Sapindaceae	Kerala (India)	Narayanan et al. (2011)
Schisandra grandiflora Hook. f. and Thoms. (Singato)	Schisandraceae	Nepal	Shrestha and Dhillion (2006)
Scleropyrum pentandrum (Dennst.) Mabb.	Santalaceae	Kerala (India)	Narayanan et al. (2011)
Smilax zeylanica L.	Smilacaceae	Kerala (India)	Yesodharan and Sujana (2007)
Saurauia napaulensis DC. (Gogan)	Actinidiaceae	Nepal	Shrestha and Dhillion (2006)
Solanum americanum Mill.	Solanaceae	Kerala (India)	Chaithanya et al. (2015)
Solanum anguivi Lam. (Chunda)	Solanaceae	Kerala (India)	Yesodharan and Sujana (2007)
Solanum torvum Sw. (Chunda)	Solanaceae	Kerala (India), West Indies, Europe	Binu (2010); Narayanan et al. (2011); Chaithanya et al. (2015); Musarella (2020)
Solanum virginianum L.	Solanaceae	Kerala (India)	Narayanan et al. (2011)
Solena amplexicaulis (Lam.) Gandhi	Cucurbitaceae	Kerala (India), dry deciduous forest and scrub jungles of Tamil Nadu	Narayanan et al. (2011); Krishnamoorthy and Subramaniam (2014)
Syzygium cuminii (Linn.) Skeels	Myrtaceae	Kerala (India)	Yesodharan and Sujana (2007); Binu (2010); Chaithanya et al. (2015)
Syzygium densiflorum Wall. ex Wight and Arn.	Myrtaceae	Kerala (India)	Narayanan et al. (2011)
Syzygium gardneri Thwaites	Myrtaceae	Kerala, part of Agasthyamala Biosphere Reserve (India)	Narayanan et al. (2011); Chand and Azeez (2021)
Syzygium laetum (Buch. -Ham.) Gandhi	Myrtaceae	Kerala, Silent Valley National Park (Western Ghats), India	Narayanan et al. (2011); Kunhikannan et al. (2011)
Syzygium palghatense Gamble	Myrtaceae	Kerala, Western Ghats (India)	Yesodharan and Sujana (2007); Snehalatha and Rasmi (2021)

(continued)

Table 1.1 (continued)

The scientific name of the plant	Family	Country/region	Reference
Helixanthera ligustrina (Wall.) Danser (Lisso)	Loranthaceae	Nepal	Shrestha and Dhillion (2006)
Tamarindus indica L.	Fabaceae	Kerala (India), Africa, Bangladesh, Pakistan, Nigeria	Yesodharan and Sujana (2007); Chaithanya et al. (2015); Bhadoriya et al. (2011); Jain et al. (2011)
Ziziphus jujuba Mill.	Rhamnaceae	Kerala (India), China	Narayanan et al. (2011); Ji et al. (2017)
Ziziphus oenopolia (L.) Mill.	Rhamnaceae	Kerala	Binu (2010); Narayanan et al. (2011)
Ziziphus rugosa Lam.	Rhamnaceae	Kerala, deciduous and semi-evergreen forest of Western Ghat, India	Yesodharan and Sujana (2007); Chaithanya et al. (2015); Prashith et al. (2011)
Utilized part: seeds			
Acacia nilotica (L.) Delile	Fabaceae	Kerala (India), Asia, Australia, Africa, America	Yesodharan and Sujana (2007); Ali (2012)
Acacia sinuata (Lour.) Merr.	Fabaceae	Kerala (India)	Yesodharan and Sujana (2007); Chaithanya et al. (2015); Vengadesan et al. (2002)
Adenanthera pavonia L.	Fabaceae	Kerala (India), Southern China, Malaysia, Western and Eastern Africa, Island of Pacific and Caribbean region	Yesodharan and Sujana (2007); Ara et al. (2010)
Artocarpus heterophyllus Lamk	Moraceae	Kerala (India), Asia, Africa, South America	Yesodharan and Sujana (2007); Narayanan et al. (2011); Chaithanya et al. (2015); Ranasinghe et al. (2019)
Artocarpus hirsutus Lamk	Moraceae	Kerala (India)	Yesodharan and Sujana (2007); Binu (2010); Narayanan et al. (2011); Chaithanya et al. (2015)
Bambusa bambos (L.) Voss	Poaceae	Kerala (India), South east Asian country	Yesodharan and Sujana (2007); Narayanan et al. (2011); Raju and Roy (2016)

(continued)

Table 1.1 (continued)

The scientific name of the plant	Family	Country/region	Reference
Coix lacryma-jobi L.	Poaceae	Kerala (India), China	Narayanan et al. (2011); Xi et al. (2016)
Cannabis sativa L. (Bhango)	Cannabaceae	India (India), Nepal	Kuddus et al. (2013); Shrestha and Dhillion (2006)
Castanopsis tribuloides (Sm.) A. DC. (Musure katus)	Fagaceae	Nepal	Shrestha and Dhillion (2006)
Cycas circinalis Linn.	Cycadaceae	Kerala (India), Tripura	Yesodharan and Sujana (2007); Binu (2010); Chaithanya et al. (2015); Bhattacharjee et al. (2019)
Delonix regia (Hook.) Raf.	Fabaceae	Kerala (India), Bangladesh, China, Nepal, Mexico, Brazil, East Africa, West Indies, South Africa, Mauritius	Chaithanya et al. (2015); Rahman et al. (2020)
Elaeocarpus serratus L.	Elaeocarpaceae	Kerala (India), Sri Lanka	Yesodharan and Sujana (2007); Raji and Siril (2021)
Entada rheedii Spreng.	Fabaceae	Kerala (India), Tanzania, South East Asia	Yesodharan and Sujana (2007); Binu (2010); Chaithanya et al. (2015); Arulkumar et al. 2017; Kumar (2017)
Firmiana simplex (L.) W. Wight	Malvaceae	Kerala (India), China, Taiwan, Ryukyus, Vietnam, Japan, SE and W of USA	Yesodharan and Sujana (2007); Upson and Cullen (2012)
Mimusops elengi L.	Sapotaceae	Kerala (India)	Narayanan et al. (2011)
Perilla frutescens (L.) Britton (Silam)	Labiatae	China, Japan, India, Nepal, Thailand, Korea	Dhyani et al. (2019); Shrestha and Dhillion (2006)
Pinus roxburghii Sarg. (Khote salla)	Pinaceae	Nepal	Shrestha and Dhillion 2006)
Schleichera oleosa (Lour.) Merr.	Sapindaceae	Kerala (India), Odisha	Yesodharan and Sujana (2007); Mandal et al. (2009)
Solanum virginianum L.	Solanaceae	Kerala (India)	Narayanan et al. (2011)
Solena amplexicaulis (Lam.) Gandhi	Cucurbitaceae	Kerala, Odisha (India)	Narayanan et al. (2011); Saravanan et al. (2020)

(continued)

Table 1.1 (continued)

The scientific name of the plant	Family	Country/region	Reference
Sterculia foetida L.	Malvaceae	Kerala, Tamil Nadu (India)	Narayanan et al. (2011); Arinathan et al. (2007)
Sterculia guttata Roxb. ex G.Don	Sterculiaceae	Kerala, Tamil Nadu (India)	Yesodharan and Sujana (2007); Binu (2010); Chaithanya et al. (2015); Arinathan et al. (2007)
Tamarindus indica L.	Fabaceae	Kerala (India), West Africa	Yesodharan and Sujana (2007); Havinga et al. (2010)
Terminalia bellirica (Gaertn.) Roxb.	Combretaceae	Kerala (India), Nepal	Yesodharan and Sujana (2007); Gahatraj et al. (2020)
Viburnum cylindricum Buch.-Ham.	Caprifoliaceae	China, Nepal	Zhao et al. (2022); Shrestha and Dhillion (2006)
Zanthoxylum armatum DC. (Timru)	Rutaceae	Warmer Valleys of Himalayas, Eastern Ghats and the lesser North Eastern Himalayan region of India and Nepal	Patade et al. (2020); Shrestha and Dhillion (2006)
Zanthoxylum oxyphyllum Edgew. (Bogya timur)	Rutaceae	Nepal	Shrestha and Dhillion (2006)

1.4 The Role of WFPs in Achieving Zero Hunger

Achievement of zero hunger or a world without hunger, malnutrition, and poverty by 2030 is the major agenda of the United Nations (UN Pathways to Zero Hunger | Zero Hunger Challenge (2022), https://www.uthen.org/zerohunger/content/challenge). It is a part of the UN's 17 sustainable development goals (UN The 17 Goals | Sustainable Development (2022), https://sdgs.un.org/goals). There exists an interlink between the sustainable development goals and food and nutritional security, i.e., the achievement of zero hunger is ultimately aiming towards the achievement of food security (Fig. 1.2). The concept of food security covers six dimensions i.e., food availability, affordability, accessibility, convenience (time cost for gathering and consumption of food), food quality, and promotional information (Karanja et al. 2022). Rapidly increasing population, climate change, and outbreaks of pandemics are rising challenges against food security (FAO 2017). Although, in the second half of the twentieth century, remarkable improvements in the field of agriculture occurred, it was mainly focused on improving and increasing the production of only a few crops (Godfray et al. 2010). In India, the green revolution brought many

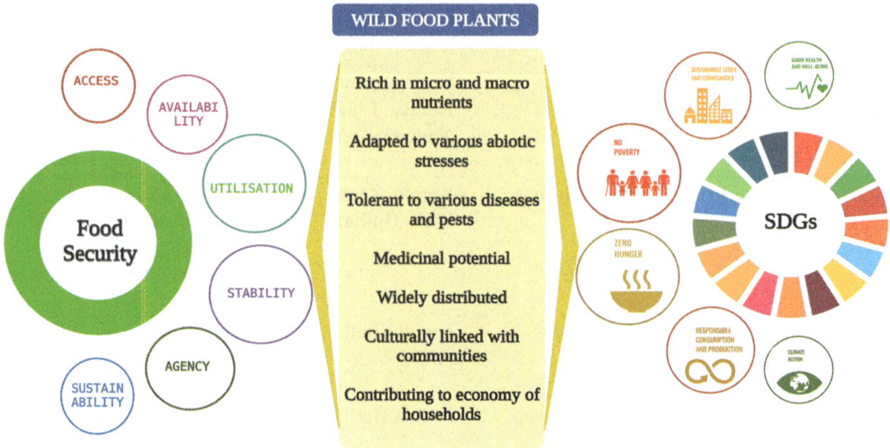

Fig. 1.2 The link between wild food plants, six dimensions of food security, and sustainable development goals (SDGs). The multiple benefits of the WFPs can help to eradicate poverty and malnutrition of consumers and thereby helps to achieve the goal of zero hunger proposed by UN (UN, https://www.uthen.org/zerohunger/content/challenge; https://www.un.org/sustainabledevelopment/news/communications-material/; Karanja et al. 2022)

revolutionary changes in the field of agriculture in terms of methods, techniques, and production scale (Chakravarti 1973; Thrupp 2000; Ameen and Raza 2018). The promotion of improved varieties of rice and wheat resulted in a significant rise in their production, but simultaneously a large decline in the production and consumption of other less popular plants occurred (Davis et al. 2019). Because of many such reasons, people selectively cultivated and consumed very few crop varieties. Modern technologies developed agriculture practices that focused on crops' high yield, and human dietary patterns shifted away from wild plants, and they became underutilized (Akinola et al. 2020; Fones et al. 2020). Nutritional, environmental, economic, and sociocultural values are considered key components of sustainable food systems (Akinola et al. 2020).

Since WFPs are locally available, low-input required, seasonally available, and less-maintenance required sources of diverse food, it is beneficial to consumers, especially poor families. Promoting the consumption of WFPs saves the money for buying food items and provides sustainable livelihood during the famine period (Shaheen et al. 2017). Compared with the WFPs, the number of food crop species that are part of the human daily diet is very less, and their nutritional quality is very poor (Chibarabada et al. 2017). Many WFPs are rich sources of protein, lipids, carbohydrates, and micronutrients (Ogle et al. 2001; Bhat and Karim 2009) and ensure greater dietary diversity (Servin and Moseley 2022). They are potential food crops for alleviating food insecurity and improving the lives of rural communities in developing or less-developed countries (Gahukar 2015). The resilience of local communities in Ethiopia depend on WFPs during seasonal insecurities exemplifies this (Guyu and Muluneh 2015). Some WFPs possess superior nutritional quality than common staples; some examples are; the vitamin C content in *Adansonia digitata* is 10 times higher than oranges (Yazzie et al. 1994), the provitamin A and vitamin C content in wild tuber *Tropaeolum tuberosum* is higher than other tubers (María Elena et al. 2019), *Achyranthes bidentata* reported vitamin C in the seeds, which is 1.6 times higher than amaranthus seeds (Marcone et al. 2003), *Plectranthus rotundifolius* recorded higher mineral content in the tubers than cassava and potato (Sethuraman et al. 2020), and *Pachira aquatica* contain P, Mg, Zn, Fe, and Cu than some starchy crops and fruits (Rodrigues et al. 2019).

Consumption of certain WFPs can help to alleviate micronutrient deficiencies up to some extent (García-Herrera and Sánchez-Mata 2016). Hunger, malnutrition, and many deficiency diseases are largely reported in underdeveloped and developing countries, especially African countries (Venter et al. 2000; Muthoni and Nyamongo 2010). So, the challenges and health issues due to the quantity and quality of food can be diminished by concentrating more on the cultivation and consumption of these neglected WFPs (Muthoni and Nyamongo 2010). Due to nutritional quality and availability in all seasons, wild leafy vegetables support the food system of rural people (Natesh et al. 2017). Wild plants are popular as rich and low-cost sources of several nutrients such as protein, essential amino acids, dietary fiber, vitamins, and minerals (e.g., Fe and Ca) (Natesh et al. 2017; Randhawa et al. 2015). Several wild plants are excellent sources of flavonoids and tannins and are popular for their antioxidant potential (Alam et al. 2019). *M. oleifera* leaves and seeds are rich in protein and lipids, respectively (Matic et al. 2018). A number of African leafy vegetables are found to be excellent sources of vitamin A, vitamin C, and minerals such as Fe, Zn, and Mg (Uusiku et al. 2010; Dube et al. 2017). Thus, including them in the diet will remedy many deficiency diseases because of micronutrients such as Fe, iodine, and vitamin A (Jarapala 2017). The richness of micronutrients and bioactive compounds makes them significantly contribute toward human nutritional security and the UN's goal for Zero hunger (Pinela et al. 2017).

1.5 Resilience of WFPs

The modern food systems are dependent on a few major crops that are associated with health problems, pollution, resource depletion, land degradation, biodiversity degradation, and environmental issues (Willett et al. 2019). Climate change is one of the important challenges that the world faces, which raises greater concerns about the future lives on earth (Mohanty and Mohanty 2009). Climate change affects the sustainability of livelihood in multiple dimensions (Kipkoech et al. 2013). While ranking staple crops based on their global production and output calories, wheat and maize come within the first five ranks (Fones et al. 2020). But these major staple crops (rice, wheat, and maize) are highly vulnerable to climatic changes, edaphic factors, and other stress conditions (Taylor et al. 2018; Lobell et al. 2011; Dabi and Khanna 2018). Monocultural systems, extensive and repeated use of fertilizers and other agrochemicals such as fungicide, weedicide and excessive water irrigation adversely affect soil biota and fertility (Thrupp 2000; Nikolskii et al. 2019). Some of the mainstream crops that we use today are highly susceptible to various diseases and abiotic stresses. The current scenario demands more abiotic and biotic stress-tolerant plants that can withstand rapid and unpredictable climate changes.

Some WFPs are capable of growing in poor soil and show environmental plasticity (Chivenge et al. 2015). It was reported that WFPs in Africa are capable of growing successfully in low humid, arid, or hot and dried areas of Africa (Rana et al. 2012). The WFPs already proved their potential to provide food security when other crops failed during severe and other stresses (Wunder et al. 2014; Shumsky et al. 2014). There are lots of WFPs, which are adapted to different environmental conditions. Biotic and abiotic tolerance of wild emmer wheat was reviewed by Huang et al. (2016). They reported that wild emmer wheat is tolerant to different pathogens such as leaf rust, powdery mildew, and head blight, and they possess different resistant genes/QTLs such as *Lr53, Yr15, YrH52, Yr35, Yr36, YrSM139-1B, Pm3k, pm42, Pm26, Pm41, Pm30, Qfhs.ndsu3AS, Qfhs, Qfhs*, and *fhs.fcu-7AL*. They also reviewed the drought-tolerant mechanism of wild emmer wheat. High-level antioxidant metabolites and tolerance to various stress were observed in wild citrus (*Citrus latipes*) and primitive citrus (*Atalantia buxifolia*) than in the cultivated crops (Rao et al. 2021). Pigeon pea is also one of the valuable crops that contribute to human nutrition in a sustainable way. The wild relative of pigeon pea reported tolerance to high precipitation, temperature, and edaphic conditions (Khoury et al. 2015). The drought tolerance potential of three landraces of wild melon (*Citrullus lanatus* L.) was explored by Zulu and Modi (2010) in South Africa. Similarly, the wild edible leafy vegetable *Reichardia picroides* displayed tolerance to salinity stress with higher deposition of proline and osmolytes in their shoot, which points toward their salt-tolerant potential (Alexopoulos et al. 2021). The stress-tolerant potential of these WFPs is a strength for the consumers, and it helps to satisfy their hunger during tough environmental conditions. Encouraging the consumption of WFPs can help to achieve sustainable food security by mitigating climate change (Wessels et al. 2021). Figure 1.3 represents the importance of WFPs for greater resilience due to their multiple benefits to the local communities.

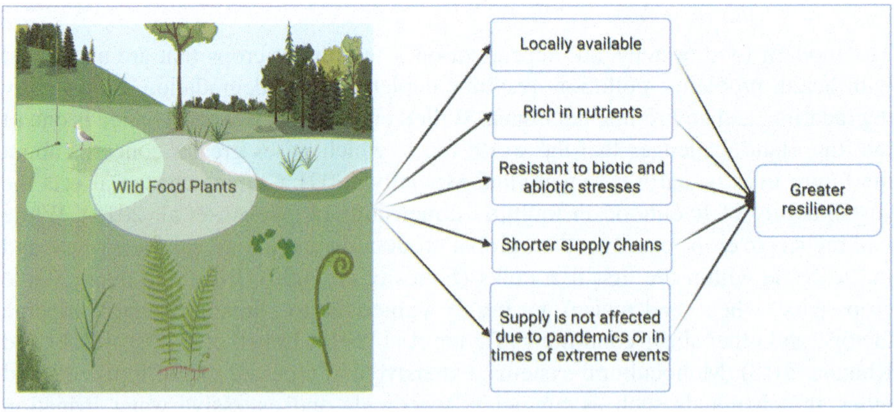

Fig. 1.3 Role of WFPs in enhancing resilience due to their multiple benefits (Created with Biorender.com)

1.6 Other Benefits

1.6.1 Medicinal Potentials

In several rural areas, WFPs are an important part of health care apart from their food value (Mujuru et al. 2020). Different communities depend on different parts, such as roots, fruits, seeds, leaves, bark, stem, and flowers of WFPs for their medicinal properties (Cohen et al. 2020). Several ethnobotanical studies recorded the medicinal potential of WFPs besides their nutritional significance (Balemie and Kebebew 2006; Della et al. 2006; Rasingam 2012; Shad et al. 2013; Cheng et al. 2022). The wild tuber *Erodium crassifolium* L'Hér recorded anti-inflammatory activity in the human cell culture in the studies conducted by Cohen et al. (2020). The antimicrobial effects of methanol extract on WFPs *Eruca sativa, Chenopodium murale, Malcolmia africana, Goldbachia laevigata, Malva neglecta, Melilotus officinalis, Medicago polymorpha*, and *Nasturtium officinale* against the microbes *Salmonella typhi, Escherichia coli, Proteus vulgaris,* and *Streptococcus pneumoniae* was reported by Khan et al. (2015). Naik et al. (2017) reported the medicinal potential of 97 wild edible plants from 48 families. The antimicrobial activity of wild yam *D. bulbifera* L. against *Escherichia coli, Klebsiella pneumoniae, Proteus vulgaris, Aspergillus niger, Staphylococcus aureus, Aspergillus fumigatus, Aspergillus flavus,* and *Rhizopus nigricans* (Seetharam et al. 2003). The anticancer activity of widely used wild *Trigonella foenum-graecum* L. was proved in mice injected with Ehrlich Ascites Carcinoma (EAC) cells (Prabhu and Krishnamoorthy 2010). Since there are lots of studies that focus on the medicinal properties of the plants, it is required to conduct detailed phytochemical and pharmacological studies on a huge diversity of WFPs (Mujuru et al. 2020).

1.6.2 Economic Potential

According to FAO (2012), nutrition is an important factor influencing economic growth, so as a source of nutrients, WFPs could also support the developing economy. The cultivation of neglected wild plants does not pose a financial burden to farmers (Muthoni and Nyamongo 2010). It is estimated that 18.8% of harvested wild plants are used as a source of income globally (Hickey et al. 2016). Forest dwellers mostly depend on WFPs to generate income. The contribution of WFPs to the household income of rural areas is significant in developing and underdeveloped countries (Asprilla-Perea and Díaz-Puente 2019). Locals of Nigeria depended on wild root and tuber crops such as *Plectranthus esculentus* N. E. Br. and *Tacca leontopetaloides* (L.) Kuntze not only for food but also for income (Safwan and Mohammed 2016). Commercially important species such as *Amaranthus* spp., *M. oleifera, I. aquatica, H. sabdariffa,* and *B. alba* can support household income (Ebert 2014). People of Africa made a yellow dye from *Cleome gynandra* L. plant (Venter et al. 2000). The population of Arasbaran region in Iran generated a household income share of 27% by selling wild edible fruits such as *Cornus mas* (Cornelian Cherry), *Ribes biebersteinii* (Reddish Blackberry), and *Rhus coriaria* (Sumac) (Ghanbari et al. 2022). Puttahariyappa et al. (2021) also reported that tribal communities from Malai Madeshawara hills wildlife sanctuary are earning an income of 15–20% of their annual income by selling various WFPs in different seasons. From another point of view, in the poorest villages of countries such as Congo, where the consumers are unable to purchase commercially available food sources from the market, they depend on WFPs for their major dietary needs, which indicates the importance of WFPs among the poorest people as a method of survival than an income generation method (Bharucha and Pretty 2010).

1.6.3 Sociocultural Potential

In addition to their use for food, WFPs are also a part of their culture for many communities (Akinola et al. 2020). Traditional eating practices have influenced the approach of people toward food safety (Kapelari et al. 2020). Traditional knowledge of communities on WFPs can help to solve present issues and can be used for community resilience (Shava et al. 2009). The agroecosystem developed by communities of a period to ensure food security is relearned and developed by later generations to respond to their environmental conditions (Holmgren and Öberg 2006). The home gardens found near dwelling places of communities comprise WFPs that they utilize for food, medicinal, rituals, and other purposes; hence this wild food plant-based agroforestry system reflects the culture of tribal communities (George and Christopher 2020). A wild species, *Feoniculum vulgare* Mill. ssp. *vulgare* used in Italy for its aroma is believed to be protect people from evil eyes (Pieroni 2001). *Emilia sonchifolia* L. is one of the 10 sacred herbs of Kerala (Varghese et al. 2010). *G. pentaphylla* leaves were used to eliminate evil beliefs (Chaithra and Thomas 2017). These examples reveal that some wild food plants are also part of rituals and have become important part of their culture.

1.7 Challenges Associated with WFPs

WFPs are useful and have the potential to contribute significantly to the food security of consumers, but they are partially marginalized (Feyssa et al. 2011; Padulosi et al. 2012). The lack of complete documentation of the diversity and utilization of WFPs and the lack of interest or negative attitude of the young generation toward the consumption of WFPs is the major reason behind their knowledge loss (Lalonde 1991; Sneyd 2013). There are many other challenges in accepting and utilizing WFPs as an alternative food system; for popularization and wide commercialization, such challenges need to be addressed (Akinola et al. 2020; Pudasaini et al. 2016). On account of reasons such as agricultural development focused on certain crops, misinterpretations of WFPs, lack of awareness about their nutritional value, and ignorance of their use have declined, this change is reflected in the food system, and some other crops, mostly cereals like rice and maize replaced the place of WFPs as the main food source (Akinola et al. 2020). It is another fact that global food policies regarding food security and nutritional value are not effectively utilized (Mbhenyane 2017). Many wild food varieties of different fruits, cereals, and legumes are almost at the edge of extinction; some are in a highly threatened stage too (Fowler and Mooney 1990). Therefore, it is necessary to document the WFP diversity in different regions of the globe and execute programs and policies for their effective utilization and consumption. The commercialization and consumption of WFPs should be promoted globally, which can support achieving zero hunger by 2030 in a scenario of global climate change and other rising food security challenges.

1.8 Conclusions and Future Prospects

A consolidated idea about the importance and characteristics of WFPs is provided in this chapter. Various categories of WFPs, the diversity and distribution of WFPs, and the multiple benefits of WFPs are discussed in detail. It was recorded that, in rural areas, WFPs are major alternative sources of proteins and carbohydrates. WFPs are an important source of food and income for several communities all over the world. WFP gene pool is enriched with a diversity of genes for abiotic and biotic stress tolerance and nutritional components. Their capability to survive in extreme environmental conditions contributes to their stress tolerance. Including WFPs in the programs to attain food security can provide sustainable outcomes and be a suitable strategy. A practice of consumption of WFPs can reduce the dependency on staple crops that require high agricultural input, high cost, and maintenance. But the diversity and significance of WFPs are not popular globally, hence widespread information regarding WFPs in different regions of the world. The consumption of regionally adapted WFPs is limited in that particular area and it should be diversified across different corners of the world. The awareness about WFPs should be spread globally, especially among the younger generation. The government should initiate public-funded projects or cooperation projects to promote the consumption of WFPs. Besides that, it is necessary to conduct scientific research and investigations to understand the physiological, biochemical, and molecular machinery of WFPs and make extensive efforts to document the WFP diversity across the globe. Research organizations should prioritize and establish novel conservation strategies for WFPs too.

References

Abate T (2012) Tropical grain legumes in Africa and South Asia: knowledge and opportunities. ICRISAT, Nairobi

Afolayan AJ, Jimoh FO (2009) Nutritional quality of some wild leafy vegetables in South Africa. Int J Food Sci Nutr 60:424–431. https://doi.org/10.1080/09637480701777928

Agarwal R, Chandra V (2019) Diversity of wild edible plants in the Mandal-Chopta forest, Uttarakhand. J Med Plants Stud 7:89–92

Akarca G (2019) Composition and antibacterial effect on food borne pathogens of *Hibiscus surrattensis* L. calyces essential oil. Ind Crop Prod 137:285–289. https://doi.org/10.1016/j.indcrop.2019.05.043

Akinola R, Pereira LM, Mabhaudhi T, de Bruin F-M, Rusch L (2020) A review of indigenous food crops in Africa and the implications for more sustainable and healthy food systems. Sustainability 12:3493. https://doi.org/10.3390/su12083493

Alam MK, Rana ZH, Islam SN, Akhtaruzzaman M (2019) Total phenolic content and antioxidant activity of methanolic extract of selected wild leafy vegetables grown in Bangladesh: a cheapest source of antioxidants. Potr S J F Sci 13:287–293. https://doi.org/10.5219/1107

Alemayehu G, Asfaw Z, Kelbessa E (2015) Plant diversity and ethnobotany in Berehet District, North Shewa Zone of Amhara Region (Ethiopia) with emphasis on wild edible plants. J Med Plants Stud 3:93–105

Aletor VA, Aladetimi OO (1989) Compositional evaluation of some cowpea varieties and some under-utilized edible legumes in Nigeria. Nahrung 33:999–1007. https://doi.org/10.1002/food.19890331023

Alexopoulos AA, Assimakopoulou A, Panagopoulos P, Bakea M, Vidalis N, Karapanos IC, Petropoulos SA (2021) Impact of salinity on the growth and chemical composition of two underutilized wild edible greens: *Taraxacum officinale* and *Reichardia picroides*. Horticulturae 7:160. https://doi.org/10.3390/horticulturae7070160

Ali A (2012) *Acacia nilotica*: a plant of multipurpose medicinal uses. J Med Plant Res 6. https://doi.org/10.5897/JMPR11.1275

Ameen A, Raza S (2018) Green revolution: a review. Int J Adv Sci Res 3:129. https://doi.org/10.7439/ijasr.v3i12.4410

Ara A, Arifuzzaman M, Ghosh CK, Hashem MA, Ahmad MU, Bachar SC, Nahar L, Sarker SD (2010) Anti-inflammatory activity of *Adenanthera pavonina* L., Fabaceae, in experimental animals. Rev Bras Farmacogn 20:929–932. https://doi.org/10.1590/S0102-695X2010005000039

Arinathan V, Mohan VR, DeBritto AJ, Murugan C (2007) Wild edibles used by Palliyars of the Western Ghats, Tamil Nadu. Indian J Tradit Knowl 6(1):163–168

Arulkumar M, Velmurugan C, Wesely EG, Bobby N, Kumar NV (2017) Phytochemical evaluation and antimicrobial efficacy of *Entada rheedii* Spreng. Glob J Res Anal 6(2):624–625

Asfaw Z, Tadesse M (2001) Prospects for sustainable use and development of wild food plants in Ethiopia. Econ Bot 55:47–62. https://doi.org/10.1007/BF02864545

Asprilla-Perea J, Díaz-Puente JM (2019) Importance of wild foods to household food security in tropical forest areas. Food Secur 11:15–22. https://doi.org/10.1007/s12571-018-0846-8

Balemie K, Kebebew F (2006) Ethnobotanical study of wild edible plants in Derashe and Kucha Districts, South Ethiopia. J Ethnobiol Ethnomed 2:53. https://doi.org/10.1186/1746-4269-2-53

Banik A, Nema S, Shankar D (2014) Wild edible tuber and root plants available in bastar region of Chhattisgarh. IJFCI 5(2):85–89. https://doi.org/10.15740/HAS/IJFCI/5.2/85-89

Bedi SJ (1978) Ethnobotany of the Ratan Mahal Hills, Gujarat, India. Econ Bot 32(3):278–284. https://doi.org/10.1007/BF02864701

Berihun T, Molla E (2017) Study on the diversity and use of wild edible plants in bullen district Northwest Ethiopia. J Bot 2017:1–10. https://doi.org/10.1155/2017/8383468

Bhadoriya SS, Ganeshpurkar A, Narwaria J, Rai G, Jain AP (2011) *Tamarindus indica*: extent of explored potential. Phcog Rev 5(9):73. https://doi.org/10.4103/0973-7847.79102

Bharucha Z, Pretty J (2010) The roles and values of wild foods in agricultural systems. Philos Trans R Soc B 365:2913–2926. https://doi.org/10.1098/rstb.2010.0123

Bhat R, Karim AA (2009) Exploring the nutritional potential of wild and underutilized legumes. Compr Rev Food Sci Food Saf 8:305–331. https://doi.org/10.1111/j.1541-4337.2009.00084.x

Bhatia H, Sharma YP, Manhas RK, Kumar K (2018) Traditionally used wild edible plants of district Udhampur, J&K, India. J Ethnobiol Ethnomed 14:73. https://doi.org/10.1186/s13002-018-0272-1

Bhattacharjee T, Sen D, Irenaeus TKS, Das SP, Chakraborty A (2019) Diversity of underutilized vegetable crops in Tripura. Acta Hortic 1241:121–134. https://doi.org/10.17660/ActaHortic.2019.1241.19

Bhusal KK, Magar SK, Thapa R, Lamsal A, Bhandari S, Maharjan R, Shrestha S, Shrestha J (2022) Nutritional and pharmacological importance of stinging nettle (*Urtica dioica* L.): a review. Heliyon 8:e09717. https://doi.org/10.1016/j.heliyon.2022.e09717

Biswas K, Das AP (2011) Documentation of wild leafy vegetables from the tribal dominated parts of Malda District of Paschimbanga, India. In: Ghosh C, Das AP (eds) Recent studies in biodiversity and traditional knowledge in India. Gour College & Sarat Book House, pp 301–306

Binu S (2010) Wild edible plants used by the tribals in Pathanamthitta district, Kerala. Indian J Tradit Knowl 9(2):309–312

Borelli T, Hunter D, Powell B, Ulian T, Mattana E, Termote C, Pawera L, Beltrame D, Penafiel D, Tan A, Taylor M, Engels J (2020) Born to eat wild: an integrated conservation approach to secure wild food plants for food security and nutrition. Plants 9:1299. https://doi.org/10.3390/plants9101299

Bvenura C, Sivakumar D (2017) The role of wild fruits and vegetables in delivering a balanced and healthy diet. Food Res Int 99:15–30. https://doi.org/10.1016/j.foodres.2017.06.046

Cambie RC, Ferguson LR (2003) Potential functional foods in the traditional Maori diet. Mutat Res 523–524:109–117. https://doi.org/10.1016/S0027-5107(02)00344-5

Chahar K (2013) *Mesua ferrea* L.: a review of the medical evidence for its phytochemistry and pharmacological actions. Afr J Pharm Pharmacol 7:211–219. https://doi.org/10.5897/AJPP12.895

Chaithanya VV, Sugiya N, Aneesh KS (2015) Wild edible plants traditionally used by Kadar tribes of Vazhachal forest division, Thrissur, Kerala. J Glob Biosci 4:3344–3349

Chaithra M, Thomas B (2017) Traditional worshiping plants from selected sacred groves of Kozhikode district, Kerala, India. Res J Recent Sci 6:7–13

Chakravarti AK (1973) Green revolution in India. Ann Assoc Am Geogr 63(3):319–330. https://doi.org/10.1111/j.1467-8306.1973.tb00929.x

Chakravarty S, Bhutia KD, Suresh CP, Shukla G, Pala NA (2016) A review on diversity, conservation and nutrition of wild edible fruits. JANS 8:2346–2353. https://doi.org/10.31018/jans.v8i4.1135

Chand R, Singh AN, Nirmala C (2017) Ethnoecological survey of underutilized plant diversity of Hamirpur District, Himachal Pradesh, India: an edibility assessment. EER 5(1):13–29. https://doi.org/10.13189/eer.2017.050103

Chand ARJ, Azeez K (2021) Underutilised fruits of Agasthyamala Biosphere Reserve: a potential source of natural antioxidants. Vegetos 34:447–452. https://doi.org/10.1007/s42535-021-00200-6

Chandrasekara A, Josheph Kumar T (2016) Roots and tuber crops as functional foods: a review on phytochemical constituents and their potential health benefits. Int J Food Sci 2016:1–15. https://doi.org/10.1155/2016/3631647

Chauhan D, Shrivastava AK, Patra S (2014) Diversity of leafy vegetables used by tribal peoples of Chhattisgarh, India. Int J Curr Microbiol App Sci 3(4):611–622

Cheikhyoussef A, Embashu W (2013) Ethnobotanical knowledge on indigenous fruits in Ohangwena and Oshikoto regions in Northern Namibia. J Ethnobiol Ethnomed 9:34. https://doi.org/10.1186/1746-4269-9-34

Cheng Z, Lu X, Lin F, Naeem A, Long C (2022) Ethnobotanical study on wild edible plants used by Dulong people in northwestern Yunnan, China. J Ethnobiol Ethnomed 18:3. https://doi.org/10.1186/s13002-022-00501-3

Chhetri HP, Yogol NS, Sherchan J, Anupa KC, Mansoor S, Thapa P (1970) Phytochemical and antimicrobial evaluations of some medicinal plants of Nepal. Kathmandu Univ J Sci Eng Technol 4:49–54. https://doi.org/10.3126/kuset.v4i1.2883

Chibarabada T, Modi A, Mabhaudhi T (2017) Expounding the value of grain legumes in the semi- and arid tropics. Sustainability 9:60. https://doi.org/10.3390/su9010060

Chivenge P, Mabhaudhi T, Modi A, Mafongoya P (2015) The potential role of neglected and underutilised crop species as future crops under water scarce conditions in Sub-Saharan Africa. IJERPH 12:5685–5711. https://doi.org/10.3390/ijerph120605685

Cohen S, Koltai H, Selvaraj G, Mazuz M, Segoli M, Bustan A, Guy O (2020) Assessment of the nutritional and medicinal potential of tubers from hairy Stork's-Bill (*Erodium crassifolium* L 'Hér), a wild plant species inhabiting arid Southeast Mediterranean Regions. Plants 9:1069. https://doi.org/10.3390/plants9091069

Cunningham AB, Brinckmann JA, Bi Y-F, Pei S-J, Schippmann U, Luo P (2018) Paris in the spring: a review of the trade, conservation and opportunities in the shift from wild harvest to cultivation of *Paris polyphylla* (Trilliaceae). J Ethnopharmacol 222:208–216. https://doi.org/10.1016/j.jep.2018.04.048

Dabi T, Khanna VK (2018) Effect of climate change on rice. Agrotechnology 07:1–7. https://doi.org/10.4172/2168-9881.1000181

Dansi A, Adjatin A, Adoukonou-Sagbadja H, Faladé V, Yedomonhan H, Odou D, Dossou B (2008) Traditional leafy vegetables and their use in the Benin Republic. Genet Resour Crop Evol 55: 1239–1256. https://doi.org/10.1007/s10722-008-9324-z

Dansi A, Vodouhè R, Azokpota P, Yedomonhan H, Assogba P, Adjatin A, Loko YL, Dossou-Aminon I, Akpagana K (2012) Diversity of the neglected and underutilized crop species of importance in Benin. Sci World J 2012:1–19. https://doi.org/10.1100/2012/932947

Das AK, Stalin N, Muthumperumal C, Swamy PS (2013) Wild plants used by Muthuvan and Kattunaikkan tribal communities of Palakkayam settlement in Nilambur of Malappuram district, Kerala. Med Plnts Int Jrnl Phyt Rela Ind 5(2):82. https://doi.org/10.5958/j.0975-6892.5.2.013

Davis KF, Chhatre A, Rao ND, Singh D, Ghosh-Jerath S, Mridul A, Poblete-Cazenave M, Pradhan N, DeFries R (2019) Assessing the sustainability of post-Green Revolution cereals in India. Proc Natl Acad Sci U S A 116:25034–25041. https://doi.org/10.1073/pnas.1910935116

Della A, Paraskeva-Hadjichambi D, Hadjichambis AC (2006) An ethnobotanical survey of wild edible plants of Paphos and Larnaca countryside of Cyprus. J Ethnobiol Ethnomed 2:34. https://doi.org/10.1186/1746-4269-2-34

Dhyani A, Chopra R, Garg M (2019) A review on nutritional value, functional properties and pharmacological application of perilla (*Perilla frutescens* L.). Biomed Pharmacol J 12:649–660

Dube P, Heijman WJM, Ihle R, Ochieng J (2017) The potential of traditional leafy vegetables for improving food security in Africa. In: Establishing food security and alternatives to international trade in emerging economies. IGI Global, pp 220–243

Duguma HT (2020) Wild edible plant nutritional contribution and consumer perception in Ethiopia. Int J Food Sci 2020:1–16. https://doi.org/10.1155/2020/2958623

Dutta B (2015) Food and medicinal values of certain species of *Dioscorea* with special reference to Assam. J Pharmacogn Phytochem 3:15–18

Ebert A (2014) Potential of underutilized traditional vegetables and legume crops to contribute to food and nutritional security, income and more sustainable production systems. Sustainability 6: 319–335. https://doi.org/10.3390/su6010319

Enyiukwu DN, Awurum AN, Nwaneri JA (2014) Potentials of Hausa Potato (*Solenostemon rotundifolius* (Poir.) J. K. Morton) and management of its tuber rot in Nigeria. GJAFH 2:027–037. https://doi.org/10.15580/GJAFH.2014.2.010314008

FAO (2002) The state of food and agriculture, 2002. Agriculture and global public goods ten years after the earth summit. https://www.fao.org/documents/card/en/c/Y6000EN

FAO (2012) The state of food insecurity in the world: economic growth is necessary but not sufficient to accelerate reduction of hunger and malnutrition. FAO, Rome

FAO (2017) The future of food and agriculture: trends and challenges. FAO, Rome, pp 11–16

FAO (2018a) The future of food and agriculture: trends and challenges. FAO, Rome

FAO (2018b) Working for zero hunger. FAO, Rome

FAO (ed) (2019) The state of food insecurity in the world: eradicating world hunger—taking stock ten years after the world food summit. FAO, Rome

FAO, IFAD, UNICEF, WFP, WHO (2020) In brief to the state of food security and nutrition in the world 2020. Transforming food systems for affordable healthy diets. FAO, Rome, pp 1–60. https://doi.org/10.4060/ca9699en

Feyssa DH, Njoka JT, Asfaw Z, Nyangito MM (2011) Seasonal availability and consumption of wild edible plants in semiarid Ethiopia: implications to food security and climate change adaptation. J Hortic For 3(5):138–149

Fones HN, Bebber DP, Chaloner TM, Kay WT, Steinberg G, Gurr SJ (2020) Threats to global food security from emerging fungal and oomycete crop pathogens. Nat Food 1:332–342. https://doi.org/10.1038/s43016-020-0075-0

Fowler C, Mooney P (1990) Shattering: food, politics, and the loss of genetic diversity. University of Arizona Press, Tucson

Gahatraj S, Bhusal B, Sapkota K, Dhami B, Gautam D (2020) Common medicinal plants of Nepal: a review of Triphala: Harro (*Terminalia chebula*), Barro (*Terminalia bellirica*), and Amala (*Emblica officinalis*). https://doi.org/10.13140/RG.2.2.15807.84643

Gahukar RT (2015) Local plants for rural food security. In: Lichtfouse E (ed) Sustainable agriculture reviews. Springer International Publishing, Cham, pp 141–158

Galil J (1984) *Ficus religiosa* L.—the tree-splitter. Bot J Linn Soc 88:185–203. https://doi.org/10.1111/j.1095-8339.1984.tb01570.x

García-Herrera P, Sánchez-Mata M d C (2016) The contribution of wild plants to dietary intakes of micronutrients (II): mineral elements. In: Sánchez-Mata M d C, Tardío J (eds) Mediterranean wild edible plants. Springer, New York, pp 141–171

George MV, Christopher G (2020) Structure, diversity and utilization of plant species in tribal homegardens of Kerala, India. Agrofor Syst 94:297–307. https://doi.org/10.1007/s10457-019-00393-5

Ghanbari S, Weiss G, Liu J, Eastin I, Fathizadeh O, Moradi G (2022) Potentials and opportunities of wild edible forest fruits for rural household's economy in Arasbaran, Iran. Forests 13:453. https://doi.org/10.3390/f13030453

Ghimeray AK, Sharma P, Ghimire B, Lamsal K, Ghimire B, Cho H (2010) Wild edible angiospermic plants of Illam Hills (Eastern Nepal) and their mode of use by the local community. Korean J Plant Taxon 40:1–4

Godfray HCJ, Garnett T (2014) Food security and sustainable intensification. Philos Trans R Soc B 369:20120273. https://doi.org/10.1098/rstb.2012.0273

Godfray HCJ, Beddington JR, Crute IR, Haddad L, Lawrence D, Muir JF, Pretty J, Robinson S, Thomas SM, Toulmin C (2010) Food security: the challenge of feeding 9 billion people. Science 327:812–818. https://doi.org/10.1126/science.1185383

Graham PH, Vance CP (2003) Legumes: importance and constraints to greater use. Plant Physiol 131:872–877. https://doi.org/10.1104/pp.017004

Guigma Y, Ouédraogo A, Zerbo P, Millogo-Rasolodimby J (2014) The use of wild plants as food in three adjoining villages in Southern Burkina Faso. J Nutr Ecol Food Res 2(2):105–115. https://doi.org/10.1166/jnef.2014.1076

Gupta S, Jyothi Lakshmi A, Manjunath MN, Prakash J (2005) Analysis of nutrient and antinutrient content of underutilized green leafy vegetables. LWT Food Sci Technol 38(4):339–345. https://doi.org/10.1016/j.lwt.2004.06.012

Gupta S, Srivastava A, Lal EP (2017) Indigenous leafy vegetables for food and nutritional security in two districts of Jharkhand, India. J Pharmacogn Phytochem 6(6):901–909

Guyu DF, Muluneh W-T (2015) Wild foods (plants and animals) in the green famine belt of Ethiopia: do they contribute to household resilience to seasonal food insecurity? For Ecosyst 2:34. https://doi.org/10.1186/s40663-015-0058-z

Haridas R, Kunhikannan C (2020) Food plants of the Cholanaikkan and Kattunaikkan communities of Nilambur taluk, Malappuram District, Kerala, India. Res J Recent Sci 9:12–18

Havinga RM, Hartl A, Putscher J, Prehsler S, Buchmann C, Vogl CR (2010) *Tamarindus indica* L. (Fabaceae): patterns of use in traditional African medicine. J Ethnopharmacol 127:573–588. https://doi.org/10.1016/j.jep.2009.11.028

Hickey GM, Pouliot M, Smith-Hall C, Wunder S, Nielsen MR (2016) Quantifying the economic contribution of wild food harvests to rural livelihoods: a global-comparative analysis. Food Policy 62:122–132. https://doi.org/10.1016/j.foodpol.2016.06.001

Holmgren K, Öberg H (2006) Climate change in Southern and Eastern Africa during the past millennium and its implications for societal development. Environ Dev Sustain 8:185–195. https://doi.org/10.1007/s10668-005-5752-5

Huai KH-Y, Pei S-J (2000) Wild plants in the diet of arhorchin mongol herdsmen in inner Mongolia. Econ Bot 54:528–536. https://doi.org/10.1007/BF02866550

Huang L, Raats D, Sela H, Klymiuk V, Lidzbarsky G, Feng L, Krugman T, Fahima T (2016) Evolution and adaptation of wild emmer wheat populations to biotic and abiotic stresses. Annu Rev Phytopathol 54:279–301. https://doi.org/10.1146/annurev-phyto-080614-120254

Jahan N, Ahmed W, Malik A (1995) New steroidal glycosides from *Mimusops elengi*. J Nat Prod 58:1244–1247. https://doi.org/10.1021/np50122a014

Jain AK, Tiwari P (2012) Nutritional value of some traditional edible plants used by tribal communities during emergency with reference to Central India. Indian J Tradit Knowl 11:51–57

Jain A, Bhadoriya S, Ganeshpurkar A, Narwaria J, Rai G (2011) *Tamarindus indica*: extent of explored potential. Phcog Rev 5:73. https://doi.org/10.4103/0973-7847.79102

Jarapala SR (2017) Nutrition science in India: green leafy vegetables: a potent food source to alleviate micronutrient deficiencies. Int Res J Basic Appl Sci 2:7–13

Ji X, Peng Q, Yuan Y, Shen J, Xie X, Wang M (2017) Isolation, structures and bioactivities of the polysaccharides from jujube fruit (*Ziziphus jujuba* Mill.): a review. Food Chem 227:349–357. https://doi.org/10.1016/j.foodchem.2017.01.074

Jordanovska S, Jovovic Z, Andjelkovic V (2020) Potential of wild species in the scenario of climate change. In: Salgotra RK, Zargar SM (eds) Rediscovery of genetic and genomic resources for future food security. Springer, Singapore, pp 263–301

Ju Y, Zhuo J, Liu B, Long C (2013) Eating from the wild: diversity of wild edible plants used by Tibetans in Shangri-la region, Yunnan, China. J Ethnobiol Ethnomed 9:28. https://doi.org/10.1186/1746-4269-9-28

Kapelari S, Alexopoulos G, Moussouri T, Sagmeister KJ, Stampfer F (2020) Food heritage makes a difference: the importance of cultural knowledge for improving education for sustainable food choices. Sustainability 12(4):1509. https://doi.org/10.3390/su12041509

Karanja A, Ickowitz A, Stadlmayr B, McMullin S (2022) Understanding drivers of food choice in low- and middle-income countries: a systematic mapping study. Glob Food Secur 32:100615. https://doi.org/10.1016/j.gfs.2022.100615

Kerr RB (2012) Lessons from the old Green Revolution for the new: social, environmental and nutritional issues for agricultural change in Africa. Prog Dev Stud 12:213–229. https://doi.org/10.1177/146499341101200308

Khan A, Gilani SS, Hussain F, Durrani MJ (2003) Ethnobotany of Gokand Valley, District Buner, Pakistan. Pak J Biol Sci 6(4):363–369. https://doi.org/10.3923/pjbs.2003.363.369

Khan H, Jan SA, Javed M, Shaheen R, Khan Z, Ahmad A, Safi SZ, Imran M (2015) Nutritional composition, antioxidant and antimicrobial activities of selected wild edible plants: composition and biological activities of plants. J Food Biochem 40(1):61–70. https://doi.org/10.1111/jfbc.12189

Khomdram SD, Fanai L, Yumkham SD (2019) Local knowledge of edible flowers used in Mizoram. Indian J Tradit Knowl 18:714–723

Khoury CK, Castañeda-Alvarez NP, Achicanoy HA, Sosa CC, Bernau V, Kassa MT, Norton SL, van der Maesen LJG, Upadhyaya HD, Ramírez-Villegas J, Jarvis A, Struik PC (2015) Crop wild relatives of pigeonpea [*Cajanus cajan* (L.) Millsp.]: distributions, ex situ conservation status, and potential genetic resources for abiotic stress tolerance. Biol Conserv 184:259–270. https://doi.org/10.1016/j.biocon.2015.01.032

Kidane L, Kejela A (2021) Food security and environment conservation through sustainable use of wild and semi-wild edible plants: a case study in Berek Natural Forest, Oromia special zone, Ethiopia. Agric Food Secur 10:29. https://doi.org/10.1186/s40066-021-00308-7

Kipkoech C, Kipkosgei D, Murgor FAC (2013) Climate change and food security. In: Silvern S (ed) Environmental change and sustainability. InTech

Konsam S, Thongam B, Handique AK (2016) Assessment of wild leafy vegetables traditionally consumed by the ethnic communities of Manipur, northeast India. J Ethnobiol Ethnomed 12:9. https://doi.org/10.1186/s13002-016-0080-4

Krishnamoorthy K, Subramaniam P (2014) Phytochemical profiling of leaf, stem, and tuber parts of *Solena amplexicaulis* (Lam.) Gandhi using GC-MS. Int Sch Res Notices 2014:e567409. https://doi.org/10.1155/2014/567409

Kuddus M, Ginawi I, AlHazimi A (2013) *Cannabis sativa*: an ancient wild edible plant of India. Emir J Food Agric 25:736. https://doi.org/10.9755/ejfa.v25i10.16400

Kumar S, Hamal IA (2009) Wild edibles of Kishtwar High Altitude National Park in Northwest Himalaya, Jammu and Kashmir (India). Ethnobotanical Leaflets:295–202

Kumar S (2017) Yam (*Dioscorea* species): future functional wild food of tribal Odisha, India. In: At the crossroads between nutrition and pharmacology, vol 2, pp 165–185

Kumar S, Kamboj J, Suman, Sharma S (2011) Overview for various aspects of the health benefits of *Piper longum* Linn. fruit. J Acupunct Meridian Stud 4:134–140. https://doi.org/10.1016/S2005-2901(11)60020-4

Kumar RP, Shammy J, Nitin G, Rinu R (2014) An inside review of *Amaranthus spinosus* Linn: a potential medicinal plant of India. Int J Res Pharm Chem 4(3):643–653

Kumari B, Solanki H (2019) The traditional knowledge of wild edible leaf used by tribal people in Chhattisgarh. IJPE 5(04):284–292. https://doi.org/10.18811/ijpen.v5i04.9

Kunhikannan C, Venkatasubramanian N, Sivalingam R, Pramod Kumar N, Thomas S, Thampan S (2011) Diversity of woody species in the Medicinal Plant Conservation Area (MPCA) of Silent Valley National Park, Kerala. Biodiversity 12:97–107. https://doi.org/10.1080/14888386.2011.602930

Ladio AH, Lozada M (2003) Comparison of wild edible plant diversity and foraging strategies in two aboriginal communities of northwestern Patagonia. Biodivers Conserv 12:937–951. https://doi.org/10.1023/A:1022873725432

Lalonde A (1991) African indigenous knowledge and its relevance to environment and development activities. Canadian International Development Agency, pp 1–12

Lem FF, Yong YS, Goh S, Chin SN, Chee FT (2022) Withanolides, the hidden gem in *Physalis minima*: a mini review on their anti-inflammatory, anti-neuroinflammatory and anti-cancer effects. Food Chem 377:132002. https://doi.org/10.1016/j.foodchem.2021.132002

Lim TK (2012) *Prunus domestica* subsp. insititia. In: Edible medicinal and non-medicinal plants. Springer, Netherlands, Dordrecht, pp 476–479

Liu Q, Liu J, Zhang P, He S (2014) Root and tuber crops. In: Encyclopedia of agriculture and food systems. Elsevier, pp 46–61

Lobell DB, Schlenker W, Costa-Roberts J (2011) Climate trends and global crop production since 1980. Science 333:616–620. https://doi.org/10.1126/science.1204531

Lokhande KS (2020) Ethnobotanical survey on wild edible plants used by tribals & rural people of Arjuni/Mor Taluka, Gondia District, Maharashtra State, India. AZB 8(3):209–217. https://doi.org/10.13189/azb.2020.080317

Lulekal E, Asfaw Z, Kelbessa E, Van Damme P (2011a) Wild edible plants in Ethiopia: a review on their potential to combat food insecurity. Afrika Focus 24. https://doi.org/10.21825/af.v24i2.4998

Lulekal E, Asfaw Z, Kelbessa E, Van Damme P (2011b) Wild edible plants in Ethiopia: a review on their potential to combat food insecurity. AFOC 24(2):71–122. https://doi.org/10.1163/2031356X-02402006

Madhurima P, Kuppast IJ, Mankani KL (2012) A review on *Amorphophallus paeoniifolius*. Int J Adv Sci Res Technol 2:99–111

Mahapatra AK, Panda PC (2012) Wild edible fruit diversity and its significance in the livelihood of indigenous tribals: evidence from eastern India. Food Sec 4:219–234. https://doi.org/10.1007/s12571-012-0186-z

Mahendra T, Krishan P, Rai SK (2017) Traditional use of wild plants for food in West Sikkim. Int J Life Sci 5:730–741

Mahklouf MH (2019) Ethnobotanical study of edible wild plants in Libya. Eur J Ecol 5:30–40. https://doi.org/10.2478/eje-2019-0011

Maikhuri RK, Rao RS, Saxena SK, Semwal RL (1999) Traditional crop-diversity based nutrition and the prospects for sustainable rural development in the central Himalaya. Himalayan Paryavaran 6:36–42

Malkanthi SHP (2017) Importance of underutilized crops in Thanamalwila Divisional Secretariat Division in Monaragala District in Sri Lanka. J Agric Sci 12:197. https://doi.org/10.4038/jas.v12i3.8266

Mallick SN, Sahoo T, Naik SK, Panda PC (2020) Ethnobotanical study of wild edible food plants used by the tribals and rural populations of Odisha, India for food and livelihood security. Plant Archives 20(1):661–669

Mandal P, Misra T, Singh I, Das K, Bhunia M (2009) Free-radical-scavenging activity in the inflorescence of European nettle/sisnu (*Urtica dioica* L.). J Young Pharm 1:129. https://doi.org/10.4103/0975-1483.55745

Maphosa Y, Jideani VA (2017) The role of legumes in human nutrition. In: Hueda MC (ed) Functional food—improve health through adequate food. InTech

Marcone MF, Jahaniaval F, Aliee H, Kakuda Y (2003) Chemical characterization of *Achyranthes bidentata* seed. Food Chem 81:7–12. https://doi.org/10.1016/S0308-8146(02)00250-9

María Elena JH, Yamilet Irene GG, Iván YG, Migdalia MM (2019) Chemical study and determination of the antioxidant activity of three varieties *Tropaeolum tuberosum* (Mashua). AJPS 10: 2279–2297. https://doi.org/10.4236/ajps.2019.1012159

Maroyi A (2011) An ethnobotanical survey of medicinal plants used by the people in Nhema communal area, Zimbabwe. J Ethnopharmacol 136(2):347–354. https://doi.org/10.1016/j.jep.2011.05.003

Mary S, Nair NV, Chaturvedi PK, Selvi A (2006) Analysis of genetic diversity among *Saccharum spontaneum* L. From four geographical regions of india, using molecular markers. Genet Resour Crop Evol 53:1221–1231. https://doi.org/10.1007/s10722-005-2433-z

Maseko I, Mabhaudhi T, Tesfay S, Araya H, Fezzehazion M, Plooy C (2017) African leafy vegetables: a review of status, production and utilization in South Africa. Sustainability 10: 16. https://doi.org/10.3390/su10010016

Mathieu G, Meissa D (2008) Traditional leafy vegetables in Senegal: diversity and medicinal uses. Afr J Tradit Complement Altern Med 4:469. https://doi.org/10.4314/ajtcam.v4i4.31239

Matic I, Guidi A, Kenzo M, Mattei M, Galgani A (2018) Investigation of medicinal plants traditionally used as dietary supplements: a review on *Moringa oleifera*. J Public Health Afr 9. https://doi.org/10.4081/jphia.2018.841

Matthews PJ, Ghanem ME (2021) Perception gaps that may explain the status of taro (*Colocasia esculenta*) as an "orphan crop". Plants People Planet 3(2):99–112. https://doi.org/10.1002/ppp3.10155

Matthews PJ, Agoo MG, Tandang DN, Madulid DA (2012) Ethnobotany and ecology of wild Taro (*Colocasia esculenta*) in the Philippines: implications for domestication and dispersal. Senri Ethnol Stud 78:307–340

Maundu PM (1997) The status of traditional vegetable utilization in Kenya. National Museums of Kenya. IPGRI

Mbhenyane XG (2017) Indigenous foods and their contribution to nutrient requirements. South Afr J Clin Nutr 30:5–7

Mc Carthy U, Uysal I, Badia-Melis R, Mercier S, O'Donnell C, Ktenioudaki A (2018) Global food security—issues, challenges and technological solutions. Trends Food Sci Technol 77:11–20. https://doi.org/10.1016/j.tifs.2018.05.002

Mehra PN, Mittal TC (1962) Pharmacognosy of Indian substitutes of male fern II. J Pharm Sci 51: 367–371. https://doi.org/10.1002/jps.2600510417

Menendez-Baceta G, Aceituno-Mata L, Tardío J, Reyes-García V, Pardo-de-Santayana M (2012) Wild edible plants traditionally gathered in Gorbeialdea (Biscay, Basque Country). Genet Resour Crop Evol 59:1329–1347. https://doi.org/10.1007/s10722-011-9760-z

Mirunalini S, Krishnaveni M (2010) Therapeutic potential of *Phyllanthus emblica* (amla): the ayurvedic wonder. J Basic Clin Physiol Pharmacol 21. https://doi.org/10.1515/JBCPP.2010.21.1.93

Modi M, Modi A, Hendriks S (2006) Potential role for wild vegetables in household food security: a preliminary case study in Kwazulu-Natal, South Africa. Afr J Food Agric Nutr Dev 6. https://doi.org/10.4314/ajfand.v6i1.19167

Mohanty S, Mohanty BP (2009) Global climate change: a cause of concern. Sci Technol Policy Issue 32:149–156

Morton S, Pencheon D, Squires N (2017) Sustainable Development Goals (SDGs), and their implementation. Br Med Bull 1–10. https://doi.org/10.1093/bmb/ldx031

Mosissa D, Abraha B (2018) Tuber wild plants of Shinasha people as food and medicine in bullen district, Northwest of Ethiopia. Biodivers Int J 2(1). https://doi.org/10.15406/bij.2018.02.00043

Motti R (2022) Wild edible plants: a challenge for future diet and health. Plants 11:344. https://doi.org/10.3390/plants11030344

Mudannayake D, Wimalasiri K, Silva K, Ajlouni S (2015) Selected Sri Lankan food plants and other herbs as potential sources of inulin-type fructans. J Natl Sci Found Sri Lanka 43(1):35. https://doi.org/10.4038/jnsfsr.v43i1.7913

Mujuru L, Jimu L, Mureva A, Mapaura A, Nyakudya IW, Muvengwi J (2020) Diversity of local knowledge on use of wild food and medicinal plants in communities around five biodiversity hotspots in Zimbabwe. Adv Tradit Med 20:663–671. https://doi.org/10.1007/s13596-020-00512-z

Mundaragi A, Thangadurai D, Appaiah KAA, Dandin CJ, Sangeetha J (2019) Physicochemical characterization, antioxidant potential and sensory quality of wine from wild edible fruits of *Flacourtia montana*. J Graham Braz Arch Biol Technol 62:e19180377. https://doi.org/10.1590/1678-4324-2019180377

Muniappan R, Reddy GVP, Raman A (2009) *Coccinia grandis* (L.) Voigt (Cucurbitaceae). In: Raman A, Reddy GVP, Muniappan R (eds) Biological control of tropical weeds using arthropods. Cambridge University Press, Cambridge, pp 175–182

Musarella CM (2020) *Solanum torvum* Sw. (Solanaceae): a new alien species for Europe. Genet Resour Crop Evol 67:515–522. https://doi.org/10.1007/s10722-019-00822-5

Musinguzi E, Kikafunda J, Kiremire B (2011) Utilization of indigenous food plants in Uganda: a case study of south-western Uganda. Afr J Food Agric Nutr Dev 6. https://doi.org/10.4314/ajfand.v6i2.71738

Musvoto C, Nortje K, de Wet B, Mahumani BK, Nahman A (2015) Imperatives for an agricultural green economy in South Africa. SAJS 111:1–8. https://doi.org/10.17159/sajs.2015/20140026

Mutaqin AZ, Kurniadie D, Iskandar J, Nurzaman M, Partasasmita R (2020) Ethnobotany of suweg, *Amorphophallus paeoniifolius*: utilization and cultivation in West Java, Indonesia. Biodiversitas 21(4). https://doi.org/10.13057/biodiv/d210444

Muthoni J, Nyamongo DO (2010) Traditional food crops and their role in food and nutritional security in Kenya. J Agric Food Inf 11:36–50. https://doi.org/10.1080/10496500903466745

Mutie FM, Rono PC, Kathambi V, Hu G-W, Wang Q-F (2020) Conservation of wild food plants and their potential for combating food insecurity in Kenya as exemplified by the drylands of Kitui County. Plants 9(8):1017. https://doi.org/10.3390/plants9081017

Naik R, Borkar SD, Bhat S, Acharya R (2017) Therapeutic potential of wild edible vegetables—a review. J Ayurveda Integr Med Sci 2:85–97

Narayanan MKR, Anilkumar N, Balakrishnan V, Sivadasan M, Alfarhan HA, Alaar AA (2011) Wild edible plants used by the Kattunaikka, Paniya and Kuruma tribes of Wayanad district, Kerala, India. J Med Plants Res 5(15):3520–3529

Natesh NH, Abbey L, Asiedu SK (2017) An overview of nutritional and anti nutritional factors in green leafy vegetables. HIJ 1:58–65. https://doi.org/10.15406/hij.2017.01.00011

Nazarudeen A, Seeni S, Koshy KC, Pushpangadan P (1996) Folk plants of food, medicine, adornment and repellent used by the paniyar community in North Kerala. J Econ Taxon Bot 12:299–305

Nguyen CH, Averyanov L, Egorov A, Nguyen CV, Tran TT (2021) Edible wild plants in the flora of Pu Luong Nature Reserve, Thanh Hoa Province, northern Vietnam. IOP Conf Ser: Earth Environ Sci 876(1):012054. https://doi.org/10.1088/1755-1315/876/1/012054

Nikolskii YN, Aidarov IP, Landeros-Sanchez C, Pchyolkin VV (2019) Impact of long-term freshwater irrigation on soil fertility. Irrig Drain 68:993–1001. https://doi.org/10.1002/ird.2381

OECD (2013) Global food security: challenges for the food and agricultural system. https://www.oecd.org/publications/globalfood-security-9789264195363-en.htm

Ogle BM, Hung PH, Tuyet HT (2001) Significance of wild vegetables in micronutrient intakes of women in Vietnam: an analysis of food variety. Asia Pac J Clin Nutr 10:21–30. https://doi.org/10.1046/j.1440-6047.2001.00206.x

Ogunjinmi AA, Ijeomah HM, Aiyeloja AA (2009) Socio-economic importance of bamboo (*Bambusa vulgaris*) in Borgu Local Government Area of Niger State, Nigeria. J Sustain Dev Afr 10:284–298

Ojelel S, Mucunguzi P, Katuura E, Kakudidi EK, Namaganda M, Kalema J (2019) Wild edible plants used by communities in and around selected forest reserves of Teso-Karamoja region, Uganda. J Ethnobiol Ethnomed 15:3. https://doi.org/10.1186/s13002-018-0278-8

Ojha R, Devkota HP (2021) Edible and medicinal pteridophytes of Nepal: a review. Ethnobot Res Appl 22:1–16

Ong HG, Kim Y-D (2017) The role of wild edible plants in household food security among transitioning hunter-gatherers: evidence from the Philippines. Food Secur 9:11–24. https://doi.org/10.1007/s12571-016-0630-6

Orech FO, Christensen DL, Larsen T, Friis H, Aagaard-Hansen J, Estambale BA (2007) Mineral content of traditional leafy vegetables from western Kenya. Int J Food Sci Nutr 58:595–602. https://doi.org/10.1080/09637480701350288

Padulosi S, Bergamini N, Lawrence T (eds) (2012) On-farm conservation of neglected and underutilized species: status, trends and novel approaches to cope with climate change. Bioversity International, Rome

Pandey A, Malav PK, Rai MK, Ahlawat SP (2021) 'Neodomesticates' of the Himalayan allium spices (*Allium* species) in Uttarakhand, India and studies on eco-geography and morphology. Genet Resour Crop Evol 68:2167–2179. https://doi.org/10.1007/s10722-021-01164-x

Patade VY, Kumar K, Gupta AK, Grover A, Negi PS, Dwivedi SK (2020) Improvement in seed germination through pretreatments in Timur (*Zanthoxylum armatum* DC.): a plant with high medicinal, economical and ecological importance. Natl Acad Sci Lett 43:295–297. https://doi.org/10.1007/s40009-019-00851-9

Patil AS, Paikrao HM, Patil SR (2013) *Passiflora foetida* Linn: a complete morphological and phytopharmacological review. Int J Pharm Bio Sci 4:285–296

Peris NW, Gacheri KM, Theophillus MM, Lucas N (eds) (2014) Morphological characterization of mulberry (*Morus* spp.) accessions grown in Kenya. Sustain Agric Res. https://doi.org/10.22004/ag.econ.230528

Pfoze NL, Kumar Y, Myrboh B (2011) Survey and assessment of floral diversity on wild edible plants from Senapati district of Manipur, Northeast India. J Biodivers Environ Sci 1:50–62

Pieroni A (2001) Evaluation of the cultural significance of wild food botanicals traditionally consumed in Northwestern Tuscany. J Ethnobiol 21:89–104

Pinela J, Carvalho AM, Ferreira ICFR (2017) Wild edible plants: nutritional and toxicological characteristics, retrieval strategies and importance for today's society. Food Chem Toxicol 110: 165–188. https://doi.org/10.1016/j.fct.2017.10.020

Pingali PL (2012) Green Revolution: impacts, limits, and the path ahead. Proc Natl Acad Sci U S A 109:12302–12308. https://doi.org/10.1073/pnas.0912953109

Prabhu A, Krishnamoorthy M (2010) Anticancer activity of *Trigonella foenum-graecum* on Ehrlich Ascites carcinoma in *Mus musculus* system. J Pharm Res 3:1181–1183

Pradhan DK, Lepcha N, Lama R (2021) Traditional lepcha hats in Sikkim Himalaya. Himalayan J Soc Sci Human 16. https://doi.org/10.51220/hjssh.v16i1.13

Pradheep K, Joseph John K, Latha M, Suma A (2021) Status of crop plants of agricultural importance in Kerala state, India: an update. Genet Resour Crop Evol 68:1849–1873. https://doi.org/10.1007/s10722-020-01100-5

Prashith KT, Vinayaka K, Mallikarjun N, Bharath A, Shailendra KB, Rakesh KM, Vinod KH (2011) Antibacterial, insecticidal and free radical scavenging activity of methanol extract of

Ziziphus rugosa Lam. (Rhamnaceae) fruit pericarp. Pharmacogn J 2:65–69. https://doi.org/10.1016/S0975-3575(11)80028-3

Pudasaini N, Sthapit SR, Gauchan D, Bhandari B, Joshi BK, Sthapit BR (2016) Baseline Survey Report: I. Jungu, Dolakha. Integrating traditional crop genetic diversity into technology: using a biodiversity portfolio approach to buffer against unpredictable environmental change in the Nepal Himalayas. Pokhara, Nepal: LI-BIRD, NARC, DOA and Bioversity International. https://doi.org/10.13140/RG.2.2.11711.51360

Puttahariyappa HR, Setty RS, Ravikanth G (2021) Dependency and economic benefits of use of wild food plants use among tribal communities in Malai Madeshawara Hills wildlife sanctuary, Southern India. https://doi.org/10.17170/KOBRA-202011192211

Rahman FB, Ahmed S, Noor P, Rahman MMM, Huq SMA, Akib MTE, Shohael AM (2020) A comprehensive multi-directional exploration of phytochemicals and bioactivities of flower extracts from *Delonix regia* (Bojer ex Hook.) Raf., *Cassia fistula* L. and *Lagerstroemia speciosa* L. Biochem Biophys Rep 24:100805. https://doi.org/10.1016/j.bbrep.2020.100805

Raj S, Jayaraj R, Kodiveri Muthukaliannan G (2022) Chemical profiling and evaluation of antioxidant and anticancer potential of tuber crop *Amorphophallus commutatus* var. *wayanadensis*. Plant Foods Hum Nutr 77:68–76. https://doi.org/10.1007/s11130-021-00942-3

Raji R, Siril EA (2021) Genetic diversity analysis of promising Ceylon olive (*Elaeocarpus serratus* L.) genotypes using morphological traits and ISSR markers. Curr Plant Biol 26:100201. https://doi.org/10.1016/j.cpb.2021.100201

Raju RI, Roy SK (2016) Mass propagation of *Bambusa bambos* (L.) Voss through in vitro culture. Jahangirnagar Univ J Biol Sci 5:15–26. https://doi.org/10.3329/jujbs.v5i2.32514

Rana JC, Pradheep K, Chaurasia OP, Sood S, Sharma RM, Singh A, Negi R (2012) Genetic resources of wild edible plants and their uses among tribal communities of cold arid region of India. Genet Resour Crop Evol 59:135–149. https://doi.org/10.1007/s10722-011-9765-7

Ranasinghe RASN, Maduwanthi SDT, Marapana RAUJ (2019) Nutritional and health benefits of jackfruit (*Artocarpus heterophyllus* Lam.): a review. Int J Food Sci 2019:1–12. https://doi.org/10.1155/2019/4327183

Randhawa MA, Khan AA, Javed MS, Sajid MW (2015) Green leafy vegetables: a health promoting source. In: Handbook of fertility. Elsevier, pp 205–220

Ranil RHG, Chamara RMSR, Pushpakumara DKNG, Bussmann RW (2021) Exploration, conservation, and utilization of ethnobotanical knowledge: Sri Lankan perspective. In: Abbasi AM, Bussmann RW (eds) Ethnobiology of mountain communities in Asia. Springer International Publishing, Cham, pp 409–431

Rao MJ, Wu S, Duan M, Wang L (2021) Antioxidant metabolites in primitive, wild, and cultivated citrus and their role in stress tolerance. Molecules 26:5801. https://doi.org/10.3390/molecules26195801

Rasingam L (2012) Ethnobotanical studies on the wild edible plants of Irula tribes of Pillur Valley, Coimbatore district, Tamil Nadu, India. Asian Pac J Trop Biomed 2:S1493–S1497. https://doi.org/10.1016/S2221-1691(12)60443-2

Ravishankar M, Kaur DP, Bhushan KB, Easdown WJ (2015) Traditional leafy vegetables of a tribal community in Jharkhand, India. Acta Hortic 1102:43–52. https://doi.org/10.17660/ActaHortic.2015.1102.4

Ray RC, Sivakumar PS (2009) Traditional and novel fermented foods and beverages from tropical root and tuber crops: review. Int J Food Sci Technol 44:1073–1087. https://doi.org/10.1111/j.1365-2621.2009.01933.x

Ray A, Ray R, Sreevidya EA (2020) How many wild edible plants do we eat—their diversity, use, and implications for sustainable food system: an exploratory analysis in India. Front Sustain Food Syst 4:56. https://doi.org/10.3389/fsufs.2020.00056

Reddy KN, Pattanaik C, Reddy CS, Raju VS (2007) Traditional knowledge on wild food plants in Andhra Pradesh. IJTK 6(1)

Rodrigues AP, Pereira GA, Tomé PHF, Arruda HS, Eberlin MN, Pastore GM (2019) Chemical composition and antioxidant activity of Monguba (*Pachira aquatica*) seeds. Food Res Int 121:880–887. https://doi.org/10.1016/j.foodres.2019.01.014

Roque RM (2004) Wood of *Gmelina arborea* in Costa Rica. New For 28:299–307. https://doi.org/10.1023/B:NEFO.0000040957.83398.e3

Sabaghnia N, Asadi-Gharneh HA, Janmohammadi M (2014) Genetic diversity of spinach (*Spinacia oleracea* L.) landraces collected in Iran using some morphological traits. AAS 103:101–111. https://doi.org/10.14720/aas.2014.103.1.11

Safwan II, Mohammed UA (2016) Review on the nutritional value, cultivation and utilization potential of some minor and underutilized indigenous root and tuber crops in Nigeria. Int J Adv Res 4(3):1298–1303

Saranraj P, Behera SS, Ray RC (2019) Traditional foods from tropical root and tuber crops. In: Innovations in traditional foods. Elsevier, pp 159–191

Saravanan R, Kannan D, Panda SP, Datta S (2020) Traditionally used wild edible plants of Kuldiha wildlife sanctuary (KWLS), Odisha, India. J Pharmacogn Phytochem 9(6S):482–488

Sardeshpande M, Shackleton C (2019) Wild edible fruits: a systematic review of an under-researched multifunctional NTFP (non-timber forest product). Forests 10:467. https://doi.org/10.3390/f10060467

Sathishsekar D, Subramanian S (2005) Antioxidant properties of *Momordica charantia* (bitter gourd) seeds on Streptozotocin induced diabetic rats. Asia Pac J Clin Nutr 14:153–158

Seetharam YN, Jyothishwaran G, Sujeeth H, Barad A, Sharanabasappa G, Shivkumar D (2003) Antimicrobial activity of *Dioscorea bulbifera* bulbils. Indian J Pharm Sci 65:195–196

Semwal P, Painuli S, Cruz-Martins N (2021) *Dioscorea deltoidea* Wall. Ex Griseb: a review of traditional uses, bioactive compounds and biological activities. Food Biosci 41:100969. https://doi.org/10.1016/j.fbio.2021.100969

Servin J, Moseley WG (2022) The hidden safety net: wild and semi-wild plant consumption and dietary diversity among women farmers in Southwestern Burkina Faso. Afr Geogr Rev 1–21. https://doi.org/10.1080/19376812.2022.2074481

Sethuraman G, Nizar MNN, Muhamad FN, Suhairi TASNM, Jahanshiri E, Gregory PJ, Azam-Ali S (2020) Nutritional composition of Black Potato (*Plectranthus rotundifolius* (Poir.) Spreng.) (Synonym: *Solenostemon rotundifolius*). Int J Sci Eng Res 11:1145–1150

Shad AA, Shah HU, Bakht J (2013) Ethnobotanical assessment and nutritive potential of wild food plants. J Anim Plant Sci 23(1):92–97

Shaheen S, Ahmad M, Haroon N (2017) Edible Wild Plants: an alternative approach to food security. Springer International Publishing, Cham

Shava S, O'Donoghue R, Krasny ME, Zazu C (2009) Traditional food crops as a source of community resilience in Zimbabwe. Int J Afr Renaiss Stud Multi- Inter- Transdiscip 4:31–48. https://doi.org/10.1080/18186870903101982

Sheikh N, Yogendra K, Misra AK, Pfoze L (2013) Phytochemical screening to validate the ethnobotanical importance of root tubers of *Dioscorea* species of Meghalaya. North East India, J Med Plants Stud, p 1

Shin T, Fujikawa K, Moe AZ, Uchiyama H (2018) Traditional knowledge of wild edible plants with special emphasis on medicinal uses in Southern Shan State, Myanmar. J Ethnobiol Ethnomed 14(1):48. https://doi.org/10.1186/s13002-018-0248-1

Shirai Y, Rambo AT (2014) Urban demand for wild foods in Northeast Thailand: a survey of edible wild species sold in the Khon Kaen municipal market. Ethnobot Res Appl 12:113–129

Shravya S, Vinod BN, Sunil C (2017) Pharmacological and phytochemical studies of *Alangium salvifolium* Wang. – a review. Bull Fac Pharm, Cairo University 55(2):217–222. https://doi.org/10.1016/j.bfopcu.2017.07.001

Shrestha PM, Dhillion SS (2006) Diversity and traditional knowledge concerning wild food species in a locally managed forest in Nepal. Agrofor Syst 66(1):55–63. https://doi.org/10.1007/s10457-005-6642-4

Shrestha R, Dawadi P, Bhusal S, Bhatt LR (2021) Nutritional value and antioxidant properties of *Diospyros malabarica* (Desr.) Kostel., fruit from midhills of western Nepal. Nepal J Sci Technol 20:113–125. https://doi.org/10.3126/njst.v20i1.43361

Shumsky SA, Hickey GM, Pelletier B, Johns T (2014) Understanding the contribution of wild edible plants to rural social-ecological resilience in semi-arid Kenya. E&S 19:art34. https://doi.org/10.5751/ES-06924-190434

Singh B, Sultan P, Hassan QP, Gairola S, Bedi YS (2016) Ethnobotany, traditional knowledge, and diversity of wild edible plants and fungi: a case study in the Bandipora District of Kashmir Himalaya, India. Int J Geogr Inf Syst 22:247–278. https://doi.org/10.1080/10496475.2016.1193833

Singh D, Siew Y-Y, Chong T-I, Yew H-C, Ho S, Lim C, Tan W-X, Neo S-Y, Koh H-L (2019) Identification of phytoconstituents in *Leea indica* (Burm. F.) Merr. leaves by high performance liquid chromatography micro time-of-flight mass spectrometry. Molecules 24:714. https://doi.org/10.3390/molecules24040714

Snehalatha VR, Rasmi AR (2021) Phytochemical evaluation and pharmacognostic standardization of *Syzygium palghatense* endemic to Western Ghats. Futur J Pharm Sci 7:147. https://doi.org/10.1186/s43094-021-00282-8

Sneyd L (2013) Wild food, prices, diets and development: sustainability and food security in Urban Cameroon. Sustainability 5:4728–4759. https://doi.org/10.3390/su5114728

Solanki S, Bhardwaj R, Vasudeva R, Chourey S, Archak S (2020) Biochemical composition of pulp and seed of wild jack (*Artocarpus hirsutus* Lam.) fruit. Plant Foods Hum Nutr 75:659–660. https://doi.org/10.1007/s11130-020-00849-5

Sthapit B, Padulosi S (2011) On-farm conservation of neglected and underutilized crops in the face of climate change. In: On-farm conservation of neglected and underutilized species: status, trends and novel approaches to cope with climate change. Proceedings of the International Conference, Friedrichsdorf, Frankfurt, Germany. Bioversity International: Rome, Italy, pp 31–48

Subedee BR, Chaudhary RP, Uprety Y, Dorji T (2020) Socio-ecological perspectives of Himalayan Giant Nettle (*Girardinia diversifolia* (Link) Friis) in Nepal. J Nat Fibers 17:9–17. https://doi.org/10.1080/15440478.2018.1458684

Subedee BR, Tripathi GR, Munankarmi NN, Chaudhary RP (2022) Genetic diversity in populations of *Girardinia diversifolia* from Nepal Himalaya using ISSR markers. Ecol Genet Genomics 23:100120. https://doi.org/10.1016/j.egg.2022.100120

Sujarwo W, Arinasa IBK, Caneva G, Guarrera PM (2016) Traditional knowledge of wild and semi-wild edible plants used in Bali (Indonesia) to maintain biological and cultural diversity. Plant Biosyst 150:971–976. https://doi.org/10.1080/11263504.2014.994577

Suwardi AB, Navia ZI, Harmawan T, Syamsuardi, Mukhtar E (2020) Ethnobotany and conservation of indigenous edible fruit plants in South Aceh, Indonesia. Biodiversitas 21. https://doi.org/10.13057/biodiv/d210511

Takooree H, Aumeeruddy MZ, Rengasamy KRR, Venugopala KN, Jeewon R, Zengin G, Mahomoodally MF (2019) A systematic review on black pepper (*Piper nigrum* L.): from folk uses to pharmacological applications. Crit Rev Food Sci Nutr 59:S210–S243. https://doi.org/10.1080/10408398.2019.1565489

Taylor M, Lebot V, McGregor A, Redden RJ (2018) Sustainable production of roots and tuber crops for food security under climate change. In: Food security and climate change. Wiley, Chichester, pp 359–376

Thrupp LA (2000) Linking agricultural biodiversity and food security: the valuable role of agrobiodiversity for sustainable agriculture. Int Aff 76:265–281. https://doi.org/10.1111/1468-2346.00133

Turan M, Kordali S, Zengin H, Dursun A, Sezen Y (2003) Macro and micro mineral content of some wild edible leaves consumed in Eastern Anatolia. Acta Agric Scand Sect B Soil Plant Sci 53:129–137. https://doi.org/10.1080/090647103100095

Turner NJ, Łuczaj ŁJ, Migliorini P, Pieroni A, Dreon AL, Sacchetti LE, Paoletti MG (2011) Edible and tended wild plants, traditional ecological knowledge and agroecology. Crit Rev Plant Sci 30(1–2):198–225. https://doi.org/10.1080/07352689.2011.554492

Ugwu FM (2009) The potentials of roots and tubers as weaning foods. Pak J Nutr 8:1701–1705. https://doi.org/10.3923/pjn.2009.1701.1705

UN Pathways to Zero Hunger | Zero Hunger Challenge (2022). https://www.un.org/zerohunger/content/challenge-hunger-can-be-eliminated-our-lifetimes#:~:text=The%20Zero%20Hunger%20vision%20of,make%20this%20vision%20a%20reality. Accessed 21 Jul 2022

UN The 17 Goals | Sustainable Development (2022). https://sdgs.un.org/goals. Accessed 21 Jul 2022

Uprety Y, Poudel RC, Shrestha KK, Rajbhandary S, Tiwari NN, Shrestha UB, Asselin H (2012) Diversity of use and local knowledge of wild edible plant resources in Nepal. J Ethnobiol Ethnomed 8:16. https://doi.org/10.1186/1746-4269-8-16

Upson TM, Cullen J (2012) 736. *Firmiana simplex*. Curtis's Bot Mag 29:170–181. https://doi.org/10.1111/j.1467-8748.2012.01783.x

Uusiku NP, Oelofse A, Duodu KG, Bester MJ, Faber M (2010) Nutritional value of leafy vegetables of sub-Saharan Africa and their potential contribution to human health: a review. J Food Compos Anal 23:499–509

van der Hoeven M, Osei J, Greeff M, Kruger A, Faber M, Smuts CM (2013) Indigenous and traditional plants: South African parents' knowledge, perceptions and uses and their children's sensory acceptance. J Ethnobiol Ethnomed 9:78. https://doi.org/10.1186/1746-4269-9-78

Varghese K, Anila J, Nagalekshmi R, Resiya S, Sonu J (2010) Dasapushpam: the traditional uses and the therapeutic potential of ten sacred plants of Kerala state in India. IJPSR 10:50–59. https://doi.org/10.13040/IJPSR.0975-8232

Vengadesan G, Ganapathi A, Prem Anand R, Ramesh Anbazhagan V (2002) In vitro propagation of *Acacia sinuata* (Lour.) Merr. via cotyledonary nodes. Agrofor Syst 55:9–15. https://doi.org/10.1023/A:1020269022363

Venter SL, Van den Heever E, Allemann J, Viljoen J (2000) Potential vegetable and medicinal uses of traditional crops in South Africa. Acta Hortic 25–28. https://doi.org/10.17660/ActaHortic.2000.523.2

Vishnu BV, Ramachandran R, Aswanilal LV, Gangaprasad A, Siril EA (2022) Genetic evaluation of *Garcinia gummi-gutta* L. (Roxb.) accessions based on inter simple sequence repeat markers. Ecol Genet Genomics 24:100130. https://doi.org/10.1016/j.egg.2022.100130

Wahidah BF, Afiati N, Jumari J (2021) Community knowledge of *Amorphophallus muelleri* Blume: cultivation and utilization in Central Java, Indonesia. Biodiversitas 22(7). https://doi.org/10.13057/biodiv/d220722

Walker DH, Thorne PJ, Sinclair FL, Thapa B, Wood CD, Subba DB (1999) A systems approach to comparing indigenous and scientific knowledge: consistency and discriminatory power of indigenous and laboratory assessment of the nutritive value of tree fodder. Agric Syst 62:87–103. https://doi.org/10.1016/S0308-521X(99)00058-X

Waseem S, Imadi SR, Gul A, Ahmad P (2017) Oilseed crops: present scenario and future prospects. In: Ahmad P (ed) Oilseed crops. Wiley, Chichester, pp 1–18

Wessels C, Merow C, Trisos CH (2021) Climate change risk to southern African wild food plants. Reg Environ Chang 21:29. https://doi.org/10.1007/s10113-021-01755-5

Willett W, Rockström J, Loken B, Springmann M, Lang T, Vermeulen S, Garnett T, Tilman D, DeClerck F, Wood A, Jonell M, Clark M, Gordon LJ, Fanzo J, Hawkes C, Zurayk R, Rivera JA, De Vries W, Majele Sibanda L, Afshin A, Chaudhary A, Herrero M, Agustina R, Branca F, Lartey A, Fan S, Crona B, Fox E, Bignet V, Troell M, Lindahl T, Singh S, Cornell SE, Srinath Reddy K, Narain S, Nishtar S, Murray CJL (2019) Food in the Anthropocene: the EAT–Lancet Commission on healthy diets from sustainable food systems. Lancet 393:447–492. https://doi.org/10.1016/S0140-6736(18)31788-4

Wunder S, Börner J, Shively G, Wyman M (2014) Safety nets, gap filling and forests: a global-comparative perspective. World Dev 64:S29–S42. https://doi.org/10.1016/j.worlddev.2014.03.005

Xi X-J, Zhu Y-G, Tong Y-P, Yang X-L, Tang N-N, Ma S-M, Li S, Cheng Z (2016) Assessment of the genetic diversity of different job's tears (*Coix lacryma-jobi* L.) Accessions and the active composition and anticancer effect of its seed oil. PLoS One 11:e0153269. https://doi.org/10.1371/journal.pone.0153269

Xu Y, Liang D, Wang G-T, Wen J, Wang R-J (2020) Nutritional and functional properties of wild food-medicine plants from the coastal region of South China. J Evid Based Complementary Altern Med 25:2515690X2091326. https://doi.org/10.1177/2515690X20913267

Yazzie D, VanderJagt DJ, Pastuszyn A, Okolo A, Glew RH (1994) The amino acid and mineral content of Baobab (*Adansonia digitata* L.) leaves. J Food Compos Anal 7:189–193. https://doi.org/10.1006/jfca.1994.1018

Yeşil Y, Çelik M, Yılmaz B (2019) Wild edible plants in Yeşilli (Mardin-Turkey), a multicultural area. J Ethnobiol Ethnomed 15:52. https://doi.org/10.1186/s13002-019-0327-y

Yesodharan K, Sujana KA (2007) Wild edible plants traditionally used by the Tribes of Parambikulam wildlife sanctuary, Kerala, India. Nat Prod Radiance 6(1):74–80

Zaman F, Sarkar C, Hossain R, Molla S, Das AK, Shuvo APR, Bappi MH, Atolani O, Islam MT, Rouf R (2022) Phytochemical evaluation and pharmacological activities of *Antidesma montanum* Blume leaf extract. Acta Biol Marisiensis 5:1–16. https://doi.org/10.2478/abmj-2022-0001

Zhao C-C, Zhang X-H, Chen J, Shao J-H, Zhao Z-Y, Tang Y-Y (2022) Lignans with α-glucosidase, protein tyrosine phosphatase 1B, and aldose reductase inhibitory activities from the fruits of *Viburnum cylindricum*. Ind Crop Prod 178:114601. https://doi.org/10.1016/j.indcrop.2022.114601

Zulhazman H, Asraf Fizree M, Muhamad Azahar A, Mohd Fadzelly AB, Nazahatul Anis A (2021) A survey on edible aroids consumed by locals in Kelantan, Peninsular Malaysia. IOP Conf Ser: Earth Environ Sci 736(1):012076. https://doi.org/10.1088/1755-1315/736/1/012076

Zulu NS, Modi AT (2010) A preliminary study to determine water stress tolerance in wild melon (*Citrullus lanatus* L.). S Afr J Plant Soil 27:334–336. https://doi.org/10.10520/EJC8798

Chapter 2
The Culture Has Not Faded: Reliance on Diverse Wild Edible Plants in Prehistory, History, and Modern Times

Avik Ray and Rajasri Ray

Abstract Food embodies our cultural identity, sense of taste, social status, and the extent of dependence on resources. It has been the most crucial element in establishing a liaison between humans and their environment. As hunter-gathers, the human race has depended on a diverse resource base procured from the environment from the early dawn of their evolution. The practice was partly lost owing to the domestication of crops and the embracing of agricultural life. The dietary diversity further shrunk in the recent century because of over-reliance on selected cereal crops, while a large suite of edible species still remains at disposal. Here, we review and synthesize the trajectory of consumption of wild edible plants, a part of edible biodiversity, to delineate the link from prehistory to the present times. I also analyze the diversity and its pattern, discuss the role of culture in use, and outline their implications on food security policies in improving dietary diversity and in creating a sustainable food system. The findings show that the consumption of wild uncultivated biota, though lessened over time, has not been completely abandoned. They remained as subsistence food, insurance crop, or key alternate resource to tide over the unfavorable period. Even now, apart from forest gathering, a plethora of plants and animals from the various anthropogenic landscapes are collected, cooked, and consumed likewise. Remarkable biodiversity (nearly 1400 species altogether) of green leafy shoots, roots and tubers, fruits, flowers, and many other edible plant organs still form a part of our dietary repertoire. However, the edible floral diversity is distributed heterogeneously across families, i.e., some families (e.g., Leguminosae, Compositae, Poaceae, Malvaceae, Rosaceae, Rubiaceae, Lamiaceae, Moraceae, etc.) contribute disproportionately to the edible species pool. Of total diversity, a maximum number of species is consumed for their leafy shoots (740 spp.), followed by fruits (657 spp.), roots and tubers (219 spp.), seeds (155 spp.), and flowers (153 spp.). The diversity also indicates their assimilation

A. Ray (✉) · R. Ray
Center for Studies in Ethnobiology, Biodiversity, and Sustainability (CEiBa), Malda, West Bengal, India
e-mail: avikray@ceibatrust.org

A. Kumar et al. (eds.), *Wild Food Plants for Zero Hunger and Resilient Agriculture*, Plant Life and Environment Dynamics,
https://doi.org/10.1007/978-981-19-6502-9_2

and wider acceptance of our food culture. The rich tradition of consumption of underutilized biota has the potential to be included in food policies to render our diet diverse and to enhance the nutritional outcome, like the inclusion under National Nutrition Mission programs, encouraging cultivation in nutrition or kitchen gardens, incorporation in mid-day meals, or related programs. Their high diversity, wider acceptance, and abundance could be a more effective way to combat micronutrient deficiency or hidden hunger than the mass-scale promotion and adoption of bio-fortified crops.

Keywords Food security · Wild edible plants · Edible biodiversity · Food policy · Biofortification · Sustainable food systems · Agricultural weeds · Green leafy shoots

2.1 Introduction

Food embodies our cultural identity, sense of taste, social status, and the extent of dependence on resources (Almerico 2014; Counihan and Van Esterik 2012; Ng and Karim 2016). It has been the most crucial element in establishing a liaison between humans and their environment—from where food had been procured or produced. Essentially of its indispensability, it has always occupied the central position throughout humanity's journey from the early dawn of human evolution, be it hunter-gatherers or agriculturists (Chevalier et al. 2014a; Gosden and Hather 2004). The role of food is quite obvious; it provides humans with the necessary energy and nutrition to survive, grow, and tide over. In the past, the hunt for obtaining food has been mediated through a thorough exploration of the environment. It required environmental learning that endowed early humans to accumulate fine-grained wisdom about their surroundings. The elements of the wisdom that was often passed on from generation to generation were revised, left out, honed up, and reconstructed depending on the context and need (Berkes et al. 2000). In this journey, humans have developed a rich body of knowledge to exploit a host of plants and animals, manipulate their availability and abundance, process and utilize them to their needs (Chevalier et al. 2014b; LaRochelle and Berkes 2003; Turner et al. 2011).

Therefore, the interaction of humanity with their environment has been quite old, from the earliest part of human history. The consequence of this interaction was a broad association of humans with biological resources that primarily consisted of edible biota. The tripartite interaction engaged early humans, their edible biological resources, and their inhabited and foraged landscape. Throughout their evolutionary history, they largely modified the landscapes (Smith 2001, 2011), often to enhance their resource base and ensure their predictable supply. As a result, the scale and magnitude of biological and cultural traces that early humans left on their catchment areas used to be prolific, and many scholars prefer to term it the domestication of landscape (Clement 2014; Terrell et al. 2003; Eriksson and Arnell 2017; Harris 2012). There are examples of profound prehistoric transformations of the low-lying Llanos de Mojos of the Bolivian Amazon by creating raised fields, forest islands, causeways, reservoirs, fish weirs, and settlement mounds to transform marginal lands into biodiverse and productive (Erickson 2006). A landscape may comprise fixed structures or landscape elements, like agricultural fields, hedgerows, groves, shorelines, hillocks, fences, and living things in the area (Johnson and Hunn 2010). It amalgamates the physical space of a bounded territory and the human values, customs, and cultural practices of a people or community (Olwig 1996).

So, an entanglement among the three partners, landscape, plant (or more generally edible biota), and people, has already been established from the very early stages of human evolution that became deep-rooted over millions of years (Chevalier et al. 2014a). It continued even during predomestication cultivation or low-level food production prior to settled and organized agriculture. It took a major turn following the domestication of crops, initiation of predominantly settled agriculturist life, and expansion of agriculture when humans became relatively less reliant on wild edible resources. Yet, settled agricultural activities went through many phases before becoming a mainstay of food procurement. Unlike specific geographic regions (e.g., fertile floodplains, riverbeds, etc.) or inertia received from social enterprise, cooperation, and agricultural activities were not equally well-adopted at the same time in most places and by many communities (Bellwood 2004; Fuller et al. 2011; Zohary et al. 2012). As a result, a large number of forest dwellers, along with urban or semi-urban inhabitants, depended on wild biota as their alternate resource, normally or during the lean season or at the time of crop failure. So, foraging for uncultivated plants or animals was never abandoned entirely during human history but remained a life-saving skill to supplement major subsistence activities (Smith 2006).

Here, I review the trajectory of consumption of wild edible plants from prehistory to present times to delineate the link, mostly focusing on their utility as an important resource. I also analyze their diversity and discuss the role of culture. Finally, I outline their implications on food security policies in improving dietary diversity and creating a sustainable food system.

2.2 Human History: Hunting and Gathering—The Spectra of Wild Edible Resources

For the last several centuries, the human race used to depend on farmlands for food; in other words, food produced through agrarian activities constitutes the major portion of our consumption. However, it has not always been like this. Almost all (99%) through their evolutionary history, humans have spent their lives as hunters and gatherers. As hunter-gatherers, their activities had been diverse, from hunting small-medium to larger games, collecting honey, eggs, seeds and nuts, catching fish, shellfish, clams and mussels, digging up roots and tubers, plucking ripe and unripe fruits, grabbing edible flowers and leafy juicy green tender shoots, and so on. So, in the relative time scale, agriculture is just a recent invention of human societies. But, it has had an overarching impact on humanity's journey ranging from supporting huge population growth in the Neolithic, stimulating sedentary lifestyle, impacting human health, social stratification, the evolution of larger urban spaces, etc. (Belfer-Cohen and Goring-Morris 2011; Bellwood 2004; Cauvin 2000; Diamond 2002; Wolfe et al. 2007). Importantly, humans armed with the newly evolving culture began to manipulate nature in their own way, perhaps a seminal turn in human cultural evolution after the discovery and management of fire (Scott 2011). However, embracing settled agriculture and ensuring the sizable supply of domesticated cereals perhaps had been achieved through nonlinear progress. It happened sporadically across many smaller regions of the world and without entirely abandoning hunting and gathering for a period. The relative importance of hunting and gathering in subsistence has been debated by many archeologists and anthropologists, putting extra emphasis on one or the other, e.g., man the hunter and women the gatherer, but their individual roles in human subsistence are not undeniable (Sterling 2014).

Prehistoric humans went through various phases before agrarian activities set off in full swing, e.g., predomestication cultivation or low-level food production, when they did perform agrarian activities but at a much lower scale and magnitude. It meant no prior domestication but tending and harvesting of wild or semi-wild proto-crops, some of which were later domesticated, and others left out from the race and may have emerged as non-crops or weeds (Harris 1989, 2012; Rindos 1989; Smith 2001). A vast body of ethnographic and historical evidence has revealed diverse resource management techniques employed by the nonagricultural societies in most environments, and they depended on a biologically diverse assemblage for their food supply, with few species as staples. The wide array of cultural practices pertained to burning, sowing, planting, pruning, tending, weeding, tilling, irrigation, taming, and herding that ensured and/or enhanced their resources (Harris 2012). The scale of resource management could have been larger and may not essentially culminate in domestication. Undomesticated carbohydrate-rich cereals are sometimes observed to have supported many large-scale feasts on porridge and stew, as in the 'rock gardens' of Göbekli Tepe (Curry 2021). These scattered findings echo the significance of resource management in an 'intermediate system' with still undomesticated or semi-wild plants and a continuum of human subsistence. In the late Paleolithic or Neolithic period, low-level food production prior to agriculture entailed the maintenance of a defined catchment area from where they procured resources and adopted various means to intensify them for better availability. Therefore, the domestication of the landscape could be a preceding state of the low-level food production in the well-defined catchment areas. This is particularly likely when foragers were practicing low-level food production with wild-plant husbandry, growing and tending wild species of cereal plants or tuber crops or fruit trees, and intensifying their resource base (Smith 2001, 2011). So intensification of resources for better availability was not an unlikely trait for hunter-gatherers. Resource intensification through landscape engineering has been documented vividly across the Amazon of South America, Cuzco of Peru, and Chinampas of Mexico (Erickson 1992, 2006, 2008), be it agro-pastoral landscapes or forested or semi-forested land, aquatic bodies like clam gardens. We can anticipate a similar scale of domestication of landscapes to exist in the old tropics. Yet, a dearth of archeological research does not directly infuse support to this hypothesis; the multifarious ways of landscape modification and enabling humans to adapt and survive in the domesticated landscape can be observed across the south and south-east Asia (Cairns and Garrity 1999; Michon 2005; Ray and Ray 2018; Ray in press; Roy 1989).

Low-level food production in domesticated landscapes or intermediate systems of subsistence may also suggest the management of wider diversity of species that happened in the larger background of broad-spectrum resource use. However, it might have undergone overall reductions in biotic diversity over time, as the preferred species made greater contributions to food production and received greater attention, thereby being manipulated within the realms of anthropogenic niches. Elevated demand from environmental changes, population growth, resource crunch, sedentary life, and other factors were instrumental drivers and sparked the process of choice of crops. It could have led to the domestication of a few, eventually.

Domestication of crops and settled agriculture was a turning point in human history that had actually made humans dependent on a few selected crops, cereals, and noncereals depending on the geography and narrowed down the resource base greatly. We continue to bear the same cultural legacy. Out of several thousand, today, nearly 30 domesticated species constitute a significant portion of dietary diversity, and only three principal cereal grains (rice, wheat, and maize) contribute to more than half of the world's calorie intake (FAO 2010). So, a diet that once comprised diverse food from various plant and animal sources and provided them with required energy and nutrition became heavily dependent on a few staple cereals which can be produced at a much larger amount with available human labor, tools and technology, capital, and state support; however great it might be. So, rather than wandering mile after mile to gather essential resources, settled and organized agriculture has undermined the necessity of gathering wild, uncultivated edible resources but cannot abandon it entirely.

2.3 A Parallel Resource to Agricultural Production and Changing Dynamics

Settled agriculture paved the way for reliance on a set of domesticated crops and their predictable supply that in turn reduced, if not abolished, the hunting and gathering of wild resources. The net outcome has been manifested in the dietary repertoire, which had been widely diverse once but drastically narrowed down to only a handful of domesticated crops, specifically carbohydrate-packed cereals or protein-rich lentils. The underlying reason, perhaps, was that they could be produced at a larger amount, stored for a relatively long time, manageable, and can be upscaled contingent on demand. However, it does not imply that hunting and gathering have been completely divorced from subsistence. Recent ethnographic research tends to suggest a continuum of subsistence strategies being contingent on dual mode but with varying levels of dependence (Morgan 2015; Fortier 2009). Therefore, foraging, hunting, and gathering remained with the agriculturists as insurance throughout but the intensity lessened since cultivation became the mainstay. At the time of crop failure due to various reasons, such as lower or untimely rainfall, pest or pathogen attack, and depredation by wild animals, the knowledge of forest resources became immensely useful to tide over unfavorable conditions (Ertug 2014; Smith 2006). If I situate the geographic context within the Indian subcontinent, it becomes pertinent since records say about Indian encounters with several *El Nino* events in history, and that should have lasting impacts on crop production, either crop loss or failure (Fisher 2018). Historical texts are not devoid of reports of famine or crisis food collected from adjacent forests landscapes to tide over the unfavorable period. Diverse forest resources, primarily consisting of uncultivated roots and tubers, ripe fruits, juicy parts of tender stems, seeds of cereals, fleshy flowers or other plant organs, etc. served as famine or crisis food (Bhandari 1974). Even in the recent past, environmental historians described the survival of indigenous communities on the forest resources even when the fangs of famines lurked everywhere. In the Ranchi district, where indigenous people comprised a major portion of the population, there were several spells of droughts in 1820, 1823, 1827, 1837, and 1866 possibly because of *El Nino* events, but the severity never reached a level of famine owing to the dependence on wild forest resources of the indigenous people (Grove et al. 1998).

Since settled agriculture had not spread uniformly throughout the world's geography and was accepted unequivocally, it left much room for other streams of subsistence, exchange, and use that necessitated forest resource extraction. South Asia had been bioculturally diverse, and the Indian subcontinent had been a home for a plethora of indigenous communities living amid the forests. These ancient forest dwellers have not settled agriculturists like the dominant group belonging to the urban or peri-urban subjects of larger centralized kingdoms or empires. However, many forest-dwelling people were primarily shifting cultivators (*Bondo, Gadava, Saora, Kondhs*) or solely hunter-gatherers (*Chenchu, Birhor*). But, many largely maintained a dual mode of subsistence, adhering to shifting cultivation as the primary means of food production with regular procurement of resources from nearby forested land (Ray in press). Although their antiquity is sketchy, we find mentions of forest-dwelling tribes as barbarians or stone-age communities in the historical texts as opposed to civilized city dwellers, and in most cases, their distinction was stark (Fisher 2018; Thapar 2012; Parasher-Sen 1998). The indigenous communities in south Asia mostly belonged to the Austroasiatic (mostly Munda-speaking people in central India and around), Tibeto-Burman (mostly in the northeast), Dravidian (*Irula* and *Kuruba*), and Indo-European (*Vedda*) language families. It is likely that the vast section of forest dwellers mentioned in the historical texts could have been from one of these groups or the other.

Although agriculturists had gradually taken firmer hold of food production in the Indian subcontinent, forest-dwelling people persisted alongside the mainstream society perhaps as early as the Mauryan kingdom or even earlier. The forest dwellers not only hunted and gathered their food but, many a time, also maintained sporadic contact with the agriculturists living in the cities (Thapar 2012; Parasher-Sen 1998). Sometimes they were subjugated to various degrees and a few have also gradually assimilated into the mainstream culture, especially those living at the confluence of urban or semi-urban and forest culture. Historians also described their ever-changing relationship with society and their traded goods. In addition to roots or tubers, leafy shoots, and flowers, they used to exchange honey, wax, and cotton from *semal* trees (*Bombyx ceiba*), medicinal plants, and bushmeat (Parasher-Sen 1998). A reading of early historical texts reveals a consistent tradition of exchange between city people and forest men, even from the pre-Mauryan era, which continued unabated throughout history (Thapar 2012; Parasher-Sen 1998). Yet, the supporting evidence is scant, and it focuses more on forest dwellers and their relational dynamics with agriculturists. We can still predict their traded goods that include a plethora of edible forest resources. The historians also suggest the gross degradation of forests in the Indian subcontinent, disturbance in the socioeconomic and cultural lives of their inhabitants, and the eviction of their inhabitants began in the pre-Mauryan period and continued thereafter. It meant the slow expansion of agrarian activities and the disruption of the material cycle of production of many of the hunting and gathering communities. They were made dependent on the produce of the plains, and it was also possible that they adopted agrarian activities or modified their shifting cultivation. On the other hand, the importance of forests as a source of revenue or as a resource base has been recognized, and it fueled the prosperity of the metropolitan society, directly or indirectly (Parasher-Sen 1998). Later, Ashoka submitted to Buddhism and restricted the killing of wild animals, which should have impacted the lives of forest dwellers.

The process of decline continued, and rights to wildly growing forest produce were further impaired during colonial times. The scale of destruction was much larger than in the early historic period. We find vivid descriptions of the forest resources in central and eastern India and their critical role in the lives of indigenous people across various accounts (Ball 1867, 1985). Also, the situation changed with the implementation of various rules regarding the usage of the forests constraining their access to resources. The wild resources which once ensured their food security or insured them from impending food scarcity became rare or inaccessible owing to the mass destruction of forest tracts, conversion to agricultural land, or infringing their rights through a fee for usage of the jungles. As a consequence, famine which was generally undetectable in the forested landscapes because of resource abundance soon became an imminent threat after the middle of the nineteenth century (Damodaran 1995).

In light of the discussion, it was quite evident that the shared history of forested landscapes and their inhabitants had been changing dramatically since the onset of agrarian life, and so did their subsistence of gathering wild edible plants. Yet, we often myopically look at the last couple of centuries for the decline of forests, forest dwellers, and hunting and gathering. In the Indian subcontinent, we find evidence of such changing dynamics and burgeoning tension from the pre-Mauryan period. Agricultural expansion through the ages has largely led to the dwindling of forested lands and the cornering of their inhabitants (Fisher 2018; Ickowitz et al. 2019). It is reasonable to assume that it intensified over time and with the changing socio-political situation of the subcontinent (Later British Colonialism, for example). Other factors, such as rapid urbanization, industrialization, and other developmental activities, were also instrumental in later centuries. The obvious consequences were reflected in the radical changes in the landscape, especially the decimation of the habitat of wild edible biota and the loss of the knowledge system linked to the landscape (Pretty 1995; Labadarios 2005).

2.4 Various Dimensions of Wild Edible Plants in the Anthropogenic Landscapes

The consumption of uncultivated flora or fauna has been well entrenched in the food culture of the country or elsewhere in many different forms (Ray et al. 2020; Ray 2021a). In most geographies, the indigenous communities living amidst forests or fringes largely adhered to these cultural traditions. In India, the forest resources collected by various groups of indigenous communities, especially around the central, eastern, south-eastern, and north-eastern ranges, constituted a greater part of the foraging communities. However, their reliance on forest resources has shrunk considerably in the last couple of centuries. Other regions were also equally affected. Not only limited to forests and forest-dwelling communities, but the collection has also been in vogue in parts of the anthropogenic landscapes as a part of the cultural legacy. This reinforces that the contribution of wild, uncultivated edible biota is integral to our energy and nutrition through agriculture-derived food still holds a larger share in most modern societies (Ray et al. 2020). I draw on a couple of examples from the perspective of human–environment interaction to demonstrate the diversity of culture of consuming uncultivated biota in the context of varied environmental settings, rituals, practices, etc.

2.4.1 Agricultural Weeds

Agricultural weeds portray one such by-product of human–environment interactions. Agricultural weeds co-grow with the main crop and are supposed to interfere with them and reduce production. Weeds are almost ubiquitous and estimated to be represented by approximately 30,000 species present across various ecosystems, notably dominant and recognized as an integral element in agroecosystems (Molina et al. 2014). Although weeds share an evolutionary history with crops, their relevance has multiplied in recent years, especially in industrial agriculture. The meteoric growth of industrial agriculture over centuries spurred increasing production of selected crops, promoting monoculture, and ignoring the importance of farmland heterogeneity (Benton et al. 2003). Hence, agricultural weeds are mostly unwanted

Fig. 2.1 Various edible species in agricultural fields, (**a**) *Marsilea quadrifolia* and (**b**) *Chenopodium album*

and are removed mechanically or chemically to lessen the interference with the sole crop. In concert with agricultural intensification, the recommendation by agricultural scientists and policymakers has compelled farmers to eradicate weeds interfering with the growth and yield of cultivated crops; it has intensified in recent times with the specialized and rampant application of weedicides, e.g., glyphosate (Myers et al. 2016). Despite this dominant prescription, the acceptance of weeds is highly variable among farmers, and so are their treatments by farming communities as recorded across cultural geographic regions (Sinha and Lakra 2007; Tanji and Nassif 1995). In many agricultural landscapes, they are well tolerated and not cleared at all, and some are even consumed as leafy greens or used as fodder crops (Mazhar et al. 2007). It has been enumerated globally that 89% of the most widespread and aggressive weeds are edible or partly edible. Many so-called 'noxious' weeds *like Ageratum conyzoides, Leucana leucocephala,* and *Synedrella nudiflora* are relished as food or used as medicine in many parts of the world (Hillocks 1998). Other co-inhabitants of agricultural fields that are regularly consumed, such as *Marsilea quadrifolia* or four-leaf clover or pigweed (*Chenopodium album*), frequent entrants in agricultural fields (Ray 2021a). There are numerous other edible herbs in the rice fields that co-grow with rice plants (Fig. 2.1a, b). Therefore, rice fields offer an example of the habitat of several edible biotas. Rice is one of the major staple crops cultivated in nearly 160 million hectares of land worldwide (Global Rice Science Partnership (GRiSP) 2013). Rice fields, either irrigated, deep water, or rain fed, are a suitable habitat for many flora and fauna. A recent study by Ray and Chakraborty (2021) has recorded a great diversity of edible flora or fauna, such as edible plants, fish, mollusks, arthropods, and insects found and gathered from the rice fields of Asia. Biodiversity tends to abound in the rice fields with its aquatic environment since they emulate the dynamics of wetland ecosystems, often comparable to natural ecosystems. Before the Green revolution, agriculture, such as abundant aquatic life in the rice fields and contiguous water bodies, used to cater to the people of south, east, and south-east Asia with food resources. The intensification of production has brought about the indiscriminate use of agrochemicals that contaminated the aquatic ecosystems, caused water pollution, nutrient imbalance, severe eutrophication, etc., which has threatened aquatic biodiversity (Ray and Chakraborty 2021) and also taken a toll on human health (Pingali and Roger 2012). The loss of biodiversity therefore not only endangers their livelihood but also narrows their dietary diversity damaging food and nutritional security.

2.4.2 Other Nondomesticated Biota

Food from ecosystems, as provisioning service, also depends on ecosystem health, where heterogeneity of anthropogenic landscapes is also crucial. Situated within the matrix, fallows, pastures, hedges, big-old trees, semi-forested land, or non-perennial water bodies quite often add to this heterogeneity and serve as temporary shelters for local biodiversity. So, rural and semi-rural landscapes are dotted with many other edible species found in the anthropogenic landscapes (Sinha and Lakra 2007), e.g., leafy green herbs, various unripe or ripe fruit trees or shrubs, various fungi, and trees with flowers are seasonally harvested. These trees, shrubs, or herbs remain scattered in the humanized landscape and are generally not tended formally like agricultural or horticultural crops. Despite that, they grow, regenerate, and propagate naturally as parts of the ecosystems until they are tolerated. Perhaps, they sprout from the seeds dispersed by insects, birds, mammals, or domestic animals or unintentionally thrown by humans, or remain in the soil as a soil seed bank. In India, there are many examples of such trees, shrubs, or herbs scattered in rural or semi-rural landscapes, yet agriculture or many developmental activities have homogenized these landscapes. Although the actual richness of species varies with geography, some of the very common fruit trees are *Tamarindus indica*, *Zizyphus jujuba*, *Bassia latifolia*, *Averrhoa bilimbi*, *Diospyros melanoxylon*, *Carissa carandas*, *Aegle marmelos, Limonia acidissima*, etc. There are a number of species whose flowers are consumed, trees (*Sesbania grandiflora, Moringa oleifera*) or climbers (*Dregea volubilis*), or aquatic members (*Nelumbo nucifera* or *Nymphaea nouchali*) (Fig. 2.2a–d). Among leafy greens, *Diplazium esculentum*, *Portulaca oleracea, Oxalis corniculata, Hygrophila spinosa, Boerhaavia diffusa, Amaranthus* spp. are very common; these herbs grow and propagate in huge numbers either in moist and shady places, around the fallow corners of human settlements or roadsides. There are many others that are characteristically aquatic, *Nasturtium officinale*, *Nymphaea nouchali, Enhydra fluctuans, Nelumbo nucifera,* or *Ipomoea aquatica* are present in high numbers near the water bodies, streams, or ponds nearer to the households. On the animal front, many clams or mollusks (*Bellamya bengalensis*) are also regularly harvested and eaten apart from fish (Sahani 2019; Singh 2018). Similarly, mammals (especially rodents), fishes, or mud-eels (*Monopterus cuchia*) are also caught near the agricultural fields or their adjacent water channels employing very simple indigenous tools (Barman et al. 2013). Altogether, their diversity, seasonal abundance, and traditional knowledge of village folks to catch, process, and cook them reflect the multifarious ways humans dynamically interact with the environment for the collection and use of common resources.

Fig. 2.2 Various edible species in anthropogenic landscapes, (**a**) *Diplazium esculentum,* (**b**) *Portulaca oleracea,* (**c**) *Nelumbo nucifera,* (**d**) *Dimocarpus longan*

2.5 The Cultural Legacy of Gathering: Consuming Various Wild Plants and Their Diversity

Uncultivated wild edible plants fall into five broad categories based on edible plant organs, grains or seeds, fleshy fruits, leaf or leafy shoots, roots or tubers, and flowers. However, there still exist miscellaneous edible plant parts eaten regularly, e.g., pith, latex, aril, bark, petiole, peduncles, gums, and various kinds of modified shoots, e.g., phylloclade and bulbils. This implies our tryst with plants, especially uncultivated edibles, has been quite diverse. Much of this is a legacy inherited from our hunter and gatherer ancestors who used to hunt and gather apart from performing agriculture. If we turn to human history, we will find early use of various seeds of grasses, nuts and acorns, roots and tubers, fungi, insects, shellfish, reptiles, birds, fleshy fruits, pollens, etc. But, archeological finds are often biased and allow one to reconstruct past events building on the evidence that has been unearthed. So, we are often able to identify specific groups of plant food, e.g., grains, more than others. Although the soft tissues of many of the gathered food degrade with time, recent advancements in the analyses of parenchyma, starch, phytolith, and macrobotanical remains allow the detection of a suite of plant usage. It makes the case of seeds of grasses and forbs or roots and tubers more prominent in the diet of early hunter and gatherer or low-level food producers because of the relative ease of detection. Fleshy fruits or green leafy shoots demonstrate less occurrence in the archeological finds. But that may not suggest their lesser presence in past human diets. Some call them the "missing link" in knowledge about ancient diets. Charred remains do not even allow one to prove the presence of green-leafy shoots in the diet though they are quite common in human coprolites (Curry 2021).

Although early hunter-gatherers survived on meat-centric diets, carbohydrates were always on their menu (Cordain et al. 2000). Apart from wild cereals, underground roots, and tubers, leafy plant parts were quite well placed in the diet of ancient hunter-gatherers (Curry 2021). Several decades of archeological and ethnohistorical research on historic and prehistoric resource management also revealed the deep-time dependence of the human race on grass or forb seeds and underground storage organs like taproots, root and stem tubers, corms, rhizomes, and bulbs (Harris 2012; Ströhle and Hahn 2011). Seeds of various herbaceous plants, primarily grasses, and legumes, have been a major source of human food after techniques for harvesting and processing were developed. The grasses and forbes mostly occur wild in monodominant stands in almost all terrestrial ecosystems and can regenerate quickly. So, spatially concentrated populations, easily harvestable seeds with readily stored supplies of food of high nutritional value facilitated the gathering of the seeds (Weiss et al. 2004). On the other hand, knowledge of processing fibrous, bitter, and toxic varieties of underground storage organs to render them digestible allowed humans to harvest them. Dependence on roots and tubers as staples was most widespread in the past for tropical and temperate foragers. These starch-rich energy-packed foods were often complemented by many animal fats and proteins from terrestrial or aquatic mammals, reptiles, mollusks, and fishes (Harris 2012). Among many edible species so exploited as food in diverse geographic locations, a few were domesticated but many others continued to remain wild or semi-wild. Building on existing evidence, it seems likely that there was a long period of management and consumption of seeds of wild grasses and forbs during which broad-spectrum gathering and pre-domestication cultivation co-existed. In some places that may have lasted even after domestication and settled agriculture.

In the Indian subcontinent, archaeobotanical studies over half a century also revealed the dimensions of prehistoric plant–people relationships (Kajale 1991; Pokharia 2008; Pokharia et al. 2009, 2015; Saraswat 1980, 2004; Sharma 1985; Singh 2001; Tewari et al. 2006, 2008). Pappu (2020) has also reviewed the state of knowledge on hunters and gatherers in Indian archeology. Similar ethnohistorical accounts also elucidated the practices of indigenous communities to harvest, consume, and manage various edible species from forests and semi-forested landscapes (von Fürer-Haimendorf and von Fürer-Haimendorf 1945; Pratap 2000; Kajale et al. 1991; Murty 1981; Ray in press; Roy 1935; Roy 1989; Sharma 1990). There were seeds of grasses, underground organs, fleshy fruits or flowers, stems, or leaves. Among the vast repertoire of species, many wild ancestors of modern crops have been commonly used even to this day, e.g., *Tinn ka chawal, Tinni* rice, *Udi dhaan* (*Oryza rufipogon*), the wild ancestor of modern-day rice (Singh et al. 2012); or *Shyama Ghas,* jungle rice, Deccan grass, or awnless barnyard grass (*Echinochloa colonum*), the wild ancestor of the cultivated cereal crop *Echinochloa frumentacea* or Sawa millet (Saha et al. 2014). Apart from the cultivation of domesticated crops, some of the grass, forbs, or other species (*Amaranthus tricolor, A. viridis,*

Bauhinia vahlii, Oryza coarctata, Nymphaea nouchali, Nelumbo nucifera) were also harvested for their seeds. Underground or aboveground storage organs (*Dioscorea* spp., *Amorphophallus* spp., *Pueraria tuberosa, Alocasia macrorrhizos, Nymphaea nouchali, Commelina benghalensis, Canna indica, Zingiber* spp., *Curcuma* spp.) were also in the menu of pre-agriculturists or transient hunter-gathers. A rich body of literature suggests a long-standing liaison between plants (and animals) and people manifested through gathering, managing, and intensifying resources found in natural or anthropogenic landscapes.

It also reinforces the notion that though a major portion of our food today is sourced from organized agriculture, the culture of consumption of uncultivated edibles has not faded; we still bear the legacy of hunter-gatherers or foragers to an extent. Several regional studies have uncovered the great wealth of wild edible flora that has not entered into the mainstream cultivation and production cycle. It implies they have been either consumed locally or regionally by various indigenous communities or people living in rural, peri-urban, or semi-forested landscapes (Singh and Arora 1978; Sinha and Lakra 2007; Angami et al. 2006; Jeyaprakash et al. 2011). A recent study has summarized and identified more than 1400 species of uncultivated wild plants belonging to nearly 200 plant families consumed in India (Ray et al. 2020). The taxonomic diversity of the edible species is quite high, which is probably owing to the enormous size of the country and its intrinsic biological and cultural diversity (Loh and Harmon 2005; Maffi and Woodley 2012). The choice and consumption of economically or culturally important species are often governed by the availability and distribution of the plants, which in turn are largely determined by environmental parameters, e.g., temperature, altitude, rainfall, edaphic factors, etc. So, the biogeographic pattern of the floral distribution translates to their heterogeneous distribution and regional or localized availability that significantly affects the peoples' choices (Turner et al. 2011; Chevalier et al. 2014b). However, among so many species, Ray et al. (2020) found the presence of some of the families was overwhelming in the species pool; for example, Leguminosae was the family with the highest number of edible species (119), followed by several others like Compositae (57), Poaceae (53), Malvaceae (49), Rosaceae (48), Rubiaceae (42), Lamiaceae (41), Moraceae (39), Amaranthaceae (38), and Araceae (32) (Fig. 2.3). Over-representation of some of the key families was also reflected when the species count was plotted against the number of families, it showed the top 17 families (9.2% of the total families) contributed to nearly 50% (646 species), and the top 44 families (23.9% of the families) contributed to 75% (971 species) of the total species count (Fig. 2.4). Reliance on a small subset from a large pool of plant families is a common trend that has been observed in the use of medicinal or wild edible species (Moerman 1996; Leonti et al. 2003; Turner et al. 2011). Similarly, the over-representation of certain families (Leguminosae, Poaceae, Rosaceae, Moraceae, Lamiaceae, Compositae, Araceae, etc.) in the spectrum of wild edible plants has also been recorded in Southeast Asia (Cruz-Garcia and Price 2011; Sujarwo et al. 2014; Shin et al. 2018). The strong biogeographic affinity between South and Southeast Asia perhaps underlies the near similar floral distribution, overlapping availability, and acceptance of resources, where the Indo-Malayan biota represented a major

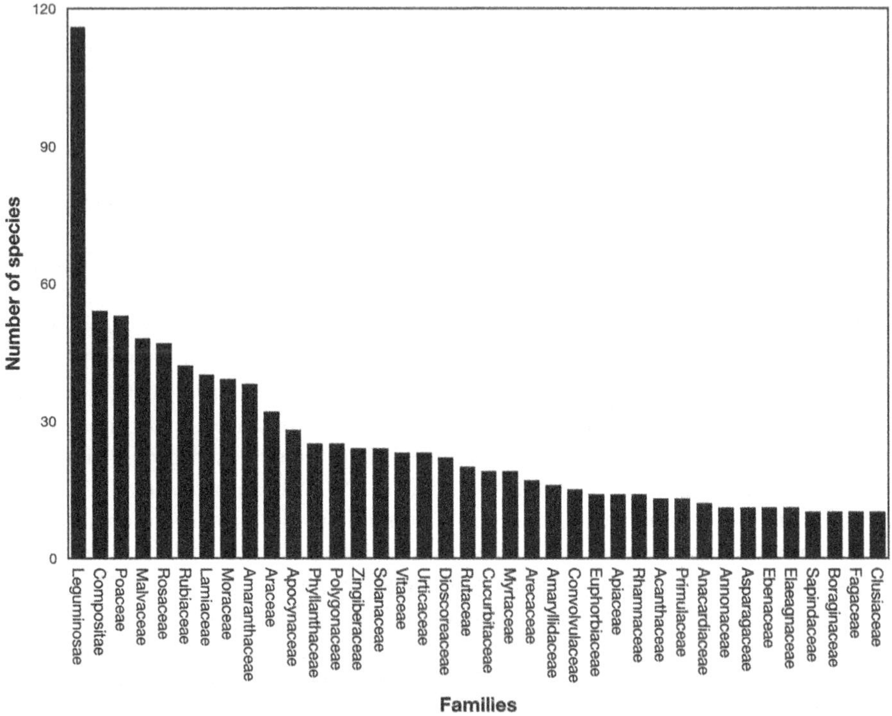

Fig. 2.3 Distribution of species across selected families

fraction (Mani 1974). Thus, it could be a plausible reason for sharing a major portion of edible floral species. Whereas wild edibles from Rosaceae, Liliaceae, Lamiaceae, Asteraceae, and Apiaceae were more frequently present in the European cuisines than the other families, perhaps owing to their abundance (Leonti et al. 2006; Tardío et al. 2006; Pardo-de-Santayana et al. 2007; Hadjichambis et al. 2008; González et al. 2011; Dogan et al. 2013). On the other hand, Ogle and Grivetti (1985a, b) found a prevalence of selected families in South Africa, such as Compositae, Myrtaceae, Moraceae, Rubiaceae, Leguminosae, and Amaranthaceae that also broadly overlap with Asian assemblage.

2.6 Heterogeneity in Use: Role of Culture

The acceptance and consumption of wild uncultivated edible species have neither been spatially nor culturally homogeneous owing to the biogeographic pattern of floral distribution, and not all plant organs are equally favored. Also, there exists a conspicuous heterogeneity in the use of edible organs in terms of species count. In other words, people's choices have demonstrated that some edible organs are

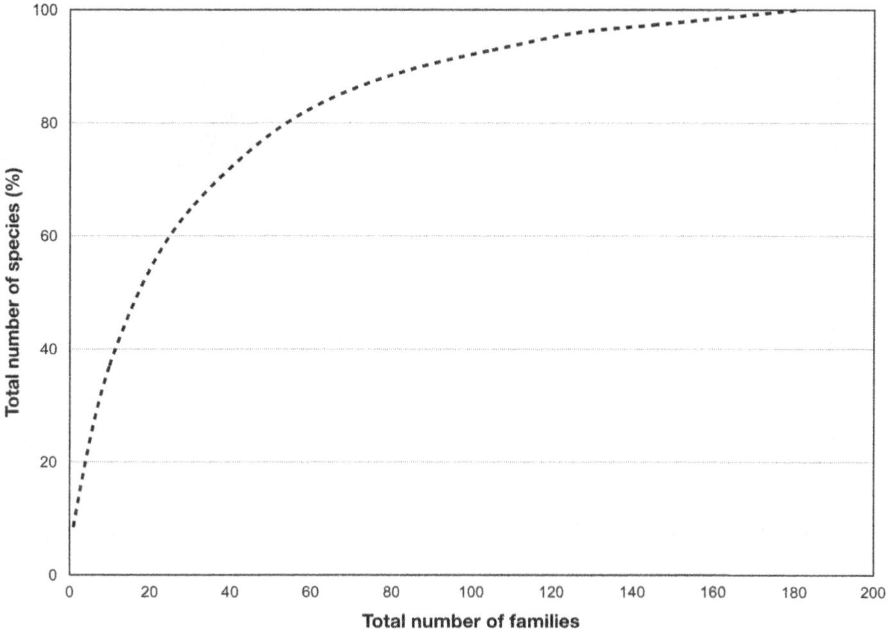

Fig. 2.4 Contribution of families to the total species pool (%) (Adapted from Ray et al. 2020. Front Sustain Food Syst 4:56. Doi: https://doi.org/10.3389/fsufs.2020.00056)

consumed more than others and are diverse in species number (Ray et al. 2020). Although the prehistoric remains are relatively scant because of preservation bias in the archeological samples, the group with edible green leafy shoots seems to be the most diverse. In India, people eat more species of leafy greens (740 species from 134 families) than fruits, flowers, or even tubers, etc. Consumption of fruits ranked next to it with 657 species from 110 different families. Underground storage organs and edible seeds (of grasses or fruits) have a fair share of species with 219 (70 families) and 155 (39 families), respectively. Despite their relatively lower diversity, these two groups are most commonly unearthed in prehistoric remains. Lastly, the diversity of edible flowers is represented by 153 species from nearly 60 plant families (Fig. 2.5a–e).

Human culture tends to play a central role in plant choice. Culture is in turn shaped by the local distribution, abundance, and availability of biological resources. So, it is often constrained by accessible resources, but it is also driven by norms, knowledge systems, taboos, biases, etc. In the past, endogamous or semi-closed small-scale societies residing in disjunct cultural geographic regions used to exercise their own way of plant use, either for edible or other purposes. The knowledge of plant use has been passed on intergenerationally through social learning. However, with the passage of time, the loop of information transfer has been disrupted as the world became increasingly connected, transportation developed, trades enhanced,

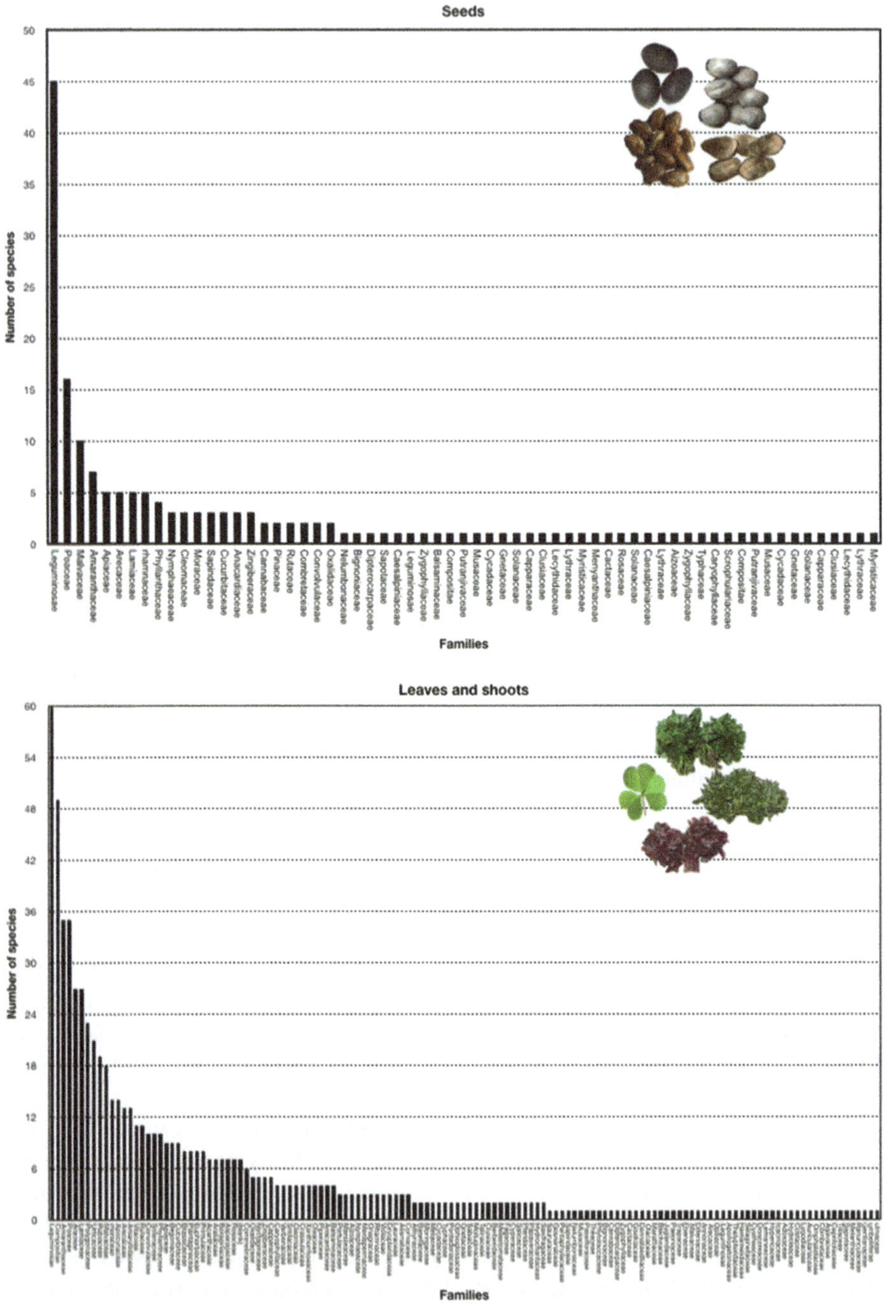

Fig. 2.5 Distribution of species across families based on individual edible plant parts: (**a**) seeds, (**b**) leaves and leafy shoots, (**c**) flowers, (**d**) fruits, and (**e**) underground parts (Adapted from Ray et al. 2020. *Front. Sustain. Food Syst.* 4:56. Doi: https://doi.org/10.3389/fsufs.2020.00056)

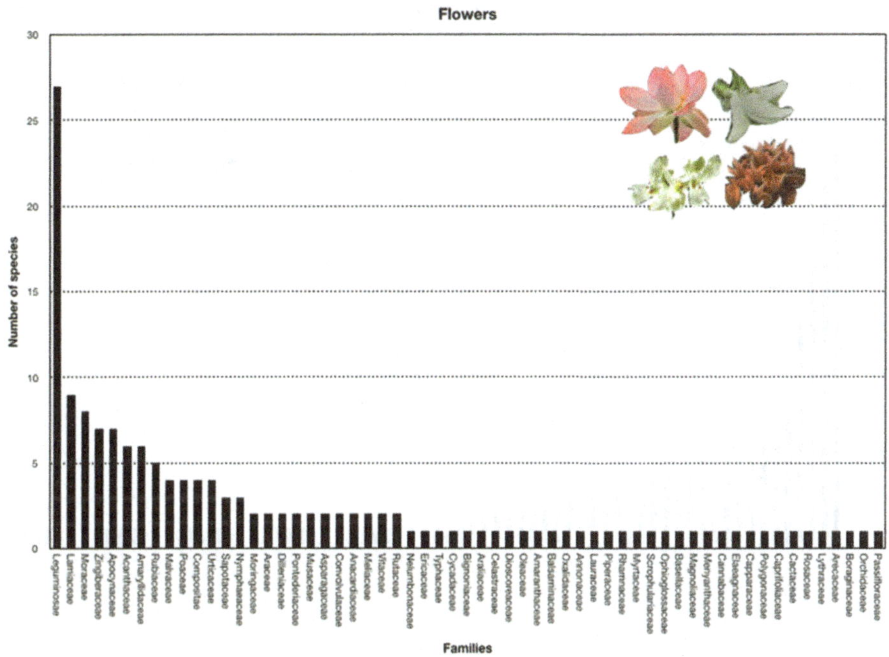

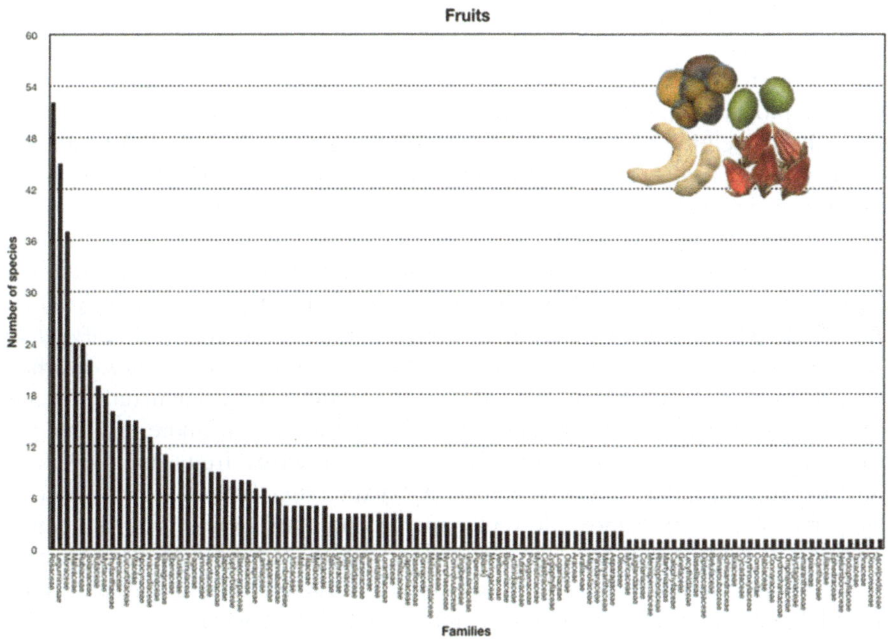

Fig. 2.5 (continued)

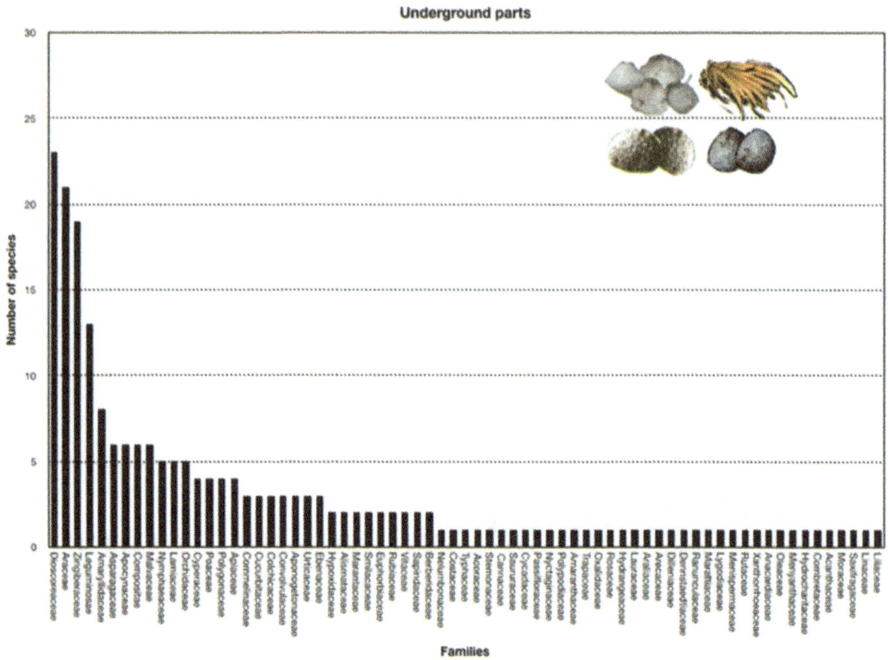

Fig. 2.5 (continued)

urban or peri-urban spaces expanded, and forests shrunk. The dwindled and marginalized biocultural landscapes still bear the legacy of the past and tend to maintain their cultural identity through their mode of plant usage. The central role of culture in wild edible use has been highlighted by Turner et al. (2011). The frequent use of leafy greens in south-east Asia is reiterated in a couple of studies (Cruz-Garcia and Price 2011; Sujarwo et al. 2014). A similar inclination toward leafy greens is also a common observation across other Asian countries, such as Korea (Pemberton and Lee 1996), Thailand (Price 2006), Vietnam (Ogle 2001), India (Ray and Ray 2022), and China (Hu 2005). In contrast, in northern Europe, where wild leafy green vegetables have been relatively less used (Luczaj 2008), or the Amazon, where they are used rarely (Katz et al. 2012). Of so many edible species in natural or semi-natural ecosystems, some plants are extremely vital because of more than one part being eaten, e.g., superfood *Moringa oleifera*, whose leaves, fruits, and flowers are eaten throughout India; similarly, aquatic *Nymphaea nouchali, N. rubra* whose all five parts (seeds, leafy shoots, flowers, underground parts, petiole, etc.) are edible. Similarly, four different parts of *Tamarindus indica, Nelumbo nucifera, Justicia adhatoda, Spondias pinnata, Ensete superbum* are widely consumed. And there are around 33 or more species with three edible organs (e.g., *Boehmeria glomerulifera, Ficus hispida, Urtica parviflora*, and several *Allium* species) (Fig. 2.6a, b). So, employment of a highly diverse assemblage of plants and their multiple plant organs not only suggests a traditional ecological knowledge base about their distribution,

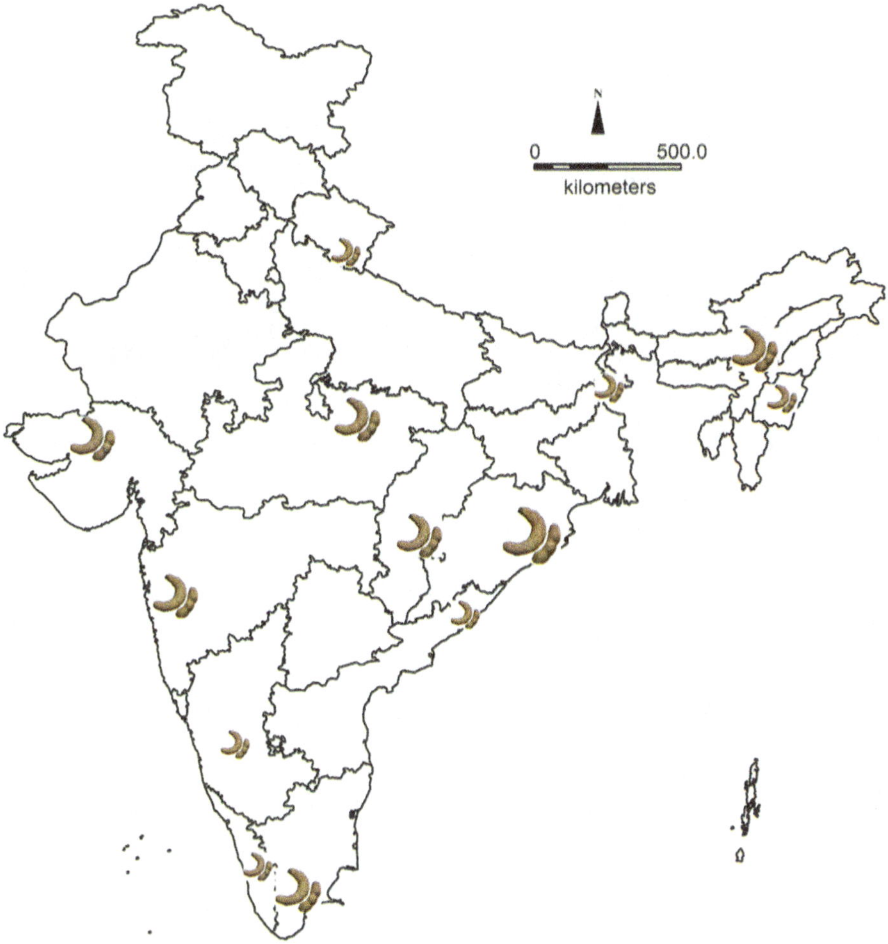

Fig. 2.6 Spatial distribution of a few species (dots represent spatial distribution and dot size is proportional to the number of plant part used): (**a**) *Tamarindus indica* and (**b**) *Spondias pinnata* (Adapted from Ray et al. 2020. *Front. Sustain. Food Syst.* 4:56. Doi: https://doi.org/10.3389/fsufs.2020.00056)

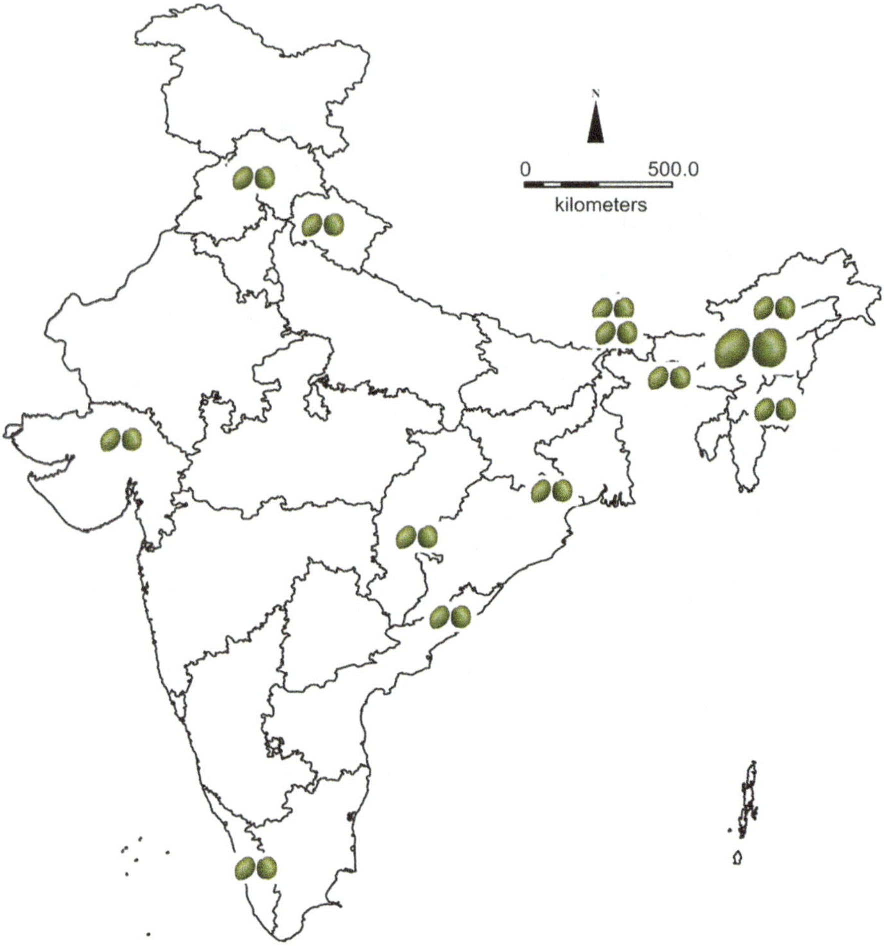

Fig. 2.6 (continued)

abundance, flowering and fruiting phenology, removal of toxins (if any), processing and cooking skills, it also validates the acumen of human resource management.

2.7 Implications for Food Security Policies and Sustainable Food System

Of 30,000 species, nearly 7000 plants have been accepted as food. This apparently large number funnels into almost 30 domesticated species that constitute our diet, and three principal cereals, rice, wheat, and maize, contribute to more than half of the world's calorie intake (FAO 2010). The addition of sugar, barley, soy, palm, and potato gets the percentage up to 76% (West et al. 2014). The figures allow us to have an idea of narrowing species diversity in our food basket. In contrast, a bewilderingly diverse range of wild uncultivated or orphan plants is still being consumed across countries and continents depending on a variety of human gastronomic choices. They tend to supplement carbohydrates, proteins, essential micronutrients, and vitamins that enrich dietary quality (Ogle 2001); and thereby provide an affordable (or in many cases no-cost) source of nutrition for rural, semi-urban, and even urban societies (Rowland et al. 2016; Jones 2017; Ray 2021a; Ray 2021b; Ray and Ray 2022). It is crucial since dietary diversity has been recommended for optimum human growth and nutrition, good health, and well-being (Oldewage-Theron and Kruger 2008; FAO et al. 2012). These underutilized foods reserve little-explored potential to transform our food systems to be diverse and nutritious, relatively sustainable, and resilient to climate change (Hunter and Fanzo 2013; Powell et al. 2015; Ray et al. 2020).

With the dire effects of climate change looming large, ensuring food security would be a critical issue. When the environmental footprints of input-intensive industrial agriculture are widening and intensifying, crop diversification is making space into vast stretches of monocropped fields, and our objective and challenge to shift toward sustainable food systems become imperative. Integration and judicious use of intrinsic resources with less negative ecological impact have become critical, and adoption of wild edible resources could emerge as a sustainable strategy (Bharucha and Pretty 2010). Although they cannot meet the demand for agriculturally produced crops by any means, their supportive or complementary role cannot be ignored. Several advantages, i.e., availability and easy access, little to no costs, high local or regional diversity and distribution, time-tested reliability, least management, etc. render them attractive as food or supplements (Mazhar et al. 2007; Bharucha and Pretty 2010; Cruz-Garcia and Price 2011; Turner et al. 2011) and may appear to effectively intervene in the crisis of hidden hunger or micronutrient deficiency. The ubiquity and diversity of wild edibles in preexisting food cultures have the potential for food policy implementation, especially in the alleviation of hidden hunger or narrowly vitamin A deficiency. Yet fortification of commercial food is widely implemented, and the production of biofortified food crops has already been flooding the market to combat hidden hunger. For example, golden rice has been sanctioned or nearly so in a few Asian countries for commercial cultivation, as was orange-fleshed sweet potato in Africa for vitamin A augmentation (Ahmad 2021; Gonzales 2021). Recently, The Govt. of India has been seeking mandatory biofortification of several food items, processed or farmed. Many crops have already been developed to fulfill this promise, e.g., zinc- or protein-rich rice, iron- and zinc-rich pearl millet, and lysine- and tryptophan-rich maize, etc. are on their way (Yadava et al. 2017; The Hindu 2017). Perhaps, many more are awaited. All of these are essentially resting on the myopic premise that subverts the importance of dietary diversity.

Here, the untapped nutritional potential of wild food to enhance dietary diversity and improve nutritional outcomes has a central role to play (Ray et al. 2020; Ray and Chakraborty 2021; Ray and Ray 2022). They are the source of many vitamins and other nutrients, e.g., leafy shoots (e.g., *Boerhaavia diffusa, Aerva lanata*) are a source of vitamin A and calcium or fruits (e.g., *Tamarindus indica*), are a source of iron and zinc, etc. Many species of the genus *Amaranthus* are regularly eaten as green leafy vegetables across the tropics, and it contains 200 times more vitamin A and 10 times more iron than the same-sized portion of cabbage (McGarry and Shackleton 2009). Compared to the widely consumed cabbage (*Brassica oleracea*), Malabar spinach (*Basella alba*) has over 13.5 times as much iron. The health benefits of hundreds of locally grown, neglected, underutilized, or orphan crops have also been widely recorded and often remain as the focus of discussion of food and nutritional security (Schönfeldt and Pretorius 2011; Ogle and Grivetti 1985a, b; Cook et al. 2000; Ogle 2001; Ogle et al. 2003; Simopoulos 2004; Uusiku et al. 2010; Ranfa et al. 2014; Mishra et al. 2015; Hama-Ba et al. 2017). In India, a couple of studies from Sikkim, Jharkhand, the north-eastern region, and South India examined a suite of edible plants also reiterates the same facts (Rajyalakshmi et al. 2001; Sundriyal and Sundriyal 2001; Bhatt et al. 2005; Ghosh-Jerath et al. 2015). While it has been argued that the contribution of wild foods to total energy intake is low (Bharucha and Pretty 2010; Schlegel and Guthrie 1973; van Liere et al. 1995; Termote et al. 2012; Powell et al. 2013), cross-country investigations have shown their higher micronutrient content and therefore critical role in enhancing the nutritional outcome (Blaney et al. 2009; Powell et al. 2013; Schlegel and Guthrie 1973).

At the confluence of sustainable food systems and wild edible flora surfaces a key aspect of food—the culture of taste. Food is not only a commodity to satisfy our hunger, but it is loaded with gustatory, economic, cultural, and spiritual ethos (Counihan and Van Esterik 2012). Therefore, a complex and dynamic interaction of various qualities finally decides the acceptance of food within a society. It signifies their embeddedness or assimilation in society and can be regarded as an indicator of their long-drawn historical entanglement with the consuming groups. That implies a raw plant material can be harvested, processed, and cooked in many different ways (boiled, pickled, mixed with other elements, or eaten raw) before finally being consumed as food. Food choice is essentially deep-rooted in culture and embodies the identity of an individual or a society or a community (Almerico 2014). However, lost in the midst of the debate on biofortification, food security, and sustainable systems is the ingrained food culture. It is a long-nurtured tradition shaped over generations—it could be difficult to be infiltrated through a top-down approach (Ray 2021b). An example can be sought from biofortified orange-fleshed sweet potato varieties of Malawi, where researchers found sensory and cultural acceptability tradeoffs with enhanced nutritional content (Hummel et al. 2018). On the other hand, there are accelerated promotions for golden rice in many parts of the world (Gonzales 2021). A sweeping answer to the biofortified crops could be the alternatives that already exist in our food culture but have eroded or lost over time. Hence, the discourses on food or nutritional security and overarching policy measures have to take into account the plurality of food cultures. Culturally embedded, widely available, and less costly alternatives are always an effective step (Wesseler and Zilberman 2017). A potential way out could be the revival of the culture and ensuring dietary diversity by promoting informal cultivation or moderate management in homesteads, fringes, pastures, or fallow lands (Broegaard et al. 2017). The resources can be integrated into various government schemes or other interventions, such as the promotion of kitchen gardens or home gardens for the propagation of their natural population and for easy access by a diversity of consumer groups (Ray and Ray 2022). On a broader scale, it can be conceived through '*Poshan Abhiyaan*' or the scheme for holistic nourishment under the National Nutrition Mission of the Government of India to improve nutritional outcomes for children, pregnant women, and lactating mothers (NITI Aayog 2019). More specifically, the Ministry of Human Resource Development's 'School Nutrition Gardens' program could be another way to encourage the growing, nurturing, and consuming of a suite of plants as part of the school's mid-day meals (Government of India 2019). There are also success stories from the other part of the world, e.g., in Western Kenya, where these leafy greens were introduced into school meals to reduce undernutrition piloted by the BFN project (UNSCN 2017). On the other hand, the Brazilian initiative of the Zero Hunger (Fome Zero) campaign ensured the exploitation of the potential of edible native species through public policies. It encouraged the farming of more than a hundred species of neglected and underutilized plants by the National Food and Nutrition Policy, Food Acquisition Program, and the National School Meals Program, etc. (Beltrame and Hunter 2015; Bioversity International 2017). These Government policies allowed the harnessing of underutilized biodiversity to improve the dietary quality and nutrition of thousands of people.

Acknowledgments We acknowledge the informal discussions with Abhra Chakraborty, Rakesh Mitra, and Sumit Sarkar in various stages of the study. We received no funding for this study and declare no conflict of interest.

References

Ahmad R (2021) Philippines becomes first country to approve Golden Rice for planting. Dhaka Tribune July 23rd, 2021. https://www.dhakatribune.com/world/2021/07/23/Philippines-Becomes-first-country-to-approve-golden-rice-for-planting

Almerico GM (2014) Food and identity: food studies, cultural, and personal identity. J Int Bus Cult Stud 8

Angami A, Gajurel PR, Rethy P, Singh B, Kalita SK (2006) Status and potential of wild edible plants of Arunachal Pradesh. Indian J Tradit Knowl 5:541–550

Ball V (1867) On the jungle products used as articles of food by the inhabitants of the districts of Manbhum & Hazaribagh. J Asiatic Soc Bengal 36:73–82

Ball V (1985) Tribal and peasant life in nineteenth century India, Usha Publications, Delhi (originally published in 1880)

Barman J, Baruah UK, Goswami UC (2013) Indigenous techniques of catching the mud eel, *Monopterus cuchia* (Ham.) in Goalpara district, Assam. Indian J Tradit Knowl 12(1):109–115

Belfer-Cohen A, Goring-Morris AN (2011) Becoming farmers: the inside story. Curr Anthropol 52 (Suppl 4):S209–S220

Bellwood PS (2004) First farmers: the origins of agricultural societies. Wiley-Blackwell, Malden, 384 p

Beltrame D, Hunter D (2015) Biodiversity for food and nutrition in Brazil. In: A4NH outcome note, November (IFPRI, 1–3). CGIAR Research Program on Agriculture for Nutrition and Health, Washington, DC

Benton TG, Vickery JA, Wilson JD (2003) Farmland biodiversity: is habitat heterogeneity the key? Trends Ecol Evol 18(4):182–188. https://doi.org/10.1016/S0169-5347(03)00011-9

Berkes F, Colding J, Folke C (2000) Rediscovery of traditional ecological knowledge as adaptive management. Ecol Appl 10(5):1251–1262. https://doi.org/10.1890/1051-0761(2000)010[1251:ROTEKA]2.0.CO;2

Bhandari MM (1974) Famine foods in the Rajasthan desert. Econ Bot 28(1):73–81. https://www.jstor.org/stable/4253471

Bharucha Z, Pretty J (2010) The roles and values of wild foods in agricultural systems. Philos Trans R Soc B 365:2913–2926. https://doi.org/10.1098/rstb.2010.0123

Bhatt BP, Singh K, Singh A (2005) Nutritional values of some commercial edible bamboo species of the North Eastern Himalayan region, India. J Bamboo Rattan 4:111–124. https://doi.org/10.1163/1569159054699317

Bioversity International (2017) Mainstreaming agrobiodiversity in sustainable food systems: scientific foundations for an agrobiodiversity index. Bioversity International, Rome

Blaney S, Beaudry M, Latham M (2009) Contribution of natural resources to nutritional status in a protected area of Gabon. Food Nutr Bull 30(1):49–62. https://doi.org/10.1177/156482650903000105

Broegaard RB, Rasmussen LV, Dawson N, Mertz O, Vongvisouk T, Grogan K (2017) Wild food collection and nutrition under commercial agriculture expansion in agriculture-forest landscapes. For Policy Econ 84:92–101. https://doi.org/10.1016/j.forpol.2016.12.012

Cairns M, Garrity DP (1999) Improving shifting cultivation in Southeast Asia by building on indigenous fallow management strategies. Agrofor Syst 47(1):37–48. https://doi.org/10.1023/A:1006248104991

Cauvin J (2000) The birth of the gods and the origins of agriculture. Cambridge University Press, Cambridge

Chevalier A, Marinova E, Peña-Chocarro L (eds) (2014a) Plants and people: choices and diversity through time, vol 1. Oxbow Books, Oxford

Chevalier A, Marinova E, Peña-Chocarro L (2014b) Factors and issues in plant choice. In: Chevalier A, Marinova E, Peña-Chocarro L (eds) Plants and people: choices and diversity through time, vol 1. Oxbow Books, Oxford; Philadelphia, pp 3–12. https://doi.org/10.2307/j.ctvh1dwfp

Clement CR (2014) Landscape domestication and archaeology. In: Smith C (ed) Encyclopedia of global archaeology. Springer, New York

Cook JA, VanderJagt DJ, Pastuszyn A, Mounkaila G, Glew RS, Millson M et al (2000) Nutrient and chemical composition of 13 wild plant foods of Niger. J Food Compos Anal 13:83–92. https://doi.org/10.1006/jfca.1999.0843

Cordain L, Miller JB, Eaton SB, Mann N, Holt SH, Speth JD (2000) Plant-animal subsistence ratios and macronutrient energy estimations in worldwide hunter-gatherer diets. Am J Clin Nutr 71(3):682–692. https://doi.org/10.1093/ajcn/71.3.682

Counihan C, Van Esterik P (eds) (2012) Food and culture: a reader. Routledge

Cruz-Garcia GS, Price LL (2011) Ethnobotanical investigation of 'wild' food plants used by rice farmers in Kalasin, Northeast Thailand. J Ethnobiol Ethnomed 7:33. https://doi.org/10.1186/1746-4269-7-33

Curry A (2021) The ancient carb revolution. Nature 594:488–491

Damodaran V (1995) Famine in a forest tract: ecological change and the causes of the 1897 famine in Chotanagpur, Northern India. Environ Hist 1(2):129–158. https://doi.org/10.3197/096734095779522636

Diamond J (2002) Evolution, consequences and future of plant and animal domestication. Nature 418:700. https://doi.org/10.1038/nature01019

Dogan Y, Ugulu I, Durkan N (2013) Wild edible plants sold in the local markets of Izmir, Turkey. Pak J Bot 45:177–184

Erickson CL (1992) Prehistoric landscape management in the Andean Highlands: raised field agriculture and its environmental impact. Popul Environ 13(4):285–300. https://doi.org/10.1007/BF01271028

Erickson CL (2006) The domesticated landscapes of the Bolivian Amazon. In: Balée W, Erickson C (eds) Time and complexity in historical ecology: studies in the neotropical lowlands. Columbia University Press, New York, pp 235–278

Erickson CL (2008) Amazonia: the historical ecology of a domesticated landscape. In: Silverman H, Isbell WH (eds) Handbook of South American archaeology. Springer, New York

Eriksson O, Arnell M (2017) Niche construction, entanglement and landscape domestication in Scandinavian infield systems. Landsc Res 42(1):78–88. https://doi.org/10.1080/01426397.2016.1255316

Ertug F (2014) Wild plant food: routine dietary supplement or famine food? In: Chevalier A, Marinova E, Peña-Chocarro L (eds) Plants and people. Choices and diversity through time. Oxbow Books, Oxford, pp 211–215

FAO (2010) The second report on the state of the world's plant genetic resources for food and agriculture. Food and Agriculture Organization of the United Nations, Rome

FAO, WFP, IFAD (2012) The state of food insecurity in the world 2012 economic growth is necessary but not sufficient to accelerate reduction of hunger and malnutrition. FAO, Rome

Fisher MH (2018) An environmental history of India: from earliest times to the twenty-first century. Cambridge University Press

Fortier J (2009) The ethnography of south Asian foragers. Annu Rev Anthropol 38:99–114. https://doi.org/10.1146/annurev-anthro-091908-164345

Fuller DQ, Willcox G, Allaby RG (2011) Cultivation and domestication had multiple origins: arguments against the core area hypothesis for the origins of agriculture in the Near East. World Archaeol 43(4):628–652. https://doi.org/10.1080/00438243.2011.624747

Ghosh-Jerath S, Singh A, Kamboj P, Goldberg G, Magsumbol MS (2015) Traditional knowledge and nutritive value of indigenous foods in the Oraon tribal community of Jharkhand: an exploratory cross-sectional study. Ecol Food Nutr 54:493–519. https://doi.org/10.1080/03670244.2015.1017758

Global Rice Science Partnership (GRiSP) (2013) Rice almanac, 4th edn. International Rice Research Institute, Los Baños (Philippines), p 283

Gonzales C (2021) "Golden Rice" tastes, smells like regular rice, but more nutritious. https://www.philrice.gov.ph/golden-rice-tastes-smells-like-regular-rice-but-more-nutritious/posted on OCT-15-2021

González JA, García-Barriuso M, Amich F (2011) The consumption of wild and semi-domesticated edible plants in the Arribes del Duero (Salamanca-Zamora, Spain): an analysis of traditional knowledge. Genet Resour Crop Evol 58:991–1006. https://doi.org/10.1007/s10722-010-9635-8

Gosden C, Hather JG (eds) (2004) The prehistory of food: appetites for change. Routledge, London

Government of India (2019) School nutrition (kitchen) gardens. Ministry of Human Resource Development, Government of India. https://samagra.education.gov.in/docs/SNG_Guidelines.pdf

Grove RH, Damodaran V, Sangwann S (1998) Nature and the orient: the environmental history of South and Southeast Asia. Oxford University Press, Chennai, p 1036

Hadjichambis AC, Paraskeva-Hadjichambi D, Della A, Elena Giusti M, De Pasquale C, Lenzarini C et al (2008) Wild and semi-domesticated food plant consumption in seven circum-Mediterranean areas. Int J Food Sci Nutr 59:383–414. https://doi.org/10.1080/09637480701566495

Hama-Ba F, Sibiria N, Powell B, Ickowitz A, Maundu P (2017) Micronutrient content of wild vegetable species harvested in forested and non-forested areas in Southwest Burkina Faso. For Res 6:213. https://doi.org/10.4172/2168-9776.100213

Harris DR (1989) An evolutionary continuum of plant-people interaction. In: Harris DR, Hillman GC (eds) Foraging and farming: the evolution of plant exploration. Unwin-Hyman, London, pp 11–26

Harris DR (2012) Evolution of agroecosystems: biodiversity, origins, and differential development. In: Biodiversity in agriculture: domestication, evolution, and sustainability. p 2156

Hillocks RJ (1998) The potential benefits of weeds with reference to smallholder agriculture in Africa. Integr Pest Manag Rev 3(3):155–167. https://doi.org/10.1023/A:1009698717015

Hu SY (2005) Food plants of China. The Chinese University Press, Hong Kong

Hummel M, Talsma EF, Van der Honing A, Gama AC, Van Vugt D, Brouwer ID et al (2018) Sensory and cultural acceptability tradeoffs with nutritional content of biofortified orange-fleshed sweet potato varieties among households with children in Malawi. PLoS One 13(10): e0204754. https://doi.org/10.1371/journal.pone.0204754

Hunter D, Fanzo J (2013) Introduction: agricultural biodiversity, diverse diets and improving nutrition. In: Fanzo J, Hunter D, Borelli T, Mattei F (eds) Diversifying food and diets. Routledge, New York, pp 33–46

Ickowitz A, Powell B, Rowland D, Jones A, Sunderland TCH (2019) Agricultural intensification, dietary diversity, and markets in the global food security narrative. Glob Food Sec 20:9–16. https://doi.org/10.1016/j.gfs.2018.11.002

Jeyaprakash K, Ayyanar M, Geetha KN, Sekar T (2011) Traditional uses of medicinal plants among the tribal people in Theni district (Western Ghats), Southern India. Asian Pac J Trop Biomed 1: S20–S25. https://doi.org/10.1016/S2221-1691(11)60115-9

Johnson LM, Hunn ES (eds) (2010) Landscape ethnoecology: concepts of biotic and physical space. Berghahn Books

Jones AD (2017) Critical review of the emerging research evidence on agricultural biodiversity, diet diversity, and nutritional status in low- and middle-income countries. Nutr Rev 75:769–782. https://doi.org/10.1093/nutrit/nux040

Kajale MD (1991) Current status of Indian palaeoethnobotany: introduced and indigenous food plants with a discussion of the historical and evolutionary development of Indian agriculture and agricultural systems in general. In: Renfrew JM (ed) New light on early farming–recent developments in palaeoethnobotany. Edinburgh University Press, Edinburgh, pp 155–189

Kajale MD, Murty MLK, Sharma BL, Dayal R, Vijendra Rao R (1991) Archaeological wood remains from the prehistoric cave site at Betamcherla (Muchchatla Chintamanu Gavi), District Kurnool, Andhra Pradesh. Man Environ 16:115–119

Katz E, Lopez CL, Fleury M, Miller RP, Payê V, Dias T et al (2012) No greens in the forest? Note on the limited consumption of greens in the Amazon. Acta Soc Bot Pol 81:283–293. https://doi.org/10.5586/asbp.2012.048

Labadarios D (2005) Malnutrition in the developing world: the triple burden. South Afr J Clin Nutr 18:119–121. https://hdl.handle.net/10520/EJC64893

LaRochelle S, Berkes F (2003) Traditional ecological knowledge and practice for edible wild plants: biodiversity use by the Rarámuri, in the Sirerra Tarahumara, Mexico. Int J Sustain Dev World Ecol 10(4):361–375. https://doi.org/10.1080/13504500309470112

Leonti M, Ramirez R, Sticher F, Heinrich OM (2003) Medicinal flora of the Popoluca, México: a botanico-systematical perspective. Econ Bot 57:218–230. https://doi.org/10.1663/0013-0001(2003)057[0218:MFOTPM]2.0.CO;2

Leonti M, Nebel S, Rivera D, Heinrich M (2006) Wild gathered food plants in the European Mediterranean: a comparative analysis. Econ Bot 60:130–142. https://doi.org/10.1663/0013-0001(2006)60[130:WGFPIT]2.0.CO;2

Loh J, Harmon D (2005) A global index of biocultural diversity. Ecol Indic 5:231–241. https://doi.org/10.1016/j.ecolind.2005.02.005

Luczaj L (2008) Archival data on wild food plants eaten in Poland in 1948. J Ethnobiol Ethnomed 4:4. https://doi.org/10.1186/1746-4269-4-4

Maffi L, Woodley E (2012) Biocultural diversity conservation: a global sourcebook. Routledge, London. https://doi.org/10.4324/9781849774697

Mani MS (1974) Biogeographical evolution in India. In: Mani MS (ed) Ecology and biogeography in India. Springer, Dordrecht, pp 698–724. https://doi.org/10.1007/978-94-010-2331-3_24

Mazhar F, Buckles D, Satheesh PV, Akhter F (2007) Food sovereignty and uncultivated biodiversity in South Asia. Academic Foundation, New Delhi

McGarry DK, Shackleton CM (2009) Children navigating rural poverty: rural children's use of wild resources to counteract food insecurity in the Eastern Cape, South Africa. J Child Poverty 15(1): 19–37. https://doi.org/10.1080/10796120802677594

Michon G (2005) Domesticating forests: how farmers manage forest resources. Institut de Recherche pour le Développement, Center for International Forestry Research, The World Agroforestry Centre, Bogor

Mishra S, Sharma S, Vasudevan P (2015) Role of edible food plants in women's nutrition and health. In: Vasudevan P, Sharma S, Sharma VP, Verma M (eds) Women, technology and development. Narosa Publishing House, New Delhi, pp 1–17

Moerman DE (1996) An analysis of the food plants and drug plants of native North America. J Ethnopharmacol 52:1–22. https://doi.org/10.1016/0378-8741(96)01393-1

Molina M, Tardío J, Aceituno-Mata L, Morales R, Reyes-García V, Pardo-de-Santayana M (2014) Weeds and food diversity: natural yield assessment and future alternatives for traditionally consumed wild vegetables. J Ethnobiol 34(1):44–67. https://doi.org/10.2993/0278-0771-34.1.44

Morgan C (2015) Is it intensification yet? Current archaeological perspectives on the evolution of hunter-gatherer economies. J Archaeol Res 23(2):163–213. https://doi.org/10.1007/s10814-014-9079-3

Murty MLK (1981) Hunter-gatherer ecosystems and archaeological patterns of subsistence behaviour on the southeast coast of India: an ethnographic model. World Archaeol 13:47–58. https://doi.org/10.1080/00438243.1981.9979813

Myers JP, Antoniou MN, Blumberg B, Carroll L, Colborn T, Everett LG, Hansen M, Landrigan PJ, Lanphear BP, Mesnage R, Vandenberg LN, vom Saal FS, Welshons WV, Benbroo CM (2016) Concerns over use of glyphosate-based herbicides and risks associated with exposures: a consensus statement. Environ Health 15(19):13. https://doi.org/10.1186/s12940-016-0117-0

Ng CY, Karim SA (2016) Historical and contemporary perspectives of the Nyonya food culture in Malaysia. J Ethnic Foods 3(2):93–106. https://doi.org/10.1016/j.jef.2016.05.004

NITI Aayog (2019) Transforming nutrition in POSHAN Abhiyaan, India

Ogle BM (2001) Wild vegetables and micronutrient nutrition: studies on the significance of wild vegetables in women's diets in Vietnam (Doctoral dissertation). Acta Universitatis Upsaliensis

Ogle BM, Grivetti LE (1985a) Legacy of the chameleon: edible wild plants in the kingdom of Swaziland, Southern Africa. A cultural, ecological, nutritional study. Part II—demographics, species availability and dietary use, analysis by ecological zone. Ecol Food Nutr 17:1–30. https://doi.org/10.1080/03670244.1985.9990879

Ogle BM, Grivetti LE (1985b) Legacy of the chameleon: edible wild plants in the kingdom of Swaziland, Southern Africa. A cultural, ecological, nutritional study. Part IV—Nutritional analysis and conclusions. Ecol Food Nutr 17:41–64. https://doi.org/10.1080/03670244.1985.9990881

Ogle BM, Tuyet HT, Duyet HN, Dung NNX (2003) Food, feed or medicine: the multiple functions of edible wild plants in Vietnam. Econ Bot 57:103–117. https://doi.org/10.1663/0013-0001(2003)057[0103:FFOMTM]2.0.CO;2

Oldewage-Theron WH, Kruger R (2008) Food variety and dietary diversity as indicators of the dietary adequacy and health status of an elderly population in Sharpeville, South Africa. J Nutr Elderly 27(1–2):101–133. https://doi.org/10.1080/01639360802060140

Olwig KR (1996) Recovering the substantive nature of landscape. Ann Assoc Am Geogr 86(4): 630–653. https://doi.org/10.1111/j.1467-8306.1996.tb01770.x

Pappu S (2020) Down ancient trails: hunter-gatherers in Indian archaeology. In: Hunter-gatherers in history, archaeology and anthropology. Routledge, pp 129–141

Parasher-Sen A (1998) Of tribes, hunters and barbarians: forest dwellers in the Mauryan period. Stud Hist 14(2):173–191. https://doi.org/10.1177/025764309801400202

Pardo-de-Santayana M, Tardío J, Blanco E, Carvalho AM, Lastra JJ, San Miguel E et al (2007) Traditional knowledge of wild edible plants used in the northwest of the Iberian Peninsula (Spain and Portugal): a comparative study. J Ethnobiol Ethnomed 3:27. https://doi.org/10.1186/1746-4269-3-27

Pemberton RW, Lee NS (1996) Wild food plants in South Korea; market presence, new crops, and exports to the United States. Econ Bot 50:57–70. https://doi.org/10.1007/BF02862113

Pingali PL, Roger PA (2012) Impact of pesticides on farmer health and the rice environment, vol 7. Springer Science & Business Media

Pokharia AK (2008) Palaeoethnobotanical record of cultivated crops and associated weeds and wild taxa from Neolithic site, Tokwa, Uttar Pradesh, India. Curr Sci 248–255. https://www.jstor.org/stable/24101866

Pokharia AK, Pal JN, Srivastava A (2009) Plant macro-remains from Neolithic Jhusi in Ganga Plain: evidence for grain-based agriculture. Curr Sci 25:564–572. https://www.jstor.org/stable/24111887

Pokharia AK, Srivastava C, Vikram B, Dimri DN, Nautiyal CM, Sharma S (2015) On the botanical findings from excavations at Ahichchhatra: a multicultural site in Upper Ganga Plain, Uttar Pradesh. Curr Sci 10:1293–1304. https://www.jstor.org/stable/24905893

Powell B, Maundu P, Harriet VK, Johns T (2013) Wild foods from farm and forest in the East Usambara Mountains, Tanzania. Ecol Food Nutr 52(6):451–478. https://doi.org/10.1080/03670244.2013.768122

Powell B, Thilsted S, Ickowitz A, Termote C, Sunderland T, Herforth A (2015) Improving diets with wild and cultivated biodiversity from across the landscape. Food Sec 7:535–554. https://doi.org/10.1007/s12571-015-0466-5

Pratap A (2000) The hoe and the axe. Oxford University Press, Oxford

Pretty JN (1995) Regenerating agriculture: policies and practice for sustainability and self-reliance. London: Earthscan Publications; Washington, DC: National Academy Press; Bangalore: ActionAid and Vikas

Price LL (2006) Wild food plants in farming environments with special reference to Northeastern Thailand, food as functional and medicinal, and the social roles of women. In: Pieroni A, Price LL (eds) Eating and healing: traditional food as medicine. Food Products Press, New York/London/Oxford, pp 65–100

Rajyalakshmi P, Venkatalaxmi K, Venkatalakshmamma K, Jyothsna Y, Devi KB, Suneetha V (2001) Total carotenoid and beta-carotene contents of forest green leafy vegetables consumed by tribals of South India. Plant Foods Hum Nutr 56:225–238. https://doi.org/10.1023/A:1011125232097

Ranfa A, Maurizi A, Romano B, Bodesmo M (2014) The importance of traditional uses and nutraceutical aspects of some edible wild plants in human nutrition: the case of Umbria (central Italy). Plant Biosyst 148:297–306. https://doi.org/10.1080/11263504.2013.770805

Ray A (2021a) Fifty shades of food—gathering no-cost nutrition. CEiBa Newslett 4(2):8–13. https://www.researchgate.net/publication/353324868_CEiBa_Newsletter_Volume_4_Issue_2_2021

Ray A (2021b) Biofortification—associated costs and alternatives? Econ Polit Wkly LVI No 35: 13–15

Ray A (in press) Domestication, cultivation, and diversification of crops in shifting cultivation systems across the highlands of India—a cross-cultural investigation. In: Cairns MF (ed) Farmer innovations and best practices by shifting cultivators in Asia-Pacific. CAB International, Wallingford

Ray A, Chakraborty A (2021) The edible biota in irrigated, deepwater, and rainfed rice fields of Asia—a neglected treasure—for sustainable food system. Environ Dev Sustain 23:17163. https://doi.org/10.1007/s10668-021-01386-0

Ray A, Ray R (2018) The birth of Aus agriculture in the south-eastern highlands of India—an exploratory synthesis. Ancient Asia 9(3):1–7. https://doi.org/10.5334/aa.146

Ray A, Ray R (2022) The leafy greens of India—their diversity, pattern of consumption, and overriding implication on food and nutrition security. Agroecol Sustain Food Syst 46(3): 432–451. https://doi.org/10.1080/21683565.2022.2025987

Ray A, Ray R, Sreevidya EA (2020) How many wild edible plants do we eat—their diversity, use, and implications for sustainable food system: an exploratory analysis in India. Front Sustain Food Syst 4:56. https://doi.org/10.3389/fsufs.2020.00056

Rindos D (1989) Darwinism and its role in the explanation of domestication. In: Harris DR, Hillman GC (eds) Foraging and farming: the evolution of plant exploration. Unwin-Hyman, London, pp 27–41

Rowland D, Ickowitz A, Powell B, Nasi R, Sunderland TCH (2016) Forest foods and healthy diets: quantifying the contributions. Environ Conserv 44:102–114. https://doi.org/10.1017/S0376892916000151

Roy SC (1935) The Hill Bhuiyas of Orissa, with comparative notes on the Plains Bhuiyas, 'Man in India' Office, Ranchi, Bihar (now Jharkhand)

Roy SK (1989) Aspects of neolithic agriculture and shifting cultivation: GaroHills, Meghalaya. Asian Perspect 15(2):193–221. https://www.jstor.org/stable/42928062

Saha A, Hoque A, Mallick SK, Panda S (2014) Medicinal uses of grasses by the tribal people in West Bengal: an overview. Int J Basic Appl Sci 3:63–70

Sahani RK (2019) Food from the Paataal Loke—fishing apple snail, field rats and more in the Sickle Economy. CEiBa Newslett 2(2):2–5. https://doi.org/10.13140/RG.2.2.18549.32482

Saraswat KS (1980) The ancient remains of the crop plants at Atranjikhera (c. 2,000–1,500 B.C.). J Indian Bot Soc 590:306–319

Saraswat KS (2004) Plant economy of early farming communities. In: Singh BP (ed) Early farming communities of the Kaimur (excavations at Senuwar). Scheme, Jaipur, pp 416–535

Schlegel SA, Guthrie HA (1973) Diet and the tiruray shift from swidden to plow farming. Ecol Food Nutr 2(3):181–191. https://doi.org/10.1080/03670244.1973.9990335

Schönfeldt HC, Pretorius B (2011) The nutrient content of five traditional South African dark green leafy vegetables—a preliminary study. J Food Compos Anal 24(8):1141–1146. https://doi.org/10.1016/j.jfca.2011.04.004

Scott J (2011) Four domestications: fire, plants, animals, and . . . us. [The Tanner Lectures on Human Values]. Harvard University, 4–6 May 2011

Sharma GR (1985) From hunting and food gathering to domestication of plants and animals in the Belan and Ganga Valleys. In: Misra VN, Bellwood P (eds) Recent advances in Indo-Pacific prehistory. Oxford & IBH Publ. Co., New Delhi, Bombay, Calcutta, pp 369–371

Sharma TC (1990) Prehistoric background of shifting cultivation. In: Majumdar DN (ed) Shifting cultivation in North-east India. Osmos Publications, Guwahati

Shin T, Fujikawa K, Moe AZ, Uchiyama H (2018) Traditional knowledge of wild edible plants with special emphasis on medicinal uses in Southern Shan State, Myanmar. J Ethnobiol Ethnomed 14:48. https://doi.org/10.1186/s13002-018-0248-1

Simopoulos AP (2004) Omega-3 fatty acids and antioxidants in edible wild plants. Biol Res 37: 263–277. https://doi.org/10.4067/S0716-97602004000200013

Singh BP (2001) Stages of culture development in the Middle Ganga Plains—a case study of Senuwar. Pragdhara 11:109–118

Singh G (2018) Not a meat of choice: many rural Bengalis can only afford snails. Quint 10th April, 2018. https://www.thequint.com/lifestyle/food/googli-snail-meat-very-popular-rural-bengal-instead-of-chicken-mutton

Singh HB, Arora RK (1978) Wild edible plants of India. ICAR, New Delhi

Singh RK, Turner NJ, Pandey CB (2012) "Tinni" Rice (*Oryza rufipogon* Griff.) production: an integrated sociocultural agroecosystem in Eastern Uttar Pradesh of India. Environ Manag 49(1): 26–43. https://doi.org/10.1007/s00267-011-9755-8

Sinha R, Lakra V (2007) Edible weeds of tribals of Jharkhand, Orissa and West Bengal. Indian J Tradit Knowl 6(1):217–222. http://hdl.handle.net/123456789/909

Smith BD (2001) Low-level food production. J Archaeol Res 9(1):1–43. https://doi.org/10.1023/A:1009436110049

Smith ML (2006) How ancient agriculturalists managed yield fluctuations through crop selection and reliance on wild plants: an example from Central India. Econ Bot 60(1):39–48. https://doi.org/10.1663/0013-0001(2006)60[39:HAAMYF]2.0.CO;2

Smith BD (2011) General patterns of niche construction and the management of 'wild' plant and animal resources by small-scale pre-industrial societies. Philos Trans R Soc B Biol Sci 366(1566):836–848. https://doi.org/10.1098/rstb.2010.0253

Sterling K (2014) Man the hunter, woman the gatherer? The impact of gender studies on hunter-gatherer research (a retrospective). In: The Oxford handbook of the archaeology and anthropology of hunter-gatherers. Oxford University Press, Oxford, pp 151–174

Ströhle A, Hahn A (2011) Diets of modern hunter-gatherers vary substantially in their carbohydrate content depending on ecoenvironments: results from an ethnographic analysis. Nutr Res 31(6): 429–435. https://doi.org/10.1016/j.nutres.2011.05.003

Sujarwo W, Arinasa IBK, Caneva G, Guarrera PM (2014) Traditional knowledge of wild and semi-wild edible plants used in Bali (Indonesia) to maintain biological and cultural diversity. Plant Biosyst 150:971–976. https://doi.org/10.1080/11263504.2014.994577

Sundriyal M, Sundriyal DC (2001) Wild edible plants of the Sikkim Himalaya: nutritive values of selected species. Econ Bot 55:377–390. https://doi.org/10.1007/BF02866561

Tanji A, Nassif F (1995) Edible weeds in Morocco. Weed Technol 617–620. https://doi.org/10.1017/S0890037X0002393

Tardío J, Pardo-de-Santayana M, Morales R (2006) Ethnobotanical review of wild edible plants in Spain. Bot J Linn Soc 152:27–71. https://doi.org/10.1111/j.1095-8339.2006.00549.x

Termote C, Meyi MB, Dhed'a Djailo B, Huybregts L, Lachat C, Kolsteren P, Van Damme P (2012) A biodiverse rich environment does not contribute to a better diet: a case study from DR Congo. PLoS One 7(1):e30533. https://doi.org/10.1371/journal.pone.0030533

Terrell JE, Hart JP, Barut S, Cellinese N, Curet A, Denham T, Kusimba CM, Latinis K, Oka R, Palka J, Pohl ME (2003) Domesticated landscapes: the subsistence ecology of plant and animal domestication. J Archaeol Method Theory 10(4):323–368. https://doi.org/10.1023/B:JARM.0000005510.54214.57

Tewari R, Srivastava RK, Singh KK, Saraswat KS, Singh IB, Chauhan MS, Pokharia AK, Saxena A, Prasad V, Sharma M (2006) Second preliminary report of the excavations at Lahuradewa, District Sant Kabir Nagar, UP: 2002–2003–2004 & 2005–06. Pragdhara 16:35–68

Tewari R, Srivastava RK, Saraswat KS, Singh IB, Singh KK (2008) Early farming at Lahuradewa. Pragdhara 18:347–373

Thapar R (2012) Perceiving the forest of early India. In: Rangarajan M, Sivaramakrishnan K (eds) India's environmental history: from ancient times to the colonial period, vol 1, Permanent Black, Ranikhet, Uttarakhand, India

The Hindu (2017) Experts raise concerns over mandatory fortification of food items. August 02, 2021. https://www.thehindu.com/news/national/experts-raise-concerns-over-mandatory-fortification-of-food-items/article35685066.ece

Turner NJ, Luczaj L, Migliorini P, Pieroni A, Dreon AL, Enrica L et al (2011) Edible and tended wild plants, traditional ecological knowledge and agroecology. Crit Rev Plant Sci 30:198–225. https://doi.org/10.1080/07352689.2011.554492

UNSCN (2017) Schools as a system to improve nutrition. A new statement for school-based food and nutrition interventions (Discussion paper). United Nations System Standing Committee on Nutrition, Rome. (Also available at www.unscn.org/uploads/web/news/document/School-Paper-EN-WEB-nov2017)

Uusiku NP, Oelofse A, Duodu KG, Bester MJ, Faber M (2010) Nutritional value of leafy vegetables of sub-Saharan Africa and their potential contribution to human health: a review. J Food Compos Anal 23:499–509. https://doi.org/10.1016/j.jfca.2010.05.002

van Liere MJ, Ategbo E-AD, Den Hartog AP, Hautvast JGAJ (1995) The consequences of seasonal food insecurity for individual food-consumption patterns in north-western Benin. Food Nutr Bull 16(2):1–8. https://doi.org/10.1177/156482659501600208

von Fürer-Haimendorf C, von Fürer-Haimendorf E (1945) The Reddis of the Bison Hills: a study in acculturation, with a Foreword by J. P. Mills. Macmillan Publishers, New York

Weiss E, Kislev ME, Simchoni O, Nadel D (2004) Small-grained wild grasses as staple food at the 23 000-year-old site of Ohalo II, Israel. Econ Bot 58(1):S125–S134. https://doi.org/10.1663/0013-0001(2004)58[S125:SWGASF]2.0.CO;2

Wesseler J, Zilberman D (2017) Golden rice: no progress to be seen. Do we still need it? Environ Dev Econ 22(2):107–109. https://doi.org/10.1017/S1355770X16000292

West PC, Gerber JS, Engstrom PM, Mueller ND, Brauman KA, Carlson KM, Cassidy ES, Johnston M, MacDonald GK, Ray DK (2014) Leverage points for improving global food security and the environment. Science 345(6194):325–328. https://doi.org/10.1126/science.1246067

Wolfe ND, Dunavan CP, Diamond J (2007) Origins of major human infectious diseases. Nature 447(7142):279. https://doi.org/10.1038/nature05775

Yadava DK, Choudhury PR, Hossain F, Dinesh Kumar D (2017) Biofortified varieties: sustainable way to alleviate malnutrition. Indian Council of Agricultural Research, New Delhi

Zohary D, Hopf M, Weiss E (2012) Domestication of Plants in the Old World: the origin and spread of domesticated plants in Southwest Asia, Europe, and the Mediterranean Basin. Oxford University Press

Chapter 3
Wild Food Plants: History, Use, and Impacts of Globalization

R. P. Harisha, R. Siddappa Setty, and G. Ravikanth

Abstract Wild food plants (WFPs) make significant contributions to food baskets and livelihoods of a large number of subsistence farming communities. Many rural households and forest-dwelling communities are dependent on WFPs as a subsistence strategy, especially during adverse conditions of food insecurity. WFPs have played an important role in livelihood opportunities and providing the required nutritional security to people enduring crop losses. In recent years, globalization has led to drastic changes in food systems/diets, which has had a major impact on health and malnutrition in many small and marginalized communities. Simplification and reduction in the diversity of diet, as expected of the globalization pattern, has led to food systems that are low in nutritional requirements but high in calorific value. Understanding the importance of WFP is critical for the region and the country. This chapter gives a broad overview of the importance of WFPs and their role in tackling food security and meeting the nutritional requirement of many marginalized communities. The WFPs are culturally deep-rooted in many communities across the globe. Detailed documentation on the nutritional and curative values, amount, part and time of collection, and phenology of WFPs is important. Re-establishing the intimate association with the use of traditional foods originating from trees and herbaceous plants could help in meeting the nutritional requirements. Incorporating indigenous knowledge may help in the sustainable management of WFPs along with meeting the needs of the communities. An integrated conservation approach is needed to document, protect, and promote WFP resources as well as ensure their accessibility for future generations.

R. P. Harisha · R. Siddappa Setty
Centre for Environment and Development, Bengaluru, India
e-mail: hari@atree.org; siddappa@atree.org

G. Ravikanth (✉)
SMS Foundation Centre for Biodiversity and Conservation, Ashoka Trust for Research in Ecology and the Environment (ATREE), Royal Enclave, Bengaluru, India
e-mail: gravikanth@atree.org

A. Kumar et al. (eds.), *Wild Food Plants for Zero Hunger and Resilient Agriculture*, Plant Life and Environment Dynamics,
https://doi.org/10.1007/978-981-19-6502-9_3

Keywords Globalization · Wild food plants · Traditional knowledge · Socioeconomics · Sustainability

3.1 Introduction

Wild food plants (WFPs) are contribute to the diets and livelihoods of millions of people worldwide (Pfoze et al. 2012). The knowledge of their use has been passed on orally from one generation to the other (Shackleton 2003; Feyssa et al. 2012). The WFPs and the biodiversity around farms have been the only source of food and nutrition for a number of marginalized communities during crop failures caused due to droughts or floods (Ravikanth et al. 2020; Aditya et al. 2020). The term "Wild Food Plants" refers to any wild edible plants which are not cultivated but gathered or collected from natural forests and agricultural land (Cruz-Garcia et al. 2016). Many WFPs are found across different habitats with wide adaptability and high reproductive potential (Heywood 1999). The WFPs do not just provide calorific value, but are also a potential source of important nutrients and also serve as medicine for a large number of communities in rural areas (de Medeiros et al. 2021). Besides food and nutrition, WFPs also have socioeconomic value and are often associated with cultural identities (Agea et al. 2011; Harisha et al. 2021a). The WFPs are an important staple food for over 1000 million people in underdeveloped as well as in developing countries and play a vital role in their life (MEA 2005). Besides the use of the term 'WFP,' 'indigenous food plants,' 'wild edible plants,' and 'traditional plants' are also used in different regions of the world.

In this chapter, we discuss the history of WFPs and the impacts of globalization resulting in the loss of food diversity. With specific examples from southern India, we discuss the role of traditional knowledge, cultural, socioeconomic, and nutritional values associated with WFP. This chapter also provides a conceptual framework that incorporates the traditional conservation practices and tools for promoting sustainable use of WFP resources and its associated knowledge. Finally, we discuss the policy interventions needed to promote and conserve the use of WFP.

3.2 History and Culture of WFPs

A number of communities have depended on wild plants for food and other needs, much before the documentation of history (Lulekal et al. 2011). Many domesticated plants, which are now commonly cultivated, have originated from wild relatives (Zapata et al. 2014). The civilizations such as the Indus valley, Mesopotamian, Greek, Egyptian, Persian, Mayan, Roman, Aztecs, Chinese, and the Incas were centers of wild plant use and domestication (Vavilov 1935). Five centers of distribution and the origin of the domestication of food plants have been recorded in the world. These include central or southwestern Asia and mountainous regions,

South-eastern Asia, the Mediterranean region, Abyssinia and adjacent parts of North Africa, and Neotropical American highlands (Vavilov 1935).

In the early sixteenth century, a European botanist was the first to record the traditional use of plants and compiled it as a book on his observations while traveling across different continents. Linnaeus in the eighteenth century had compiled the writings of a number of herbalists and documented the collection as well as use of the plants. It was in 1855 that Alphonse de Candolle introduced the use of archaeological evidence in taxonomy and historical descriptions of plant morphology and domestication. It was later in 1952 that Hills described in chronological order, the historical use of wild plants.

In India, the Indus valley civilization was well known to be the center of the origin of plant domestication and use. A number of archaeological evidence suggests that the domestication of a number of wild plants occurred across the region around the river valleys. Apart from rice, barley, and wheat which were a staple diet, horse gram, black gram, wild mung bean, and pigeon pea were cultivated much before 5000 BC (Nene 2006). Through most of the subcontinent, agriculture was well-established by 6000–5000 BC. In 3000 BC, the central Indian plains across the Ganga River valley were the center of origin of the domestication of plants (Gupta 2004). In southern India, the domestication of legumes and finger millet has been documented independently in the Tamil cultural texts (Fuller 2011).

The Rigveda and the Atharvaveda, the ancient Indian manuscripts, have documented wild plant usage. However, many of these have largely focused on the therapeutic values and the medicinal properties of wild plants (Jain 1991). The Indian ethnobotany documented by Jain in1991 has described over 1500 wild plants and related traditional knowledge on their use (Jain 1991). Similarly, in Ayurveda, the history of plant usage has been documented, which also refers to the center of agro-diversity (Reddy and Vijaya 2002).

Around the world, many cultures have been identified by their affiliations to a particular region or to an animal or plant species (Cocks 2006; Madegowda 2009; Kothari et al. 2012). The dependence of humans and their survival has depended on the biodiversity of the region. People were largely dependent on the plant species that were found around them for food, shelter, and medicine, and this had direct relevance in recognition of peoples' life (Gadgil et al. 2000; Shackleton and Shackleton 2006; Sommano et al. 2011). The plant species around them have also played a significant role in shaping the cultural identities of the people and these are also called 'cultural keystone' species (Cristancho and Vining 2004). These cultural keystone species vary from one region to another and from one culture to another. However, many indigenous communities are closely associated with few species and depend upon them most extensively to meet their daily needs (Garibaldi and Turne 2004). Many of these species are rooted in their cultural traditions and have been inevitable components of social activities (Smith et al. 2019).

3.3 Documenting the Use of WFPs

Many WFPs have high cultural significance, indicating their major role in the traditional food systems of many indigenous communities. The critical role played by the WFP to rural livelihoods on a daily basis has been well documented. These WFPs have often served as a safety net in times of famine/distress to many indigenous communities. India is a biodiversity hotspot with a rich culture and indigenous knowledge among its 550 indigenous communities and 227 linguistic groups, who inhabit varied climatic and geographic zones (Grover et al. 2002; Scoones et al. 1992). Like other many south and southeast Asian countries, the indigenous communities in India depend on WFPs resources to meet their food need (Jain 1991; Ghosh 2003; Muthu et al. 2006; Shrestha and Dhillion 2006). For example, in a study conducted across eight villages in MM Hills Wildlife Sanctuary in Southern India, Harisha and Padmavathy (2013) documented a large number of WFPs being used by two forest-dwelling Soliga and Bedagampana communities. Their study using free listing exercises, household surveys, focused group discussions, and key informant interviews obtained both qualitative and quantitative data on WFP across these communities (Harisha and Padmavathy 2013; Harisha et al. 2021b). Their study revealed that these communities use over 126 plants for food and about 68 species for medicine, apart from using 32 species for making agricultural implements and in-house construction; they also trade 14 species for obtaining cash income and 26 species for cultural and spiritual activities (Harisha and Padmavathy 2013). Among the 126 WFP species recorded, 103 species (83.7%) were common across land use and forest types. There were 96 (78.5%) and 91 (73.9%) species distributed within the non-forest and forest habitats, respectively (Harisha and Padmavathy 2013). The largest number of leafy vegetables and fruits (81 species) were available and collected during the monsoon (May to October). Most tubers were collected in summer (February to April), and shoots and flowers were obtained in the winter season (November to January). The usage and seasonal availability of WFP were positively correlated (Harisha et al. 2021b). Thus, dependency and use of knowledge on WFP not only provide a safety net for rural households but also provides an opportunity for agro-forestry-based species conservation (Srivastava 2008; Ravikanth et al. 2020).

In another study on the cultural importance value of WFP using quotations, availability, and utilization frequency, Harisha et al. (2015) documented the parts used and multifunction food use indices in the same landscape. They documented the cultural index value, which captures the theoretical importance of a plant for two forest-dwelling Soliga and Bedagampana communities, whereas the index of relative importance value documents the multiple uses and intensity of plant use in daily life (Harisha and Padmavathy 2013; Puri 2015). However, the cultural index score of a plant might change independently from the relative importance of a plant or the economic value index. For example, depending on the availability of a species, there could be differences between the economic and cultural indices. Among the two species, *Acacia farnesiana* and *Dacalepsis hamiltoni* are culturally more valuable than economic terms. The leafy vegetables scored high cultural index values followed by fruits and tubers. *Jasminum ritchiei* scored very high in cultural index value (2086.6), whereas *Blepharis maderaspatensis* (3.6) had the minimum cultural index value (Harisha et al. 2015). The mean cultural index was 365.0 ± 4.7 and more than 93 species scored greater than 100 cultural index value. Thus, it was seen that most WFP species used by the community had significant cultural importance (Harisha et al. 2015).

Based on quantitative relative cultural importance indices, the contrasting cultural significance of plant species to human cultures has been recognized (Garibaldi and Turne 2004; Reyes-Garcia et al. 2006). The acceptance of WFP with a very high to moderate cultural index could be delineated based on the relative high quotation, availability, partly used as well as multifunctional food use indices. For instance, many WFP collected and consumed as leafy vegetables have a bitter taste (*Senna hirsuta*), but scored high taste appreciation in MM Hills (Harisha et al. 2015). The older generation tended to appreciate their bitter taste and often related it to their medicinal value and health (Reyes-Garcia et al. 2006; Agea et al. 2011). On the other hand, the younger generation tends to ignore such bitter-tasting plants.

Many local communities have a number of ways of using WFP resources to meet their day-to-day requirements and overcome the ramifications by trial and error (Donovan and Puri 2004). As a result, they have rich knowledge of the usage of plant resources around their habitat (Bussmann 2006). Local traditional knowledge and practices can help in understanding and adaptive management and socio-ecological systems (Narayanan and Anil Kumar 2007; FAO 2014). However, local traditions of medicinal plant use and knowledge have been devalued and replaced by the modern allopathic system and state-sponsored practices of conservation. The institutional takeover of resources by the state/private sector, including the displacement or infiltration of local village institutions and market forces, further erodes traditional knowledge systems (Peters 1996). The erosion of traditional knowledge on WFP directly leads to negative consequences on biodiversity, especially in the case of common property resources like forests, wetlands, and sacred species (Ayyanar and Ignacimuthu 2005). In recent years, a number of efforts are being made to protect and revive traditional knowledge. For instance, traditional knowledge digital library has been created in India to document traditional ecological knowledge systems through nonspatial means.

3.4 Socioeconomic Status of WFPs

Rural livelihoods are characterized by the extent of contribution of each livelihood activity in the form of monetary or nonmonetary benefits. A monetary contribution to household livelihood comes from formal cash income through local wages, pensions, livestock, NTFPs collection, and seasonal migration as laborers to urban areas, etc. (Puri 1997; Lele 1998). A number of studies have shown that nonfarm activities contribute to more than 70% of the total livelihood of households in the country (Rathore 2009). Historically, subsistence farming, collection of WFP for own consumption, and other forest resources used in day-to-day needs constitute nonmonetary benefits. The contribution of nonmonetary benefits to the total household's income is critically important (Reyes-Garcia et al. 2006). Specific linkages between a household's social, economic, and cultural factors play a significant role in resource use and have large implications for policy and livelihood opportunities for poor households (Shackleton and Shackleton 2006; De Laucena et al. 2007). However, there is a lack of detailed knowledge, resulting in a large number of WFPs being ignored in socioeconomic valuations and in the policy framework.

In India, WFPs are an important source to many rural households; unfortunately, their values are not accounted in for the economic analysis of natural resources (Puri 2015; Harisha et al. 2015). In a recent study, Harisha et al. (2021a) estimated the source of income at the household level for both the Soliga and Bedagampana communities. The income per capita per year from WFP income was ₹1459.6 for the Soliga communities and ₹1508.5 for the Bedagampana communities. It shows that income through WFP is as important as income from agriculture. The study also divulged that both communities equally rely on WFP for their dietary, therapeutic, and nutritional needs. Statistically, there was no significant difference in per capita income across both communities to the total household income (Mann–Whitney pairwise comparison, $p > 0.05$). Harisha et al. (2021a, b) study also revealed that WFPs are mainly used to supplement the staple diet and fill dietary gaps. The greater number of plant citations by both factions indicates a large consumption level and knowledge of these plants. At times of seasonal food shortages when household stocks are vacant and the new crop is still in the field, are the times of intensive collection and consumption of WFP.

The economic values of WFP have not been documented at the international or national or state level (Dovie et al. 2008). Many attempts have been made to assess the economic benefits of WFP (Bharucha and Pretty 2010). However, only a few studies have fully documented the economic valuation of WFP with the complexity of the quantitative assessment. Most of the data that is available has been secured from case studies carried out with individual local communities or community groups (Bussmann 2006; Eyong 2007). Moreover, many plant species collected are consumed through sharing or bartering and offering (Sundriyal and Sundriyal 2003; Scherrer et al. 2005). Therefore, it is essential to assess the economic value of WFP consumption in forested communities (Reyes-Garcia et al. 2006; Agea et al. 2011).

WFPs have been providing economic supplements for many generations in rural households. As agricultural yields have failed to meet dietary needs, increasing the use of WFP in the diet becomes crucial. The importance of evaluating the dependency and the economic value of lesser-known WFP is crucial and has been realized to a greater extent by the scientific world in recent years.

3.5 Impacts of Globalization and WFPs

Globalization refers to the increasing homogenization of the world's economy through the removal of barriers in international trade. Through globalization, regional economies, cultures, and societies have become integrated through transportation, communication, and trade. For the largest part of human history, information and knowledge about the WFP have been transmitted orally and by observation. However, globalization has changed the ways in which food is consumed (or what type of food is consumed), and this has profoundly affected ecosystems and human health. Globalization has led to accelerated loss of food biodiversity and degraded most ecosystems. A large proportion of the world's population in the tropical region suffers from hunger, and twice as many are devoid of a nutrient-filled diet. Simplification of diets, distinctive of the globalization pattern, has led to diets that are low in nutritional diversity but high in calorific value. While this food provides the required energy, they do not solve the problems of malnutrition and micronutrient deficiencies, particularly among poor segments of the population in developing countries. Intimate associations with the use of traditional foods originating from trees and herbaceous plants were lost because of globalization and industrialized approaches to farming led by corporate agribusiness. The loss of WFP consumption could also be attributed to social migration and rapid urbanization, which has had a severe impact on the dislocation of families and a significant increase in hunger, malnutrition, environmental degradation, and climate change. This has also led to the loss of traditional knowledge associated with the collection, processing, and consumption of WFPs.

Globalization has led to a significant change in food habits, lifestyle, and, more importantly, perceptions of using WFP. In a recent study, 89.5 ± 2.3% of people expressed a lack of sharing traditional knowledge across generations (Harish et al. 2021). Changes in lifestyle (84.5 ± 1.8% of people) and changes in food habits (71.0 ± 3.6% of people) were the major reasons for the change in the use of WFP. Despite the benefit of WFP in meeting food and nutritional as well as financial needs, changing lifestyles have resulted in a negative perception of WFP consumption. The younger generation has a negative perception with respect to WFP; consumption of it is considered uncivilized and a display of their poverty. Socioeconomic changes, in turn, have influenced consumption habit in cities and rural areas. Other studies too have reported a similar trend of the use of WFP as an indication of social backwardness (Agea et al. 2011). Several studies across the globe have shown higher reliance on store-bought food and marginalization of wild foods in many regions of the world (Orech et al. 2007; Centinkaya 2009).

In another study, both the Soliga and Bedagampana communities in Southern India depend on WFP resources that have been a part of their social and economic supplement (Harisha and Padmavathy 2013). However, changing socioeconomic conditions have resulted in a significant difference in traditional knowledge on the usage of wild plants across age class (68%), and gender (32.6%). There is also a significant difference in traditional knowledge on the usage of wild plants between educated (61.5%) and uneducated people (38.5%), as well as between farming (58.4%) and nonfarming households (41.6%) in both communities. However, there was no difference in usage between low-income and high-income households. Their study showed a significant positive correlation between the number of WFP listed by people and age ($r = 0.696$, $p < 0.05$) indicating that the younger generation is less fond of WFP than the older generation. The local community gained knowledge about WFP utilization, processing, and management through experience. However, the younger generation seems to have less interest both in acquiring knowledge and in imparting the knowledge across generations.

Food sharing is another social characteristic of the indigenous food system that has been severely impacted by globalization. Harisha et al. (2021b) reported that the traditional communities, on average, shared 15 WFP with their neighbors in the village, eight species between villages, and five species between relatives or friends living in distant villages/towns. Regularly, eight species of greens, three species of fruits, three species of tuber, and one species of the shoot were shared. During drought or failure of food crops and economic or social or environmental crisis, sharing of WFP would happen more frequently (especially during rainy and summer seasons) than on normal days. Most members of the community (89%) felt that sharing is a part of their culture which was practiced by their ancestors and 11% of them reported that sharing WFP provides an opportunity to express their love and care between them. However, in recent years, this culture of sharing among the communities has gradually decreased despite their increased connectedness.

These prosocial interactions not only influence the welfare of the community but are also encouraged as social and ethical obligations (Agea et al. 2011). Food sharing has traditionally been considered a characteristic feature of human societies and morality since early hominids to modern humans, from hunting and gathering to agricultural practices (Agea et al. 2011). Sharing web is a social fabric that served as a safety net between families, households, relatives, and even across communities (Srivastava 2008). Sharing resources and information has been identified as a long-term strategy to balance and manage risk in the traditional knowledge system (Madegowda and Usha Rao 2014; Chaubey et al. 2015). However, over the years, the tradition of sharing has gradually decreased.

3.6 Health and Nutritional Benefits

Despite over a thousand species of plants being listed as edible globally, only 120 species are cultivated today. It is well known that only nine plants meet 75% of the human diet and over half of the human diet is composed of only three species such as rice, wheat, and maize (Redzic 2006; FAO 2014). This decreasing diversity and reliance on very few species has largely contributed to malnutrition and is making communities more vulnerable to climate change impacts (Pinela et al. 2017). Half of the 35 biodiversity hotspots are located in regions with over 20%

malnutrition representing one-quarter of the malnourished people in developing countries.

Over centuries, people have relied on WFP resources as a source of several important micronutrients. WFPs are nutritionally loftier than some of the cultivated crops (Parvati and Kumar 2002; Toledo and Burlingame 2006). Globally, insufficient use of fruit and vegetables has led to malnutrition causing 1.7 million deaths (Vishwakarma and Dubey 2011; Sansanelli and Tassoni 2014). There is a strong and direct relationship between a low intake of fruits and vegetables to higher mortality as well as a higher risk to major chronic diseases such as cardiovascular diabetes and others (Grivetti and Ogle 2000; Addis et al. 2005). WFPs provide low cost but quality nutrition and high food therapeutic value for large parts of the population in rural areas.

In the MM Hills region, Harisha et al. (2015) estimated 65% of the WFP as critical supplements of micronutrients. The majority of the WFP provided nutrient supplements of zinc and iron, especially in the leaves and shoots (Sudarshan 1998). Most fruits are rich in macronutrients and provide rare micronutrients such as calcium, phosphorus, and vitamins B and C (Grivetti and Ogle 2000; Parvati and Kumar 2002). Similarly, in summer, tubers provide the required carbohydrates (starch) and other micronutrients at the time of shortage of nutritive foods. The relative frequency citation index value of WFP ranged from 0.250 to 0.598 (Harisha et al. 2015). For 19 species, they evaluated the leaves and shoots, fruits from eight species, and tubers, flowers, and bark were evaluated from four species. *Boerhavia diffusa, Acacia farnesiana, Alternanthera sissilis,* etc. frequently used as vegetables, were reported to increase iron in the blood, reduce blood pressure, and improve eyesight, and in addition, were laxative and diuretic. Both communities have adopted WFP as a common ailment for many common sicknesses like fever, cold, cough, headache, stomach ache, ulcer, and skin allergies. Many researchers across India and other countries disclosed that WFPs are an important source of nutritional security (Dovie et al. 2007; Bhattarai et al. 2009; Sneyd 2013; Aditya et al. 2020). Harisha et al. (2015) also found that many fruits, leafy shoots, and tubers are vital nutritional supplements to the people's diets, especially during food scarcity. Their study revealed that many of these provide critical nutritional supplements, especially to women during their puberty as well as during childbirth.

3.7 Diversity and Food Security

Of 300,000 known higher plant species, 5000 are being used and of these only 20–30 species are regarded as a staple food for all of the human population (Cotton 1996; Heywood 1999). On the other hand, thousands of WFP species are used by humans. However, food security policies all over the world have not completely acknowledged the significance of the diversity of WFP, and this has often limited food security and biodiversity conservation. Recent studies have stressed that the drastic decline in biodiversity and the increase in global food insecurity should be addressed together (Puri et al. 2006; Ravikanth et al. 2020). Tropical countries have potential genetic diversity of WFP, which can serve as food for the entire world if they are conserved and a systematic domestication process is undertaken (Delang 2006). The decline in biodiversity has led to lower dietary diversity and the elimination of essential food and nutrient sources, particularly for rural people (Gadgil and Guha 1995; Malik et al. 2001; Burlingame 2000). For example, in the MM Hills region, previous studies have documented 126 WFP species, belonging to 94 genera and 58 families (Harisha and Padmavathy 2013; Harisha et al. 2021a). An average of 50 species of WFP collected are from farmland, 64 species from forests, and 12 species from other land use. The six plant species which have the highest relative frequency citation (RFC) scoring are *Jasminum ritchiei*, *Cocculus villosus*, *Canthium parviflorum*, *Holostemma annulare*, *Celosia argentea*, and *Solanum nigrum* (Harisha et al. 2021a). Both Bedagampana and the Soliga communities consume a high proportion of wild leaves as greens which fall under the category of weeds (83%) and are often collected from forests, farms, wasteland, and kitchen gardens. Though weeds unlike invasive species are a human-perceived ecological concept, 88% reported that these plants have high nutritional value and since they have a high reproductive capacity, rapid growth, and a high range of adaptation to different environmental conditions, they could be harvested multiple times in a season.

Their study also indicated that fruits and tender leafy shoots were the most commonly used parts of WFP (78%), while flowers, tubers, and roots were the least used parts (Harisha et al. 2015). The WFP species belong to different lifeforms: trees, shrubs, herbs, climbers, and grass. The trees contributed 48%, while the grasses contributed only 1.5% of all the identified species. Herbs comprised 42.8% of WFP, most of which are seasonal leafy vegetables. Trees comprised around 26.9% of which are seasonal fruit-bearing trees, which are regarded as healthy by the tribal communities and a few are marketable. Shrubs (17.4%) were largely for seasonal fruits, followed by 11.1% climbers, which are seasonal tubers, greens, and fruits. The raw fruits were eaten, and others (herbs/shrubs and leafy shoots) were cooked before consumption. Trees (34%), shrubs (22%), and herbs (54%) were often the most frequently used as WFP (Fig. 3.1).

The diversity in the diet or dietary diversity increases the probability of consuming sufficient amounts of all food components essential to health (Bhattacharjee 2006; Pinela et al. 2017). Dietary diversity is defined as the total number of food groups consumed by an individual or household in a given period (Godfray et al. 2010; Muller and Almedom 2008). As human societies in developing countries suffer from malnutrition, enhancing dietary diversity would enhance health (Pardo-De-Santayana et al. 2005; Godfray et al. 2010). The community perceived dietary diversity practices in the indigenous food system as the most common adaptive mechanism that served as therapeutic ailments or medicine to many common diseases.

Other than the environmental factors and biology of the species, anthropogenic factors have a large influence on the distribution and diversity of WFP (Saha et al. 2014). Local communities have learned to use local WFP species, which are hardly

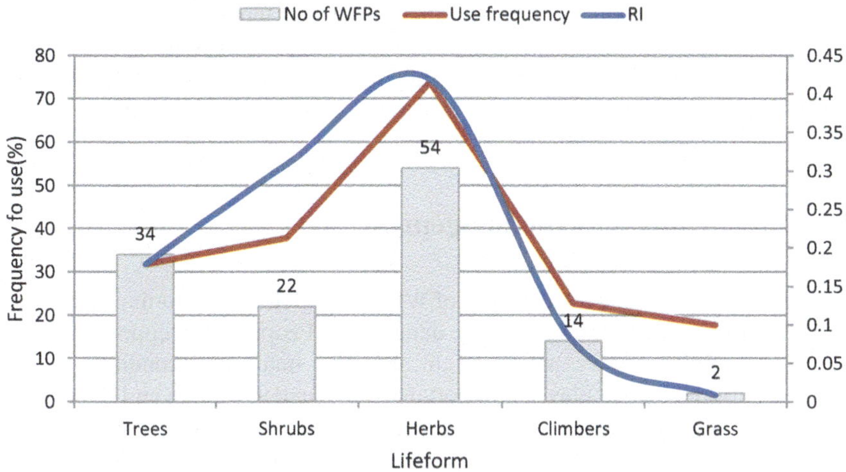

Fig. 3.1 The use frequency and relative importance across lifeforms of WFP species (Adapted and modified from Harisha et al. 2021b)

studied; even their identification and classification are in considerable confusion in the scientific community. The importance of these resources in a complex livelihood network, agroecosystems, economics, and restoration potentials are little known (Bawa et al. 2007; Bharucha and Pretty 2010). Therefore, most of the species which are documented have never been cultivated or domesticated, and agronomy is virtually unknown.

3.8 Conservation and Sustainable Use

Since WFPs are easily accessible, a lack of study on the community perceptions, practices, and monitoring has resulted in a poor understanding of the knowledge of WFP resources and their importance in the livelihood of the local community (Ghosh et al. 2007; Hamilton et al. 2016). There is a real danger of genetic erosion, which in turn calls for the need to assess the conservation status and threat of WFP. For the conservation of vulnerable WFP species, cultivation is often considered an alternative to wild collection (IUCN 2001). There is considerable interest in bringing endangered WFP and important therapeutic species into cultivation to reduce the pressure on wild populations.

A combination of scientific and traditional knowledge is required for the effective conservation of wild plant resources (Malik et al. 2001; Chaubey et al. 2015). It is important to involve the stakeholders and recognize their knowledge and practices to achieve sustainable use of natural resources and conservation. The intervention of new technologies and the neoliberal economy have been motivating exotic genetic resources' conservation rather than indigenous genetic resources. Indigenous knowledge of WFP production, preservation, use, and therapeutic values are no longer transmitted to the next generation and are disappearing drastically. Overlooking of WFP is unfortunate since they have historically been better adapted to the local climatic condition than the introduced exotic vegetable crops.

3.9 Monitoring and Co-management

Local people gain knowledge on the use of WFP through their day-to-day interaction and experience with the resources they depend on. People's perception and their indigenous knowledge give more insight into the status and condition of the resources. This knowledge plays an important role in understanding and developing conservation protocols and management practices. There is a lack of research on the monetization aspect of WFP and the importance of local people's traditional practices and knowledge in the management of these resources (Mandal et al. 2010). There is no sufficient database on the conservation status, issues, and management practices of WFP species, which are common and open sources subject to over-utilization and poor management (Bawa et al. 2007; Christensen et al. 2008). Moreover, these resources are highly neglected and ignored in the conservation and socioeconomic assessment. These species are disappearing at an alarming rate in many parts of the world because of climatic variation, invasive proliferation, and anthropogenic and biological impacts (Castro and Espinosa 2015).

Many studies revealed that the intimate knowledge held by the forest-dwelling communities had been gained from generations with continuous interaction and observation (Rao et al. 2003; Donovan and Puri 2004). Many studies have suggested that communities' knowledge has been useful in the monitoring and management of the species under threat and in developing management practices (Acharya and Acharya 2010; Sinu 2013). It is important to engage the local people and their intimate knowledge in addressing conservation issues and better management of WFP resources. The fallow land and cattle sheds in the forest are in-situ conservation areas where the communities have been conserving WFP species by traditional practices. The backyard and farmland could be considered ex-situ conservation areas.

3.10 Challenges Ahead

Changed perception in the local community from a subsistence agriculture system to an economic and intended business system, and rapid development activities have led to changes in a microhabitat that has affected WFP resources in the landscape. Major changes in occupation, from traditional agricultural practices to migration and business, have increased the erosion of indigenous knowledge on the use of WFP. The loss of knowledge and WFP resources has resulted in reliance on store-bought foods that have changed the diet composition of the local communities. These changes have led to the phenomenon of 'nutrition transition', which contributes to several public health problems and further burdens the food security of the local community, aggravating malnutrition.

There has been definite erosion of traditional knowledge and the disappearance of many WFPs due to modernization, globalization, and climate change. The WFP and related traditional knowledge systems have an important role in biodiversity conservation and sustainability in an era of global climate change. Nevertheless, it is underestimated and neglected in the economic assessment at the state and national levels. The local knowledge derived from long-term nature-society interactions has been extremely useful in validating scientific hypotheses and suggesting new research directions. The combined potential of traditional and scientific knowledge should be harnessed to validate, protect, promote, and develop a sustainable use strategy for WFP resources and related traditional knowledge. It would be an important tool to develop adaptation strategies to mitigate climate change impact and enhance the health of the environment and human well-being.

Understanding the distribution patterns, ecology, and ecosystem service of WFP is crucial for the region and the country. WFPs are the potential sources to tackle food security and potential sources of rare micronutrients apart from their established role in the sociocultural system. The local communities and policymakers together must lay down integrated scientific and traditional principles for a holistic approach to the conservation of WFP and ecosystem management. In-depth documentation on the therapeutic values, collection methods, phenology, and monitoring and non-monitoring values of WFP is important. Incorporating indigenous knowledge expands human understanding and may enhance biodiversity and resource sustainability for future generations.

3.11 Policy and Interventions

Until the neoliberal era, rural people had a simple linear integrated life with available natural resources and were satisfied with minimum components of human well-being. The bare necessities were fulfilled by natural resources using their traditional knowledge, which had been passed through generations. After globalization and privatization, traditional knowledge was opened up to the global market, making an impact on rural livelihoods that became inter-reliant and complex. In India, many policies about natural resource management and conservation strategies are based on separating the local people from their environment. Many conservation policies fail to address the linkages between the local people and their indigenous knowledge of ecosystem functioning, development, and human well-being. Policies that integrate scientific and local knowledge systems are vital for conserving and managing WFP resources. Effective implementation of such policies through local institutions is crucial and the best management strategy. Paying attention to the linkages and knowledge systems, which exist between local communities and the government, would be the best practice to address the needs of the local communities successfully and to achieve sustainability goals in ecosystem management. Retrieving traditional knowledge systems through the provisions provided in the inclusive policies at the national level, and as indicated globally in the CBD, SDGs, and MEAs, would help conserve and promote WFPs. At present, there are three policy tools such as the Biological Diversity Act, the 2002 Forest Rights Act, 2006, and the legal framework of the Intellectual Property Rights (IPR) Act. These can be used to discuss ways in which traditional knowledge about the collection, use, and cultivation of WFP can be protected and promoted.

3.12 Conclusions

The dependency on WFP resources is inevitable and critically important to meet the dietary, therapeutic, socioeconomic, and cultural practices. The economic value of WFP for households is as important as the crops grown for subsistence. It revealed that households, which follow the traditional occupation, were still dependent on WFPs for food and other uses than households with nontraditional occupations. The reliance on WFPs is, therefore, a safety net and potential source of nutrition. Emphasizing the improvement of nutrition and health through initiatives that protect WFP diversity and related traditional knowledge systems is critically important. Towards this, a systematic review of WFP resources at the regional and national level with local consultation is necessary. Understanding the role of WFPs in the food, nutrition, culture, and economics of the local communities is very important. The benefits of WFP resources to the forest-dependent community in the semi-arid tropical region of the country are enormous and can no longer be neglected in national and regional resource accounting. Most importantly, to address the challenges posed by recent climate change issues, the financial crisis, and their implications on food and nutritional security, the use of WFP in the diet becomes crucial. Documenting the traditional knowledge of WFP use is necessary to gain information, facilitate its sustainable use, and increase its positive impact on community resilience. A better understanding of the degree of their significance in the new globalization, climate change, and policies has major implications for the socio-ecological system to achieve sustainable development goals (SDGs).

References

Acharya KP, Acharya M (2010) Traditional knowledge on medicinal plants used for the treatment of livestock diseases in Sardikhola VDC, Kaski, Nepal. J Med Plant Res 4(2):235–239

Addis G, Urga K, Dikasso D (2005) Ethnobotanical study of edible wild plants in some selected districts of Ethiopia. J Hum Ecol 33(1):83–118

Aditya V, Sumashini PS, Aravind NA, Ravikanth G, Çhandrashekara K, Uma Shaanker R (2020) Reconciling biodiversity conservation with agricultural intensification: challenges and opportunities for India. Curr Sci 118(12):1870–1873

Agea JG, Okia CA, Obua J, Hall J, Teklehamanot Z (2011) Wild and semi-wild food plants in Bunyoro-Kitara Kingdom, Uganda: cultural significances, local perceptions and social implications of their consumption. Int J Med Aromat Plants 1(2):137–152

Ayyanar M, Ignacimuthu S (2005) Traditional knowledge of Kani tribals in Kouthalai of Tirunelveli hills, Tamil Nadu, India. J Ethnopharmacol 102:246–255

Bawa KS, Joseph G, Setty S (2007) Poverty, biodiversity and institutions in forest-agriculture ecotones in the Western Ghats and Eastern Himalaya ranges of India. Agric Ecosyst Environ 121(3):287–295. Elsevier, Science Direct

Bharucha Z, Pretty J (2010) The roles and values of wild foods in agricultural systems. Philos Trans R Soc B 365:2913–2926. https://doi.org/10.1098/rstb.2010.0123

Bhattacharjee L (2006) Indigenous peoples' food systems for health: finding interventions that work. Public Health Nutr 9:1013–1019

Bhattarai S, Chaudhary RP, Taylor RSL (2009) Wild edible plants used by the people of Manang District, Central Nepal. Ecol Food Nutr 48(1):1–20

Burlingame B (2000) Wild nutrition. J Food Compos Anal 13:99–100

Bussmann RW (2006) Ethnobotany of the Samburu of Mt. Nyiru, South Turkana, Kenya. J Ethnobiol Ethnomed 2:35

Castro A, Espinosa CI (2015) Seasonal diversity of butterflies and its relationship with woody-plant resources availability in an Ecuadorian tropical dry forest. Trop Conserv Sci 8(2):333–351

Centinkaya G (2009) Challenges for the maintenance of traditional knowledge in the Satoyama and Satoumi ecosystems, Noto Peninsula, Japan. Hum Ecol Rev 6(1):27–40

Chaubey OP, Sharma A, Krishnamurthy G (2015) *Ex-situ* conservation of indigenous, threatened and ethno-medicinal diversity of forest species. Int J Bio-Sci Bio-Technol 7(3):9–22

Christensen M, Bhattarai S, Devkota S, Larsen HO (2008) Collection and use of wild edible fungi in Nepal. Econ Bot 62(1):12–23

Cocks ML (2006) Bio-cultural diversity: moving beyond the realm of indigenous and local people. Hum Ecol 34(2):185–200

Cotton CM (1996) Ethnobotany. Principles and applications. Wiley, Chichester

Cristancho S, Vining J (2004) Culturally defined keystone species. Soc Hum Ecol 11(2)

Cruz-Garcia GS, Struik PC, Johnson DE (2016) Wild harvest: distribution and diversity of wild food plants in rice ecosystems of Northeast Thailand. NJAS: Wageningen J Life Sci 78:1–11. https://doi.org/10.1016/j.njas.2015.12.003

De Laucena RFP, de Lima Araujo E, De Albuquerque UP (2007) Does the local availability of woody Caatinga plants (Northeastern Brazil) explain their use value? Econ Bot 61:347–361

de Medeiros PM, Figueiredo KF, Gonçalves PHS, de Almeida Caetano R, da Costa Santos ÉM, dos Santos GMC, Barbosa DM, de Paula M, Mapeli AM (2021) Wild plants and the food-medicine continuum an ethnobotanical survey in Chapada Diamantina (Northeastern Brazil). J Ethnobiol Ethnomed 17:37. https://doi.org/10.1186/s13002-021-00463-y

Delang CO (2006) The role of wild plants in poverty alleviation and biodiversity conservation in tropical countries. Prog Dev Stud 4:275–286

Donovan D, Puri R (2004) Learning from traditional knowledge of non-timber forest products: Penan Benalui and the autecology of Aquilaria in Indonesian Borneo. Ecol Soc 9(3):3

Dovie DBK, Shackleton CM, Witkowski ETF (2007) Conceptualizing the human use of wild edible herbs for conservation in South African communal areas. J Environ Manag 84(2):146–156

Dovie DBK, Witkowski ETF, Shackleton CM (2008) Knowledge of plant resource use based on location, gender and generation. Appl Geogr 28:311–322

Eyong CT (2007) Indigenous knowledge and sustainable development in Africa: Case Study on Central Africa. Tribes Tribals Special 1:121–139

FAO (2014) State of the world's forests. Food and Agriculture Organization of the United Nations, Rome. http://www.fao.org/3/a-i3710e.pdf

Feyssa DH, Njoka JT, Asfaw Z, Nyangito MM (2012) Comparative analysis of indigenous knowledge on use and management of wild edible plants: the case of central East Shewa of Ethiopia. Ethnobot Res Appl 10:287–304. https://ethnobotanyjournal.org/index.php/era/article/view/650

Fuller DQ (2011) Finding plant domestication in the Indian subcontinent; the origins of agriculture. Curr Anthropol 52(S4):S347–S362

Gadgil M, Guha R (1995) Ecology and equity: the use and abuse of nature in contemporary India. Routledge, London and New York

Gadgil M, Seshagiri Rao PR, Utkarsh G, Pramod P, Ashwini Chhatre and Members of the People's Biodiversity Initiative (2000) New meanings for old knowledge: the people biodiversity registers program. Ecol Appl 10(5):1307–1317

Garibaldi A, Turne N (2004) Cultural keystone species: implications for ecological conservation and restoration. Ecol Soc 9(3):1

Ghosh A (2003) Herbal folk remedies of Bankura and Medinipur districts, West Bengal (India). Indian J Tradit Knowl 2:393–396

Ghosh B, Barbhuiya AR, Chowdahury C (2007) Changes in health status of the Soliga tribe at BRT due to modern interventions. Curr Sci 92(12):1688–1689

Godfray HCJ, Beddington JR, Crute IR, Haddad L, Lawrence D, Muir JF, Pretty J, Robinson S, Thomas SM, Toulmin C (2010) Food security: the challenge of feeding 9 billion people. Science 327(5967):812–818. https://doi.org/10.1126/science.1185383

Grivetti LE, Ogle BM (2000) Value of traditional foods in meeting macro- and micronutrient needs: the wild plant connection. Nutr Res Rev 13:31–46

Grover JK, Yadav S, Vats V (2002) Medicinal plants of India with anti-diabetic potential. J Ethnopharmacol 81:81–100

Gupta AK (2004) Origin of agriculture and domestication plants and animals linked to early Holocene climate amelioration. Curr Sci 87(1):54–59

Hamilton AC, Karamura D, Kakudidi E (2016) History and conservation of wild and cultivated plant diversity in Uganda: forest species and banana varieties as case studies. Plant Divers. 38: 23–44

Harisha RP, Padmavathy S (2013) Knowledge and use of wild edible plant in two communities in Malai Madeshwara Hills, Southern India. Int J Bot 9(2):64–72

Harisha RP, Padmavathy S, Nagaraja BC (2015) Traditional Ecological Knowledge (TEK) and its importance in South India: perspective from local community. Appl Ecol Environ Res 14(1): 311–326

Harisha RP, Siddappa Setty R, Ravikanth G (2021a) Dependency and economic benefits of the use of wild food plants among tribal communities in Malai Madeshawara Hills wildlife sanctuary, Southern India

Harisha RP, Gowthami R, Siddappa Setty R (2021b) Vocal to local: indigenous dietary practices and diversity of wild food plants in Malai Mahadeswara wildlife sanctuary, South India. Ethnobot Res Appl 22:22

Heywood V (1999) Use and potential of wild plants in farm households. FAO, Rome, p 113

IUCN (2001). https://www.iucn.org/content/iucn-red-list-categories-and-criteria-version-31

Jain SK (1991) Dictionary of Indian folk medicine and ethnobotany. Deep Publications, New Delhi

Kothari A, Rai N, Madegowda C (2012) Green approach, an alternative conservation model for the BRT Sanctuary is a step closer to becoming reality. Conserv Frontline 29(1):99–102

Lele S (1998) Socio-economic factors influencing Soliga-Forest relationship. In: Ganeshaiah KN, Uma Shaanker R (eds) Biligiri Rangaswamy Temple Wildlife Sanctuary, natural history, biodiversity and conservation. Ashoka Trust for Research in Ecology and the Environment (ATREE) and Vivekananda Girijana Kalayana Kendra (VGKK), B. R. Hills, pp 21–22

Lulekal E, Asfaw Z, Kelbessa E, Van Damme P (2011) Wild edible plants in Ethiopia: a review on their potential to combat food insecurity. Afrika Focus 24(2):71–121

Madegowda C (2009) Traditional knowledge and conservation. Econ Polit Wkly 14:21

Madegowda C, Usha Rao C (2014) The traditional ecological knowledge of Soliga tribe on eradication of *Lantana camara* and their livelihood through Mahatma Gandhi National Rural Employment Guarantee Act at Biligiri Rangaswamy Temple Wildlife Sanctuary, South India. Cult Anthropol 163–173

Malik SS, Umesh Srivastava B, Tomar DC, Bhandari, Pandey A, Hore DK, Dikshit N (2001) Plant exploration and germplasm collection. In: Dhillon BS et al (eds) National Bureau of Plant Genetic Resources: A compendium of achievements. National Bureau of Plant Genetic Resources, New Delhi, pp 31–68

Mandal S, Rai ND, Madegowda (2010) Culture, conservation and co-management: strengthening Soliga stake in biodiversity conservation in Biligiri Rangaswamy Temple wildlife sanctuary, India. In: Verschuuren B, Wild R, McNeely JA, Oviedo G (eds) Sacred natural sites: conserving nature and culture. Earthscan and IUCN, London, pp 263–271

Millennium Ecosystem Assessment (MEA) (2005) Ecosystem and human well-being: policy responses, vol 3. http://www.millenniumassessemnt.org/en/Responses.aspx

Muller J, Almedom AM (2008) What is "famine food"? Distinguishing between traditional vegetables and special foods for times of hunger/scarcity (Boumba, Niger). Hum Ecol 36(4): 599–607. https://doi.org/10.1007/s10745-008-9179-0

Muthu C, Ayyanar M, Raja N, Ignacimuthu S (2006) Medicinal plants used by traditional healers in Kancheepuram district of Tamil nadu, India. J Ethnobiol Ethnomed 2:43

Narayanan MKR, Anil Kumar N (2007) Gendered knowledge and changing trends in utilization of wild edible greens in Western Ghats. Indian J Tradit Knowl 6:204–216

Nene YL (2006) Indian pulses through the Millennia 1. Asian Agric Hist 10(3):179–202

Orech FO, Aagaard-Hansen J, Friis H (2007) Ethnoecology of traditional leafy vegetables of the Luo people of Bondo district, western Kenya. Int J Food Sci Nutr 58(7):522–530

Pardo-De-Santayana M, Tardi J, Morales RN (2005) The gathering consumption of wild edible plants in the Campoo (Cantabria, Spain). Int J Food Sci Nutr 56(7):529/542

Parvati S, Kumar VJ (2002) Studies on chemical composition and utilization of the wild edible vegetable athalakkai (*Momordica tuberose*). Plant Foods Hum Nutr 57(3–4):215–220

Peters CM (1996) The ecology and management of non-timber forest resources. World Bank Technical Paper 322. The World Bank, Washington, DC

Pfoze NL, Kumar Y, Sheikh N, Myrboh B (2012) Assessment of local dependency on selected wild edible plants and fruits from Senapati District, Manipur, Northeast India. Ethnobot Res Appl 10: 357–367

Pinela J, Carvalho AM, Ferreira ICFR (2017) Wild edible plants: nutritional and toxicological characteristics, retrieval strategies, and importance for today's society. Food Chem Toxicol 110(2017):165–188

Puri RK (1997) Penan Benalui knowledge and use of tree palms. In: Sorensen KW, Morris B (eds) People and plants of the, Kayan Mentarang. UNESCO and WWF, Indonesia, Jakarta, pp 195–226

Puri RK (2015) Chapter 10: The uniqueness of the everyday: herders and invasive species in India. In: Barnes J, Dove M (eds) Climate cultures: anthropological perspectives on climate change. Agrarian Studies Series. Yale University Press, New Haven, pp 249–272

Puri RK, Sheil C, Van Heist M, Liswanti N, Rukmiyati R, Samsoedin I (2006) Local people's priorities for biodiversity: examples from the forests of Indonesian Borneo. Ambio 35(1):17–24

Rao KS, Nautiyal S, Maikhuri RK, Saxena KG (2003) Indigenous ecological knowledge, biodiversity and sustainable development in the Central Himalayas. Trop Ecol 44(1):93–111

Rathore M (2009) Nutrient content of important fruit trees from arid zone of Rajasthan. J Hortic For 1(7):103–108

Ravikanth G, Aditya V, Uma Shaanker R (2020) Biodiversity in and around farmlands: food and nutritional security and rural livelihoods. Econ Polit Wkly 55(49):34–39

Reddy KC, Vijaya Prakash P (2002) Socio-economic profile of Adivasis in India. India International Centre, New Delhi

Redzic SJ (2006) Wild edible plants and their traditional use in the human nutrition in Bosnia-Herzegovina. Ecol Food Nutr 45:189–232

Reyes-Garcia V, Huanca T, Vadez V, Leonard W, Wilkie D (2006) Cultural, practical and economic value of wild plants: a quantitative study in the Bolivian Amazon. Econ Bot 60(1): 62–74

Saha D, Sundriyal M, Sundriyal RC (2014) Diversity of food composition and nutritive analysis of edible wild plants in a multi-ethnic tribal land, Northeast India: an important facet for food supply. Indian J Tradit Knowl 13(4):698–705

Sansanelli S, Tassoni A (2014) Wild food plants traditionally consumed in the area of Bologna (Emilia Romagna region, Italy). J Ethnobot Ethnomed 10:69

Scherrer AM, Motti R, Weckerle CS (2005) Traditional plant use in the areas of Monte Vesole and Ascea, Cilento National Park (Campania, Southern Italy). J Ethnopharmacol 97:129–143

Scoones I, Melnyk M, Pretty JN (1992) The hidden harvest-wild foods and agricultural systems: a literature review and annotated bibliography. Sustainable Agriculture Programme, International Institute for Environment and Development, London, p 256. ISBN-13: 978-0905347936

Shackleton CM (2003) The prevalence of use and value of wild edible herbs in South Africa. S Afr J Sci 99:23–25

Shackleton CM, Shackleton SE (2006) Household wealth status and natural resource use in the Kat River Valley, South Africa. Ecol Econ 57:306–317

Shrestha PM, Dhillion SS (2006) Diversity and traditional knowledge concerning wild food species in a locally managed forest in Nepal. Agrofor Syst 66:55–63

Sinu E (2013) Social-welfare measures for children of tribes in India. Int Res Anal 2(3):169–171

Smith E, Ahmed S, Dupuis V, Running Crane M, Eggers M, Pierre M, Flagg K, Byker Shanks C (2019) Contribution of wild foods to diet, food security, and cultural values amidst climate change. J Agric Food Syst Commun Dev 9:1–24

Sneyd LQ (2013) Wild food, prices, diets and development: sustainability and food security in Urban Cameroon. Sustainability 5:4728–4759

Sommano S, Caffin N, Mcdonald J, Cocksedge R (2011) Food safety and standard of Australian native plants. Qual Assur Saf Crops Foods 3:176–184

Srivastava VK (2008) Tribal economy and society. Eastern Anthropol 61(4):533–543

Sudarshan H (1998) Traditional medicine and health care system of Soligas. In: Ganeshaiah KN, Uma Shaanker R (eds) Biligiri Rangaswamy Temple wildlife sanctuary, natural history, biodiversity and conservation. Ashoka Trust for Research in Ecology and the Environment (ATREE) and Vivekananda Girijana Kalayana Kendra (VGKK), B. R. Hills, pp 17–19

Sundriyal M, Sundriyal RC (2003) Underutilized edible plants of the Sikkim Himalaya: need for domestication. Curr Sci 85(6):731–736

Toledo A, Burlingame B (2006) Biodiversity and nutrition; a common path toward global food security and sustainable development. J Food Compos Anal 19:477–483

Vavilov NI (1935) Theoretical basis for plant breeding, vol 1. Moscow. Origin and geography of cultivated plants. In: The phytogeographical basis for plant breeding (D. Love, transl.). Cambridge Univ. Press, Cambridge, pp 316–366

Vishwakarma KL, Dubey V (2011) Nutritional analysis of indigenous wild edible herbs used in eastern Chhattisgarh, India. Emir J Food Agric 23(6):554–560

Zapata L, Cummings LS, Calderon JLM, Russel M, Sigaut F, Reyero SG, Cruz-Garcia G (2014) Choices and diversity through time. In: Chevalier A, Pena-Chocarro L, Marinova E (eds) Plants and people: choices and diversity through time, pp 1–432

Chapter 4
A Cross-continental Survey of Traditional Food Systems That Are Based on Wild Food Plants

Sajana Sreedharan, Vaishnav Suresh Babu, Golla Nagaraju Gari Saritha, Sarang Viswanathan Sherly, and Ajay Kumar

Abstract Food systems comprising items derived from local or indigenous and naturally available sources were followed by the human race all over the world. Such traditional food systems (TFS) are generally linked with specific communities and hence are known as the reflection of their culture and symbols of their heritage. The utilization of a food crop and preparation of dishes would not always be similar all over the world, and it depends on various factors such as traditional knowledge and food habits of the communities. Different communities may utilize different edible parts of the same plant, follow different preparation methods, and products may be entirely different in look and taste. Some traditional food crops may be only consumed in a small region or by one or a few communities. In Africa, Asia, and South America, wide diversity of wild traditional food plants, which were part of traditional food systems is found. This diet combination included root and tuber vegetables, cereals, leafy vegetables, pulses, seeds, fruits, tree barks, beverages, and many other plant foods. Healthy and nutritious food were one of the necessary needs humans ever demanded. Indigenous people of various regions in the world faced many challenges in their community history, such as food shortage, nutritional insecurity, crop failure, malnutrition, natural disasters, unfavorable weather patterns, and hunger. They developed a diverse, nutritional, and multipotential dietary system by integrating local food items to ensure their food security for the large populations. Food systems reflect the environmental specialties of the dwelling places; the locally available food crops or traditionally cultivated wild food crops were most apt for the existing climate conditions. In many parts of the world where climatic and edaphic

Sajana Sreedharan, Vaishnav Suresh Babu, Golla Nagaraju Gari Saritha, and Sarang Viswanathan Sherly contributed equally to this work.

S. Sreedharan · V. S. Babu · G. N. G. Saritha · S. V. Sherly
Department of Plant Science, School of Biological Sciences, Central University of Kerala, Kasaragod, Kerala, India

A. Kumar (✉)
Department of Plant Science, Central University of Kerala, Kasaragod, Kerala, India

A. Kumar et al. (eds.), *Wild Food Plants for Zero Hunger and Resilient Agriculture*, Plant Life and Environment Dynamics,
https://doi.org/10.1007/978-981-19-6502-9_4

conditions were very challenging for agriculture, many of the crops which were part of TFS were cultivated or naturally grown. They nourished and sustained the people over generations. This chapter summarises traditional food systems based on wild food plants from across the globe that are based on wild food plants. It also discusses the preparation, significance and medicinal importance of some of the traditional foods prepared from wild food plants.

Keywords Wild food plants · Traditional food systems · Traditional food crops · Food security

4.1 Introduction

Tradition is a wider term that can be expressed in many ways, as a part of a religious program or an event in a family or as a function of time scale. The traditional food system (TFS) is sometimes related to the local food, indicating their ingredients and how they are made (Cayot 2007). Wide varieties of locally available food plants are the basis of culturally acceptable TFS followed by various indigenous groups across the world (Kuhnlein and Receveur 1996). TFS not only meet the requirement of a healthy diet but also act as medicines due to the presense of important bioactive compounds. According to Indian philosophy, food is the gift of Paramathma (God), and it should be for the survival, strengthening, and nourishment of the body (Bhat 2012). The Siddha system of medicine (a traditional medicinal system in India) was based on the theory that "**Food is Medicine and Medicine is Food**" (Sivaraman and Rajalakshmi 2005). Traditional food systems possess multidimensional aspects, which include cultural and economic significance (Trichopoulou et al. 2006). The usage of wild food plants (WFPs) by humans has a long history (Zohary et al. 2012). With the changing lifestyles and standards, attitude towards food and its ingredients also changes (Sproesser et al. 2019).

The identity and characteristics of the TFS of each indigenous group are preserved by the verbal transfer of traditional knowledge from generation to generation (Kuhnlein et al. 2009). The proper usage of underutilized and orphan wild plants can contribute to the well-being of a society in terms of health, economy, and livelihood (Romojaro et al. 2013). The food culture of a society greatly depends on the locality of origin, and their livelihood depends on the crops they cultivate (Wahlqvist and Lee 2007). Traditional food crops are plants that have historically remained an integral part of the farming system and dietary patterns in the world for long years but were later neglected and underutilized over time (Muthoni and Nyamongo 2010). Being more connected to nature, tribal communities have developed immense knowledge about traditional crops and their uses. There are many tribal communities with vast diversity in their cultures, environments, and in their traditions. These communities and numerous indigenous foods that still exist in their environment showcase and reflect the rich biodiversity which can be used to promote food security, nutrition, and health (Ghosh-Jerath et al. 2016). Nutritional, environmental, economic, and sociocultural values are considered key components of sustainable food systems (Akinola et al. 2020). Nowadays, wild traditional food crops are greatly accepted by the common people because of their significance in terms of key components of food security. The large variety of wild food crops is part of rural because of their affordability and nutritional security (Ray et al. 2020). Wild food plants are still widely used by indigenous communities across the globe. They prepare special dishes from the wild food plants that comprise their wild food plant based traditional food systems. As the WFPs are directly collected from the wild, these TFSs are not very common in the urban areas. This chapter summarises traditional food systems from across the globe that are based on wild food plants. This chapter is an attempt to provide a glimpse of the huge diversity and richness of the wild food plant based traditional food systems from across the globe. Examples of various TFS are taken from various different countries. This chapter further provides details about the preparation and the significance of the traditional foods (TFs) from the WFPs. This chapter also explores the health benefits of some of the TFs obtained from WFPs with examples.

4.2 Wild Food Plants

Wild foods have been consumed by humans since prehistoric times (Grivetti and Ogle 2000), but their importance in the human diet declined after the introduction of agricultural practices and the industrial revolution (Łuczaj 2010; Łuczaj et al. 2013). The low presence of wild food crops in urban diet makes their importance less in urban diets (Bharucha and Pretty 2010). Wheat, maize, and rice contribute to majority of the human calorie requirements (Fanzo et al. 2020). Wild food plants are used for many purposes other than a food source in many countries. Khatri et al. (2017) have emphasised the need to recognise the importance of the wild foods for nutritional security. In Nepal, of 62 wild food plants, 80% are used for other purposes like thirst quenchers, flavourants, and honey beers (Shrestha and Dhillon 2006). Many of the wild plants are also used for the treatment of livestock diseases (Mafimisebi et al. 2012). Huge diversity of wild food plants is still consumed by the indigenous communities (Ray et al. 2020). These wild food plants are mostly collected from the wild seasonally and sometimes they are also processed and used offseason. Several studies have documented the wild food plants from various regions of the world. Guarrera and Savo (2016) have documented 276 wild taxa distributed in 40 families in Italy. Another study also reported 357 taxa belonging to 72 families in Tuscany, Italy (Baldi et al. 2022). Wild food plants are also sold in the markets in some countries. A study by Łuczaj et al. (2021) recorded 110 species of wild edible plants being sold in the markets of Luang Prabang, Lao PDR. Several WFPs are also used ad spices and herbs (Motti 2021). Pawera et al. (2020) have documented usage of 106 WFPs in West Sumatra, Indonesia. This study further found that there is a decline in the usage of WFPs in the past. A study by Thakur et al. (2020) recoreded usage of 49 edible plants belonging to 24 families among

Gaddis (ethnic, tribal group spread across Himachal Pradesh and Jammu and Kashmir, India) of Himachal Pradesh.

4.3 Traditional Food Systems Based on Wild Food Plants

A wide variety of TFS based on wild food plants exists among different tribal groups in the world (Devi and Suresh 2012). Traditional foods have their own significance in current dietary patterns (Trichopoulou et al. 2007). By definition, traditional foods mean 'the food product with specific raw materials and recipes which is known for a long time and made through specific processes' (Cayot 2007). The TFS is not only significant due to its health benefits but also plays a pivotal role in socioeconomic and cultural aspects (Kuhnlein et al. 2006). The diversity in the traditional foods among the indigenous group is proportional to their nutrient intake. A study done by Roche et al. (2008) among people of Awajun in the Peruvian Amazon indicates the positive relationship between traditional food diversity and nutrient intake by considering the Traditional Food Diversity Score (TFDS) (Roche et al. 2008). The studies on traditional foods enable the understanding of how much knowledge the tribal groups possess and how much effort they made to make each traditional diet (Inamdar et al. 2005). Women of tribal communities have a significant role in the protection and promotion of their indigenous foods. The production of indigenous foods of Northeast India, like *Kinema, Tungrymbai, Hawaijar, Aakhone, Bekang,* and *Peruyyan* are exclusively done by women of different tribal groups of Northeast India. The indigenous knowledge of the production of these foods is inherited from mother to daughter only (Tamang et al. 2009). In some countries like Nepal, the technologies and methods behind the production of traditional foods are exclusively done by specific ethnic communities (Dahal et al. 2005). Table 4.1 presents a list of traditional foods that are mainly prepared from wild food plants or uses any ingredient of wild food plant origin. This chapter provides traditional food names and their preparation and the ingredients used along with the significance of these traditional foods. Some examples of non wild food plants but greater importance as traditional food systems are also included in the table to emphasise on the importance and diversity of the local traditional food systems that have huge regional importance. Some traditional foods presented in the table also show meat as an important part but uses some plants either as spice or flavour or for any other purposes. This table is not extensive but a representative of the diversity of traditional food systems. There are many such examples from various parts of the world. Many databases can be created to document the wild food plant based traditional foods.

Table 4.1 A cross-continental overview of traditional food systems based on wild food plants

Sl. no.	Country/ state	Traditional food	Preparation	Significance	Reference
Asia					
1	Indonesia	Brongkos	Combination of meat stew with tofu and black-eyed peas, cooked with coconut oil	Important dish of Indonesian cuisine	Harmayani et al. (2019)
		Pecel	A mixture of boiled carrot, spinach, and sprout with peanut sauce	Contain bio-active compounds	
		Sayur Tempe Lombok lijo	The soup consists of tempe, green chili, and coconut oil	An alternative source of protein	
		Gulei rebung	Fermented bamboo shoots are mixed with coconut milk and spices		Tamang et al. (2009)
		Sayur ladeh	Fermented bamboo shoots are mixed with vegetables		
		Lun-pia	Fried fermented bamboo shoots are mixed with vegetables		
2	India	Ushoi	Pieces of the inner core of bamboo soaked in water	Can be used to make other bamboo-based food and can be used in off season by drying	Nongdam and Tikendra (2014)
		Iromba	Chutney made up of processed bamboo shoots		Nongdam and Tikendra (2014)
		Sobium	Fermentation of thin sliced juvenile bamboo shoots	Traditional food of Manipur	Premlata et al. (2020)
		Hawaijar	Fermented soybeans	Alternative source of proteins	Devi and Suresh (2012)
		Ambali	Fermentation of finger millet flour	Rich in calcium and amino acids	Sarkar et al. (2015)
		Tambittu	Made from roasted and milled foxtail	Traditional festival sweet of North Karnataka	Inamdar et al. (2005)

(continued)

Table 4.1 (continued)

Sl. no.	Country/ state	Traditional food	Preparation	Significance	Reference
		Vatte pudi/Esluli pudi	Fine powder of *Artocarpus gomezianus* Wall. ex Trecul (Monkey Jack) with 2–3% of cooking salt	Powder is used as an alternative to Tamarind	Sarala and Krishnamurthy (2014)
		Rabdi	A mixture of flour of barley, pearl millet, and corn with buttermilk	A cereal-based fermented food, rich in lactic acid bacteria	Blandino et al. (2003)
		Sinki	Fermentation of radish roots	Traditional local food of Sikkim	Sundriyal et al. (2004)
		Kinema	Fermented soybean seeds	Traditional food prepared by women of *Limboo* and *Rai* caste of Sikkim	Tamang et al. (2009)
		Aakhone	Fermented soybean seeds	Traditional food of Nagaland	Mao and Odyuo (2007)
		Anishi	Dried cakes made by grinding fermented *Colocasia* leaves	Traditional food of *Ao* tribe of Nagaland	Mao and Odyuo (2007)
		Dawl rep	Air or fire dried leaves of Taro (*Colocasia esculenta*)	Traditional foods of Mizoram	Lalmuanpuii (2021)
		Antumbu Mung	Young shoots and leaves of *Brassaiopsis hainla* are boiled in water for about 15–30 min, water removed, served along with chilies and salt		
		Tawkte buhpawk	Green fruits of *Solanum indicum* is mixed with boiled rice, along with salt and baking soda		

(continued)

Table 4.1 (continued)

Sl. no.	Country/ state	Traditional food	Preparation	Significance	Reference
		Hulhu Zeu	Boiled tender leaves and shoots of *Aganope thyrsifolia* is deep fried and mixed with salt, onion, and garlic		
		Meihle Tlak	Boiled tender shoots of *Caryota mitis* are mixed with smoked roselle leaves, and served along with salt and chili		
3	China	Sufu	Prepared from fermented soya milk by using *Actinomucor elagans*	High amount of vitamin B12	Li et al. (2004)
		Douchi	Fermentation of steamed soybean	Antioxidant activity	
4	Nepal	Masyaura	Sun-dried fermented vegetable balls made with black gram and *Colocasia* tuber	Legume and tuber-based fermented food. Rich in vitamin B complex	Dahal et al. (2003)
		Mesu	Pickle made from tender edible bamboo. Mixing of sliced and fermented bamboo shoots with chilies and salt	Sour taste and strong ammonical odor	Tamang et al. (1988)
5	Turkey	Boza	Millet semolina is boiled with water, added sugar and fermented to obtain a thick yellow liquid	Nutritious and healthy beverage useful in digestion	Arici and Daglioglu (2002)
6	Japan	Miso	Soya beans or rice or barley are fermented with *Aspergillus oryzae* (koji mold) and salt is added to make a thick paste	Traditional and staple food of Japanese diet	Murooka and Yamshita (2008)
7	Thailand	Kapi	Small fish or shrimp is mixed with salt in a bamboo basket and left overnight and later dried in broad	Healthy traditional and low calorie dish	Yongsmith and Malaphan (2016)

(continued)

Table 4.1 (continued)

Sl. no.	Country/ state	Traditional food	Preparation	Significance	Reference
			daylight on a bamboo mat. It is then grinded to fine paste and fermented in clay pot for about 4 months		
8	Sri Lanka	Kiribath or milk rice	Rice is cooked with coconut milk until a thick paste is formed. It can be made into shapes and cut into pieces	Traditional and ethnic food	Mihiranie et al. (2020)
9	South Korea	Chungkookjang	Soaked soybeans are steamed for 5–7 h. After cooling, it is fermented for 2 days with boiled rice straw by covering on the soybeans	Traditional Korean food with high amounts of fatty acids and essential amino acids	Hong et al. (2012)
10	Philippines	Bagoong	Shrimp or fish fry is mixed with salt and fermented for 3–12 months to obtain a fine paste	Traditional and regular dish	Montaño et al. (2001)
11	Vietnam	Thit lon chua	Grilled pork is chopped into thin slices to which roasted rice powder, salt, sugar and spices are added and fermented for 3–5 days at 30 °C to obtain a sour tasting yellowish brown product	Traditional fermented meat product	La Anh (2015)
12	Singapore	Laksa	Thick vermicelli made of rice is cooked with a gravy curry containing laksa leaves, coconut milk, tiny dried prawns and spices	Traditional and popular dish	Catherine Henderson (2014)
13	Malaysia	Otak	Fish flesh added with tapioca flour, eggs, little coconut milk, sugar, salt, kaffir lime leaves and other spices is	Traditional orange colored fish cake which is nutritious and has antioxidant properties	Harmayani et al. (2019)

(continued)

Table 4.1 (continued)

Sl. no.	Country/ state	Traditional food	Preparation	Significance	Reference
			wrapped in a banana leaf and grilled		
14	Pakistan	Barian	Meat, gram flour, and salt are mixed with spices and air dried followed by sun drying	Local dish popular in rural areas	Mushtaq et al. (2012)
15	Maldives	Rihaakuru	Gutted tuna fish is boiled in salt water for 45 min and after removing cooked fish the leftover soup is cooked until a thick a paste is obtained	Traditional side dish	Naila et al. (2011)
16	Qatar	Laban or leben or lban	Fermented milk is churned to separate butter and the left-over liquid is consumed as laban	Traditional dairy product	Hassan et al. (2018)
17	Iran	Shir-Berenj	Washed rice is boiled with water, sugar, milk, and cardamom seeds for 20–30 min and then cooled and served	Traditional rice pudding	Gharibzahedi (2018)
18	Bangladesh	Jilapi	Curd fermented wheat flour is put in hot oil in spiral shape and deep fried followed by soaking in sugar syrup	Traditional dessert	Hossain and Kabir (2016)
19	Myanmar	Laphet	Steamed tea leaves are packed in clay pots and fermented with limited air for 3–4 months	Traditional snack and side dish	Bo et al. (2020)
20	Iraq	Khobz	Wheat flour dough is added with starter culture (ripened dates soaked in water overnight) and fermented overnight and then baked to form a sour bread	Traditional thick and sour bread	Muhialdin et al. (2022)
21	Bhutan	Datshi	Fermented milk/dahi is churned to obtain	Traditional dairy product	Rai et al. (2016)

(continued)

Table 4.1 (continued)

Sl. no.	Country/ state	Traditional food	Preparation	Significance	Reference
			mohi which is then boiled till formation of clumps to obtain datshi		
Africa					
22	Ethiopia	Injera	Cooking the batter made up of primary and secondary fermented teff flour (*Eragrostis tef* Zucc) in a hot clay plate	Source of minerals and vitamins	Neela and Fanta (2020)
23	Nigeria	Efo riro	Soup made up of Spinach (*Amaranthus hybridus*) and Tomatoes (*Solanum lycopersicum*)	Antioxidant capacity	Tchokouaha et al. (2015)
		Akara	Cowpea flour is pasted and deep fried into ball like shape	Source of micronutrients like iron and zinc	McWatters (1983)
		Ayaraya oka	Mixture of corn, pigeon pea, vegetables, and palm oil	Contain vitamin A, iron, and zinc	Okeke et al. (2008, 2009)
24	Algeria	Klila	Fermented cheese prepared by mild boiling of Lben (a fermented milk beverage) until it becomes curdled and then drained using a cotton cloth	Popular, authentic and traditional cheese made by artisanal processes	Leksir et al. (2019)
		Smen/Dhan	Fermented butter made either by salting Zebda (butter) or by modera	Ethnic and traditional dairy product, flavoring agent, also used for therapeutic purpose for treating flu, burns, headache, migraine, and eczema	Boussekine et al. (2020)
		Traditional couscous or seksu	Durum wheat semolina is steamed and dried which is then	Traditional and staple food of north African regions usually	Chemache et al. (2018)

(continued)

Table 4.1 (continued)

Sl. no.	Country/ state	Traditional food	Preparation	Significance	Reference
			cooked with oils and vegetables	made during festivals and gatherings	
25	Libya	Fasolia	White beans are boiled with onion, garlic, tomato sauce, few spices, and sandwiched in wheat bread	Traditional fast food	Buzgeia et al. (2021)
26	Morocco	Bubanita/ boubanita	Small cut pieces of lamb meat is seasoned with spices and dried and fermented in the lumen of lamb in shade	Traditional meat product eaten along with pasta or vegetables	Benkerroum (2013)
27	Tunisia	Kadid	Small slices of beef are added with olive oil, salt and garlic, and sun dried	Traditional meat product usually used in winters	Essid et al. (2007)
28	Egypt	Falafel	Faba beans, onion, garlic, cumin seeds, and salt are made into paste balls and deep fried	Traditional and staple diet	Hefni and Witthöft (2014)
		Rayeb	Raw milk of buffalo is fermented at room temperature until it is coagulated	Traditional milk product especially in rural areas	Abd El Gawad et al. (2010)
29	Ghana	Yakayake	Cassava roots are grated, dried, sieved, salted and made into a spongy ball or cake by steaming	Traditional food of Ghana	Lancaster et al. (1982)
30	Namibia	Oskikundu	Flour of pearl millet and sorghum malt are cooked and fermented to obtain a faintly viscous greenish-brown beverage	Nutritious and nonalcoholic beverage	Taylor and Emmambux (2008)
31	Niger	Kilichi	Thin and flat straps of boneless meat is sun dried and marinated with spices and coloring agents which is followed by	Popular and valued meat product both in urban rural areas	Boubacar Seydou et al. (2021)

(continued)

Table 4.1 (continued)

Sl. no.	Country/ state	Traditional food	Preparation	Significance	Reference
			another round of sun drying and at last grilled in wood fire		
32	Chad	Kawal	A paste made by pounding leaves of *Cassia obtusifolia* is put in a clay pot and covered with leaves of sorghum. The pot is tucked under ground except for neck and left for fermentation for 14 days (hand stirring once in 3 days) and then sundried	Substitute or replacement for meat	Mbaiguinam et al. (2005)
33	Sudan	Kisra	Flat, thin bread made from fermented sorghum, roasted/ pounded chickpeas and sugar/salt	Staple and traditional food of Sudan and nutritious for infants	Badi et al. (1990)
34	South Africa	Mahewu (Magou)	Maize flour is made into thin porridge and boiled. After cooling, a small amount of wheat flour is put and left for fermentation in a sunny place	Traditional, fermented and nonalcoholic beverage	Hesseltine (1979)
35	Kenya	Kimyet (ugali)	Maize or sorghum or millet flour is mixed with water and boiled to make a thin porridge	Staple and traditional food and recommended for pregnant women	Riang'a et al. (2017)
36	Zambia	Munkoyo	Maize flour is boiled with water for several hours and roots of *Rhynchosia* are added and left for fermentation for about 2 days	Traditional, fermented and nonalcoholic beverage	Chileshe et al. (2020)
37	Uganda	Inya	Inyaasa (dried and powdered fermented cassava) is boiled in water and stirred until it becomes a hard cake	Traditional food of Lughbara ethnic group of Uganda	Amone (2014)

(continued)

Table 4.1 (continued)

Sl. no.	Country/ state	Traditional food	Preparation	Significance	Reference
		Akalo	Cassava and millet flour mixed in hot water in a pan and then served in baskets	Energy food of Banyoro and Batooro groups in Uganda	Tumuhe et al. (2020)
38	Zimbabwe	Masau	The ripened fruits of *Ziziphus mauritiana* (buffalo thorn) are pounded and left for fermentation after adding some water	Traditional fruit-based fermented beverage	Gadaga (1999)
39	Tanzania	Porridge	Whorl pearl millet flour, salt, water or Whorl finger millet flour, sugar, water	Less fat comparatively good fiber content	Kulwa et al. (2015)
Australia					
40	Fiji	Duruka	Unopened flower of sugarcane is cooked with coconut milk	Traditional Fijian dish	Lim (2014)
41	Papua New Guinea	Mone	Sago starch is made into dumpling and cooked with coconut cream	Traditional dessert or dumpling	Macintyre (1987)
42	New Caledonia	Bougna	Meat, yam, coconut milk and sweet potatoes are wrapped in banana leaves and cooked in a pit	Traditional dish of kanak community	Anderson (2013)
Europe					
43	Germany	Sauerkraut	White cabbage leaves are sliced into pieces, salted and fermented	Traditional fermented vegetable dish	Peñas et al. (2010)
44	Ukraine	Uzvar	Fruits like apples, plums, and cherries are dried and boiled with water and honey	Traditional fruit based dish	Costa et al. (2013)
45	France	Camembert	After adding rennet to cow milk, whey is drained and coagulam is collected which is then added with salt and *Penicillium*	Traditional cheese from cow milk	Boisard (1991)

(continued)

Table 4.1 (continued)

Sl. no.	Country/ state	Traditional food	Preparation	Significance	Reference
			candidum and left for ripening for 2 weeks		
46	Austria	Krautfleckerl	Cabbage, noodles, onions, and bacon are cooked with added flavors of car-away and white wine	Traditional noodle recipe	Weichselbaum et al. (2009)
47	Bulgaria	Dko ot rozi	Rose petals are dried and boiled with water, sugar and citric acid	Traditional rose jam	Albuquerque et al. (2013)
48	Georgia	Churchkhela	Walnuts are put on a string and dipped in paste made out of white grapes and wheat flour and dried	Traditional sweet dish	Costa et al. (2013)
49	Russia	Kefir	Milk is added with kefir grains and fermented for 1 day	Nonalcoholic beverage hav-ing wound healing and anti-cancerous properties	John and Deeseenthum (2015)
50	Italy	Brasato al Barolo	Ox meat soaked for 12 h in 2 L of Barolo wine is cooked with herbs and vegetables	Traditional cuisine of Italy	Weichselbaum et al. (2009)
51	Romania	Socata	Elderberry flowers are fermented with sugar, yeast, water, and lemon juice	Traditional fruit based beverage	Albuquerque et al. (2013)
North America					
52	Mexico	Sopes	Corn flour dough cake deep fried and topped with meat and vegetables	Traditional snack	Chiu et al. (2021)
53	Canada	Poutine	Potato fries and cheese curd are mixed with a special gravy and tossed	Traditional Canadian recipe	Fabien-Ouellet (2016)
54	New York	Pastrami	Beef brisket is cured for 6 days and smoked for 4 h and flavored with black pepper	Traditional meat dish	Cardoso et al. (2020)

(continued)

Table 4.1 (continued)

Sl. no.	Country/ state	Traditional food	Preparation	Significance	Reference
South America					
55	Venezuela	Arepa	Corn flour is mixed with water and salt, dough either baked at 400 °C for 4 min or fried in soybean oil	Traditional food of Venezuelans	Alvarez (1981)
56	Brazil	Feijoada	Bacon, garlic, and onions are fried until brown and ingredients such as black beans, different types of meat, orange slices, and cassava flour are added and cooked for about 4 h	Traditional dish rich in protein	Faller et al. (2012)
57	Chile	Curanto	Nalca leaves are layered with ingredients like shellfish, chicken, potatoes, and other spices and cooked slowly for hours in a clay vessel	Traditional fish recipe	Daughters (2016)
58	Argentina	Asado	Different types of meat cuts are grilled with sausages	Traditional meat dish usually prepared by men	Zycherman (2008)

4.4 Wild Food Plant-Based Traditional Foods as a Source of Healthy and Nutritious Diets

The transition from traditional foods to fewer nutrient-based foods significantly increased the percentage of people who have been suffering from noncommunicable chronic diseases (Batal et al. 2005). The addition of a small number of traditional foods to the diet can considerably increase the nutritional quality of the diet (Schuster et al. 2011). Many wild food plants are also used as medicines (de Medeiros et al. 2021). The traditional foods which are produced by microbial fermentation have their own nutritional benefits because of higher concentrations of proteins, amino acids, and vitamins (Sha et al. 2013). Traditional fermented foods effectively work against diarrhea through their antimicrobial properties (Watson et al. 1996). The underutilization of traditional foods among indigenous people created many health issues like malnutrition, underweight, stunting, and micronutrient deficiencies (Kuhnlein et al. 2009). Even though traditional foods function as a source of a healthy and nutritious diet, the environmental contaminants in traditional foods create problems related to health. The contaminants like heavy metals reach traditional foods mainly through local and long-range transport avenues (Kuhnlein and Chan 2000). Table 4.2. provides examples of the health benefits of some of the wild food plants that are part of traditionall foods. This table also provides the part of the plant used and the important nutritional component of the plant. Figure 4.1 provides an overview of the various activities associated with the wild food plants such as collection, marketing and processing. This figure also depicts the major roles of WFPs and the WFP based traditional food systems.

4.5 Traditional Food Systems and Food Security

"Food security exists when all people, at all times, have physical, social, and economic access to sufficient, safe, and nutritious food that meets their dietary needs and food preferences for an active and healthy life" (FAO 2002). Physical availability of food, economic and physical access to food, food utilization, and stable food access are four dimensions of food security. The disturbance of any of these or all results in food insecurity (FAO 2008). The utilization of indigenous knowledge of the TFS, and its in integration with the current food systems, can effectively improve food security (Elliott et al. 2012). For the survival of human beings, the consideration of agricultural biodiversity along with the conventional systems is very critical (Millennium Ecosystem Assessment 2008). Several countries are very rich in their traditional foods. For example, South Korea's Kimchi is very famous traditional food. Nutritionally, kimchi is rich in vitamins, fiber, minerals and other important nutrients (Cheigh et al. 1994).

The TFSs are getting more attention nowadays, which contributes to food nutrition and health (Chyne et al. 2019). Compared with common cultivars, traditional food serves as a good source of macro and micronutrients (vitamins A, D, C, and minerals such as iron, calcium, and zinc) (Ghosh-Jerath et al. 2015). The reports state that traditional foods have a higher nutritional value than market foods (Elliott et al. 2012). According to the FAO (Food Agricultural Organization), the world produces sufficient amounts of food for the global population. Wheat, barley, and rice are the main food crops in the international market. A comparison of modern dietary guidelines and agricultural production statistics reveals that although we are consuming energy rich crops, they lack micro and macronutrients (Borelli et al. 2020a, b). Diversification of the diets with traditional plant-based foods is one way to address the challenges of rapidly emerging diet- and lifestyle-related noncommunicable chronic diseases (NCDs) in indigenous communities around the world. Indigenous communities must rebuild native ecosystems, use traditional food crop cultivation methods, and revitalize traditional knowledge of food preparation, processing, and preservation in order to combat an NCD epidemic. A number of

Table 4.2 Health benefits of wild food plants that are used for the preparation of traditional foods

Sl. no.	Country	Wild food plant	Parts used	Health benefits	Remarks on nutritional/ phytochemical composition	References
Asia						
1.	Indonesia	*Arenga pinnata* (Wrumb) Merr.	Fruit	Beneficial to gastrointestinal tract function	High fiber content	Harmayani et al. (2019)
		Arachis hypogaea L.	Seeds	Reduce cholesterol level, prevent bacterial infections	Omega fatty acids, arginine	Harmayani et al. (2019)
		Cyclea barbata Miers	Leaves	Treatments for fever, gastric inflammation, nausea, and hypertension. Antioxidant and anticancer property	Carotenoids and flavonoids	Harmayani et al. (2019)
		Kaempferia galanga L.	Rhizome	Prevent dry cough, fungal infections, diphtheria, and gonorrhea	Pinostrobin, hydroxy panduratin A, panduratin A	Wang et al. (2021)
2.	Thailand	*Alpinia galanga* L.	Rhizome	Used for osteoarthritis, has antitumor and anti-inflammatory property	Acetoxychavicol acetate, ρ-hydroxycinnamaldehyde	Pengseng et al. (2010)
		Centella asiatica L.	Leaves	Wound healing, skin treatment, and antimicrobial	Pentacyclic terpene derivatives, asiaticoside, and madecassoside	Brinkhaus et al. (2000)
		Aegle marmelos L.	Leaves, root and fruit	Used for gastrointestinal diseases, gynecological dysfunctions	Steroids, terpenoids, flavonoids	Patkar et al. (2012)
		Parkia speciosa Hassk	Seeds	Antioxidant	Polyphenol, phytosterol, and flavonoids	Kamisah et al. (2013)
3.	India	*Solanum spirale* Roxb.	Leaves and fruits	Treating fever and stomach ache	Rich in vitamin C and crude lipid	Kalita et al. (2014)
		Acacia catechu (L. f.) Willd	Bark	Antibacterial, antidiabetic and antioxidant properties	Polyphenol and flavonoids	Aryal et al. (2021)
		Oxalis corniculata L.	Leaves	Antiepileptic, antioxidant and antitumorigenic		Kathiriya et al. (2010)

(continued)

Table 4.2 (continued)

Sl. no.	Country	Wild food plant	Parts used	Health benefits	Remarks on nutritional/ phytochemical composition	References
		Basella rubra L.	Leaves	Antidiabetic, hypoglycemic, and antioxidant	Rich in fiber	Mensah et al. (2008); Nirmala et al. (2011)
		Cocculus hirsutus (L.) W.Theob.	Leaves	Antidiabetic and antihepatotoxic	Isoquinoline alkaloids and triterpenes	Badole et al. (2006); Sangameswaran and Jayakar (2007)
Africa						
4.	Kenya	*Vigna unguiculata* L. Walp.	Seed and leaf	Used for gastrointestinal disorders 21, cardiovascular diseases, hypercholesterolemia	High protein content	Jayathilake et al. (2018)
		Cajanus cajan L.	Seed	Flower used for treating ailments such as bronchitis, coughs, and pneumonia	Rich in protein calcium and iron	Singh (2016)
5.	Morocco	*Opuntia ficus-indica* (L.) Mill.	Seeds and fruits	Antioxidant activity	Rich in macroelements and minerals	Ghazi et al. (2015)
		Opuntia dillenii (Ker Gawl.) Haw.	Seeds and fruits	Antioxidant activity	High Zn and linoleic acid content	Ghazi et al. (2015)
6.	South Africa	*Kedrostis africana* (L) Cogn	Tuber	Antidiuretic	Rich with minerals	Unuofin et al. (2017)
Australia						
7.	New zealand	*Sonchus asper* (L.) Hill	Leaves and stem	Protection against infectious disease	High vitamin C and phenolic compounds	Rush et al. (2010)
8.	Papua New Guinea	*Pandanus brosimos* Merr. & L. M.Perry	Nut	Antidiuretic, antidiabetic, and antioxidant	Isoflavones and coumestrol	Borelli et al. (2020a, b); Gurmeet and Amrita (2015)
9.	Fiji	*Colacasia esculeta* (L.) Schott	Tuber	Antimicrobial, antioxidant, and anticancer		Chand et al. (2018)
			Fruit	Anxiolytic, antioxidant, and antidiabetic		Chand et al. (2018)

Europe						
10.	Italy	*Achillea erba-rotta* All.	Flower	Anti-inflammatory and antibacterial		Mattalia et al. (2013)
		Malva sylvestris L.	Flower and leaves	Used for digestive problems		di Tizio et al. (2012)
		Parietaria judaica L.	Aerial parts	Used for treating hemorrhoids		di Tizio et al. (2012)
11.	Ukraine	*Helichrysum arenarium* (L.) Moench	Flower	Antiseptic, used in the treatment of tuberculosis and hepatitis		Stryamets et al. (2015)
12.	Russia	*Vaccinium myrtillus* L.	Fruit	Treatment of eyes and stomach diseases	Vitamin rich	Stryamets et al. (2015)
13.	Poland	*Corylus avellana* L.	Nuts	Antioxidant, cardioprotective, anti-inflammatory and hypolipidemic	Rich in flavonoids and polyphenols especially *p*-coumaric acid	Król et al. (2020)
14.	Spain	*Anchusa azurea* Mill.	Leaves	Antidiarrheal and used in treatment of urinary tract infections		Rivera et al. (2005)
15.	Portugal	*Foeniculum vulgare* Mill.	Leaves	Helps in digestion		Carvalho and Morales (2010)
16.	Greece	*Origanum dubium* Boiss.	Aerial parts	Antistress, antipyretic and good for digestion		Della et al. (2006)
17.	Austria	*Rubus idaeus* L.	Twigs and leaves	Good for hematopoiesis, cold, and fever		Sõukand and Kalle (2013)
North America						
18.	Mexico	*Portulaca oleracea* L.	Leaves, stem and flowers	Antifungal, antibacterial, antiscorbutic, and contraceptive properties	Ascorbic acid and iron content is present	Angelica et al. (2011)
		Anoda cristata (L.) Schltdl	Leaves	Reduce balding, heal lacerations, and stomach infections	Rich in iron, retinol, and ascorbic acid	Angelica et al. (2011)
19.	Canada	*Acer pensylvanicum* L.	Bark	Cold, cough, and bronchitis treatment	Allantoin phytochemical	Arnason et al. (1981)

(continued)

Table 4.2 (continued)

Sl. no.	Country	Wild food plant	Parts used	Health benefits	Remarks on nutritional/ phytochemical composition	References
		Juglans cinerea Liebm.	Nut	Emetic and purgative	High amounts of iron	Arnason et al. (1981)
20.	Cuba	*Bromelia pinguin* L.	Fruit	Anthelmintic		Volpato and Godinez (2006)
21.	Panama	*Chamaedorea tepejilote* Liebm.	Male inflorescence	Antidiabetic and anti tussive	Rich in vitamin C and iron	Andrews et al. (2018); Michon (2012)
South America						
22.	Argentina	*Citrus aurantium* L.	Fruit	Used in treating respiratory infections		Kujawska and Łuczaj (2015)
23.	Brazil	*Anredera cordifolia* (Ten.) Steenis	Leaves	Heals intestinal and stomach ailments		de Medeiros et al. (2021)
24.	Chile	*Sarcocornia neei* (Lag.) M.A. Alonso & M.B. Crespo	Leaves	Antioxidant properties	Rich in proteins and poly-unsaturated fatty acids	Riquelme et al. (2016)
25.	Peru	*Bejaria resinosa* Mutis ex L.f.	Leaves	Anti-inflammatory and used in controlling female matrix infections		Torres-Guevara et al. (2020)

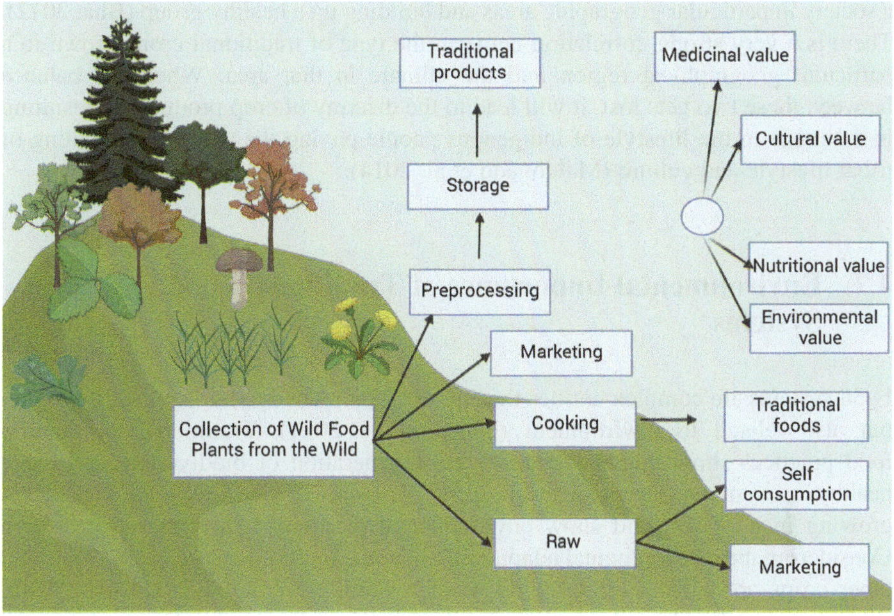

Fig. 4.1 Main activities associated with wild food plants such as collection, processing and marketing and the roles of the wild food plants and the traditional food systems based on them

traditional plant-based foods are rich in bioactive compounds with diverse health benefits that are relevant to human health (Sarkar et al. 2020).

4.6 Cultural Importance of Traditional Food Systems

The food available from local resources that are culturally accepted are termed traditional foods; they may contain a wide range of plants and animals (Kuhnlein and Receveur 1996). These foods provide the essential nutrients for the sustainability of that community. Traditional food can be accessed by the indigenous people through their local resources. The knowledge about the TFS arises by various means like folk taxonomy, folk medicine, and sociocultural activities (Kuhnlein and Receveur 1996). Activities like harvesting and preparation of foods enable social relationships and knowledge transfer through the gathering of individuals in a community; it also facilitates a spiritual connection with the land (Lambden et al. 2007). TFS allows us to identify the major food in a culture which is obtained through local resources and their importance within the indigenous people. They play an important role in the health of indigenous people. The cultural identity of indigenous people can be identified through their traditional food system (Kwik 2008). Knowledge regarding traditional food systems had a greater role in sustaining a society in particular geographic areas and building up a healthy group (Bhat 2012). There is a very strong correlation between the type of traditional crops grown in a particular geographical region and the climate in that area. When the balance between these two gets lost, it will lead to the disarray of crop production, resulting in a change in the lifestyle of indigenous people paving the way to remodeling of tribal lifestyle and culture (Maldonado et al. 2014).

4.7 Environmental Importance of Traditional Food Systems

Food systems are complex entities that are not only related to human health and diet but also related to environment (Fanzo et al. 2021). Traditional agricultural food practices show positive signs in the regeneration of biodiversity and make healthy ecosystems (Deaconu et al. 2021). Traditional food crops are capable of growing in poor soil and show environmental plasticity (Chivenge et al. 2015). Considering the environmental adaptability of some traditional food crops, roots and tuber crops are capable of growing in high humidity conditions (Pearce 1990), cowpea is widely adapted to poorly fertile soil (Carvalho et al. 2017) and adapted to high temperature and drought conditions (Hall 2012). Indigenous traditional crops such as cowpea and pigeon pea can improve and restore the fertility of the soil by biological nitrogen fixation (Elowad and Hall 1987). Coarse cereals can be referred to as a good alternative crop for cultivation in stressed climatic and edaphic conditions (Rai et al. 2008; Kaur et al. 2014; Eliazer Nelson et al. 2019). Therefore, we can say that traditional food plants are important in the context of increasing environmental stresses.

4.8 Traditional Food Systems and Climate Change

Climate change makes adverse impacts that have cascading effects on various sectors of human life (Lawrence et al. 2020). The climate change negatively affects the lives of indigenous people around the globe. The disappearance of plant species, alternation of water quality and quantity, and changes in weather and soil all create visible impacts on the traditional food systems of indigenous people (Guyot et al. 2006). Indigenous people depend on a wide variety of plants, animals, and fungi as their food source, and they are also used for medicine, ceremony, and economic benefit. Even the water is also held as sacred by indigenous people. Climate change, along with other unfavorable conditions, affects the relationship between tribes and traditional foods. Traditional food production declines when the ecosystem and water resources are excessively used for other purposes. The ecosystem gets exploited through the pollution and introduction of invasive species and poor management. Climate change affects the total life cycle of traditional crops - by the affecting pollinators, timing in flowering time, and shift in harvesting time. These changes not only affect the production but also impact the food web (Maldonado et al. 2014).

4.9 Conclusions and Future Prospects

Rural communities across the world are generally identified by their locality, lifestyles, culture, and traditional beliefs they follow. But when considering eating habits, clear differences are visible even in different tribal groups in the same country or nearby states. As we all know, human civilization is closely related to agriculture, and it arises on land suitable for agriculture. It is clear evidence that our ancestors were aware of how food is essential for the our survival. Likewise, the indigenous communities depended on various food crops suitable for cultivating in their locality or available in their forests to balance their energy needs for their hard work. Though the used traditional crop species vary, it can be concluded that all the groups were capable of including proximate components, antioxidants, and various protective secondary metabolites through a diverse diet of tuber, legumes, leaves, fruit, oil seeds, flowers, and other edible parts in their diets. Knowingly or unknowingly, the diet they followed made them healthy, and these natural resources kept away the lifestyle diseases such as diabetes, cancer, increased blood pressure, and cholesterol among them. According to WHO reports in 2005, lifestyle diseases or internationally so-called chronic diseases is the main cause of 61% of global death. An unhealthy diet is one of the reasons for this. So, healthy diet sources are very much important. In the recent years, we have witnessed the increasing importance of traditional foods. Identification, documentation, and improvement of these food crops and integrating them into our diet can be helpful for improving our existing food systems and making it more diverse. Identification of the plants with significant properties such as anti-cancer, anti-inflammatory, and antimicrobial can open new opportunities for drug development. In the future, more studies on the nutritional and pharmacological properties of these wild plants and the traditional foods based on them are needed.

References

Abd El Gawad IA, Abd El Fatah AM, Al Rubayyi KA (2010) Identification and characterization of dominant lactic acid bacteria isolated from traditional Rayeb milk in Egypt. J Am Sci 6:728–735

Akinola R, Pereira LM, Mabhaudhi T, de Bruin F-M, Rusch L (2020) A review of indigenous food crops in Africa and the implications for more sustainable and healthy food systems. Sustainability 12:3493. https://doi.org/10.3390/su12083493

Albuquerque TG, Costa HS, Sanches-Silva A, Santos M, Trichopoulou A, D'Antuono F, Alexieva I, Boyko N, Costea C, Fedosova K, Karpenko D, Kilasonia Z, Koçaoglu B, Finglas P (2013) Traditional foods from the Black Sea region as a potential source of minerals:

traditional foods from the Black Sea region as a source of minerals. J Sci Food Agric 93:3535–3544. https://doi.org/10.1002/jsfa.6164
Alvarez RJ (1981) Selected nutritional and biochemical changes during the production of baked and fried arepas. Ecol Food Nutr 11:201–205. https://doi.org/10.1080/03670244.1981.9990676
Amone C (2014) We are strong because of our millet bread: staple foods and the growth of ethnic identities in Uganda. Trames 18(2):159. https://doi.org/10.3176/tr.2014.2.04
Anderson J (2013) The empire bites back?: writing food in Oceania. Portal: J Multidiscip Int Stud 10:1–16
Andrews CM, Wyne K, Svenson JE (2018) The use of traditional and complementary medicine for diabetes in rural Guatemala. J Health Care Poor Underserved 29:1188–1208. https://doi.org/10.1353/hpu.2018.0092
Angelica BC, Maria RAVA, Gabino AMG, Patricia SSM, Rafael PP (2011) The traditional medicinal and food uses of four plants in Oaxaca, Mexico. J Med Plants Res 5:3404–3411
Arici M, Daglioglu O (2002) Boza: a lactic acid fermented cereal beverage as a traditional Turkish food. Food Rev Intl 18:39–48. https://doi.org/10.1081/FRI-120003416
Arnason T, Hebda RJ, Johns T (1981) Use of plants for food and medicine by Native Peoples of eastern Canada. Can J Bot 59:2189–2325
Aryal B, Adhikari B, Aryal N, Bhattarai BR, Khadayat K, Parajuli N (2021) LC-HRMS profiling and antidiabetic, antioxidant, and antibacterial activities of *Acacia catechu* (L.f.) Willd. Biomed Res Int 2021:1–16. https://doi.org/10.1155/2021/7588711
Badi S, Pedersen B, Monowar L, Eggum BO (1990) The nutritive value of new and traditional sorghum and millet foods from Sudan. Plant Food Hum Nutr 40:5–19. https://doi.org/10.1007/BF02193775
Badole S, Patel N, Bodhankar S, Jain B, Bhardwaj S (2006) Antihyperglycemic activity of aqueous extract of leaves of *Cocculus hirsutus* (L.) Diels in alloxan-induced diabetic mice. Indian J Pharmacol 38:49
Baldi A, Bruschi P, Campeggi S, Egea T, Rivera D, Obón C, Lenzi A (2022) The renaissance of wild food plants: insights from Tuscany (Italy). Foods 11(3):300. https://doi.org/10.3390/foods11030300
Batal M, Gray-Donald K, Kuhnlein HV, Receveur O (2005) Estimation of traditional food intake in indigenous communities in Denendeh and the Yukon. Int J Circumpolar Health 64:46–54. https://doi.org/10.3402/ijch.v64i1.17953
Benkerroum N (2013) Traditional fermented foods of North African countries: technology and food safety challenges with regard to microbiological risks: safety of North African traditional foods .... Compr Rev Food Sci Food Saf 12:54–89. https://doi.org/10.1111/j.1541-4337.2012.00215.x
Bharucha Z, Pretty J (2010) The roles and values of wild foods in agricultural systems. Philos Trans R Soc B 365(1554):2913–2926. https://doi.org/10.1098/rstb.2010.0123
Bhat S (2012) Importance of traditional food system. In: Recent trends in food science and technology. The Registrar, Tumkur University
Blandino A, Al-Aseeri ME, Pandiella SS, Cantero D, Webb C (2003) Cereal-based fermented foods and beverages. Food Res Int 36:527–543. https://doi.org/10.1016/S0963-9969(03)00009-7
Bo B, Kim S-A, Han NS (2020) Bacterial and fungal diversity in Laphet, traditional fermented tea leaves in Myanmar, analyzed by culturing, DNA amplicon-based sequencing, and PCR-DGGE methods. Int J Food Microbiol 320:108508. https://doi.org/10.1016/j.ijfoodmicro.2020.108508
Boisard P (1991) The future of a tradition: two ways of making camembert, the foremost cheese of France. Food Foodways 4:173–207. https://doi.org/10.1080/07409710.1991.9961981
Borelli T, Hunter D, Padulosi S, Amaya N, Meldrum G, de Oliveira Beltrame DM, Samarasinghe G, Wasike VW, Güner B, Tan A, Koreissi Dembélé Y, Lochetti G, Sidibé A, Tartanac F (2020a) Local solutions for sustainable food systems: the contribution of orphan crops and wild edible species. Agronomy 10:231. https://doi.org/10.3390/agronomy10020231
Borelli T, Hunter D, Powell B, Ulian T, Mattana E, Termote C, Pawera L, Beltrame D, Penafiel D, Tan A, Taylor M, Engels J (2020b) Born to eat wild: an integrated conservation approach to

secure wild food plants for food security and nutrition. Plants 9:1299. https://doi.org/10.3390/plants9101299

Boubacar Seydou R, Douny C, Mahillon J, Anihouvi VB, Scippo M, Hounhouigan DJ (2021) Technological performance and effects of traditional grilling on the physicochemical characteristics and polycyclic aromatic hydrocarbon contamination of *kilichi*, a grilled meat produced in Niger. J Food Process Preserv 45:e15627. https://doi.org/10.1111/jfpp.15627

Boussekine R, Merabti R, Barkat M, Becila F-Z, Belhoula N, Mounier J, Bekhouche F (2020) Traditional fermented butter Smen/Dhan: current knowledge, production and consumption in Algeria. J Food Res 9:71. https://doi.org/10.5539/jfr.v9n4p71

Brinkhaus B, Lindner M, Schuppan D, Hahn EG (2000) Chemical, pharmacological and clinical profile of the East Asian medical plant *Centella asiatica*. Phytomedicine 7:427–448. https://doi.org/10.1016/S0944-7113(00)80065-3

Buzgeia MH, Madi MF, Hamza MS, EL-Taguri HS (2021) Modern and Traditional fast foods consumption in Benghazi, Libya. J Int Med Res Health Sci 01:26–35. https://doi.org/10.53626/JIMRHS.2021.1204

Cardoso P d S, Fagundes JM, Couto DS, Pires E d M, Guimarães CED, Ribeiro CDF, Otero DM (2020) From curing to smoking: processes and techniques for the production of pastrami. Braz J Dev 6:61511–61520. https://doi.org/10.34117/bjdv6n8-538

Carvalho AM, Morales R (2010) Northeastern region of Portugal. Ethnobotany in the New Europe: people, health, and wild plant resources. p 14

Carvalho M, Lino-Neto T, Rosa E, Carnide V (2017) Cowpea: a legume crop for a challenging environment: cowpea for a challenging environment. J Sci Food Agric 97:4273–4284. https://doi.org/10.1002/jsfa.8250

Catherine Henderson J (2014) Food and culture: in search of a Singapore cuisine. Br Food J 116: 904–917. https://doi.org/10.1108/BFJ-12-2012-0291

Cayot N (2007) Sensory quality of traditional foods. Food Chem 101:154–162. https://doi.org/10.1016/j.foodchem.2006.01.012

Chand RR, Devi SS, Kumari SS, Kumar SS, Goundar NN, Naranyan N, Chandra P (2018) Traditional use of medicinal plants among selected villages in Fiji Islands: a review perspective. Pac Med Stud J 1:10–20

Cheigh H, Park K, Lee CY (1994) Biochemical, microbiological, and nutritional aspects of kimchi (Korean fermented vegetable products). Crit Rev Food Sci Nutr 34:175–203. https://doi.org/10.1080/10408399409527656

Chemache L, Kehal F, Namoune H, Chaalal M, Gagaoua M (2018) Couscous: ethnic making and consumption patterns in the Northeast of Algeria. J Ethnic Foods 5:211–219. https://doi.org/10.1016/j.jef.2018.08.002

Chileshe J, van den Heuvel J, Handema R, Zwaan BJ, Talsma EF, Schoustra S (2020) Nutritional composition and microbial communities of two non-alcoholic traditional fermented beverages from Zambia: a study of Mabisi and Munkoyo. Nutrients 12:1628. https://doi.org/10.3390/nu12061628

Chiu BC, Olson ME, Fahey JW (2021) Exploring the use of *Moringa oleifera* as a vegetable in Agua Caliente Nueva, Jalisco, Mexico: a qualitative study. Food Front 2(3):294–304. https://doi.org/10.1002/fft2.10

Chivenge P, Mabhaudhi T, Modi A, Mafongoya P (2015) The potential role of neglected and underutilised crop species as future crops under water scarce conditions in Sub-Saharan Africa. IJERPH 12:5685–5711. https://doi.org/10.3390/ijerph120605685

Chyne DAL, Ananthan R, Longvah T (2019) Food compositional analysis of Indigenous foods consumed by the Khasi of Meghalaya, North-East India. J Food Compos Anal 77:91–100. https://doi.org/10.1016/j.jfca.2019.01.008

Costa HS, Albuquerque TG, Sanches-Silva A, Vasilopoulou E, Trichopoulou A, D'Antuono LF, Alexieva I, Boyko N, Costea C, Fedosova K, Hayran O, Karpenko D, Kilasonia Z, Finglas P (2013) New nutritional composition data on selected traditional foods consumed in Black Sea

Area countries: nutritional composition of traditional foods from the Black Sea region. J Sci Food Agric 93:3524–3534. https://doi.org/10.1002/jsfa.6192

Dahal NR, Rao ER, Swamylingappa B (2003) Biochemical and nutritional evaluation of Masyaura—a legume-based traditional savory of Nepal. J Food Sci Technol 40:17–22

Dahal NR, Karki TB, Swamylingappa B, Li Q, Gu G (2005) Traditional foods and beverages of Nepal—a review. Food Rev Intl 21:1–25. https://doi.org/10.1081/FRI-200040579

Daughters A (2016) Southern Chile's archipelago of Chiloé: shifting identities in a new economy: Southern Chile's archipelago of Chiloé. J Latin Am Caribb Anthropol 21:317–335. https://doi.org/10.1111/jlca.12199

de Medeiros PM, Figueiredo KF, Gonçalves PHS, de Almeida Caetano R, da Costa Santos ÉM, Santos GMCD, Barbosa DM, de Paula M, Mapeli AM (2021) Wild plants and the food-medicine continuum—an ethnobotanical survey in Chapada Diamantina (Northeastern Brazil). J Ethnobiol Ethnomed 17(1):37. https://doi.org/10.1186/s13002-021-00463-y

Deaconu A, Ekomer, Mercille G, Batal M (2021) Promoting traditional foods for human and environmental health: lessons from agroecology and Indigenous communities in Ecuador. BMC Nutr 7:1. https://doi.org/10.1186/s40795-020-00395-y

Della A, Paraskeva-Hadjichambi D, Hadjichambis AC (2006) An ethnobotanical survey of wild edible plants of Paphos and Larnaca countryside of Cyprus. J Ethnobiol Ethnomed 2:1–9

Devi P, Suresh Kumar P (2012) Traditional ethnic and fermented foods of different tribes of Manipur. Indian J Tradit Knowl 11:70–77

di Tizio A, Łuczaj ŁJ, Quave CL, Redžić S, Pieroni A (2012) Traditional food and herbal uses of wild plants in the ancient South-Slavic diaspora of Mundimitar/Montemitro (Southern Italy). J Ethnobiol Ethnomed 8(1):21. https://doi.org/10.1186/1746-4269-8-21

Eliazer Nelson ARL, Ravichandran K, Antony U (2019) The impact of the Green Revolution on indigenous crops of India. J Ethnic Food 6:8. https://doi.org/10.1186/s42779-019-0011-9

Elliott B, Jayatilaka D, Brown C, Varley L, Corbett KK (2012) "We are not being heard": aboriginal perspectives on traditional foods access and food security. J Environ Public Health 2012:1–9. https://doi.org/10.1155/2012/130945

Elowad HOA, Hall AE (1987) Influences of early and late nitrogen fertilization on yield and nitrogen fixation of cowpea under well-watered and dry field conditions. Field Crop Res 15: 229–244. https://doi.org/10.1016/0378-4290(87)90012-8

Essid I, Ben Ismail H, Bel Hadj Ahmed S, Ghedamsi R, Hassouna M (2007) Characterization and technological properties of *Staphylococcus xylosus* strains isolated from a Tunisian traditional salted meat. Meat Sci 77:204–212. https://doi.org/10.1016/j.meatsci.2007.03.003

Fabien-Ouellet N (2016) Poutine dynamics. Cuizine 7. https://doi.org/10.7202/1038479ar

Faller ALK, Fialho E, Liu RH (2012) Cellular antioxidant activity of *Feijoada* whole meal coupled with an in vitro digestion. J Agric Food Chem 60:4826–4832. https://doi.org/10.1021/jf300602w

Fanzo J, Bellows AL, Spiker ML et al (2021) The importance of food systems and the environment for nutrition. Am J Clin Nutr 113:7–16

FAO (2008) An introduction to the basic concepts of food security. Food security information for action practical guides. EC-FAO Food Security Programme

Fanzo J, Drewnowski A, Blumberg J, Miller G, Kraemer K, Kennedy E (2020) Nutrients, foods, diets, people: promoting healthy eating. Curr Dev Nutrition 4(6):nzaa069. https://doi.org/10.1093/cdn/nzaa069

Food Agricultural Organization of the United Nations (FAO) (2002) The state of food security in the world 2001

Gadaga T (1999) A review of traditional fermented foods and beverages of Zimbabwe. Int J Food Microbiol 53:1–11. https://doi.org/10.1016/S0168-1605(99)00154-3

Gharibzahedi SMT (2018) Favorite and traditional rice flour–based puddings, breads, and pastries in the north of Iran: a review. J Ethnic Foods 5:105–113. https://doi.org/10.1016/j.jef.2018.03.001

Ghazi Z, Ramdani M, Tahri M, Rmili R, Elmsellem H, El Mahi B, Fauconnier M-L (2015) Chemical composition and antioxidant activity of seeds oils and fruit juice of *Opuntia ficus-indica* and *Opuntia dillenii* from Morocco. J Mater Environ Sci 6(8):2338–2345

Ghosh-Jerath S, Singh A, Kamboj P, Goldberg G, Magsumbol MS (2015) Traditional knowledge and nutritive value of indigenous foods in the Oraon tribal community of Jharkhand: an exploratory cross-sectional study. Ecol Food Nutr 54:493–519. https://doi.org/10.1080/03670244.2015.1017758

Ghosh-Jerath S, Singh A, Magsumbol MS, Kamboj P, Goldberg G (2016) Exploring the potential of indigenous foods to address hidden hunger: nutritive value of indigenous foods of Santhal Tribal Community of Jharkhand, India. J Hung Environ Nutr 11:548–568. https://doi.org/10.1080/19320248.2016.1157545

Grivetti LE, Ogle BM (2000) Value of traditional foods in meeting macro- and micronutrient needs: the wild plant connection. Nutr Res Rev 13(1):31–46. https://doi.org/10.1079/095442200108728990

Guarrera PM, Savo V (2016) Wild food plants used in traditional vegetable mixtures in Italy. J Ethnopharmacol 185:202–234. https://doi.org/10.1016/j.jep.2016.02.050

Gurmeet S, Amrita P (2015) Unique pandanus-flavour, food and medicine. J Pharmacogn Phytochem 5:8–14

Guyot M, Dickson C, Paci C, Furgal C, Chan HM (2006) Local observations of climate change and impacts on traditional food security in two northern Aboriginal communities. Int J Circumpolar Health 65:403–415. https://doi.org/10.3402/ijch.v65i5.18135

Hall AE (2012) Phenotyping cowpeas for adaptation to drought. Front Physiol 3. https://doi.org/10.3389/fphys.2012.00155

Harmayani E, Anal AK, Wichienchot S, Bhat R, Gardjito M, Santoso U, Siripongvutikorn S, Puripaatanavong J, Payyappallimana U (2019) Healthy food traditions of Asia: exploratory case studies from Indonesia, Thailand, Malaysia, and Nepal. J Ethnic Food 6:1. https://doi.org/10.1186/s42779-019-0002-x

Hassan ZU, Al-Thani R, Atia FA, Almeer S, Balmas V, Migheli Q, Jaoua S (2018) Evidence of low levels of aflatoxin M1 in milk and dairy products marketed in Qatar. Food Control 92:25–29. https://doi.org/10.1016/j.foodcont.2018.04.038

Hefni M, Witthöft CM (2014) Folate content in processed legume foods commonly consumed in Egypt. LWT Food Sci Technol 57:337–343. https://doi.org/10.1016/j.lwt.2013.12.026

Hesseltine CW (1979) Some important fermented foods of Mid-Asia, the Middle East, and Africa. J Am Oil Chem Soc 56:367–374. https://doi.org/10.1007/BF02671501

Hong SW, Choi JY, Chung KS (2012) Culture-based and denaturing gradient gel electrophoresis analysis of the bacterial community from *Chungkookjang*, a traditional Korean fermented soybean food. J Food Sci 77:M572–M578. https://doi.org/10.1111/j.1750-3841.2012.02901.x

Hossain M, Kabir Y (2016) Ethnic fermented foods and beverages of Bangladesh. In: Tamang JP (ed) Ethnic fermented foods and alcoholic beverages of Asia. Springer India, New Delhi, pp 73–89

Inamdar V, Chimmad BV, Naik R (2005) Nutrient composition of traditional festival foods of North Karnataka. J Hum Ecol 18:43–48. https://doi.org/10.1080/09709274.2005.11905805

Jayathilake C, Visvanathan R, Deen A, Bangamuwage R, Jayawardana BC, Nammi S, Liyanage R (2018) Cowpea: an overview on its nutritional facts and health benefits: nutritional and health properties of cowpea. J Sci Food Agric 98(13):4793–4806. https://doi.org/10.1002/jsfa.9074

John SM, Deeseenthum S (2015) Properties and benefits of kefir—a review. Songklanakarin J Sci Technol 37:275–282

Kalita P, Tag H, Sarma HN, Das AK (2014) Evaluation of nutritional potential of five unexplored wild edible food plants from Eastern Himalayan Biodiversity Hotspot Region (India). https://doi.org/10.5281/ZENODO.1091224

Kamisah Y, Othman F, Qodriyah HMS, Jaarin K (2013) *Parkia speciosa* Hassk.: a potential phytomedicine. Evid Based Complement Alternat Med 2013:1–9. https://doi.org/10.1155/2013/709028

Kathiriya A, Das K, Kumar EP, Mathai KB (2010) Evaluation of antitumor and antioxidant activity of *Oxalis corniculata* Linn. against ehrlich ascites carcinoma on mice. Iran J Cancer Prevent 3: 157–165

Kaur KD, Jha A, Sabikhi L, Singh AK (2014) Significance of coarse cereals in health and nutrition: a review. J Food Sci Technol 51:1429–1441. https://doi.org/10.1007/s13197-011-0612-9

Khatri D, Shrestha K, Ojha H, Paudel G, Paudel N, Pain A (2017) Reframing community forest governance for food security in Nepal. Environ Conserv 44(2):174–182. https://doi.org/10.1017/S0376892916000369

Król K, Gantner M, Piotrowska A, Hallmann E (2020) Effect of climate and roasting on polyphenols and tocopherols in the kernels and skin of six hazelnut cultivars (*Corylus avellana* L.). Agriculture 10:36. https://doi.org/10.3390/agriculture10020036

Kuhnlein HV, Chan HM (2000) Environment and contaminants in traditional food systems of northern indigenous peoples. Annu Rev Nutr 20:595–626. https://doi.org/10.1146/annurev.nutr.20.1.595

Kuhnlein HV, Receveur O (1996) Dietary change and traditional food systems of indigenous peoples. Annu Rev Nutr 16:417–442. https://doi.org/10.1146/annurev.nu.16.070196.002221

Kuhnlein H, Erasmus B, Creed-Kanashiro H, Englberger L, Okeke C, Turner N, Allen L, Bhattacharjee L (2006) Indigenous peoples' food systems for health: finding interventions that work. Public Health Nutr 9:1013–1019. https://doi.org/10.1017/PHN2006987

Kuhnlein HV, McDonald M, Spigelski D (2009) Gwich' in traditional food for health: phase 1. Indigenous Peoples' food systems: the many dimensions of culture, diversity and environment for nutrition and health. Reprinted. Food and Agriculture Organization of the United Nations, Rome, p 45–58

Kujawska M, Łuczaj Ł (2015) Wild edible plants used by the polish Community in Misiones, Argentina. Hum Ecol 43:855–869. https://doi.org/10.1007/s10745-015-9790-9

Kulwa KBM, Mamiro PS, Kimanya ME, Mziray R, Kolsteren PW (2015) Feeding practices and nutrient content of complementary meals in rural central Tanzania: implications for dietary adequacy and nutritional status. BMC Pediatr 15:171. https://doi.org/10.1186/s12887-015-0489-2

Kwik JC (2008) Traditional food knowledge: renewing culture and restoring health. Master Thesis. University of Waterloo

La Anh N (2015) Health-promoting microbes in traditional Vietnamese fermented foods: a review. Food Sci Human Wellness 4:147–161. https://doi.org/10.1016/j.fshw.2015.08.004

Lalmuanpuii R (2021) Diversity and phytochemical study of some wild edible vegetables in Aizawl district of Mizoram. Thesis, Mizoram University

Lambden J, Receveur O, Kuhnlein HV (2007) Traditional food attributes must be included in studies of food security in the Canadian Arctic. Int J Circumpolar Health 66:308–319. https://doi.org/10.3402/ijch.v66i4.18272

Lancaster PA, Ingram JS, Lim MY, Coursey DG (1982) Traditional cassava-based foods: survey of processing techniques. Econ Bot 36:12–45. https://doi.org/10.1007/BF02858697

Lawrence J, Blackett P, Cradock-Henry NA (2020) Cascading climate change impacts and implications. Clim Risk Manag 29:100234. https://doi.org/10.1016/j.crm.2020.100234

Leksir C, Boudalia S, Moujahed N, Chemmam M (2019) Traditional dairy products in Algeria: case of Klila cheese. J Ethnic Food 6:7. https://doi.org/10.1186/s42779-019-0008-4

Li L-T, Yin L-J, Saito M (2004) Function of traditional foods and food culture in China. Jpn Agric Res Q 38(4):213–220. https://doi.org/10.6090/jarq.38.213

Lim TK (2014) *Saccharum spontaneum* var. *edulis*. In: Edible medicinal and non medicinal plants. Springer, Netherlands, Dordrecht, pp 640–642

Łuczaj Ł (2010) Changes in the utilization of wild green vegetables in Poland since the 19th century: a comparison of four ethnobotanical surveys. J Ethnopharmacol 128(2):395–404. https://doi.org/10.1016/j.jep.2010.01.038

Łuczaj Ł, Zovko Končić M, Miličević T, Dolina K, Pandža M (2013) Wild vegetable mixes sold in the markets of Dalmatia (southern Croatia). J Ethnobiol Ethnomed 9(1):2. https://doi.org/10.1186/1746-4269-9-2

Łuczaj Ł, Lamxay V, Tongchan K, Xayphakatsa K, Phimmakong K, Radavanh S, Kanyasone V, Pietras M, Karbarz M (2021) Wild food plants and fungi sold in the markets of Luang Prabang, Lao PDR. J Ethnobiol Ethnomed 17(1):6. https://doi.org/10.1186/s13002-020-00423-y

Macintyre M (1987) Nurturance and nutrition: change and continuity in concepts of food and feasting in a Southern Massim community. J Soc Océan 84:51–59. https://doi.org/10.3406/jso.1987.2561

Mafimisebi TE, Oguntade AE, Fajemisin AN, Aiyelari OP (2012) Local knowledge and socio-economic determinants of traditional medicines' utilization in livestock health management in Southwest Nigeria. J Ethnobiol Ethnomed 8:2. https://doi.org/10.1186/1746-4269-8-2

Maldonado JK, Colombi B, Pandya R (eds) (2014) Climate change and indigenous peoples in the United States: impacts, experiences and actions. Springer International Publishing, Cham. https://doi.org/10.1007/978-3-319-05266-3

Mao AA, Odyuo N (2007) Traditional fermented foods of the Naga tribes of Northeastern, India. Indian J Tradit Knowl 6:37–41

Mattalia G, Quave CL, Pieroni A (2013) Traditional uses of wild food and medicinal plants among Brigasc, Kyé, and Provençal communities on the Western Italian Alps. Genet Resour Crop Evol 60:587–603. https://doi.org/10.1007/s10722-012-9859-x

Mbaiguinam M, Mahmout Y, Tarkodjiel M, Delobel B, Bessiere J-M (2005) Constituents of Kawal, fermented *Cassia obtusifolia* leaves, a traditional food from Chad. Afr J Biotechnol 4:1080–1083

McWatters KH (1983) Compositional, physical and sensory characteristics of Akara processed from cowpea paste and Nigerian cowpea flour. Cereal Chem 60:333–336

Mensah JK, Okoli RI, Ohaju-Obodo JO, Eifediyi K (2008) Phytochemical, nutritional and medical properties of some leafy vegetables consumed by Edo people of Nigeria. Afr J Biotechnol 7

Michon A (2012) Conservation of natural resources within mature tropical forests: how an indigenous community uses and manages wild plants in the Comarca Ngöbe-Buglé, Panama. Doctoral dissertation, Carleton University

Mihiranie S, Jayasinghe JK, Jayasinghe CVL, Wanasundara JPD (2020) Indigenous and traditional foods of Sri Lanka. J Ethnic Food 7:42. https://doi.org/10.1186/s42779-020-00075-z

Millennium Ecosystem Assessment (2008) Ecosystems and human well-being: biodiversity synthesis. World Resources Institute, Washington, DC

Montaño N, Gavino G, Gavino VC (2001) Polyunsaturated fatty acid contents of some traditional fish and shrimp paste condiments of the Philippines. Food Chem 75:155–158. https://doi.org/10.1016/S0308-8146(01)00126-1

Motti R (2021) Wild plants used as herbs and spices in Italy: an ethnobotanical review. Plants 10(3):563. https://doi.org/10.3390/plants10030563

Muhialdin BJ, Filimonau V, Qasem JM, Ibrahim SA, Algboory HL (2022) Traditional fermented foods and beverages in Iraq and their potential for large-scale commercialization. J Ethnic Food 9:18. https://doi.org/10.1186/s42779-022-00133-8

Murooka Y, Yamshita M (2008) Traditional healthful fermented products of Japan. J Ind Microbiol Biotechnol 35:791–798. https://doi.org/10.1007/s10295-008-0362-5

Mushtaq M, Sultana B, Anwar F, Khan MZ, Ashrafuzzaman M (2012) Occurrence of aflatoxins in selected processed foods from Pakistan. Int J Mol Sci 13:8324–8337. https://doi.org/10.3390/ijms13078324

Muthoni J, Nyamongo DO (2010) Traditional food crops and their role in food and nutritional security in Kenya. J Agric Food Inf 11:36–50. https://doi.org/10.1080/10496500903466745

Naila A, Flint S, Fletcher GC, Bremer PJ, Meerdink G (2011) Chemistry and microbiology of traditional Rihaakuru (fish paste) from the Maldives. Int J Food Sci Nutr 62:139–147. https://doi.org/10.3109/09637486.2010.515566

Neela S, Fanta SW (2020) Injera (an ethnic, traditional staple food of Ethiopia): a review on traditional practice to scientific developments. J Ethnic Food 7:32. https://doi.org/10.1186/s42779-020-00069-x

Nirmala A, Saroja S, Devi GG (2011) Antidiabetic activity of *Basella rubra* and its relationship with the antioxidant property. Br Biotechnol J 1(1):1–9

Nongdam P, Tikendra L (2014) The nutritional facts of bamboo shoots and their usage as important traditional foods of Northeast India. Int Sch Res Notices 2014:1–17. https://doi.org/10.1155/2014/679073

Okeke EC, Eneobong HN, Uzuegbunam AO, Ozioko AO, Kuhnlein H (2008) Igbo traditional food system: documentation, uses and research needs. Pak J Nutr 7(2):365–376. https://doi.org/10.3923/pjn.2008.365.376

Okeke EC, Eneobong HN, Uzuegbunam AO, Ozioko AO, Umeh SI, Kuhnlein H (2009) Nutrient composition of traditional foods and their contribution to energy and nutrient intakes of children and women in rural households in Igbo culture area. Pak J Nutr 8(4):304–312. https://doi.org/10.3923/pjn.2009.304.312

Paoletti MG (ed) (2005) Ecological implications of minilivestock: potential of insects, rodents, frogs, and snails. Science Publishers, Enfield, (NH)

Patkar AN, Desai NV, Ranage AA, Kalekar KS (2012) A review on *Aegle marmelos*: a potential medicinal tree. Int Res J Pharm 3(8):86–91

Pawera L, Khomsan A, Zuhud EAM, Hunter D, Ickowitz A, Polesny Z (2020) Wild food plants and trends in their use: from knowledge and perceptions to drivers of change in West Sumatra, Indonesia. Foods 9(9):1240. https://doi.org/10.3390/foods9091240

Pearce R (1990) Traditional food crops in sub-Saharan Africa. Food Policy 15:374–382. https://doi.org/10.1016/0306-9192(90)90054-4

Peñas E, Frias J, Sidro B, Vidal-Valverde C (2010) Impact of fermentation conditions and refrigerated storage on microbial quality and biogenic amine content of sauerkraut. Food Chem 123:143–150. https://doi.org/10.1016/j.foodchem.2010.04.021

Pengseng N, Siripongvutikorn S, Usawakesmanee W, Wattanachant S (2010) Effect of thermal processing and protein nutrients on antioxidant activity of Tom-Kha paste extract. Asian J Food Agro-Ind 3:389–399

Premlata T, Sharma V, Bisht MS, Nirmala C (2020) Edible bamboo resources of Manipur: consumption pattern of young shoots, processing techniques and their commercial status in the local market. Indian J Tradit Knowl 19:73–82

Rai KN, Gowda CLL, Reddy BVS, Sehgal S (2008) Adaptation and potential uses of sorghum and pearl millet in alternative and health foods. Compr Rev Food Sci Food Saf 7:340–352. https://doi.org/10.1111/j.1541-4337.2008.00049.x

Rai R, Shangpliang HNJ, Tamang JP (2016) Naturally fermented milk products of the Eastern Himalayas. J Ethnic Foods 3:270–275. https://doi.org/10.1016/j.jef.2016.11.006

Ray A, Ray R, Sreevidya EA (2020) How many wild edible plants do we eat—their diversity, use, and implications for sustainable food system: an exploratory analysis in India. Front Sustain Food Syst 4:56. https://doi.org/10.3389/fsufs.2020.00056

Riang'a RM, Broerse J, Nangulu AK (2017) Food beliefs and practices among the Kalenjin pregnant women in rural Uasin Gishu County, Kenya. J Ethnobiol Ethnomed 13:29. https://doi.org/10.1186/s13002-017-0157-8

Riquelme J, Olaeta JA, Gálvez L, Undurraga P, Fuentealba C, Osses A, Orellana J, Gallardo J, Pedreschi R (2016) Nutritional and functional characterization of wild and cultivated *Sarcocornia neei* grown in Chile. Cienc Inv Agr 43:11–11. https://doi.org/10.4067/S0718-16202016000200011

Rivera D, Obon C, Inocencio C, Heinrich M, Verde A, Fajardo J, Llorach R (2005) The ethnobotanical study of local Mediterranean food plants as medicinal resources in Southern Spain. J Physiol Pharmacol 56(Suppl 1):97–114

Roche M, Creed-Kanashiro H, Tuesta I, Kuhnlein H (2008) Traditional food diversity predicts dietary quality for the Awajún in the Peruvian Amazon. Public Health Nutr 11:457–465. https://doi.org/10.1017/S1368980007000560

Romojaro A, Botella MÁ, Obón C, Pretel MT (2013) Nutritional and antioxidant properties of wild edible plants and their use as potential ingredients in the modern diet. Int J Food Sci Nutr 64(8): 944–952. https://doi.org/10.3109/09637486.2013.821695

Rush EC, Hsi E, Ferguson LR, Williams MH, Simmons D (2010) Traditional foods reported by a Māori community in 2004. Mai Review 2. Retrieved from http://www.review.mai.ac.nz/index.php/MR/article/view/279/478

Sangameswaran B, Jayakar B (2007) Anti-diabetic and spermatogenic activity of *Cocculus hirsutus* (L) Diels. Afr J Biotechnol 6

Sarala P, Krishnamurthy SR (2014) Monkey jack: underutilized edible medicinal plant, nutritional attributes and traditional foods of Western Ghat, Karnataka, India. Indian J Tradit Knowl 13: 508–518

Sarkar P, Lohith Kumar DH, Dhumal C, Panigrahi SS, Choudhary R (2015) Traditional and Ayurvedic foods of Indian origin. J Ethnic Foods 2:97–109. https://doi.org/10.1016/j.jef.2015.08.003

Sarkar D, Walker-Swaney J, Shetty K (2020) Food diversity and indigenous food systems to combat diet-linked chronic diseases. Curr Dev Nutr 4:3–11. https://doi.org/10.1093/cdn/nzz099

Schuster RC, Wein EE, Dickson C, Chan HM (2011) Importance of traditional foods for the food security of two First Nations communities in the Yukon, Canada. Int J Circumpolar Health 70: 286–300. https://doi.org/10.3402/ijch.v70i3.17833

Sha SP, Ghatani K, Tamang JP (2013) Dalbari, a Traditional Pulses based fermented food of West Bengal. Int J Agric Food Sci Technol 4:6–10

Shrestha PM, Dhillion SS (2006) Diversity and traditional knowledge concerning wild food species in a locally managed forest in Nepal. Agroforest Syst 66(1):55–63. https://doi.org/10.1007/s10457-005-6642-4

Singh IP (2016) Nutritional benefits of pigeonpea. In: Singh U, Praharaj CS, Singh SS, Singh NP (eds) Biofortification of food crops. Springer, New Delhi, pp 73–81

Sivaraman G, Rajalakshmi K (2005) Concept of siddha nutraceuticals. Acta Hortic 153–159. https://doi.org/10.17660/ActaHortic.2005.680.23

Sõukand R, Kalle R (2013) Where does the border lie: locally grown plants used for making tea for recreation and/or healing, 1970s–1990s Estonia. J Ethnopharmacol 150:162–174. https://doi.org/10.1016/j.jep.2013.08.031

Sproesser G, Ruby MB, Arbit N, Akotia CS, Alvarenga M d S, Bhangaokar R, Furumitsu I, Hu X, Imada S, Kaptan G, Kaufer-Horwitz M, Menon U, Fischler C, Rozin P, Schupp HT, Renner B (2019) Understanding traditional and modern eating: the TEP10 framework. BMC Public Health 19:1606. https://doi.org/10.1186/s12889-019-7844-4

Stryamets N, Elbakidze M, Ceuterick M, Angelstam P, Axelsson R (2015) From economic survival to recreation: contemporary uses of wild food and medicine in rural Sweden, Ukraine and NW Russia. J Ethnobiol Ethnomed 11:53. https://doi.org/10.1186/s13002-015-0036-0

Sundriyal M, Sundriyal RC, Sharma E (2004) Dietary use of wild plant resources in the Sikkim Himalaya, India. Econ Bot 58:626–638. https://doi.org/10.1663/0013-0001(2004)058[0626:DUOWPR]2.0.CO;2

Tamang JP, Sarkar PK, Hesseltine CW (1988) Traditional fermented foods and beverages of Darjeeling and Sikkim—a review. J Sci Food Agric 44:375–385. https://doi.org/10.1002/jsfa.2740440410

Tamang JP, Chettri R, Sharma RM (2009) Indigenous knowledge of Northeast women on production of ethnic fermented soybean foods. Indian J Tradit Knowl 8:122–126

Taylor JRN, Emmambux MN (2008) Gluten-free foods and beverages from millets. In: Gluten-free cereal products and beverages. Elsevier, pp 119–V

Tchokouaha LRY, Tchamgoue AD, Ann OM, Ogochukwu O, Daniel F, Ifeanyi O, Ifunanya O, Priscilla N, Agbor GA (2015) Comparative antioxidant capacity of the spices of some traditional soups in Nigeria. Scholars Acad J Biosci 3:79–84

Thakur A, Singh S, Puri S (2020) Exploration of wild edible plants used as food by Gaddis—a tribal community of the Western Himalaya. Scientific World J 2020:1–6. https://doi.org/10.1155/2020/6280153

Torres-Guevara FA, Ganoza-Yupanqui ML, Suárez-Rebaza LA, Malca-García GR, Bussmann RW (2020) Wild plants of Northern Peru. Wild plants: the treasure of natural healers. p 37

Trichopoulou A, Vasilopoulou E, Georga K, Soukara S, Dilis V (2006) Traditional foods: why and how to sustain them. Trends Food Sci Technol 17:498–504. https://doi.org/10.1016/j.tifs.2006.03.005

Trichopoulou A, Soukara S, Vasilopoulou E (2007) Traditional foods: a science and society perspective. Trends Food Sci Technol 18:420–427. https://doi.org/10.1016/j.tifs.2007.03.007

Tumuhe CL, Ategeka K, Sunday C, Tibaijuka D, Muhindo CB (2020) The utilization of traditional and indigenous foods and seeds in Uganda. J Food Secur 8:11–21. https://doi.org/10.12691/jfs-8-1-2

Unuofin JO, Otunola GA, Afolayan AJ (2017) Nutritional evaluation of *Kedrostis africana* (L.) Cogn: an edible wild plant of South Africa. Asian Pac J Trop Biomed 7(5):443–449. https://doi.org/10.1016/j.apjtb.2017.01.016

Volpato G, Godinez D (2006) Medicinal foods in Cuba: promoting health in the household. Eating and healing: traditional food as medicine. pp 213–235

Wahlqvist ML, Lee MS (2007) Regional food culture and development. Asia Pac J Clin Nutr 16:2–7

Wang S-Y, Zhao H, Xu H-T, Han X-D, Wu Y-S, Xu F-F, Yang X-B, Göransson U, Liu B (2021) *Kaempferia galanga* L.: progresses in phytochemistry, pharmacology, toxicology and ethnomedicinal uses. Front Pharmacol 12:675350. https://doi.org/10.3389/fphar.2021.675350

Watson FE, Ngesa A, Onyang'o J, Alnwick D, Tomkins AM (1996) Fermentation—a traditional anti-diarrhoeal practice lost? The use of fermented foods in urban and rural Kenya. Int J Food Sci Nutr 47:171–179. https://doi.org/10.3109/09637489609012579

Weichselbaum E, Benelam B, Soares Costa H (2009) Traditional foods in Europe. EuroFIR Project, Norwich

Yongsmith B, Malaphan W (2016) Traditional fermented foods in Thailand. In: Kristbergsson K, Oliveira J (eds) Traditional foods. Springer, Boston, pp 31–59

Zohary D, Hopf M, Wiess E (2012) Domestication of plants in the Old World: the origin and spread of domesticated plants in South West Asia, Europe, and the Mediterranean Basin. Oxford University Press

Zycherman A (2008) To beef or not to beef: defining food security and insecurity in Tucumán Argentina. Ecol Environ Anthropol (University of Georgia) 30

Chapter 5
Usage of Wild Edible Plants Among Upland Indigenous Communities of Northeastern States of India

Nilay Kumar, Sunandani Chandel, Dileep Kumar Pandey, Praveen Kumar Singh, and Kalkame Cheran Momin

Abstract The use of wild plants as food is an integral part of the culture and tradition of many indigenous communities around the world. These wild plants are the precious gift of our nature, and most ethnic communities depend on them for their day-to-day life. There are about 800 different species of wild edible plants (WEPs) in India, of which 300 species are used mostly by the tribal and rural population of the Northeastern region alone. Wild plants as food constitute a significant portion of the daily diet among the people of the Northeast region. Such edible plants obtained from the wild occupy a significant place among the indigenous community and are found to play an important role in supplementing the diet of rural people. These wild edible resources are to be explored in such a manner so that tribal people can get their due share as they have conserved it over the centuries based on their traditional knowledge. The Northeast region of India falls in the global hotspot of biodiversity, whereas WEPs have high potential. Its usage is much visible in their traditional foods and rituals. A comprehensive assessment covering six hilly states of the region, the present study indicated that the majority of these plants have high value and maximum usage of wild resources as food. With the view of reducing the gap in traditional knowledge and tapping the potential hidden resources for proper utilization, exploitation, and sustainable management of WEPs are crucial. The study discusses these issues in-depth and suggests biodiversity-based intervention to sustain the livelihood, nutritional security, and economic prosperity of the people in the NE region.

N. Kumar (✉) · D. K. Pandey · K. C. Momin
College of Horticulture and Forestry, Central Agricultural University, Pasighat, Arunachal Pradesh, India

S. Chandel
ASPEE College of Horticulture and Forestry, Navsari Agricultural University, Navsari, Gujarat, India

P. K. Singh
Center for Protected Cultivation Technology, ICAR - Indian Agricultural Research Institute, Pusa, New Delhi, India

A. Kumar et al. (eds.), *Wild Food Plants for Zero Hunger and Resilient Agriculture*, Plant Life and Environment Dynamics,
https://doi.org/10.1007/978-981-19-6502-9_5

Keywords Ethnic community · Ethnobotany · Wild edible plants · Nutritional security · Northeast India

5.1 Introduction

The plant species that are untamed in nature and are consumed widely for nourishment are known as wild edible plants (WEPs) (Beluhan and Ranogajec 2010). In the distant past, a wide variety of WEPs are consumed owing to cultural significance and traditional food habits by tribals and rural populations living in the close vicinity of forests (Satter et al. 2016). While in the National and International market spectrum, the WEPs have not gained much significance, but locally, they are of great importance (Yesodharan and Sujana 2007). Several folkloric communities consider these WEPs as their staple food. These natural food sources not only enhance their nutrition but also play an important role in uplifting the economy of the tribal people. North East India had a great diversity of ethnic communities encompassing about 50% of the tribal population of India. A decent amount of around 130 tribal networks dwell in North East area. It comprises well-known ethnic groups viz. Assamese, Bodo, Chakma, Chakma, Deuri, Hmar, Karbi, Khasi, Kuki, Meiteis, Mizo, Paite, Rabha, Tangkhul, and Tripuris (Deb et al. 2013; Dutt and Dutt 2005; Yumkham et al. 2016). In today's world, most of the cultivated crops such as Oats (*Avena sativa*), foxtail millet (*Setaria italica* derived from *S. viridis*), chicory (*Cichorium intybus*), pak choi (*Brassica rapa*), spinach beet (*Beta vulgaris*) owes its origin from the wild but humans cultured them for food. However, with the advent of time, the propensity for gathering wild food plants has completely vanished. These untamed natural resources encompass a great genetic pool that can be harbored to sustain the growing nutrition demand for developing new, improved resistant crops (Rapoport and Drausal 2013; Ju et al. 2013).

5.2 Potential Usage of Wild Edible Resources in Different States in Northeast India

India epitomizes immense heterogeneity in plant diversity owing to varied climates and biogeography. It encapsulates 7% of the universe's verdure (45,000 plants) and is among the 25 most plant-rich nations in the world. Approximately, it harbors 800 diverse species of WEPs (Hazarika and Pongener 2018). Northeast India acknowledged as the "Cradle of Flowering Plants" is home to numerous botanical resources (Das et al. 2016). The region encompasses eight states i.e. Arunachal Pradesh, Assam, Manipur, Meghalaya, Mizoram, Nagaland, Sikkim, and Tripura forms a part of Indo-Malayan and Himalayan biodiversity hotspot and harbors 50% of the biodiversity of India. The region has a rich repository of about 300 WEPs (Tripathi et al. 2016). North East India and Arunachal Himalayan Region are considered as one of the center of origin for several cultivable crops and their wild relatives (Vavilov 1926). Each community in the region have its own unique identity, tradition, and culture and, more specifically, different dietary patterns and eating habit. The Northeast region is blessed with a rich biodiversity of flora and fauna and is known to be one of the biodiversity hotspots in the world. Several ethnobotanical studies have been undertaken to document the wild edible resources in North East India.

5.2.1 Arunachal Pradesh

Arunachal Pradesh illustrates a rich divergence of vegetation in the eastern Himalayas. It is the largest state among all the other states in the locale viz. 26.28°–29.30° North latitude and 91.20°–97.30° East longitude covering a topographical zone of 83,743 sq. km (Khongsai et al. 2011). The state holds a 9.2 lakhs tribal population, which accounts for 0.9% of the total tribal population of India. The state encompasses several tribes, viz. *Abor, Adi, Aka, Apatani, Galo, Hrusso, Idu, Khampti, Khowa, Khamba, Mishmi, Monpa, Naga, Nyishi, Sherdukpen, Singpho, Taroan,* and *Tagin* (Census 2011). The *Adi* tribe residing in the east and east-west Siang, lower Dibang valley in the west, and *Monpa* tribe in the west, the north-west part of the state are among the prime tribes in the state owing to their distinct culture, traditional practices, and greater kinship towards the wild resources for their nourishment. The tribal ethnic communities residing in the state represent diverse utilization of wild resources for their livelihood as well as nutritional enhancement. The state holds 346 diverse WEPs representing 215 genera belonging to 102 different families. Among all the wild resources, 165 species are utilized as medicinal plants, followed by 97 species of wild edible vegetables and 39 species of wild edible fruits. The maximum number of wild edibles consumed by the tribal's belongs to the Fabaceae family (Angami et al. 2006; Das et al. 2019). Baruah and Bhattacharyya (2019) reported that the ethnic people belonging to *Galo* tribe consume several wild plants viz. *Clerodendrum glandulosum, Diplazium esculenium, Elatostema sublaxum, Ganostegia hirta, Litsea cubeba, Solanum americanum* as a vegetable after boiling in their day-to-day food. The production of several fermented drinks from wild plants is a traditional ongoing practice by the *Monpa* tribe. The several wild plants extensively used for the same are *Alnus nepalensis, Acer campbellii, Asparagus racemosus, Scurrula elata*, and *Hypericum uralum* (Tsering et al. 2017). A sweetened fruit wine having stimulant properties is locally prepared using the *Elaeagnus umbellata* fruits by the tribal community (Tag et al. 2014). Wild edible mushrooms are often consumed in different parts of the state for nourishment and food security. Singh et al. (2017) reported 32 species of WEPs in the west Siang district represented by 20 families and 32 genera. The commonly used species in local cuisines are *Auricularia polytricha* and *A. delicata.*

5.2.2 Assam

Assam is widely recognized as the gateway towards the northeastern region, encompassing 78,438 sq. km area representing 38.84 lakhs tribal population constituting 3.7% of total tribal inhabitants in the nation. The major tribal's dwellings in the state are *Bhoi, Chakma, Dimasa, Garo, Hajong, Hmar, Jaintia, Kachari, Khasi, Kuki, Lyngngam, Pnar, Synteng*, and *War* (Census 2011; Dhar 2014). The state holds 286 diverse WEPs representing 192 genera belonging to 93 different families, which accounts for about 7.27% of the total plant diversity in the state (Sarma et al. 2010). A high rainfall rate in North East India promotes the growth of wild edible ferns. *Dhekia Saag*, a popular local cuisine in Assam, is prepared using a wild fern *Diplazium esculentum* (Yumkham et al. 2016). Another study was conducted to document the availability of wild edible resources in the local market throughout the state. Chaudhury et al. (2021) documented a wide spectrum of edible resources from the wild i.e. about 143 wild edible species representing 115 genera and 57 families. Several wild edible mushrooms, namely *Auricularia delicata, Termitomyces* sp., *Lentinula edodes, Lentinus polychrous* and *L. squarrosulus,* are widely consumed by the various ethnic communities residing in close vicinity of forests. Nath (2015) recognized several wild edible resources from western Assam, which are used as leafy vegetables and preparation of several local cuisines, chutneys, traditional delicacies, etc. About 75 plants belonging to 62 genera and 44 different families are documented. Consumption of wild edible flowers, either in fresh or dry form, in several regional delicacies is a long-ago tradition in the state. Deka and Nath (2014) cataloged 23 families of edible flowers representing 31 genera and 35 species. Several important species are *Lasia spinosa, Alocasia acuminateare, Sesbania grandiflora* (popularly known as Bor)*, Monochoria hastate, Oroxylum indicum*, etc. The ethnic groups, viz. *Bodo* and *Garos*, are mostly situated in the close vicinity of the forest and are traditionally consuming several wood mushrooms collected from the wild to meet their nutritional necessities. The important wild edible mushrooms belong to *Termitomyces* and *Lentinus* genera (Sarma et al. 2010). Narzary et al. (2014) reported 29 wild edible flower species represented by 29 genera widely consumed as local food by the *Bodo* tribe residing in the state. The inflorescence including the calyx, corolla, and ovary are consumed, whereas the other parts of the flower are discarded. The several wild flowers consumed by the ethnic communities are *Adhatoda vasica, Anthocephalus cadamba, Ficus lepidosa, Gmelina arborea, Leucas aspera, Oroxylum indicum, Phlogacanthus thyrsiflorus, Sesbania grandiflora, Tragia involcurata,* etc. The leaves of *Colocasia esculanta, Houttuyania cordata*, and *Maihot esculanta* are widely consumed as wild edible vegetables after cooking by the *Karbi* tribals (Kar and Borthakur 2007).

5.2.3 Manipur

Manipur is positioned between 23°27′ to 25°41′ N latitude and 93°61′ and 94°48′ E longitude constitute a fundamental part of the biodiversity hotspot between India and Myanmar. The state incorporates 22,327 sq. km area representing an 11.67 lakhs tribal population, which accounts for 1.1% of the total tribal population of India (Census 2011; Yumnam et al. 2012). The state holds 38 different tribal communities imperatively categorized into four groups viz. *Kuki, Mizo, Naga*, and others. Several recognized ethnic groups residing in the state are *Anal, Hmar, Kabui, Kacha, Kom, Maring, Monsang, Moyon, Naga, Paite, Simte, Tangkhul, Thadou, Vaiphei*, and *Zou* (Devi et al. 2013). Konsam et al. (2016) documented the rich diversity of wild leafy vegetables having 68 species represented by 42 families utilized by several ethnic groups in the state. These wild leafy vegetables are found in several local markets, and one of the popular local markets is run by the women's belonging to several ethnic communities, viz. "Ima Keithel." The highly recognized wild leafy vegetables used by the tribal's as a source of nourishment are *Alocasia cucullata, Alpinia nigra, Amomum aromaticum, Chimonobambusa callosa, Eryngium foetidum, Euryale ferox, Ficus benghalensis, Hedychium coronarium, Houttuynia cordata, Ipomoea aquatica, Neptunia oleracea, Oenanthe javanica*, and *Zanthoxylum budranga. Singju* and *Iromba*, the popular local cuisine, are prepared from the young green pods of *Parkia timoriana* collected from the wild by the tribal's (Sinha 2001). Several ethnic communities, namely *Garo, Khasi, Mizo Naga*, and *Reang*, dwelling in forest-inhabited areas also consume the young green pods of *Parkia roxburghii* as a vegetable (Singha et al. 2021). The flower buds of *Crotalaria tetragona* are traditionally consumed as a vegetable by ethnic communities (Bhatt et al. 2009). Apshahana and Sharma (2018) concluded that different species of wild edible mushrooms are popularly consumed and marketed in the state by several ethnic groups, viz. *Kalapahar, Kanglatombi, Kangpokpi, Mao Gate, Motbung*, and *Senapati*. The popularly wild edible species of mushroom in the state are *Auricularia delicate, Cantharellus cibarius, Lactarius volemus, Lentinula edodes, Pleurotus citrinopileatus, Pleurotus ostreatus, Schizophllum commune*, and *Termitomyces eurhizus.*

5.2.4 Meghalaya

Meghalaya is widely acknowledged as the "*Adobe of cloud*". It is bounded by Assam in three directions and in the south by Myanmar. The state clutches a rich diversity of vegetation composition extending from tropical to temperate. The geographical expansion viz. 22,429 sq. km extents from latitude 25°47′–26°10′ North and longitude 89°45′–92°45′ East. The heaviest precipitation throughout the world is received by the *Mawsynram* of about 1169 cm on the southern slopes of Khasi hills (Jeeva 2009). The state embraces a 25.65 lakhs tribal population constituting 2.4%

of the total tribal population of the country. The imperative ethnic communities residing in the state are *Garo, Jaintia*, and *Khasi*. Among these, the *Jaintia*, locally known as "*Ka Ri Ki Khadar Doloi*," is the pioneering resident of Jaintia hills (Census 2011; Phawa et al. 2019). A total of 249 plants belonging to 82 distinctive families and 153 genera are documented (Sawian et al. 2007). Raw fruits of *Elaeagnus latifolia* are often consumed by the local inhabitants. The fruits are expansively used in the preparation of jam, jelly, and local fruit juice. Several local chutneys are prepared using the fruits of *Dillenia pentagyna* and *Elaeagnus latifolia* (Seal 2012). Wild edible mushrooms are an important edible resource among the various ethnic communities in the state. Several traditional food recipes are prepared by the *Khasi* tribe using *Cantharellus cibarius, Clavulina* spp., *Gomphus floccosus, Laccaria lateritia, Lactarius volemus, Ramaria* spp., *Tricholoma* spp. etc. are *Syrwa tit, Tit tung bad doh sniang, Tit tyngab bad tungtap*, and *Tit tyngab bad jadoh* (Khaund and Joshi 2013).

5.2.5 *Mizoram*

Mizoram is adjacent to Assam and Manipur in the North-Northeast, Tripura Bangladesh in the West, followed by Myanmar in the Southeast. It encompasses undulating topography bounded by Chin hill and Chittagong hill in the east and west, respectively. It lies between the latitude 21°58′ and 24°35′ North and longitude 92°15′ and 93°29′ East (Kar et al. 2013). The state holds 10.36 lakhs of tribal communities, which represents 1% of the total tribal population of the state. The major proportion of the tribal community is represented by *Chakma, Dimasa, Garo, Hajong, Hmar, Jaintia, Khasi, Kuki, Lakher, Man (Tai-speaking), Mizo, Mikir, Naga, Paite, Pawi*, and *Synteng* (Lalramnghinglova 2003; Census 2011). In the Aizawl district, Hazarika and Nautiyal (2012) documented 60 wild edible fruits representing 35 families and 50 different genera among the *Mizo* tribe. The several wild edible fruits were not only a rich source of nutrition but also used as ethnomedicines. The several important wild edible fruits found in the state are *Ardisia macrocarpa, Averrhoa carambola, Bruinsmia polysperma, Choerospondias axillaris, Ficus* species, *Rhus semialata, Sapindus mukorossi*, and *Syzygium grandis*. Incorporation of wild edible flowers in the daily diet by women among the ethnic communities is an age-old practice in the state. Several edible flowers are used for the preparation of traditional food items. A total of 59 edible flowers have been discovered in the districts of Aizawl and Champhai. The highest numbers of edible flowers are reported among the Lamiaceae, Apiaceae, followed by Leguminosae. Some of the distinctive wild edible flowers used in the state are *Livistoni chinensis Aeschynanthus maculata, Trevesia palmata, Begonia longifolia, Mollugo pentaphylla, Viburnum* species, *Osbeckia stellate,* and *Vaccinium sprengelii* (Khomdram et al. 2019). Mizoram represents a rich diversity of wild edible mushrooms. A total of 27 wild edible mushroom species were reported. The prominent species used in traditional recipes are *Auricularia auricula, Lentinus polychrous, L. tigrinis, Termitomyces heimi*, and *Volvariella volvacea* (Lalrinawmi et al. 2017).

5.2.6 Nagaland

Nagaland is sandwiched between Arunachal Pradesh, Assam, Manipur, and Myanmar in the North, West, South, and East, respectively. It is positioned between latitude 25°06′–27°04′ North and longitude 93°20′–95°15′ East covering a geographical area of 16,579 sq. km. A total of 17.11 lakhs tribal population constituting 1.6% of total tribal inhabitants in the nation resides in the state. The distinctive ethnic communities residing in the state are *Naga, Kuki, Kachari, Mikir*, and *Garo* including several subtribes viz. *Angami, Ao, Lotha, Sumi, Sangtam, Chang, Khiamniungan, Konyak,* etc. (Census 2011; Hazarika and Pongener 2018). Wild edible fruits form a constituent part of daily nutrition among local inhabitants of the state (Das et al. 2016). Khruomo and Deb (2018) enumerated 47 wild edible fruit species represented by 39 genera and 29 families in the three districts of Nagaland viz. Kohima, Phek, and Tuensang. The prominent wild edible fruits used by the ethnic communities in the state are *Castanopsis indica, Choerospondias axillaris, Docynia indica, Elaeocarpus floribundus, Emblica officinalis, Ficus semicordata, Hodgsonia macrocarpa, Myrica esculenta, Rhus semialata, Spondias pinnata* and *Terminalia chebula*. Several wild fruits viz. *Docynia indica, Musa babisiana*, and *Phyllanthus emblica* are used in the preparation of local drinks by the ethnic communities. The seeds of *Castanopsis indica* and *Spondias pinnata* are chewed after drying (Chase and Singh 2016). The consumption of fermented nuts of *Hadgsonia heteroclite* by the tribal inhabitants is a common practice in the state (Semwal et al. 2015). Wild edible mushroom shapes a characteristic source of nourishment among the primitive ethnic communities within the state. Toshinungla et al. (2016) chronicled 33 species of wild edible mushrooms represented by 28 genera and 19 families in the state. The rampant species consumed by the local inhabitants are *Auricularia auricula-judae, Lactarius edodes, L. volemus, Lentinus edodes, L. squarrosulus Pleurotus pulmonarius, Schizophyllum commune, Termitomyces eurhizus*, and *T. heimi*. *Gynura cusimbua*, a popular seasonal leafy vegetable, is extensively consumed in mid-day meal in combination with other leafy vegetables (Nakro 2011).

5.2.7 Sikkim

Sikkim is positioned between 27°4′46″–28°7′48″ north and 88°58″–88°55′25″ east longitude and covers a 7096 sq. km area. The state shapes the whole upper bowl of the Tista Waterway. A total of 2.06 lakhs tribal population constituting 0.2% of total tribal inhabitants in the nation, resides in the state. The chief tribal communities are

Bhutias, Lepchas, Limbus, and *Nepalese* (Census 2011). The state harbors a rich diversity of WEPs viz. 190 species (Sundriyal and Sundriyal 2001). A pragmatic study conducted in the western part of the state chronicled 124 wild edibles represented by 61 families in 100 genera. Among partitioning of edible plant parts, 38% was represented by fruits, 14% by seeds, 30% by young shoots and leaves, 7% by tubers, 4% by flowers, 2.5% by nuts, and 2% by bark among the several WEPs (Mahendra et al. 2017). Suresh et al. (2014) reported 21 wild edible fruit species belonging to 19 genera and 15 families. The important wild fruits used by the ethnic communities in their daily nutrition are *Baccaurea sapida, Calamus flagellum, Diploknema butyracea, Elaeocarpus sikkimensis, Eleagnus latifolia, Eriolobus indica, Ficus roxburghii, Machilus edulis, Rubus ellipticus,* and *Spondias axillaris.* The fruits of *Castanopsis hystrix* and *Ficus* species are frequently consumed by the ethnic communities after drying. However, the high revenue-generating fat is extracted from the fruits of *Diploknema butyraceae* and *Machilus edulis.* The fresh fruits of *Spandias axillaris* and *Choerospondias axillaris* are profusely used in pickle making by ethnic communities (Sundriyal and Sundriyal 2004). The leaves of the wild edible plant *Cardamine macrophylla* (Locally known as goyang) are used by local inhabitants of the *Sherpa* tribe for the preparation of a *Ziang dui*, a fermented food extract (Tamang and Tamang 2009). The local inhabitants belonging to the *Lachungpa* community specifically use the seed of *Chenoposium album* to enhance the flavor of locally made alcoholic brew using finger millets viz. *Chhyang.* The leaves of *Nasturtium officinale,* locally known as *Simrayo,* are often consumed by several ethnic communities in the state. The consumption of *Urtica dioca* (*Sisnoo*) leaves and flowering spikes in several local food delicacies is a traditional custom among the *Bhutias.* The *Sisnoo* is profusely used during cultural ceremonies, such as marriages, etc., by the *Bhutias* and *Nepalese* (Pradhan and Tamang 2015). Ethnic communities play an important role in utilizing wild mushroom species for their livelihood. The several important wild edible mushrooms recorded in the state viz. *Agaricus* species, *Amantia* species, *Auricularia polytricha, Boletus* species, *Catharellus* species, *Clavulina* species, and *Termitomyces* species are frequently collected by the *Karbis* (Borah et al. 2018).

5.2.8 Tripura

Tripura is a mountainous landlocked state located at the southwestern tip of northeastern India. It is placed between 22°56′ to 24°32′ N latitude and 91°10′ and 92°21′ E longitude, covering a land area of 10, 491 sq. km, 70% of which is covered with hills (Biswas et al. 2018). About 11.67 lakhs of tribals exist in the state, which constitutes 1.1% of the total tribal population of India. The state embraces immense heterogeneity among the ethnic communities; more than 19 tribes are discovered, of which people speaking *Kokborok* form the key cluster. Several other communities residing in the state are *Bengshel, Bhil, Bhutia, Chaimal, Chakma, Dub, Garoo, Halam, Jamatia, Kaipeng, Kalai, Karbong, Kaur, Khasia, Kuki, Lengui, Lepcha,*

Lushai, Mag, Munda, Murashing, Mussum, Noatia, Orang, Riang, Rupini, Santal, Sukuchep, Thangchep, Tippera, Tripura, Tripuri, and *Uchai* (Census 2011; Chanda et al. 2018). The WEPs are extensively used as a vegetable by the ethnic communities, namely *Tripuri* in the Agartala and Khowai district. The commonly used wild plants are *Alocasia odora, Canavalia gladiata, Dioscorea hamiltonii, Ipomoea aquatica, Lasia spinosa, Monochoria hastata, Solanum indicum,* and *Solanum torvum* (Choudhury et al. 2010). *Diplazium esculentum*, a wild edible fern, is customarily utilized in preparation of a local cuisine known as *Palai Sak* in the state (Yumkham et al. 2016). The numerous traditional cuisines prepared by ethnic communities *Malsom, Rupini*, and *Tripuri* using WEPs are *Awandru, Berma bwtwi, Chakhwi, Gudak, Moisdeng, Mwiborok, Ser*, and *Thokmwi* (Deb et al. 2013).

5.3 Wild Edible Plants: Alternate Source of Nutrition and Food Security

Around 2 billion individuals endure micronutrient insufficiencies all around the world, which is a critical deterrent to financial development (Thompson and Amoroso 2011). India has a large population and is home to the largest number of food-insecure people. According to the reports, the food insecurity level has risen in many parts of the world (FAO et al. 2020). Approximately 190 million Indians are malnourished. India is positioned 102 of 117 nations within the Global Hunger Index (GHI 2019). In developing countries, malnutrition is a preeminent comprehensive issue. In India, malnutrition is a key issue among children under the age of 5; nearly 43% of them are underweight (IFPRI 2016). Nearly 90% of the world's dietary intake comes from 30 domesticated food crops, but the consumption of WEPs for nutritional and food security is an enduring custom in developing countries (Hammer et al. 2003; Shelef et al. 2017). The indigenous ethnic communities on an average use 120 wild plants for their sustenance in both developing and developed countries (Bharucha and Pretty 2010). Northeast India harbors a rich diversity of wild edibles, which has immense potential to anchorage the nutritional requirements of the whole country. These contribute significantly to the dietary pattern of the tribal population, both as famine and healthy nutritious food. Many reports have revealed that these neglected crops are known to have good amounts of nutrients like the other available commercial crops (Fernandes et al. 2016; Hunter et al. 2019). They are also the sole custodians of the rich and diverse knowledge of plant uses and traditional food systems, and to local food biodiversity existing within the ecosystems they inhabit (Kuhnlein 2017). WEPs include several edible fruits, leaves, roots, stems, flowers, nuts, mushrooms, and ferns (Turner et al. 2011).

5.3.1 Wild Edible Fruits

The edible fruits collected from the wild play a crucial role in the lives of ethnic communities. The fruits from different tree species are collected throughout the year. It has been reported that wild edible fruits are wealthy sources of fibers, minerals (both macro and micro), vitamins, polyphenols, and antioxidants (Biswas et al. 2018). Several researchers have found higher nutrients among the fruit trees growing in the wild viz. *Cornus capitata* (higher phosphorus content) and *Phoebe lanceolata* (higher potassium content) are in contrast to commercial fruit crops (Saha et al. 2014). The wild consumable fruits encompass a critical spot in the life and customary acts of the native individuals living in Northeast India to fulfil nutritional requirements (Table 5.1).

The wild edible fruits also contribute towards income generation. Several wild edible fruit species are sold in the local markets (Fig. 5.1) by ethnic communities ranging from 10 to 200 Rs. per kilogram in the different states of Northeast India (Kar et al. 2008; Pfoze et al. 2012).

Table 5.1 Nutritive value of wild edible fruits in Northeastern states India

Scientific name	Nutritive value per 100 g	Reference
Baccaurea sapida	236.50 g	Seal (2011)
Eleagnus latifolia	332.10 g	Seal (2011)
Gomphogyne cissiformis	361.36 g	Seal et al. (2014)
Gymnopetalum cochinchinense	343.01 g	Seal et al. (2014)
Morus alba	374.10 g	Seal (2011)
Parkia roxburghii	380.93 g	Seal (2011)
Terminalia bellerica	381.87 g	Seal (2011)
Terminalia chebula	362.50 g	Seal (2011)
Zanthozylum armatum	374.42 g	Seal et al. (2014)

Fig. 5.1 Tribal women selling wild leafy vegetables in the local market

5.3.2 Wild Edible Leafy Vegetables

Wild edible leafy vegetables are found abundantly in the forest and agricultural lands during the rainy season. These leafy vegetables are mostly obtained from herbaceous plants, and leaves from some trees are also edible. The consumption of these wild leafy vegetables is an integral part of daily food intake throughout the world. Wild leafy vegetables are often consumed after boiling or cooked to prepare several local delicacies (Chauhan et al. 2014). Wild vegetables comprehend high fiber, protein, and several micronutrients viz. copper, manganese, iron, and zinc. The leafy vegetables are also considered a rich source of natural antioxidants consumed by tribal to meet their daily requirement (Wada and Ou 2002; Saikia and Deka 2013; Ebert 2014). Wild leafy vegetables contain plentiful secondary metabolites, which impart specific therapeutic properties such as antibacterial and anti-cancerous (Kumar et al. 2013). They are often considered a rich source of natural antioxidants (Salam et al. 2011). Wild leafy vegetables provide a rich source of nourishment (Table 5.2) and increase the diversity of food. Apart from being nutritious these leafy vegetables also complement the income for the ethnic communities in the local market and also contribute to overcoming the famine periods (Maxwell and Caldwell 2008; McGarry and Shackleton 2009).

The income generated through selling wild leafy vegetables in local markets (Fig. 5.2) by the ethnic community ranges from 20 to 240 Rs per kilogram depending upon the season and popularity of the vegetable (Salam et al. 2012; Bhutia et al. 2021).

Table 5.2 Nutritive value of wild edible leafy vegetables in Northeastern states of India

Scientific name	Nutritive value per 100 g	Reference
Amaranthus viridis	368.20 g	Pradhan et al. (2015)
Artimisia vulgaris	100.16 g	Seal et al. (2020)
Basella alba	402.66 g	Saha et al. (2015)
Cardamine hirsuta	112.38 g	Seal et al. (2020)
Chenopodium album	162.32 g	Chandra et al. (2016)
Emblica floribunda	139.55 g	Seal et al. (2020)
Eryngium foetidium	167.82 g	Gogoi et al. (2014)
Gnetum gnenom	168.44 g	Gogoi et al. (2014)
Leucas plukenetti	215.48 g	Chandra et al. (2016)
Lycianthus pachypetala	164.32 g	Gogoi et al. (2014)
Moringa oleifera	293.66 g	Saha et al. (2015)
Nasturtium officinale	348.40 g	Pradhan et al. (2015)
Phytolacca acinosa	325.83 g	Seal et al. (2017)
Pouzolzia hirta	330.51 g	Seal et al. (2017)
Rhus semialata	166.28 g	Gogoi et al. (2014)
Talinum triangulare	169.90 g	Chandra et al. (2016)
Utica dioca	353.40 g	Pradhan et al. (2015)

Fig. 5.2 Tribal women selling wild edible fruits and pods in the local market

5.3.3 *Wild Edible Ferns*

The fern alludes to green vascular cryptograms, which reproduce through spores. A high rainfall rate in North East India promotes the growth of wild edible ferns. The whole of northeast India comprehends 23 species of wild edible ferns representing 18 families and 19 genera (Yumkham et al. 2016). Wild ferns used in several cuisines are a rich source of a variety of nutrients viz. copper, iron, magnesium, sulfur, and zinc including high protein, crude fiber, and vitamin C (Chettri et al. 2018). The nutritive value of frequently consumed wild edible fern viz. *Diplazium esculatum* varies from 195 to 376.70 g per 100 g in the different parts of Northeast India (Pradhan et al. 2015; Saha et al. 2015). The wild edible fern is also an imperative component of the rural economy among ethnic communities. Fresh ferns are sold at the rate of 3–30 Rs per bundle in various local markets (Salam et al. 2012; Bhutia et al. 2021).

5.3.4 *Wild Edible Mushroom*

The numerous species of mushrooms growing in the wild are a promising source of food among the local inhabitants for their nourishment. More than 2000 species of wild edible mushrooms are found throughout the world (Rai et al. 2005). Wild mushrooms are consumed as a vegetable and are considered a rich source of fiber, phenols, protein vitamin D2, macronutrients (magnesium and potassium), micronutrients as well as natural antioxidants (Kalac 2013; Yildiz et al. 2014; Valverde et al. 2015). In developing countries, FAO and United Nations have recognized the necessity of incorporating wild edible mushrooms for human nourishment as a sustainable practice to meet the growing nutritional demands and food insecurity (Toshinungla et al. 2016). Murugkar and Subbulakshmi (2005) conducted

Table 5.3 Nutritive value of wild edible mushrooms in northeastern states of India

Scientific name	Nutritive value per 100 g	Reference
Lentinus squarrosulus	307.19 g	Das et al. (2017)
Lentinus tuber-regium	291.04 g	Das et al. (2017)
Macrocybe gigantea	336.81 g	Das et al. (2017)

an observational study in Manipur to look at the dietary status of wild-eatable mushrooms. They found that wild mushrooms contained protein (6.12 g per 250 g), fat (0.712 g per 250 g), several nutrients viz. calcium (287 mg per 250 g), iron (9.3 mg per 250 g), sodium (0.077 mg per 250 g), and zinc (3.72 mg per 250 g). The wild edible mushrooms are an important source of nutrition and livelihood among the imperative ethnic communities residing in the northeast region (Table 5.3).

The rural income generated by selling the wild edible species of mushroom ranges from 120 to 1200 Rs per kilogram depending upon the type of species in different regional markets of northeast India (Bhaben et al. 2011; Khaund and Joshi 2013; Das et al. 2017; Swargiari and Buragohain 2018).

5.3.5 Wild Edible Flowers

The inclusion of wild edible flowers in the human diet has been a common practice since the commencement of human civilization. The use of flowers in traditional food has been relished in Asian cuisine for centuries (Kelley et al. 2002). They are also used for garnishing, decorating food items, cocktails, canned in sugar, preserved in distilleries, tea, pickles, liquors, dyes, stir-fried dishes, etc. (Rop et al. 2012; Gupta et al. 2018). Wild edible flowers are considered a natural source of antioxidants and are known to have biochemical properties, viz. antibacterial, antifungal, antimicrobial, anti-cancerous, anti-inflammatory, etc., are also reported (Kumari and Bhargava 2021). During famine periods, the wild edible flowers extend the diversity of edible food resources and also secure food insecurity. Many flowers are known to be edible; however, proper identification and documentation are still lacking, and this is essential to know their utility (Rop et al. 2012; Gupta et al. 2018). Several wild edible flowers are sold in the local markets of the northeast region (Table 5.4) to generate an economy. The income produced through the marketing of wild edible flowers ranges from 10 to 50 Rs per bundle (Khomdram et al. 2019). The use of edible flowers for culinary purposes has regained its popularity in recent days owing to the changing lifestyle of people all over the world.

Table 5.4 Wild edible flowers sold in the local markets of northeastern states of India

Scientific name	Family	Vernacular name	Uses	Reference
Aeschynanthus maculata Lindl.	Gesneraceae	Bawlte-hlan-tai (Mizo)	Flowers eaten with meat	Kar et al. (2013)
Allium ascalonicum L.	Liliaceae	Meiti Tilhou (Mn)	Fried inflorescence is taken for 'Kangou' (a traditional Manipuri recipe)	Khomdram et al. (2019); Singh and Mathew (2016)
Allium hookeri Thw.	Liliaceae	Maroi Napakpi (Mn)	Fried inflorescence is taken for 'Kangou' (a traditional Manipuri recipe), as a salad dressing	Khomdram et al. (2019); Singh and Mathew (2016)
Adhatoda vasica Nees.	Acanthaceae	Bahaka (As; M)	Flowers are eaten fried. The plant has medicinal value	Nath (2015)
Alpinia galangal (L.) Swartz	Zingiberaceae	Kanghoo (Mn)	Young inflorescence is taken raw or boiled	Devi and Salam (2016)
Ammomum aromaticum Roxb.	Zingiberaceae	Papia (A), Khuk, Aairi (K)	Flowers eaten cooked and as chutney	Das et al. (2019)
Amorphophallus bulbilfer (Roxb.) Bl.	Araceae	Batema (K), Batet (M)	Used as a vegetable	Deb et al. (2013)
Bauhinia variegata L.	Caesalpinaceae	Bah (As), Koirala (Sk), Chingthrao Angaoba (Mn), Vaube (Mizo)	Flowers used as vegetable and as a chutney	Kar et al. (2013); Sundriyal and Sundriyal (2001)
Berberis aristata DC.	Berberidaceae	Kanchan (TM)	Flowers eaten raw	Tsering et al. (2017)
Berberis angulosa Wall.	Berberidaceae	Lae-kanchan (TM)	Flowers eaten raw	Tsering et al. (2017)
Berberis asiatica Roxb. Ex. DC.	Berberidaceae	Kanchan (TM)	Flowers eaten raw	Tsering et al. (2017)
Bombax ceiba L.	Bombaceae	Semi Phul (C), Semolu (As), Tera (Mn)	Vegetable	Deka and Nath (2014)
Buddleja asiatica Lour.	Buddlejaceae	Dieng-tuti-mynneng (Khasi)	Flowers are eaten as vegetable	
Callicarpa arborca Roxb.	Verbenaceae	Hnah-kiah (Mizo)	Flowers fried with meat	Kar et al. (2013)
Cassia fistula Linn.	Caesalpinaceae	Phung-ril (Mizo)	Flowers are eaten as fried vegetable	Kar et al. (2013)
Carica papaya L.	Caricaceae	Amita (As)	Flowers eaten cooked	Deka and Nath (2014);

(continued)

Table 5.4 (continued)

Scientific name	Family	Vernacular name	Uses	Reference
				Kar and Borthakur (2007)
Cirsium verutum (D. Don) Spreng.	Asteraceae	Brongzom (TM)	Inflorescence edible, medicinal	Tsering et al. (2017)
Curcuma aromatica Salisb.	Zingiberaceae	Jangali Haladi (A)	Vegetable	Angami et al. (2006)
Curcuma angustifolia Roxb.	Zingiberaceae	Yaipal (Mn)	Flowers used as vegetable	Singh and Mathew (2016); Khan et al. (2015); Konsam et al. (2016)
Clerodendrum serratum (L.) Moon	Verbenaceae	Moirang Khanum (Mn), Rilong-phlang (Khasi)	Raw flowers eaten as chutney and as fried vegetable	Khan et al. (2015); Konsam et al. (2016)
Crotolaria juncea L.	Papilionaceae	Tum-thang (Mizo)	Flower cooked with meat and fish	Kar et al. (2013)
Cymbidium hookerianum Rchb. f.	Orchidaceae	–	Flowers cooked as vegetable	Das et al. (2019)
Dillenia pentagyna Roxb.	Dilleniaceae	Thing-se-lei (Mizo), Dieng-soh-bar (Khasi), Oxi (As)	Flowers are eaten fried	Sawian et al. (2007); Nath (2015)
Dracaena spicata Roxb.	Dracaenaceae	Sa thang-dai (Mizo)	Flowers are cooked with meat	Kar et al. (2013)
Erythrina stricta Roxb.	Fabaceae	Chemroy (TM)	Vegetable	Angami et al. (2006)
Gentiana kurroo Royle	Gentiaceae	Panginmento (TM)	Flowers cooked as vegetable	Tsering et al. (2017)
Girardinia diversifolia (Link) Friis	Urticaceae	Patle sisnu (Sk)	Consumed as vegetable, used to make pickles	Sundriyal and Sundriyal (2001); Mahendra et al. (2017)
Gmelina arborea Roxb.	Lamiaceae	Gamare (As), Thlam-vawng (Mizo)	Flowers are eaten fried or boiled	Kar et al. (2013)
Laportia terminalis Wight	Urticaceae	Bhangre sisnu (Sk)	Consumed as vegetable, used to make pickles	Sundriyal and Sundriyal (2001); Mahendra et al. (2017)

(continued)

Table 5.4 (continued)

Scientific name	Family	Vernacular name	Uses	Reference
Meconopsis paniculata (D. Don) Prain	Papaveraceae	–	Flowers cooked as vegetable	Das et al. (2017)
Moringa pterygosperma Gaetern	Moingaceae	Sajina (As; M), Diengtoh (Khasi)	Flowers cooked as vegetable	Sawian et al. (2007)
Monochoria hastata L.	Pontederiaceae	Jonakiful (As), Ooren (Mn)	Flower bud is eaten cooked as vegetable or fried	Deka and Nath (2014); Nath (2015)
Musa balbisiana Colla.	Musaceae	Kopak/Colon (A), Wegoin (Ak), Cha Ngok (T)	Vegetable, salad	Angami et al. (2006)
Musa aurantiaca G. Mann ex Baker	Musaceae	Nyaprom (T)	Flowers cooked as vegetable	Angami et al. (2006)
Musa acuminate Colla	Musaceae	Thailik Bwfang (K),	Flowers cooked as vegetable	Deb et al. (2013); Chanda et al. (2018)
Nyctanthes arbour-tristis L.	Oleaceae	Sewali phul (As)	Flowers are used to enhance appetite	Kar and Borthakur (2007)
Oenanthe javanica (Blume) DC.	Apiaceae	Komprek (Mn)	Inflorescence used in chutneys and salad	Konsam et al. (2016); Khomdram et al. (2019)
Oroxylum indicum (L.) Kunz	Bignoniaceae	Salsali (TW), Totola (Sk), Dingori (As)	Flowers cooked as vegetable	Deka and Nath (2014); Nath (2015)
Phlogacanthus tubiflorus (Buch.-Ham ex Wall) Nees	Acanthaceae	Basak patta (C), Nongmangkha Anganba (Mn)	Vegetable, medicinal	Angami et al. (2006); Khomdram et al. (2019)
Phlogacanthus thyrsiformis (Roxb.) Nees	Acanthaceae	Phikhip (T), Ronga bahka tita (As), Nongmangkha (Mn), Allot (Garo)	Flowers cooked as vegetable, as medicine	Sawian et al. (2007); Singh and Mathew (2016); Momin et al. (2016)
Piper mullesua Buch.-Ham.ex D. Don	Piperaceae	Chabo (Sk)	Flowers added to vegetables for flavoring	Mahendra et al. (2017)
Rhododendron arboreum Smith	Ericaeae	Samu (W), Lali guran (Sk), Ching Leihao (Mn)	Flowers eaten raw or cooked, used to prepare local wine	Thokchom et al. (2016); Mahendra et al. (2017)

(continued)

Table 5.4 (continued)

Scientific name	Family	Vernacular name	Uses	Reference
Rhododendron keysii Nutt.	Ericaeae	Taam (TM)	Flowers eaten raw	Tsering et al. (2017)
Sesbania grandiflora (L.) Poir.	Fabaceae	Bogbibar (As)	Flowers are eaten as vegetable either fried or with grinded rice or gram	Kar and Borthakur (2007); Deka and Nath (2014)
Spilanthes acmella Murr.	Asteraceae	Maresa (I), Maanja-lei (Mn)	Vegetable, Medicinal	Thokchom et al. (2016); Chanda et al. (2018)
Taxus wallichiana Zucc.	Taxaceae	Tae-sheng (T)	Flowers eaten raw	Das et al. (2019)
Tupistra nutans Wall.ex Lindl.	Asperagaceae	Nakima (Sk)	Inflorescence is eaten as vegetable	Sundriyal and Sundriyal (2001)

Abbreviation used: *A* Adi, *As* Assamese, *C* Chakma, *I* Idu-Mishmi, *M* Mishing, *N* Nyishi, *S* Sinpho, *T* Tangsa, *TM* Tawang Monpa, *W* West Kameng Monpa, *Sk* Sikkimese, *K* Kokborok (Tripuri and Rupini), *MN* Manipuri

5.4 Conclusion and Future Perspectives

Wild edible resources play a significant role in nutrition, food security, and economical generation among the rural tribal population residing in Northeast India. The area is one of the richest reservoirs of genetic resources and wild relatives of several domesticated crops. The traditional knowledge among the diverse ethnic groups forms the multi-ethnic knowledge base. Ethnic communities play an indispensable role in conserving, maintaining, and utilizing these traditional bioresources. With the growing concern of food security, these plants can be introduced in the national food and nutrition security and sovereignty strategies that focus on nutrient adequacy rather than quantity of staples while being culturally acceptable. At the local level, several institutional mechanisms can be planned for sustainable harvesting and utilization of these valuable resources to boost the economy of tribals. Setting up marketing channels for selling these WEPs to generate a higher revenue can prove to be an economic strategy. Special camps should be organized on a monthly basis to popularize the consumption of wild edible resources among the community. Considering the value of the WEPs, the cultivation of species can be disseminated amongst the growers for larger-scale cultivation and marketing which will definitely benefit the indigenous communities at large. Further, these plants may also have substantial potential for the development of new crops through domestication and provide a genetic resource pool for hybridization and selection. Domestication of such WEPs should be encouraged with proper conservative measures, sustainable utilization, and harvesting of resources to preserve the local gene pool.

Many underutilized plant species possess multiple uses, including their edible nature. There is still a need to identify such plant species to work out suitable measures to identify their antioxidant perspective, phytochemical analysis for essential components, etc. More research should be conducted to fully utilize wild edible resources, which will ultimately increase their acceptability in the community. However, with the advent of modernization, a long gap has been created between the tribal's and urban communities. Over time the indigenous knowledge regarding these bioresources has also declined. The changing environmental conditions and growing population have led to several health problems and food insecurity. Thus, to meet the needs of the growing population, it is important to diversify food resources to eradicate nutrition insufficiency and food insecurity.

References

Angami A, Gajurel PR, Rethy P, Singh B, Kalita SK (2006) Status and potential of wild edible plants of Arunachal Pradesh. Indian J Tradit Knowl 5(4):541–550

Apshahana K, Sharma A (2018) Wild edible mushrooms traded by ethnic communities of Senapati and Kangpokpi district of Manipur, India. J Pharmacogn Phytochem 7(1):2303–2306

Baruah U, Bhattacharyya R (2019) Diverse ethnic food practices of the Galo tribe in Arunachal Pradesh. Food Sci 10(2):197–202

Beluhan S, Ranogajec A (2010) Chemical composition and non-volatile components of Croatian wild edible mushrooms. Food Chem 124:1076–1082

Bhaben T, Lisha G, Chandra SG (2011) Wild edible fungal resources used by ethnic tribes of Nagaland, India. Indian J Tradit Knowl 10(3):512–515

Bharucha Z, Pretty J (2010) The roles and values of wild foods in agricultural systems. Philos Trans R Soc Lond Ser B Biol Sci 365:2913–2926

Bhatt KC, Pandey A, Dhariwal OP, Panwar NS, Bhandari DC (2009) "Tum-thang" (*Crotalaria tetragona* Roxb. ex Andr.): a little known wild edible species in the north-eastern hill region of India. Genet Resour Crop Evol 56(5):729–733

Bhutia R, Rai YK, Rai LK (2021) Wild edible plant used by ethnic peoples on the boundary of subtropical & temperate natural forest, Mamlay watershed, South Sikkim. Nat Prod Radiance 11(4):350–357

Biswas SC, Majumdar M, Das S, Misra TK (2018) Diversity of wild edible minor fruits used by the ethnic communities of Tripura, India. Indian J Tradit Knowl 17(2):282–289

Borah N, Semwal RL, Garkoti SC (2018) Ethnomycological knowledge of three indigenous communities of Assam, India. Indian J Tradit Knowl 17(2):327–335

Census (2011) Census of India 2011. Provisional population totals. Government of India, New Delhi

Chanda M, Pal I, Majumder S, Datta P (2018) Assessment of traditionally used wild edible plants and its impact on health and livelihood among the ethnic communities of Tripura. Asian J Agric Hortic Res 2(3):1–13

Chandra KS, Gogoi D, Gautam KH, Handique AK (2016) Nutritive values of some non-conventional leafy vegetables and scarcity food plants of north east India. Afr J Food Sci 10(11):340–343

Chase P, Singh OP (2016) Bioresources of Nagaland: a case of wild edible fruits in Khonoma Village Forest. In: Purkayastha J (ed) Bioprospecting of indigenous bioresources of North-East India. Springer, Singapore, pp 45–60

Chaudhury G, Basumatari M, Darji C, Ahmed AF, Borah D, Sah RK, Devi A, Hazarika N, Dutta G (2021) Economic significance of wild bioresources to rural communities in the Eastern Himalayan state of Assam, Northeast India. Trees Forests People 5:100102

Chauhan D, Shrivastava AK, Patra S (2014) Diversity of leafy vegetables used by tribal peoples of Chhattisgarh, India. Int J Curr Microbiol Appl Sci 3(4):611–622

Chettri S, Manivannan S, Muddarsu VR (2018) Nutrient and elemental composition of wild edible ferns of the Himalaya. Am Fern J 108(3):95–106

Choudhury R, Choudhury MD, De B, Paul SB (2010) Importance of certain tribal edible plants of Tripura. Indian J Tradit Knowl 9(2):300–302

Das G, Patra JK, Singdevsachan SK, Gouda S, Shin HS (2016) Diversity of traditional and fermented foods of the Seven Sister states of India and their nutritional and nutraceutical potential: a review. Front Life Sci 9(4):292–312

Das AR, Saha AK, Joshi SR, Das P (2017) Wild edible macrofungi consumed by ethnic tribes of Tripura in Northeast India with special reference to antibacterial activity of *Pleurotus djamor* (Rumph. ex Fr.) Boedijn. Int Food Res J 24(2):834–838

Das AK, Mylliemngap W, Laling N, Arya OP, Sundriyal RC (2019) Investigation of plants utilization by tribal communities of Arunachal Himalayas in India. In: Singh B (ed) Plants for human survival and medicine. New India Publishing Agency, New Delhi, pp 283–210

Deb D, Sarkar A, Debbarma B, Datta BK, Majumdar K (2013) Wild edible plants and their utilization in traditional recipes of Tripura, Northeast India. Adv Biol Res 7(5):203–211

Deka K, Nath N (2014) Documentation of edible flowers of western Assam. Am J Phytomed Clin Ther 2(10):1124–1140

Devi MR, Salam S (2016) Wild edible plants used by the Monsang Naga Tribe of Manipur, India. Pleione 10(1):90–96

Devi MR, Salam S, Dutta BK, Singh PK (2013) Biodiversity of medicinal plants used by the Monsang Naga Tribe of Manipur, India. In: Singh KK, Das KC, Lalrutsanga H (eds) Bioresources and traditional knowledge of Northeast India. Mizo Post Graduate Science Society, India, pp 299–302

Dhar S (2014) Socio-economic and demographic status of Assam: a comparative analysis of Assam with India. Population (crore). Int J Humanit Soc Sci Stud 1(3):108–117

Dutt BK, Dutt PK (2005) Potential of ethnobotanical studies in North East India: an overview. Indian J Tradit Knowl 4(1):7–14

Ebert AW (2014) Potential of underutilized traditional vegetables and legume crops to contribute to food and nutritional security, income and more sustainable production systems. Sustainability 6(1):319–335

FAO, IFAD, UNICEF, WFP, WHO (2020) The state of food security and nutrition in the world 2020. Transforming food systems for affordable healthy diets. FAO, Rome

Fernandes VR, Morales P, Ruiz-Rodriguez BM, Isasa ET (2016) Nutrients and bioactive compounds in wild fruits through different continents. In: Ferreira ICFR, Morales P, Barros L (eds) Wild plants, mushrooms and nuts. Wiley, Chichester, pp 263–314

GHI (2019) 2019 Global hunger index results by severity. Global hunger index—peer-reviewed annual publication designed to comprehensively measure and track hunger at the global, regional, and country levels. https://www.globalhungerindex.org/results.html

Gogoi D, Rajkhowa RC, Handique AK (2014) Nutritive values of some non-conventional leafy vegetables of ethnic sources from Nagaland, NE India. Int J Emerg Sci Eng 2(5):78–79

Gupta YC, Sharma P, Sharma G, Agnihotri R (2018) Edible flowers. In: Abstracts of national conference on floriculture for rural and urban prosperity in the scenario of climate change. CAE & PHT Central Agriculture University (CAU), Ranipool, Sikkim, 16–18 Feb 2018

Hammer K, Arrowsmith N, Gladis T (2003) Agrobiodiversity with emphasis on plant genetic resources. Naturwissenschaften 90:241–250

Hazarika TK, Nautiyal BP (2012) Studies on wild edible fruits of Mizoram, India used as ethnomedicine. Genet Resour Crop Evol 59(8):1767–1776

Hazarika TK, Pongener M (2018) Potential wild edible fruits of Nagaland, North-east India and its significance in the livelihood and nutritional security of rural, indigenous people. Genet Resour Crop Evol 65(1):199–215

Hunter D, Borelli T, Beltrame DMO, Oliveira CNS, Coradin L, Wasike VW, Wasilwa L, Mwai J, Manjella A, Samarasinghe GWL, Madhujith T, Nadeeshani HVH, Tan A, Ay ST, Guzelsoy N, Lauridsen N, Gee E, Tartanac F (2019) The potential of neglected and underutilized species for improving diets and nutrition. Planta 250:709–729

IFPRI (2016) International Food Policy Research Institute. Global nutrition from promise to impact: ending malnutrition by 2030, Washington, DC

Jeeva S (2009) Horticultural potential of wild edible fruits used by the Khasi tribes of Meghalaya. J Hortic For 1(9):182–192

Ju Y, Zhuo J, Liu B, Long C (2013) Eating from the wild: diversity of wild edible plants used by Tibetans in Shangri-la region, Yunnan, China. J Ethnobiol Ethnomed 9(1):1–22

Kalac P (2013) A review of chemical composition and nutritional value of wild-growing and cultivated mushrooms. J Sci Food Agric 93(2):209–218

Kar A, Borthakur SK (2007) Wild vegetables sold in local markets of Karbi Anglong, Assam. Indian J Tradit Knowl 6(1):169–172

Kar A, Borkataki S, Borthakur SK (2008) Wild edible fruits of the Karbi's of Karbi Anglong district of Assam, India. Pleione. 2(2):175–181

Kar A, Bora D, Borthakur SK, Goswami NK, Saharia D (2013) Wild edible plant resources used by the Mizos of Mizoram, India. KUSET 9(1):106–126

Kelley KM, Behe BK, Biernbaum JA, Poff KL (2002) Combinations of colors and species of containerized edible flowers: effect on consumer preferences. HortScience 37:218–221

Khan MR, Kikim A, Yadava PS (2015) Conservation of indigenous wild edible plants used by different communities of Kangchup Hills, Senapati, North East India. Int J Bio-resour Stress Manag 6(6):680–689

Khaund P, Joshi SR (2013) Wild edible macrofungal species consumed by the Khasi tribe of Meghalaya, India. Indian J Nat Prod Resour 4(2):194–204

Khomdram SD, Fanai L, Yumkham SD (2019) Local knowledge of edible flowers used in Mizoram. Indian J Tradit Knowl 18(4):714–723

Khongsai M, Saikia SP, Kayang H (2011) Ethnomedicinal plants used by different tribes of Arunachal Pradesh. Indian J Tradit Knowl 10(3):541–546

Khruomo N, Deb CR (2018) Indigenous Wild Edible Fruits: sustainable resources for food, medicine and income generation—a study from Nagaland, India. J Exp Biol Agric Sci 6(2): 405–413

Konsam S, Thongam B, Handique AK (2016) Assessment of wild leafy vegetables traditionally consumed by the ethnic communities of Manipur, northeast India. J Ethnobiol Ethnomed 12(1): 1–15

Kuhnlein HV (2017) Holding on to agrobiodiversity: human nutrition and health of indigenous peoples. In: Hunter D, Guarino L, Spillane C, McKeown PC (eds) Routledge handbook of agricultural biodiversity. Routledge, pp 388–400

Kumar S, Prasad AK, Iyer SV, Vaidya SK (2013) Systematic pharmacognostical, phytochemical and pharmacological review on an ethno medicinal plant, *Basella alba* L. J Pharmacogn Phytother 5(4):53–58

Kumari P, Bhargava B (2021) Phytochemicals from edible flowers: opening a new arena for healthy lifestyle. J Funct Foods 78:104375

Lalramnghinglova H (2003) Ethno-medicinal plants of Mizoram. Bishen Singh Mahendra Pal Singh, Dehra Dun

Lalrinawmi Z, Vabeikhokhei J, Zothanzama J (2017) Edible mushrooms of Mizoram. Sci Vis 17: 172–181

Mahendra T, Krishan P, Rai SK (2017) Traditional use of wild plants for food in West Sikkim, India. Int J Life Sci 5(4):730–741

Maxwell D, Caldwell R (2008) The Coping Strategies Index: a tool for rapid measurement of household food security and the impact of food aid programs in humanitarian emergencies. Field methods manual

McGarry DK, Shackleton CM (2009) Children navigating rural poverty: rural children's use of wild resources to counteract food insecurity in the Eastern Cape, South Africa. J Child Poverty 15(1): 19–37

Momin KC, Suresh CP, Momin BC, Singh YS, Singh SK (2016) An ethno-botanical study of wild plants in Garo Hills region of Meghalaya and their usage. Int J Minor Fruits Med Aromat Plants 2(1):47–53

Murugkar DA, Subbulakshmi GJFC (2005) Nutritional value of edible wild mushrooms collected from the Khasi hills of Meghalaya. Food Chem 89(4):599–603

Nakro V (2011) Traditional agricultural practices and sustainable livelihood, a thematic report. Department of Planning and Coordination, Government of Nagaland, Nagaland

Narzary D, Brahma BK, Goyal AK (2014) The flower and inflorescence of wild plants used as vegetables by the Bodo and allied tribes of Kokrajhar district, Assam, India. Adv Res Pharm Biol 4(1):618–621

Nath N (2015) Research article wild edible vegetables from Western Assam. Sch Acad J Biosci 3(12):1044–1050

Pfoze NL, Kumar Y, Sheikh N, Myrboh B (2012) Assessment of local dependency on selected wild edible plants and fruits from Senapati district, Manipur, Northeast India. Ethnobot Res Appl 10: 357–367

Phawa GM, Dkhar E, Marbaniang D (2019) Indigenous wild edible plants of Bataw Village, East Jaintia Hills District, Meghalaya. Int J Arts Sci Humanit 7(2):17–23

Pradhan S, Tamang JP (2015) Ethnobiology of wild leafy vegetables of Sikkim. Indian J Tradit Knowl 14(2):290–297

Pradhan S, Manivannan S, Tamang JP (2015) Proximate, mineral composition and antioxidant properties of some wild leafy vegetables. J Sci Ind Res 74:155–159

Rai M, Tidke G, Wasser SP (2005) Therapeutic potential of mushrooms. Nat Prod Radiance 4(4): 246–257

Rapoport EH, Drausal BS (2013) Edible plants. In: Encyclopedia of biodiversity, vol 3. Elsevier, pp 3127–3132

Rop O, Mlcek J, Jurikova T, Neugebauerova J, Vabkova J (2012) Edible flowers—a new promising source of mineral elements in human nutrition. Molecules 17(6):6672–6683

Saha D, Sundriyal M, Sundriyal RC (2014) Diversity of food composition and nutritive analysis of edible wild plants in a multi-ethnic tribal land, Northeast India: an important facet for food supply. Indian J Tradit Knowl 13(4):698–705

Saha J, Biswal AK, Deka SC (2015) Chemical composition of some underutilized green leafy vegetables of Sonitpur district of Assam, India. Int Food Res J 22(4):1466–1473

Saikia P, Deka DC (2013) Mineral content of some wild green leafy vegetables of North-East India. J Chem Pharm Res 5(3):117–121

Salam JS, Joylani SD, Rebika ND, Priyadarshini S (2011) Secondary metabolites, antioxidant status and nutritive composition of two non-conventional leafy vegetables-*Stellaria media* L. and *Chenopodium album* L. Indian J Agric Biochem 24(2):136–140

Salam S, Jamir NS, Singh PK (2012) Wild leafy vegetables sold in local markets of Ukhrul District of Manipur, India. Pleione 6(2):298–303

Sarma TC, Sarma I, Patiri BN (2010) Wild edible mushrooms used by some ethnic tribes of Western Assam. Bioscan 3:613–625

Satter MMA, Khan MMRL, Jabin SA, Abedin N, Islam MF, Shaha B (2016) Nutritional quality and safety aspects of wild vegetables consume in Bangladesh. Asian Pac J Trop Biomed 6(2): 125–131

Sawian JT, Jeeva S, Lyndem FG, Mishra BP, Laloo RC (2007) Wild edible plants of Meghalaya, North-east India. Nat Prod Radiance 6(5):410–426

Seal T (2011) Nutritional composition of wild edible fruits in Meghalaya state of India and their Ethno-botanical importance. J Bot 6(2):58–67
Seal T (2012) Evaluation of nutritional potential of wild edible plants, traditionally used by the tribal people of Meghalaya state in India. Am J Plant Nutr Fertil Technol 2(1):19–26
Seal T, Pillai B, Chaudhuri K (2014) Nutritional potential of wild edible fruits, traditionally used by the local people of Meghalaya state in India. Indian J Tradit Knowl 5(4):359–364
Seal T, Pillai B, Chaudhuri K (2017) Evaluation of nutritional potential of five unexplored wild edible plants consumed by the tribal people of Arunachal Pradesh State in India. J Food Nutr Res 5(1):1–5
Seal T, Chaudhuri K, Pillai B, Chakrabarti S, Auddy B, Mondal T (2020) Wild-edible plants of Meghalaya State in India: nutritional, minerals, antinutritional, vitamin content and toxicity studies. Pharmacogn Mag 16(68):142–151
Semwal DP, Bhatt KC, Bhandari DC, Panwar NS (2015) A note on distribution, ethnobotany and economic potential of *Hodgsonia heteroclita* (Roxb.) Hook. f. & Thoms. in North-eastern India. Nat Prod Radiance 5(1):88–91
Shelef O, Weisberg PJ, Provenza FD (2017) The value of native plants and local production in an era of global agriculture. Front Plant Sci 8:2069
Singh D, Mathew B (2016) Wild edible plants used by Meitei Community of Eastern Himalayas, India. Int J Agric Sci 8(53):2699–2702
Singh R, Bhagawati R, Ngachan SV (2017) Ethno-mycological studies and biodiversity of edible and medicinal mushrooms in West-Siang District of Arunachal Pradesh, India. Ecol Environ 35(1):102–105
Singha WR, Kurmi B, Sahoo UK, Sileshi GW, Nath AJ, Das AK (2021) *Parkia roxburghii*, an underutilized tree bean for food, nutritional and regional climate security. Trees Forests People 4:100065
Sinha SC (2001) Medicinal plants of Manipur. Mass Publication, Imphal
Sundriyal M, Sundriyal DC (2001) Wild edible plants of the Sikkim Himalaya: nutritive values of selected species. Econ Bot 55(3):377–390
Sundriyal M, Sundriyal RC (2004) Structure, phenology, fruit yield, and future prospects of some prominent wild edible plant species of the Sikkim Himalaya, India. J Ethnobiol 24(1):113–138
Suresh CP, Bhutia KD, Shukla G, Pradhan K, Chakravarty S (2014) Wild edible tree fruits of Sikkim Himalayas. J Tree Sci 33(1):43–48
Swargiari BN, Buragohain R (2018) Wild edible mushroom trade in the local markets by tribal people of Kokrajhar, Assam. NeBIO 9(1):109–116
Tag H, Tsering J, Hui PK, Gogoi BJ, Veer V (2014) Nutritional potential and traditional uses of high-altitude wild edible plants in Eastern Himalayas, India. WASET 8(3):395–400
Tamang B, Tamang JP (2009) Traditional knowledge of bio preservation of perishable vegetables and bamboo shoots in Northeast India as food resources. Indian J Tradit Knowl 8:89–95
Thokchom R, Hanglem A, Zimisai S, Thongam CA, Devi YR, Thokchom J, Singh SS (2016) Documentation and assessment of wild medicinal and edible flowers of valley districts of Manipur. Int J Res Appl 4(11):13–20
Thompson B, Amoroso L (2011) Combating micronutrient deficiencies: food-based approaches. FAO and CABI, Rome
Toshinungla AO, Deb CR, Neilazonuo K (2016) Wild edible mushrooms of Nagaland, India: a potential food resource. J Exp Biol Agric Sci 4(1):59–65
Tripathi RS, Roy A, Kushwaha D, Lalnunmawia F, Lalnundanga LH, Lalnunzira C, Roy PS (2016) Perspectives of forest biodiversity conservation in Northeast India. J Biodivers Bioprospect Dev 3(2):1–9
Tsering J, Gogoi BJ, Hui PK, Tam N, Tag H (2017) Ethnobotanical appraisal on wild edible plants used by the Monpa community of Arunachal Pradesh. Indian J Tradit Knowl 16(4):626–637
Turner NJ, Luczaj LJ, Migliorini P, Pieroni A, Dreon AL, Sacchetti LE, Paoletti MG (2011) Edible and tended wild plants, traditional ecological knowledge and agroecology. Crit Rev Plant Sci 30(1–2):198–225

Valverde ME, Hernandez-Perez T, Paredes-Lopez O (2015) Edible mushrooms: improving human health and promoting quality life. Int J Microbiol 2015:376387

Vavilov NI (1926) Studies in the origin of cultivated plants. Institute of Applied Botany and Plant Breeding, Leningrad

Wada L, Ou B (2002) Antioxidant activity and phenolic content of Oregon cranberries. J Agric Food Chem 50:3495–3500

Yesodharan K, Sujana KA (2007) Wild edible plants traditionally used by the tribes in the Parambikulam Wildlife Sanctuary, Kerala, India. Nat Prod Radiance 6(1):74–80

Yildiz O, Can Z, Laghari AQ, Şahin H, Malkoç M (2014) Wild edible mushrooms as a natural source of phenolics and antioxidants. J Food Biochem 39(2):148–154

Yumkham SD, Chakpram L, Salam S, Bhattacharya MK, Singh PK (2016) Edible ferns and fern–allies of North East India: a study on potential wild vegetables. Genet Resour Crop Evol 64(3): 467–477

Yumnam RS, Devi CO, Abujam SS, Chetia D (2012) Study on the ethnomedicinal system of Manipur. Int J Pharm Biol Arch 3(3):587–591

Chapter 6
Nutritional and Health Benefits of High Altitude Wild Food Plant, *Hippophae rhamnoides* for the Himalayan Communities

Stanzin Ladol, Ajay Kumar, and Deepak Sharma

Abstract *Hippophae rhamnoides* L. belongs to the family Elaeagnaceae and is commonly known as Seabuckthorn (Sbt). It is an essential, multipurpose plant that grows widely in Asia and Europe. It has been observed that this plant is consumed as a wild food plant by the Himalayan communities in India. Various parts of Sbt are consumed, and the traditional communities prepare various items from it. The berries are also dried for off-season use. Recent scientific studies have demonstrated its exceptional usage in therapy, which might be because of the high amounts of many bioactive compounds. Sbt berries, leaves, and bark are good sources of nutritional and medicinal compounds. The high levels of antioxidants, antibacterial, antiviral, anti-radiation, anti-inflammatory, cardioprotective, and neuroprotective potential make Sbt unique and valuable. *Hippophae rhamnoides* contain more carotenoids, vitamins, and minerals compared to other *Hippophae* species. Because of its multiple medicinal and nutritional properties, Sbt has become one of the most important superfoods. In this context, the chapter highlights the nutritional and health benefits of Sbt for the Himalayan communities.

Keywords *Hippophae rhamnoides* · Seabuckthorn · Antioxidants · Flavonoids · Polyphenols · Therapeutic foods

S. Ladol (✉)
Department of Zoology, Central University of Jammu, Jammu and Kashmir, India
e-mail: stanzin.zool@cujammu.ac.in

A. Kumar
Department of Plant Science, Central University of Kerala, Kasaragod, Kerala, India

D. Sharma
School of Life Sciences, Jawaharlal Nehru University, New Delhi, India

A. Kumar et al. (eds.), *Wild Food Plants for Zero Hunger and Resilient Agriculture*, Plant Life and Environment Dynamics,
https://doi.org/10.1007/978-981-19-6502-9_6

6.1 Introduction

Hippophae rhamnoides L. (Seabuckthorn/Sbt) from the family Elaeagnaceae is an important multipurpose plant. *Hippophae* is a Latin term, with 'hippo' meaning horse and 'phaos' meaning flare or gloss (Michel et al. 2012). It is a shrub that can grow wild or be cultivated and is primarily found in cold, dry areas all across Europe and Asia. The Sbt is a dioecious and anemophilous plant; female flower pollination is carried out by the wind (Li and Beveridge 2003). It survives well on poor soils and tolerates extreme temperatures ±40 °C. Sbt has low growth provisions and is invasive when grown in low moisture, alluvial gravel, wet landslides, and on river banks (Li and Beveridge 2003). It is native to China, Russia, and widespread in Eurasia at 2000–4500 m above sea level (Rousi 1971). *H. rhamnoides* is typically the most widespread *Hippophae* species found in high-altitude regions of India, northwest Himalayas from Ladakh to Lahaul-Spiti. The characteristic thorns on the twigs are a peculiarity of *H. rhamnoides*, which markedly distinguishes it from other *Hippophae* species (Fig. 6.1). According to morphological variations, there are a total of four species, including *H. rhamnoides* L., *H. salicifolia*, *H. neurocarpa*, *H. tibetana*. *H. rhamnoides* L. is further categorised into nine subspecies (spp.): *rhamnoides*, *yunnanensis*, *fluviatilis*, *gyantsensis*, *sinensis*, *mongolica*, *caucasia*, *turkestanica*, and *carpatica*. According to Lu (1992), there are 6 species and 12 sub-species of *Hippophae*.

Sbt has garnered global notice, serving as a resource for researchers in the fields of nutritional supplements, pharmaceuticals, cosmetics, environmental studies, and

Fig. 6.1 A *Hippophae rhamnoides* plant growing in its natural habitat. This plant is nutritionally and medicinally important for the Himalayan communities

other scientific fields. It is suggested to harvest the ripe berries at the right time and process them or freeze them instantly for the finest flavor and retention of the bioactive components (Liu et al. 2017). The chemical constituents of Sbt berries strongly rely on parameters such as cultivar or species, plant part, growing area, soil conditions, fertilizer use, and degree of ripeness (Ciesarová et al. 2020). Species of Sbt differ in their secondary metabolite ranges, leading to variability in the effectiveness and hence the therapy. A study has shown the differential antioxidant, and neuroprotective effects of Sbt leaves that were harvested at different times. Additionally, various species and subspecies have peculiar salient features, including quality, ease of harvesting and processing, drought and cold resistance, and various diseases treatment (Ciesarová et al. 2020). Sbt cultivation has no negative environmental impact because no chemicals are used in the cultivation of seabuckthorn nurseries and orchards. Moreover, Sbt is a nitrogen-fixing plant (with Frankia), therefore it improves soil nitrogen levels and hence promotes biodiversity.

6.2 Nutritional Composition

The Sbt berries and seeds comprise 190 bioactive compounds and are exceptionally good sources of different types of vitamins, such as vitamins C, A, E, D, and K (Zeb 2004). Its bioactive compounds comprise flavonoids, phytosterols, serotonin, carotene, and essential amino acids such as, isoleucine, leucine, lysine, methionine, phenylalanine, threonine, tryptophan, and valine (Shah et al. 2007), and minerals such as iron, zinc, calcium, magnesium, selenium, and iodine (Shah et al. 2007). It is an abundant source of omega 3, 6, 7, and 9; these are essential fatty acids that improve brain function and play an essential role in cognitive processes (Yang and Kallio 2002) and are also rich in antioxidant enzymes like superoxide dismutase. Various metabolites present in Sbt are alanine, quinic acid, saturated fatty acids, unsaturated fatty acids, quercetin-3-*O*-β-D-rutinoside, isorhamnetin-7-*O*-α-L-rhamnoside, trigonelline, D-fructose, sucrose, uridine, malic acid, quercetin-3-*O*-β-D-glucoside, quercetin-3-*O*-β-D-glucoside-7-*O*-α-L-rhamnoside, isorhamnetin-3-*O*-β-D-rutinoside, oleanolic acid, β-D-glucose, α-D-glucose, isorhamnetin-3-*O*-β-D-glucoside, isorhamnetin-3-*O*-β-D-glucoside-7-*O*-α-L-rhamnoside, L-quebrachitol, tryptophan, histidine, dehydroascorbic acid, sterols, isorhamnetin-3-*O*-β-D-galactoside-7-*O*-α-L-rhamnoside, quercetin-7-*O*-α-L-rhamnoside, quercetin, kaempferol, isorhamnetin-3-*O*-α-L-arabinopyranoside-7-*O*-α-L-rhamnoside, kaempferol-3-*O*-β-D-sophoroside-7-*O*-α-L-rhamnoside, asparagine, and isorhamnetin, isorhamnetin-3-*O*-β-D-sophoroside-7-*O*-α-L-rhamnoside (Liu et al. 2017). Numerous studies have shown that Sbt berries contain flavonoids and other polyphenols. Flavonoids have gained importance in the pharmaceutical field in recent years because of their positive effects on human health. Another study also reported flavonoids such as flavonols, isorhamnetin, quercetin, and kaempferol in Sbt (Lehtonen et al. 2010). The flavonoid quercetin is receiving attention for its potential to prevent neurodegenerative diseases and the brain oxidative stress and plasma corticosterone regulation (Ishisaka et al. 2011). The main phenolics of Sbt are catechin, resveratrol, protocatechuic acids, ferulic, procyanidins B1 and B2, hyperoside, hydroxybenzoic acids, chlorogenic acid, epicatechin (Ghendov-Mosanu et al. 2020). Another study also reported five important phenolic compounds in Sbt, such as ellagic acid, gallic acid, isorhamnetin, kaempferol, and quercetin (Cho et al. 2017). Additionally, it was reported that harvest time might affect the total flavonoid and phenolic content along with the antioxidative potential of Sbt (Cho et al. 2017).

6.3 Ethnomedicinal Uses of Sbt

Hippophae rhamnoides is an ethnomedicinal plant that has become famous worldwide for its biological effects. Phytochemical studies on the shrub indicate the presence of multiple primary and secondary metabolites (Pundir et al. 2021). As it has been shown in scientific and clinical studies, the Sbt is a rich source of numerous bioactive compounds. Besides being used as a food source since the Tang Dynasty (618–907 AD), Sbt berries were reported to be used in Chinese folk medicine for the treatment of different human ailments, including skin wounds, mucosal wounds of the stomach and lung diseases, as well as cardiac disease and altitude sickness (Pundir et al. 2021).

The traditional use of Sbt dates back to the Tang Dynasty, and the records have been found in many ancient literature. Different parts of the plant, such as berries, leaves, seeds, and twigs, have been used as treatments for different disorders. Already around 900 AD, it was used as a therapeutic shrub in Tibet (Li and Schroeder 1996). In traditional Tibetan Amchi medicine, various parts of Sbt are used to prepare multiple formulations for various ailments. This school of medicine is still practiced in many countries, including China, India, and Europe. Several therapeutic effects have been reported in classics, such as Jing Zhu Ben Cao from the Qing Dynasty and Sibu Yidian from the Tang Dynasty of China. It is believed that the legendary Genghis Khan used it to improve his army's stamina, strength, memory, and immunity during wartime (Stobdan et al. 2010).

6.4 Importance of Sbt for Himalayan Communities

Sbt grows in the cold deserts of Ladakh, Himachal Pradesh, and Arunachal Pradesh in India (Suryakumar and Gupta 2011). Sbt berries are locally known as Drilbu or Chharma in Himachal and tsester in Ladakh. The indigenous Amchi Medical System in Ladakh has utilised Sbt for centuries. And the records of medicinal uses are available from the eighth century Tibetan medical text (Rgyud Bzi) (Stobdan et al. 2010, 2013). In 2010, the Ministry of the Environment and Forests (MoEF) and the Defence Research and Development Organization (DRDO) started an initiative to cultivate Sbt in the five districts of the high-altitude, cold Himalayan desert regions to promote Sbt cultivation. Since then, many such initiatives have been taken by various organizations, such as the Indian Council of Agricultural Research (ICAR) and Defence Institute of High Altitude Research (DIHAR). The socioeconomic status of rural areas is comparatively poorer than that of urban areas. In the countryside, there are few opportunities to earn money, and people are mainly dependent on government jobs alone. As a result, employability in rural areas is also lower. With awareness of Sbt cultivation and its health benefits, families in rural areas may participate in Sbt berry and leaf collection and earn a living in the Himalayan regions. The region's horticulture department, through workshops and awareness programs skill, shared the ability to produce various formulations of Sbt, including juicing, herbal tea and confectionery making and the local unemployed self-help female groups are learning these skills and making Sbt formulations locally that may contribute to the local economy. Value-added products derived from Sbt, such as herbal teas, antioxidant food supplements, beverages, jam, jelly, soft gel capsules, UV protective oil, baked goods, and animal feed are now commercially available (Stobdan et al. 2010, 2013). Patents have been filed for the methods of Sbt squash, syrup, fruit leather, and tea and feed for cows and poultry by DIHAR, DRDO, and ICAR.

6.5 Pharmacotherapeutic Uses of Sbt

Hippophae rhamnoides appear to be a promising plant that plays a positive role in improving human health. This chapter mainly highlights the potential uses of different parts of Sbt (Table 6.1). Many in vivo and in vitro studies have shown that Sbt is effective in treating various ailments. Table 6.1 presents various protective properties of *Hippophae rhamnoides* against various diseases.

6.5.1 Antioxidant Capacity

Sbt berries (Fig. 6.2) are rich in antioxidants and contain compounds with anticoagulant properties. It can prevent and heal diseases related to oxidative stress and cause alteration of platelet activity. The phenolic compounds of the Sbt fraction are responsible for the antioxidant and antiplatelet properties (Olas et al. 2016). Data showed that nutritional supplements of Sbt extract and vitamin E prevent (nicotine-induced) oxidative stress in smokers (Taysi et al. 2010). Another study reported that treatment of neuronal PC-12 cells with Sbt increased their membrane integrity and viability and reduced intracellular oxidative stress. These findings suggest that Sbt is abundant in antioxidants and medicinal compounds that mitigate oxidative stress. Sbt antioxidants can neutralize oxidative stress-induced mitochondrial dysfunction and neuronal cell apoptosis (Cho et al. 2017). Moreover, ethanol and water extracts of Sbt have been shown to be non-toxic (Taysi et al. 2010). Sbt berry oil has been

Table 6.1 Protective properties of *Hippophae rhamnoides* against various diseases

Sbt part used	Property	Disorder	Study	References
Berries	Antioxidants, anticoagulant	Oxidative stress, liver damage	In vivo study	Olas et al. (2016), Taysi et al. (2010), Cho et al. (2017), Turan et al. (2021), Zeb (2004), Ran et al. (2021)
Berries	Antiaging	Cognitive impairment	In vivo study	Sethi et al. (2009), Ahn et al. (2014)
Berries	Neuroprotective	Memory dysfunction	In vivo study	Ahn et al. (2014), Ladol and Sharma (2021), Ma et al. (2020)
Berries	Radioprotective	Lethal whole-body irradiation	In vivo study	Sureshbabu et al. (2008)
Berries	Regulate gut microbiota	Obesity and obesity-associated lipid disorders	In vivo study	Guo et al. (2020)
Sbt oil and leaf	Antiproliferative and antileukemic	Cancer	In vivo study	Solà Marsiñach and Cuenca (2019), Sharma and Deswal (2018), Zhamanbaeva et al. (2014)
Berries	Antibacterial	Skin infections	In vivo study	Ivanišová et al. (2020)
Dry berries	Regulate metabolic processes	Metabolic disorders, hypertensive load	In vivo study	Koyama et al. (2009)
Berries	Anti-inflammatory	Tissue infections	In vivo study	Zhao et al. (2020), Rédei et al. (2018)

tested for genotoxicity and teratogenicity and reported safe for potential food consumption or use as a dietary supplement (Wen et al. 2020).

6.5.2 Anti-inflammatory Capacity

In plants, proteins are involved in many biological processes, such as photosynthesis, signalling pathways, and antioxidant systems, and hence play a crucial role in their survival. In addition, nutritional intake of phenolic compounds can be considered a crucial strategy to treat or delay pathological conditions, as well as a means to prevent disease in multiple ailments. Sbt prevents acrylamide-related oxidants and cytokines and a decrease in antioxidants. Further, Sbt has been reported to inhibit acrylamide-induced oxidative and inflammation in the brain. Hence, Sbt has shown efficiency in preventing acrylamide-related neurotoxicity (Turan et al. 2021).

Fig. 6.2 Berries of *Hippophae rhamnoides* are used as a fruit. The fruits are rich in many phytochemicals with multiple health benefits

6.5.3 Antiaging Effects

Consumption of fruits and vegetables rich in polyphenols leads to a decrease in the formation of free radicals, which leads to a reduction in the occurrence of neurological disorders and age-related neurodegenerative diseases. Research has been done in the past few decades to understand the age-associated changes in the normal brain and to define the relationship with the alteration and development of neuropathological diseases, such as dementia, that are common in the aging brain. Although aging is inevitable, age-related symptoms, including cognitive ability, can be improved with antioxidant-rich supplements.

The berries are an abundant source of antioxidants (Zeb 2004), and antioxidants have been shown to be effective in reducing the onset of aging (Sethi et al. 2009). Age-associated diseases and mild cognitive impairment are common in the older population. These conditions correlate with a poor ability of progenitor cells to proliferate and differentiate into brain cells. Brain-derived neurotrophic factor (BDNF) fosters the proliferation and survival of new neuronal cells in the brain, particularly in hippocampal neurogenesis, which in turn aids in memory processing. Other antioxidant enzymes, such as superoxide dismutases (SODs), also contribute to neurogenesis. In an earlier study in the aged gerbil DG, the administration of Sbt significantly enhanced cell proliferation and neuroblast differentiation. This shows that enhanced neurogenesis is directly associated with the upregulation of intrinsic factors, such as BDNF, SOD1, SOD2, and p-GSK-3β in the aged brain, which suggests a potential role for Sbt as a neuroprotectant and antiaging candidate (Ahn et al. 2014).

6.5.4 Neuroprotective Effects

Another study shows that Sbt administration decreases the epileptiform activity in iron-induced epileptic rats. Sbt administration has significantly prevented memory dysfunction, lessened anxiety-like behavior, and reduced morphological damage in rats with iron-induced epilepsy. This finding indicates the pharmacotherapeutic potential of Sbt towards post-traumatic epilepsy, which may have been exercised through its neuroprotective ability (Ladol and Sharma 2021). Another study also reported the immunosuppressive and neuroprotective effects of the Sbt berry (Ma et al. 2020).

6.5.5 Radioprotective Capacity

An extract of Sbt berries (RH-3) is reported to have significant radiation protection ability against lethal whole-body irradiation. Radiation results in DNA damage, chromosomal aberrations, and cell death. The study reported that RH-3 protects DNA from radiation-induced strand breaks owing to its ability to attach to and stabilize DNA, which may be ascribed to its ability to scavenge free radicals. The significant role of RH-3 in ameliorating radiation-induced DNA damage supports its role in contributing to overall radioprotective capability (Sureshbabu et al. 2008). Sbt was effective in stabilizing the production of ROS, IL-1β, and TNF-α hence, showed beneficial effects in acrylamide-induced neurotoxicity (Turan et al. 2021).

6.5.6 Influence Gut Microbiota

A study reported that freeze-dried Sbt powder regulates the constituents of the gut microbiota. The change is such that it promotes a relative increase in loads of helpful bacteria, e.g. Akkermansia and Bacteroides, and a relative decrease in loads of harmful bacteria, e.g. Desulfovibrio (Guo et al. 2020). Altering helpful bacteria had a beneficial effect on genes encoding for lipolysis and a harmful effect on genes encoding for lipid lipogenesis and storage. Sbt affects the formation of short-chain fatty acids (SCFAs), particularly propionic acid, as it is the main metabolite. Therefore, freeze-dried Sbt powder may alter the constituents and form of the gut microbiota hence alleviating obesity and obesity-associated lipid disorders (Guo et al. 2020).

6.5.7 Antifungal Capacity

The composition of fatty acids contained in Sbt oil is unique and offers a multitude of benefits for human health and is therefore highly valued by both biomedicine and the cosmetics industries. The clinical action of mono-, poly-unsaturated, and saturated fatty acids in Sbt oil contributes to the proper functioning of the organism (Solà Marsiñach and Cuenca 2019). The addition of Sbt berry flour to wheat flour inhibits the growth of rope spoilage in bread and increases the bread's antioxidant activity and shelf life by 120 h. The results suggest adding 1% Sbt berry flour in wheat flour to get bread rich in biomolecules with better flavor and rich in antioxidants as well as health benefits (Ghendov-Mosanu et al. 2020).

6.5.8 Anticancer Activity

Gold nanoparticles synthesized in a single pot with Sbt berry and leaf extract showed shape-dependent diverse nanobiotechnology applications. The Sbt-mediated green synthesis of AuNPs as antioxidants, nanocatalysts, and anticancer for treating dye-polluted wastewater and medicinal applications is highly recommended (Sharma and Deswal 2018). The antiproliferative effect of Sbt is not related to the initiation of cell differentiation. However, cell growth inhibition is caused by DNA damage, which results in arrest in the cell cycle (S phase) and the initiation of apoptosis. Sbt produced antiproliferative and cytotoxic effects on human AML cells. Nonetheless, the antileukemic mechanism of action of Sbt required further investigation (Zhamanbaeva et al. 2014).

6.5.9 Antibacterial Capacity

Sbt has antioxidant and antibacterial properties that have applications in health promotion and agricultural practices (Ivanišová et al. 2020). Sbt berry extract partially activates muscarinic receptors and, in turn, has laxative, prokinetic and enteric effects confirming its therapeutic usage in gastrointestinal conditions such as delayed gastric emptying and constipation (Hanif et al. 2019). The dried berry of Sbt has effects on the metabolic processes and leads to a reduction in hypertensive load on ventricular microvessels in stroke-prone rats (Koyama et al. 2009). Earlier studies linked inflammation to various diseases. Data suggest that pretreatment with Sbt polysaccharide protects against lipopolysaccharide-induced apoptosis in porcine intestinal epithelial cells damaged through its anti-inflammatory activity (Zhao et al. 2020).

6.5.10 Hepatoprotective Effects

Long-term heavy drinking causes liver damage, known as alcoholic liver disease (ALD), which is characterized by oxidative stress and changes in the gut microbiota. Sbt fermentated liquid has a hepatoprotective effect on ALD in mice (Ran et al. 2021). Sbt is effective in treating various allergy symptoms. An earlier study reported that the activity is likely due to a membrane-stabilizing effect caused by the inhibition of mast cell degranulation (Rédei et al. 2018).

6.5.11 Other Uses

Sbt grows under salt and drought stress, and the transcription factors such as WRKY were found to play an important role in Sbt resisting stress. Knowledge of such molecular mechanisms promotes improved tolerance to abiotic stress in cultivar development (Wang et al. 2019). The 2D-PAGE profile of Sbt in low temperature (LT) stress offered beneficial protein candidates for genetic manipulation and provided an understanding of plants' ability to respond to different environmental stress. The response of Sbt to cold stress is complicated and involves multiple cellular, molecular and physiological adaptations (He et al. 2016). Whereas another group of studies deals with the assessment of different combinations of fatty acids used for various conditions. In a study, rats were supplemented with a cocktail of omega-6 and omega-3, resulting in reduced expression of MHC-II in conjunctival epithelium. The cocktail of omega-3 and GLA fatty acids supplementation has been shown to be successful in treating the dryness of the eyes (Viau et al. 2009). The oil extracted from Sbt berries and leaves is widely used in the cosmetics industry.

6.6 Conclusions

Hippophae rhamnoides appear to be a promising plant that may have a beneficial role in improving human health. This chapter mainly highlighted the potential uses of different parts of Sbt. Many in vitro and in vivo studies have shown that Sbt is an effective turning point in the treatment of various ailments. And it is scientifically proven to be safe and with no toxic effects or side effects. Numerous studies have conducted the profiling of different bioactive compounds of Sbt and have suggested their use as dietary supplements for their therapeutic potential. Additionally, people in India living in areas with a large prevalence of Sbt are unaware of the medicinal properties of Sbt. Therefore, it is important to conduct an awareness program and workshop in such areas to promote knowledge about the health benefits of Sbt. Further studies are needed to clearly understand the exact mechanism of action responsible for medical and nutritional properties.

References

Ahn JH, Chen BH, Park JH, Kim IH, Cho JH, Lee JC, Yan BC, Choi JH, Hwang IK, Park JH, Han SN, Lee YL, Kim MJ, Won MH (2014) *Hippophae rhamnoides* L. leaves extract enhances cell proliferation and neuroblast differentiation through upregulation of intrinsic factors in the dentate gyrus of the aged gerbil. Chin Med J (Engl) 127(23):4006–4011. PMID: 25430440

Cho CH, Jang H, Lee M, Kang H, Heo HJ, Kim DO (2017) Sea buckthorn (*Hippophae rhamnoides* L.) leaf extracts protect neuronal PC-12 cells from oxidative stress. J Microbiol Biotechnol 27(7):1257–1265. https://doi.org/10.4014/jmb.1704.04033. PMID: 28535611

Ciesarová Z, Murkovic M, Cejpek K, Kreps F, Tobolková B, Koplík R, Belajová E, Kukurová K, Daško Ľ, Panovská Z, Revenco D, Burčová Z (2020) Why is sea buckthorn (*Hippophae rhamnoides* L.) so exceptional? A review. Food Res Int 133:109170. https://doi.org/10.1016/j.foodres.2020.109170. Epub 2020 Mar 17 PMID: 32466930

Ghendov-Mosanu A, Cristea E, Patras A, Sturza R, Padureanu S, Deseatnicova O, Turculet N, Boestean O, Niculaua M (2020) Potential application of *Hippophae rhamnoides* in wheat bread production. Molecules 25(6):1272. https://doi.org/10.3390/molecules25061272. PMID: 32168868; PMCID: PMC7144010

Guo C, Han L, Li M, Yu L (2020) Seabuckthorn (*Hippophaë rhamnoides*) freeze-dried powder protects against high-fat diet-induced obesity, lipid metabolism disorders by modulating the gut microbiota of mice. Nutrients 12(1):265. https://doi.org/10.3390/nu12010265. PMID: 31968607; PMCID: PMC7020008

Hanif M, Mehmood MH, Ishrat G, Abdullah A, Sohail S, Ahmed M, Gilani AH (2019) Evaluation of prokinetic and laxative effects of *Hippophae rhamnoides* in rodents. Pak J Pharm Sci 32(5 (Special)):2527–2533. PMID: 31894044.

He C, Gao G, Zhang J, Duan A, Luo H (2016) Proteome profiling reveals insights into cold-tolerant growth in sea buckthorn. Proteome Sci 14:14. https://doi.org/10.1186/s12953-016-0103-z. PMID: 27761102; PMCID: PMC5054542

Ishisaka A, Ichikawa S, Sakakibara H, Piskula MK, Nakamura T, Kato Y, Ito M, Miyamoto K, Tsuji A, Kawai Y, Terao J (2011) Accumulation of orally administered quercetin in brain tissue and its antioxidative effects in rats. Free Radic Biol Med 51(7):1329–1336. https://doi.org/10.1016/j.freeradbiomed.2011.06.017. Epub 2011 Jun 23 PMID: 21741473

Ivanišová E, Blašková M, Terentjeva M, Grygorieva O, Vergun O, Brindza J, Kačániová M (2020) Biological properties of sea buckthorn (*Hippophae rhamnoides* L.) derived products. Acta Sci Pol Technol Aliment 19(2):195–205. https://doi.org/10.17306/J.AFS.0809. PMID: 32600016

Koyama T, Taka A, Togashi H (2009) Effects of a herbal medicine, *Hippophae rhamnoides*, on cardiovascular functions and coronary microvessels in the spontaneously hypertensive stroke-prone rat. Clin Hemorheol Microcirc 41(1):17–26. https://doi.org/10.3233/CH-2009-1148. PMID: 19136738

Ladol S, Sharma D (2021) The effects of *Hippophae rhamnoides* in neuroprotection and behavioral alterations against iron-induced epilepsy. Epilepsy Res 175:106695. https://doi.org/10.1016/j.eplepsyres.2021.106695. Epub 2021 Jun 24. PMID: 34186382

Lehtonen HM, Lehtinen O, Suomela JP, Viitanen M, Kallio H (2010) Flavonol glycosides of sea buckthorn (*Hippophaë rhamnoides* ssp. *sinensis*) and lingonberry (*Vaccinium vitis-idaea*) are bioavailable in humans and monoglucuronidated for excretion. J Agric Food Chem 58(1):620–627. https://doi.org/10.1021/jf9029942. PMID: 20050706

Li TS, Beveridge TH (2003) Sea buckthorn (*Hippophae rhamnoides* L.): production and utilization. NRC Research Press

Li TS, Schroeder WR (1996) Sea buckthorn (*Hippophae rhamnoides* L.): a multipurpose plant. HortTechnology 6:370–380

Liu Y, Fan G, Zhang J, Zhang Y, Li J, Xiong C, Zhang Q, Li X, Lai X (2017) Metabolic discrimination of sea buckthorn from different *Hippophaë* species by ^{1}H NMR based metabolomics. Sci Rep 7(1):1585. https://doi.org/10.1038/s41598-017-01722-3. PMID: 28484246; PMCID: PMC5431470

Lu R (1992) Seabuckthorn–a multipurpose plant for fragile mountains. International Center for Integrated Mountain Development ICIMOD. Occasional Paper (20)

Ma QG, Wei RR, Shang DL, Sang ZP, Dong JH (2020) Structurally diverse Flavonolignans with immunosuppressive and neuroprotective activities from the fruits of *Hippophae rhamnoides* L. J Agric Food Chem 68(24):6564–6575. https://doi.org/10.1021/acs.jafc.0c01432. Epub 2020 Jun 3 PMID: 32437606

Michel T, Destandau E, Le Floch G, Lucchesi ME, Elfakir C (2012) Antimicrobial, antioxidant and phytochemical investigations of sea buckthorn (*Hippophaë rhamnoides* L.) leaf, stem, root and seed. Food Chem 131:754–760

Olas B, Kontek B, Malinowska P, Żuchowski J, Stochmal A (2016) *Hippophae rhamnoides* L. fruits reduce the oxidative stress in human blood platelets and plasma. Oxid Med Cell Longev 4692486. https://doi.org/10.1155/2016/4692486. Epub 2016 Jan 10. PMID: 26933473; PMCID: PMC4737000

Pundir S, Garg P, Dviwedi A, Ali A, Kapoor VK, Kapoor D, Kulshrestha S, Lal UR, Negi P (2021) Ethnomedicinal uses, phytochemistry and dermatological effects of *Hippophae rhamnoides* L.: a review. J Ethnopharmacol 10(266):113434. https://doi.org/10.1016/j.jep.2020.113434. Epub 2020 Oct 2 PMID: 33017636

Ran B, Guo CE, Li W, Li W, Wang Q, Qian J, Li H (2021) Sea buckthorn (*Hippophae rhamnoides* L.) fermentation liquid protects against alcoholic liver disease linked to regulation of liver metabolome and the abundance of gut microbiota. J Sci Food Agric 101(7):2846–2854. https://doi.org/10.1002/jsfa.10915. Epub 2020 Nov 20 PMID: 33145761

Rédei D, Kúsz N, Jedlinszki N, Blazsó G, Zupkó I, Hohmann J (2018) Bioactivity-guided investigation of the anti-inflammatory activity of *Hippophae rhamnoides* fruits. Planta Med 84(1):26–33. https://doi.org/10.1055/s-0043-114424. Epub 2017 Jun 29 PMID: 28662529

Rousi A (1971) The genus *Hippophae* L. A taxonomic study. Societas Biologica Fennica Vanamo. Annales Botanici Fennici 1:177–227

Sethi P, Jyoti A, Hussain E, Sharma D (2009) Curcumin attenuates aluminium-induced functional neurotoxicity in rats. Pharmacol Biochem Behav 93(1):31–39. https://doi.org/10.1016/j.pbb.2009.04.005. Epub 2009 Apr 17. PMID: 19376155

Shah AH, Ahmed DI, Sabir M, Arif SH, Khaliq IS, Batool FA (2007) Biochemical and nutritional evaluations of sea buckthorn (*Hippophae rhamnoides* L. Spp. *Turkestanica*) from different locations of Pakistan. Pak J Bot 39:2059–2065

Sharma B, Deswal R (2018) Single pot synthesized gold nanoparticles using *Hippophae rhamnoides* leaf and berry extract showed shape-dependent differential nanobiotechnological applications. Artif Cells Nanomed Biotechnol 46(Suppl 2):408–418. https://doi.org/10.1080/21691401.2018.1458034. Epub 2018 Apr 4 PMID: 29616833

Solà Marsiñach M, Cuenca AP (2019) The impact of sea buckthorn oil fatty acids on human health. Lipids Health Dis 18(1):145. https://doi.org/10.1186/s12944-019-1065-9. PMID: 31228942; PMCID: PMC6589177

Stobdan T, Chaurasia OP, Korekar G, Mundra S, Ali Z, Yadav A, Singh SB (2010) Attributes of Seabuckthorn (*Hippophae rhamnoides* L.) to meet nutritional requirements in high altitudes

Stobdan T, Korekar G, Srivastava RB (2013) Nutritional attributes and health application of seabuckthorn (*Hippophae rhamnoides* L.)—a review. Curr Nutr Food Sci 9(2):151–165

Sureshbabu AV, Barik TK, Namita I, Prem Kumar I (2008) Radioprotective properties of *Hippophae rhamnoides* (sea buckthorn) extract in vitro. Int J Health Sci (Qassim) 2(2):45–62. PMID: 21475487; PMCID: PMC3068731

Suryakumar G, Gupta A (2011) Medicinal and therapeutic potential of Sea buckthorn (*Hippophae rhamnoides* L.). J Ethnopharmacol 138(2):268–78. https://doi.org/10.1016/j.jep.2011.09.024. Epub 2011 Sep 22 PMID: 21963559.

Taysi S, Gumustekin K, Demircan B, Aktas O, Oztasan N, Akcay F, Suleyman H, Akar S, Dane S, Gul M (2010) *Hippophae rhamnoides* attenuates nicotine-induced oxidative stress in rat liver. Pharm Biol 48(5):488–493. https://doi.org/10.3109/13880200903179707. PMID: 20645788

Turan MI, Aktaş M, Gundogdu B, Yilmaz SK, Suleyman H (2021) The effect of *Hippophae rhamnoides* L. extract on acrylamide-induced brain injury in rats. Acta Cir Bras 36(10): e361005. https://doi.org/10.1590/ACB361005. PMID: 34817026; PMCID: PMC8610209

Viau S, Maire MA, Pasquis B, Grégoire S, Acar N, Bron AM, Bretillon L, Creuzot-Garcher CP, Joffre C (2009) Efficacy of a 2-month dietary supplementation with polyunsaturated fatty acids in dry eye induced by scopolamine in a rat model. Graefes Arch Clin Exp Ophthalmol 247(8): 1039–1050. https://doi.org/10.1007/s00417-009-1080-z. Epub 2009 May 5 PMID: 19415319

Wang Z, Feng R, Zhang X, Su Z, Wei J, Liu J (2019) Characterization of the *Hippophae rhamnoides* WRKY gene family and functional analysis of the role of the HrWRKY21 gene in resistance to abiotic stresses. Genome 62(10):689–703. https://doi.org/10.1139/gen-2019-0024. Epub 2019 Jul 17 PMID: 31315001

Wen P, Zhao P, Qin G, Tang S, Li B, Zhang J, Peng L (2020) Genotoxicity and teratogenicity of seabuckthorn (*Hippophae rhamnoides* L.) berry oil. Drug Chem Toxicol 43(4):391–397. https://doi.org/10.1080/01480545.2018.1497047. Epub 2018 Aug 6. PMID: 30081653

Yang B, Kallio H (2002) Composition and physiological effects of sea buckthorn (*Hippophae*) lipids. Trends Food Sci Technol 1(13):160–167

Zeb A (2004) Important therapeutic uses of sea buckthorn (*Hippophae*): a review. J Biol Sci 4:687–693

Zhamanbaeva GT, Murzakhmetova MK, Tuleukhanov ST, Danilenko MP (2014) Antitumor activity of ethanol extract from *Hippophae rhamnoides* L. leaves towards human acute myeloid leukemia cells in vitro. Bull Exp Biol Med 158(2):252–255. https://doi.org/10.1007/s10517-014-2734-3. Epub 2014 Nov 30 PMID: 25432283

Zhao L, Li M, Sun K, Su S, Geng T, Sun H (2020) *Hippophae rhamnoides* polysaccharides protect IPEC-J2 cells from LPS-induced inflammation, apoptosis and barrier dysfunction in vitro via inhibiting TLR4/NF-κB signaling pathway. Int J Biol Macromol 15(155):1202–1215. https://doi.org/10.1016/j.ijbiomac.2019.11.088. Epub 2019 Nov 12 PMID: 31730993

Chapter 7
Nutritional Potential of Wild Edible Rose Hips in India for Food Security

Kanwaljeet Singh and Sumeet Gairola

Abstract Wild fruits possess important bioactive chemicals that may prove useful in alleviating several health ailments. In many places around the world, wild fruits have played an essential role in providing nutrition and dietary requirements to the poor. Nutraceuticals are gaining wide acceptance for their flavor and nutritional benefits. One such resource in India with untapped potential is the fruits of wild *Rosa* species. Rose hips are used to make a variety of foodstuffs and beverages, including jams, jellies, teas, and alcoholic beverages. In traditional medicine, the rose hips are used to treat eczema, cold, cough, bronchitis, stomachic diseases, kidney diseases, peptic ulcer, cardiac diseases, asthma, rheumatism, cancer, and hemorrhoids. Biochemically, the fruits of these species are rich in carbohydrates, phenolics, flavonoids, tocopherols, terpenes, carotenoids, fatty acids, galactolipids, vitamins, minerals, and tannins. The rich phytochemical composition attributed several pharmacological properties to the rose hips such as anti-inflammatory, antioxidant, immunomodulator, cardioprotective, anticancer, antidiabetic, neuroprotective, and antibacterial properties are attributed to the rich phytochemical composition of rose hips. In addition to nutritional benefits, these may assist in the alleviation of hunger in hilly areas of India, particularly in the Himalayan region if explored scientifically. This chapter attempts to provide detailed information on the phytochemical and nutritional composition of the wild species of *Rosa* found in India.

Keywords Nutritional composition · Therapeutic applications · Ethnomedicinal uses · Wild roses

K. Singh · S. Gairola (✉)
Plant Sciences and Agrotechnology Division, CSIR-Indian Institute of Integrative Medicine, Jammu, Jammu and Kashmir, India
e-mail: sumeetgairola@iiim.res.in

A. Kumar et al. (eds.), *Wild Food Plants for Zero Hunger and Resilient Agriculture*, Plant Life and Environment Dynamics,
https://doi.org/10.1007/978-981-19-6502-9_7

7.1 Introduction

With the increasing population and depletion of natural resources, securing enough healthy food for everyone is becoming a critical concern (Karhagomba et al. 2013). Addressing these concerns will require an increase in the overall production of food, which cannot be achieved merely by industrial agriculture through land use changes at the expense of the ecological system and afore dwindling biodiversity (Jacobsen et al. 2013; Sunderland 2011; Padulosi et al. 2011), along with a switch to healthy foods (Abarca-Gómez et al. 2010). Prospecting for natural food supplies from biodiversity is one solution in this regard (Patel 2015). Many underutilized food sources have been identified over the years and nutritional and safety assessments are being done (Patel 2017). Neglected and underexploited plant species are the ones that have the potential to improve sovereignty, food security, and livelihood of people but are not completely realized due to their low cutthroat with crops grown in prevailing agriculture (Ulian et al. 2020). Nonetheless, they are essential to people on a local level and are frequently tuned to distinct environmental and climatic conditions (Padulosi et al. 2011) Wild roses, which are members of the Rosaceae family, are one such underutilized food resource. The species of wild roses are broadly distributed in the Northern Hemisphere with nearly 200 species (Wissemann and Ritz 2005). In India, 11 species of wild *Rosa* L. have been reported (Chopra and Singh 2013). Among the growing wild rose species of this region, *Rosa webbiana, Rosa moschata* have a wide distributional range (Singh et al. 2020).

Many products, including rosehip tea, jelly, jam, nectar, marmalade, and pestil, as well as alcoholic beverages, are made from it (Grochowski 1990; Ilyasoğlu 2014). Wild rose hips are recognized to be high in phenolic acids, flavonoids, tocopherols, carotenoids, and vitamin C (Hodisan et al. 1997; Ercisli 2007; Fascella et al. 2019; Singh et al. 2021). Researchers have recently become interested in the hips (pseudo fruits) of *Rosa* species due to the presence of essential phytochemicals in them. In several European countries rose hips are employed in culinary items (Gao et al. 2005), while several extracts of rose hips are available in the market as health supplements, cosmetics, and as nutritional aid. Various extracts of *Rosa* species have been connected to potential pharmacological properties in the last decade, including anti-inflammatory, antioxidant, immunomodulator, cardioprotective, anti-cancer, antidiabetic, neuroprotective, and antibacterial properties (Patel 2017).

Furthermore, due to the preventive and therapeutic effects of rosehip against cold, cough, bronchitis, stomachic diseases, kidney diseases, eczema, peptic ulcer, rheumatism, cardiac diseases, cancer, asthma, and hemorrhoids, the fruits of the rosehip have been used traditional medicine (Ugulu et al. 2009; Genc and Ozhatay 2006). Rose hips are used to make a variety of drinks and food items such as jams, jellies, alcoholic drinks, and teas (Grochowski 1990). Wild *Rosa* species such as *Rosa webbiana, Rosa moschata*, *Rosa sericea*, and *Rosa macrophylla* have traditional medicinal value in India and are used to treat cardiac, ophthalmic, hepatic, and gastrointestinal disorders (Kumar and Bhagat 2012; Sharma et al. 2015; Kumar et al. 2015; Balodi et al. 2018). Figure 7.1 provides photographs of a few wild species of

Fig. 7.1 Pictures of some of the wild species of *Rosa* with their hips which are edible and consumed as food in various culinary preparations. (**a**, **b**) *Rosa canina* and its hips, (**c**, **d**) *Rosa macrophylla* and its hips, (**e**, **f**) *Rosa moschata* and its hips and (**g**, **h**) *Rosa webbiana* and its hips

rose along with their hips which are used in various traditional foods. In other traditional medicinal systems around the world, roses are a well-documented plant and are used to treat liver, kidney, lung, heart, pain, and stomach ailments (Ercisli

2005; Genc and Ozhatay 2006; Ayati et al. 2018). This chapter attempts to provide concise information on the phytochemical and nutritional composition of wild rose species in India.

7.2 Ethnomedicinal Uses

Rosa species have a long history of being employed in treating a broad spectrum of ailments in different traditional systems of medicine all around the world. In the Turkish medicine system, the hips of *R. canina* are used to treat colds and flu, kidney stones, hemorrhoids, inflammation, breath shortness, and diarrhea and as an immunotonic, expectorant, analgesic, and appetizer (Polat et al. 2013; Uzun and Kaya 2016). In Italy, diarrhea and urinary problems, sore throat, cold, and inflammation are treated by employing the hips of *R. canina* (Vitalini et al. 2013). In Iranian folk medicine, the hips of *R. canina* are used in treating kidney stones, stomach cramps, dandruff, and fatty liver and to regulate blood pressure and sugar level (Moghanloo et al. 2019). In Coratia, respiratory, digestive, urogenital, and cardiovascular problems are reported to be treated using the hip powder of *R. canina* (Žuna Pfeiffer et al. 2020). The fruit infusion of *Rosa foetida* is used to treat cardialgia, and cancer in Turkey (Akgul et al. 2018). In the Chinese medicinal system, the fruits and leaves are utilized to treat bone fractures, traumatic injury, appendicitis, diarrhea, enteritis, and stomachache (Hong et al. 2015). Health ailments such as fever, weakness, diarrhea, and bile disorder are reported to be treated using the fruits of *R. macrophylla* in Nepal (Ghimire and Aumeeruddy-Thomas 2009). People in Uttarakhand, India use the fruits of *R. macrophylla* to treat dermatological problems (Adhikari et al. 2007). In Indian Western Himalaya regions, the hips of *R. moschata* are employed to treat pregnancy termination and bleeding (Kumari et al. 2009), leucorrhea gastrointestinal problems (Balodi et al. 2018) and as vermicide (Dutt et al. (2011). The hips of *R. rubiginosa* are used in dermatological problems and as an antitussive in Argentina (Lozada et al. 2006). In Turkish traditional medicine, the hips of *R. sempervirens* are employed in ophthalmic problems (Maxia et al. 2008). In Uttarakhand region of India, headache and liver complaints are treated using hips of *R. sericea* (Singh and Rawat 2011). In the Indian state of Himachal Pradesh and the Union Territory of Ladakh, the hips of *R. webbiana* are used in sexual dysfunction, liver disorder, bleeding, jaundice, and liver problem (Singh and Lal 2008; Angmo et al. 2012). In Afganistan, people uses hips of *R. webbiana* to treat high blood pressure, bloody cough, ear-ache, stomach trouble, and fever (Soelberg and Jäger 2016). Figure 7.2 provides an outline of the important pharmacological properties of wild species of *Rosa*.

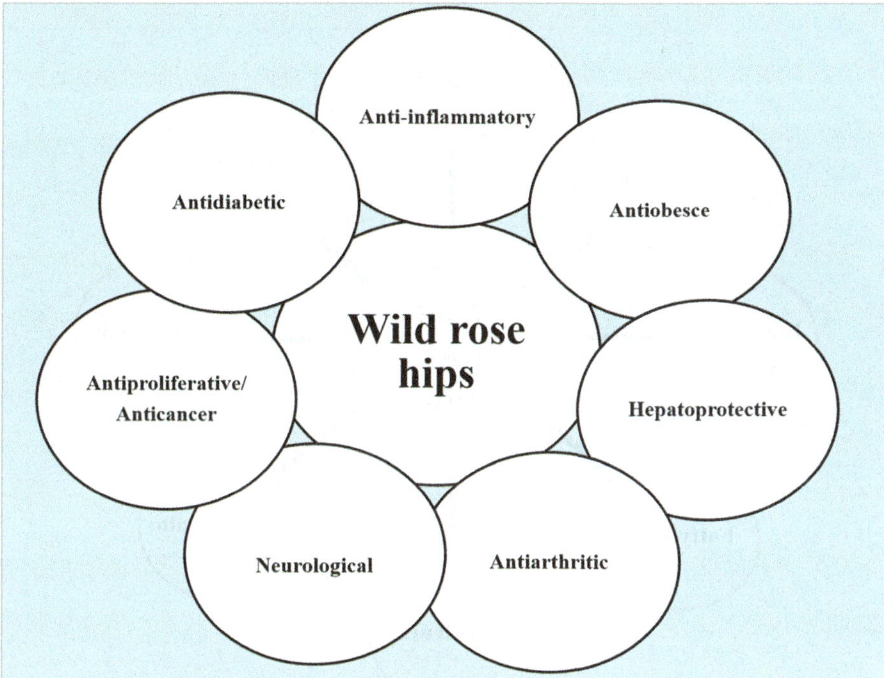

Fig. 7.2 Pharmacological properties of wild rose hips

7.3 Nutritional Composition

7.3.1 Vitamins and Minerals

Vitamins are dietary ingredients due to their health-promoting properties (Shakoor et al. 2020). Similarly, minerals perform a wide range of functions, including bone formation, muscle and neuron function, and water balance regulation in the body (Kim and Choi 2013). Wild rose species provide a high concentration of vitamins and minerals. The most well-known are *R. canina* hips, which are high in ascorbic acid, minerals, carotenoids, and other bioactive compounds (Ahmad et al. 2016; Javanmard et al. 2018). Sadia (2020) reported 669.17 mg/100 g FW and 740.72 mg/100 g FW ascorbic acid concentration in *R. webbiana* and *R. moschata*, respectively. Singh et al. (2021) discovered vitamin C levels in the wild species of *Rosa* from Western Himalaya, India, ranging from 23.44 to 143.80 mg/110 g DW. As per Ercisli (2007), ascorbic acid concentration varies from 300 mg/100 g DW to 4000 mg/100 g DW, the likely reasons identified being the geographic origin of the plant, ecological characteristics, light intensity, and oxygen level of the environment (Dogan and Kazankaya 2006; Roman et al. 2013). Ascorbic acid's immune-boosting properties have led to its widespread use in the continuing COVID-19 pandemic when no specific treatment is available. Rose species, which are high in

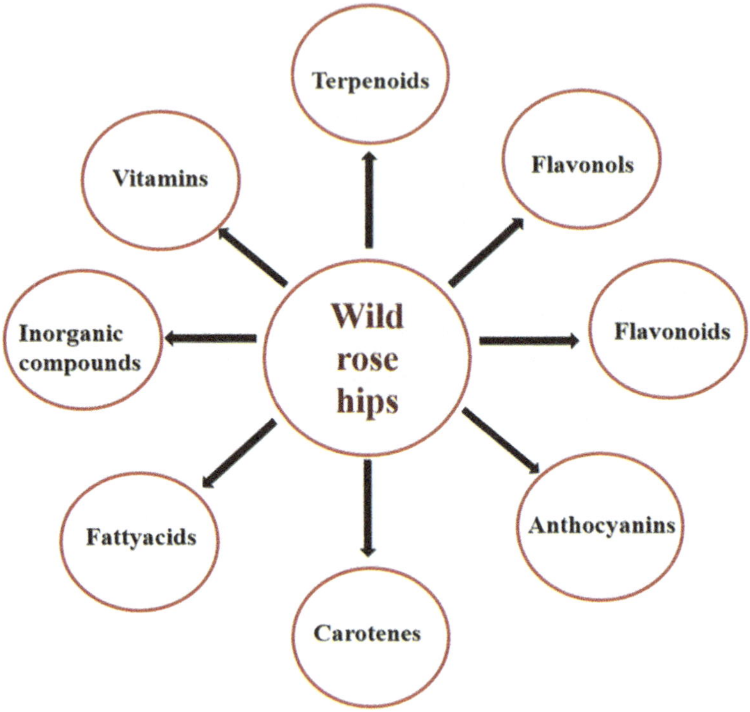

Fig. 7.3 Important phytochemical constituents of rose hips with health benefits

nutritional components and ascorbic acid, should be promoted for increased human intake in order to boost nutrition and reinforce immune processes. Aside from the phytochemicals described above, the rose hip also contains flavonoid, pectin, retinol, vitamin B-complex, tocopherols, and minerals such as Ca, K, Mg, S, Fe, Si, Mn, and Se (Barros et al. 2011; Dubtsova et al. 2012; Singh et al. 2021). Figure 7.3 provides an overview of the important phytochemicals that are found in wild species of *Rosa* with health benefits.

7.3.2 Carbohydrates and Proteins

The hips of wild rose species contain a high concentration of carbohydrates and proteins. Sadia (2020) discovered 80.73% and 80.74% carbohydrates in the hips of *R. webbiana* and *R. moschata* from Pakistan, respectively. In India, the carbohydrate proportion ranged from 74.13% to 81.43% (Singh et al. 2021). Similarly, Najda and Buczkowska (2013) discovered a sugar concentration of 17.1–27.5% in *R. canina* rose hips. Furthermore, *Rosa webbiana*, *Rosa canina*, and *Rosa moschata* have significant protein content (1.1–9.01%) in their hips (Sadia 2020; Singh et al.

2021). Xie et al. (2016) found protein content in the range of 11.50–19.36 mg BAE/100 g) in the hips of *Rosa laevigata.* In *Rosa sempervirens,* the protein content was 0.27 g/100 FW (Rosu et al. 2011).

7.3.3 Carotenoids and Tocopherols

Carotenoids' primary benefits can be explained by their antioxidant effect. Carotenoids have been shown to improve cardiovascular health, boost cognitive performance, and may assist to avert many forms of cancer, in addition to promoting eyesight (Eggersdorfer and Wyss 2018). Tocopherols, on the other hand, improve human health by safeguarding the cardiac and nervous systems and have anti-diabetic properties (Almagro et al. 2021). Rosehips are a high source of carotenoids, the most interesting of which are carotenes. Beta-carotene and lycopene are the most abundant (Hodisan et al. 1997). Hornero-Méndez and Mínguez-Mosquera (2000) found 390 mg/kg DW of lycopene in Chile rosehips of Chile, which is higher than found in tomatoes. They suggested that rosehips may be used as a substitute for tomatoes in the food sector. Lycopene is known to have cardiac and cancer-protective potential (Wan et al. 2014; Cheng et al. 2019), whereas lutein, beta-carotene, and zeaxanthin protect the eyes from oxidative damage, inflammation by quenching the free radicals generated during the complex physiological reactions (Johra et al. 2020).

Tocopherols have been found in the hips of various rose species. *R. canina* hips contain α, β and γ-tocopherols (Fan et al. 2014). The concentration varies depending on the fruit portion and harvesting period. The fruit flesh and seed parts of *R. canina* contain alpha-tocopherols in concentrations of 34.20 g/g, 21.62 g/g, and 8.05 g/g, respectively. In different harvesting times, concentration ranges of 126.9–169.5 and 60.1–77.8 for and alpha-tocopherols was recorded in *R. rubiginosa* hips (Andersson et al. 2011). Tocopherols are particularly interesting since they contribute significantly to vitamin E activity. Kazaz et al. (2009) found 34 g/g DW of tocopherols in the hips of *R. canina* and 8 g/g DW in *R. canina* seeds. However, the nutritional content of rose fruits may be affected by storage conditions (Rosu et al. 2011).

7.3.4 Phenolics, Flavonoids, and Tannins

Phytonutrients, such as phenolic compounds, have been shown to improve health and well-being (Valdés et al. 2015). *R. canina, R. corymbifera, R. moschata, and R. webbiana* hips contain a high concentration of phenolic chemicals (Demir and Özcan 2001; Kovacs et al. 2004; Ercisli 2007; Patel 2013; Fascella et al. 2019; Shameh et al. 2019). Anthocyanin, tannins, flavonoids flavanols, stilbenoid, carotenoids, chlorins, sugars, organic acids, fatty acids, tocopherols, and galactolipid are also abundant in wild roses (Ayati et al. 2018). Anthocyanins are well-acknowledged as an important nutraceutical for promoting good health by lowering oxidative stress, inflammation, and age-related health issues (Speer et al. 2020). Isoflavonoids appear to influence many signaling pathways implicated in cancer formation and progression (Rizeq et al. 2020). The total phenolic content in accessions of wild rose species in the Western Himalayas ranges from 119.68 to 529.75 mg/100 g (Singh et al. 2021). More than 50 different phenolic compounds have been reported from the species of *Rosa*, including *R. canina, R. rubiginosa, R. glauca,* and *R. sempervirens.* These compounds include flavonols, flavanols, phenolic acids, and their derivatives (Cunja et al. 2014). Polyphenols contained in food can protect against oxidative damage by neutralizing ROS and inducing an innate defense response. The fruits of *R. sempervirens* contain significant amounts of quinic acid as well as the polyphenols astragalin, ()-epicatechin, (+)-catechin, and hyperoside (Kerasioti et al. 2019).

Ferulic acid in the fruits of *R. sempervirens* is known to limit the activity of α-amylase by modifying the secondary structures of these enzymes via mixed and non-competitive methods, respectively (Zheng et al. 2020). As a result, it can be used in the development of novel functional food in the times to come. Türkben et al. (2010) found that the predominant phenolic components in *Rosa canina* rose hips were quercetin and (+) -catechin, but their quantities varied greatly depending on the ripeness of the fruits, as well as processing conditions and manipulation techniques. In their studies on hips of *R. canina*, Festa et al. (2001) and Fujii and Saito (2009) have examined anti-mutagenic and anti-carcinogenic action of ellagic acid and quercetin.

The leaves of *R. moschata, R. canina*, and *R. sempervirens* contains catechin, quercetin di-*O*-hexoside,epicatechin gallate, epicatechin, rutin, quercetin 3-*O*-rhamnoside, kaempferol 7-*O*-glucoside, Kaempferol *O*-hexoside-deoxyhexoside, quercetin 3-*O*-glucoside, aromadendrin *O*-hexoside-deoxyhexoside, quercetin *O*-pentoside 1, kaempferol 3-*O*-glucoside, quercetin *O*-pentoside 2, kaempferol 3-*O*-rutinoside, kaempferol *O*-rhamnoside (Ouerghemmi et al. 2016). Li et al. (2013) isolated new flavonoids viz., kaempferol-3-*O*-β-D-xylopyranosyl-(1 → 2)-α-D-ribopyranoside and quercetin-3-*O*-β-D-xylopyranosyl-(1 → 2)-α-D-ribopyranoside and one phenolic derivative gallicin-*p*-*O*-(6′-*O*-caffeoyl)-β-D-glucoside along with some other known bioactive compounds from the *R. sericea* leaf. Furthermore, ellagitannins are also reported in the hips of *R. canina* (Fecka 2009). Ellagitannins, ellagic acid, are known to display several health-promoting properties including antioxidant, anti-inflammatory, cardioprotective, anticancer, and prebiotic properties (Landete 2011; Cortés-Martín et al. 2019). Similarly, quercetin and quercetin-3-*O*-glucuronide reduce ROS generation, providing chemoprotection to mitochondria via antioxidative effects (Guo et al. 2013).

Festa et al. (2001) and Fujii and Saito (2009) investigated the anti-mutagenic and anti-carcinogenic effects of ellagic acid and quercetin in *R. canina* hips. Rose hips contain quercetin, which have antithrombotic activities, to preserve the oxidation of low-density lipoprotein, and to calm the smooth muscles of heart (Comalada et al. 2005). The glycosidic flavonoid 'tiliroside' (kaempferol 3--D-(6″-*p*-coumaroyl-glucopyranoside)) found in *R. canina* hips (Ninomiya et al. 2007) has been shown to reduce the LDL oxidation in vitro and have strong antioxidant, cytotoxic, antiobesity and anti-complement effects (Rao et al. 2007; Nagatomo et al. 2015). *R. sempervirens* leaves are rich in antioxidants and contain quercetin, quercetin 3-xyloside, quercitrin, and hyperoside (Bitis et al. 2017). This species' extracts have also been proven to be cytotoxic to HeLa and HT-29 cell lines (Nađpal et al. 2018). *R. rubiginosa* contains several phenolics and flavonoids, primarily quercetin derivatives (Jimenez-Lopez et al. 2017). In vivo studies of quercetin and its derivatives on inflammation reveal that it is an effective anti-inflammatory drug. An in vivo investigation found that mice fed a quercetin-rich diet had lower expression of genes associated with inflammation (Ferry et al. 1996). Quercetin has anti-cancer (Tang et al. 2020) and cardioprotective properties (Ferenczyova et al. 2020). *R. rubiginosa* seeds are rich in polyphenols and have strong antioxidant properties (Moure et al. 2001). Oleanolic acid and ursolic acid have been isolated from the hips of *R. laevigata*, *R. multiflora*, and *R. canina* (Fang et al. 1991; Pang et al. 2019; Saaby et al. 2011). The roots of R. *laevigata* have been found to contain a variety of flavonoids and stilbenes (Yan et al. 2014). The hips of *R. corymbifera* fruits contain gallic acid, vanillic acid, ellagic acid, caffeic acid, coumaric acid, and protocatechuic acid (Jimenez-Lopez et al. 2017) while that of *R. hemisphaerica* possesses cinnamic acid, caffeic acid, gallic acid, and *p*-coumaric acid (Shameh et al. 2019).

7.3.5 Terpenes

Triterpenes are among the most widespread and diversified natural phytochemical groups. Triterpenes have a wide range of applications in the medical, food, and industrial realms (Thimmappa et al. 2014). *R. canina* hips include ursolic acid, betulinic acid, and oleanic acid (Saaby et al. 2011). Moreover, various additional terpenoid compounds have been found in rose hips (Özdemir et al. 2022). So far, three types of triterpenoids, oleanolic acid, ursolic acid, and lupinic acid have been reported in *R. laevigata* (Li et al. 2021). Ursolic acid is a natural nutraceutical with anti-inflammatory properties due to its suppression of cysteine-aspartic acid protease 3 (Ma et al. 2017).

7.3.6 FAs and Galactolipids

In their study, Nowak (2005) found 97% of wild rose hip seed oil with palmitic, linoleic, stearic, and oleic acid and the remaining 3% composed of other fatty acids (Nowak 2005). The fatty acid study performed by Ercisli (2007) in *R. canina* found lauric acid (4.80%), palmitic acid (16.4%), linoleic acid (16.0%), linolenic acid (40.5%), and nonadecylic acid (4.74%). In addition to fatty acids, rose oil (seed) has been reported to be high in phytosterols, tocopherols, carotenoids, and phenolic compounds (Grajzer et al. 2015), indicating that rosehip oil has the potential to compete in the food industry. Linoleic, linolenic acid, oleic, and palmitic acid are the major fatty acids found in the seeds of *R. moschata* and *R. multiflora* (Sharma et al. 2012). *R. canina* seeds contain a high amount of PUFAs such as omega-3 and omega-6 (Ilyasoğlu 2014). Result studies have revealed several health benefits of these PUFAs such as dilatation of blood arteries, lower cholesterol and triglycerides, prevention of thrombosis, increased blood fluidity, enhanced fluidity of erythrocytes, and a lower risk for heart disease (Calder 2006; Leaf and Weber 1988).

Rose hips and seed oil have also been demonstrated to have cosmeceutical potential (Franco et al. 2007). Because of its therapeutic benefits and lovely odor, *Rosa moschata* is widely utilized in the perfume and cosmetics industries. Palmitic, arachidic, linolenic, linoleic oleic, stearic, and erucic acid are all found in *R. rubiginosa* hips (Adamczak et al. 2011). Rosehip powder, in particular, has been proven in clinical trials to alleviate symptoms associated with rheumatoid inflammation (Cohen 2012; Kharazmi and Winther 1999; Rein et al. 2004) and is related to (2S)-1,2-di-*O*-[(9Z,12Z,15Z)-octadeca-9,12,15-trienoyl]-3-*O*-D galactopyranosyl glycerol and was isolated from *R. canina* (Cohen 2012; Larsen et al. 2003).

7.3.7 Dietary Fibers

The crude fiber content of wild rose species from the Indian Himalayan region ranged from 21.6 to 46.03% (Singh et al. 2021). In other research, the crude fiber content of rose hips ranged from 3.27 to 4.03% (Demir and Özcan 2001; Sadia 2020). The higher crude fiber level stimulates the immune system and can help improve intestinal motility, cholesterol absorption, and prevent atherosclerosis (Abuajah et al. 2015; McRorie Jr and McKeown 2017).

7.4 Conclusions and Future Prospects

The summarized data showed the therapeutic use of different Indian *Rosa* species for treating various ailments all over the world in different traditional medicinal systems, including India. Various traditional uses include respiratory disorders, indigestion, impotency, jaundice, fever, leucorrhea, bleeding, pregnancy termination, and gastrointestinal disorders. The biochemical investigations provide useful information for screening accessions with a high concentration of bioactive chemicals. Nutritional research on rose hips must accelerate to generate affordable and widely available health supplements.

The wild species of *Rosa* in India have a wide range of variability in terms of important agronomic attributes and genetic diversity, making them a valuable source of information for crop improvement and a future food source. Different wild *Rosa* species may help broaden the genetic base of cultivated roses depending on the trait.

The important traits observed in *Rosa* species are the yellow flower pigmentation of *R. foetida* and *R. ecae* (Chowdhary and Wadhwa 1984; eFI 2022), flower color variation in *R. webbiana* and *R. macrophylla* (Singh et al. 2020), pleasant flower fragrance of *R. moschata* (Singh et al. 2020), and continuous flowering in *R. leschenaultiana* (eFI 2022). Although studies have been performed on the essential oil composition in roses, however, wild species such as *R. webbiana* and *R. macrophylla* remained to be characterized for their aromatic potential, and further studies are required to prove the economic potential of the oil of these *Rosa* species. Thornlessness in some of the population's *R. macrophylla* (Singh et al. 2020) can be used to manipulate rose quality in the future rationally. Several species, including *R. webbiana*, *R. macrophylla*, *R. moschata*, *R. rubiginosa*, and *R. canina*, can improve hip quality for specific medicinal or food applications. *R. webbiana* and *R. macrophylla* thrive well in high altitudinal areas of J&K, Ladakh, and Uttarakhand (Upreti et al. 2010; Singh et al. 2020). Therefore, studies aiming to clarify the mechanisms of adaptation to these extreme environments represent great potential for enhancing world food production. Therefore, the inventorization of such morphological traits is important for obtaining elite phenotypes with high morphological variability.

The development of disease-resistant roses always remains the highest priority in rose breeding research. Therefore, species such as *R. laevigata* and *R. multiflora,* which have been previously reported to be resistant to black spots (Drewes-Alvarez 2003) and are present in J&K, India can be exploited to develop resistant varieties. Similarly, *R. multiflora*, moderately resistant to powdery mildew can be exploited from developing resistant varieties.

The distribution of *Rosa* species in the wide altitudinal ranges indicates their robustness and adaptability to broad climatic conditions. For instance, *R. webbiana* is distributed in Western Himalaya's cold higher altitudinal areas, including the cold deserts of the Union Territory of Ladakh, and found growing luxuriantly in the barren land, roadside, and steep rocky areas (Singh et al. 2020). The presence of small-sized, hairy leaflets in *R. webbiana* may be an adaptation to the xeric conditions in which this species flourishes, thereby reducing the transpiration rate. In addition, the presence of hairs may also be an adaptation against herbivory. These characteristics can be incorporated for developing drought-stress rose varieties. It is also noted that studies in India on the morphological characterization of the agronomic traits of individual wild rose species, their genetic diversity, their relationship with the nutritional composition, and their commercial utility and acceptability by the farmers are lacking hitherto. Present growing demands of rose hips (pseudo fruits) in the food, pharmaceutical, and cosmetic industries, necessitates in-depth characterization of their chemical constituents and related biological activities. The biochemical analysis provides valuable information for screening accessions harboring a high amount of bioactive compounds.

References

Abarca-Gómez L, Abdeen ZA, Hamid ZA, Abu-Rmeileh NM, Acosta-Cazares B, Acuin C, Adams RJ, Aekplakorn W, Afsana K, Aguilar-Salinas CA, Agyemang C (2010) Worldwide trends in body-mass index, underweight, overweight, and obesity from 1975 to 2016: a pooled analysis of 2416 population-based measurement studies in 128· 9 million children, adolescents, and adults. Lancet 390:2627–2642

Abuajah CI, Ogbonna AC, Osuji CM (2015) Functional components and medicinal properties of food: a review. J Food Sci Technol 52:2522–2529

Adamczak A, Grys A, Buchwald W, Zielinski J (2011) Content of oil and main fatty acids in hips of rose species native in Poland. Dendrobiology 66:55–62

Adhikari BS, Babu MM, Saklani PL, Rawat GS (2007) Distribution, use pattern and prospects for conservation of medicinal shrubs in Uttaranchal State, India. J Mt Sci 4:155–180

Ahmad N, Anwar F, Gilani AU (2016) Rose hip (*Rosa canina* L.) oils. In: Preedy VR (ed) Essential oils in food preservation, flavor and safety. Academic, pp 667–675

Akgul A, Akgul A, Senol SG, Yildirim H, Secmen O, Dogan Y (2018) An ethnobotanical study in Midyat (Turkey), a city on the silk road where cultures meet. J Ethnobiol Ethnomed 14:12

Almagro L, Sabater-Jara AB, Belchí-Navarro S, Pedreño MÁ (2021) Recent trends in the biotechnological production of tocopherols using in vitro cultures. Phytochem Rev 20:1193–1207

Andersson SC, Rumpunen K, Johansson E, Olsson ME (2011) Carotenoid content and composition in rose hips (*Rosa* spp.) during ripening, determination of suitable maturity marker and implications for health promoting food products. Food Chem 128:689–696

Angmo K, Adhikari BS, Rawat GS (2012) Changing aspects of traditional healthcare system in Western Ladakh, India. J Ethnopharmacol 143:621–630

Ayati Z, Amiri MS, Ramezani M, Delshad E, Sahebkar A, Emami SA (2018) Phytochemistry, traditional uses and pharmacological profile of rose hip: a review. Curr Pharm Des 24:4101–4124

Balodi KN, Purohit MV, Shridhar V, Arunachalam K (2018) Ethno-medicinal uses of various plants species among the Jaad Bhotiya community of Uttarakhand, Western Himalaya. Stud Ethnomed 12:189–197

Barros L, Carvalho AM, Ferreira IC (2011) Exotic fruits as a source of important phytochemicals: improving the traditional use of *Rosa canina* fruits in Portugal. Food Res Int 44:2233–2236

Bitis L, Sen A, Ozsoy N, Birteksoz-Tan S, Kultur S, Melikoglu G (2017) Flavonoids and biological activities of various extracts from *Rosa sempervirens* leaves. Biotechnol Biotechnol Equip 31: 299–303

Calder PC (2006) n-3 Polyunsaturated fatty acids, inflammation, and inflammatory diseases. Am J Clin Nutr 83:1505–1519

Cheng HM, Koutsidis G, Lodge JK, Ashor AW, Siervo M, Lara J (2019) Lycopene and tomato and risk of cardiovascular diseases: a systematic review and meta-analysis of epidemiological evidence. Crit Rev Food Sci Nutr 59:141–158

Chopra VL, Singh M (2013) Ornamental plants for gardening. Scientific Publishers, Jodhpur

Chowdhary HJ, Wadhwa BM (1984) Flora of Himachal Pradesh analysis. Botanical Survey of India, Calcutta

Cohen M (2012) Rosehip: an evidence based herbal medicine for inflammation and arthritis. Aust Fam Physician 41:495–498

Comalada M, Camuesco D, Sierra S, Ballester I, Xaus J, Gálvez J, Zarzuelo A (2005) In vivo quercitrin anti-inflammatory effect involves release of quercetin, which inhibits inflammation through down-regulation of the NF-κB pathway. Eur J Immunol 35:584–592

Cortés-Martín A, Romo-Vaquero M, García-Mantrana I, Rodríguez-Varela A, Collado MC, Espín JC, Selma MV (2019) Urolithin metabotypes can anticipate the different restoration of the gut microbiota and anthropometric profiles during the first year postpartum. Nutrients 11:2079

Cunja V, Mikulic-Petkovsek M, Stampar F, Schmitzer V (2014) Compound identification of selected rose species and cultivars: an insight to petal and leaf phenolic profiles. J Am Soc Hortic Sci 139:157–166

Demir F, Özcan M (2001) Chemical and technological properties of rose (*Rosa canina* L.) fruits grown wild in Turkey. J Food Eng 47:333–336

Dogan A, Kazankaya A (2006) Fruit properties of rose hip species grown in Lake Van Basin (Eastern Anatolia Region). Asian J Plant Sci 5:120–122

Drewes-Alvarez R (2003) Disease—black spot. In: Roberts AV, Debener T, Gudin S (eds) Encyclopedia of rose science. Elsevier, Academic, Oxford, pp 148–153

Dubtsova GN, Negmatulloeva RN, Bessonov VV, Baĭkov VG, Sheviakova LV, Makhova NN, Perederiaev OI, Bogachuk MN, Baĭgarin EK (2012) Composition and content of biologically active substances in rose hips. Vopr Pitan 81:84–88

Dutt B, Sharma S, Sharma K, Gupta A, Singh H (2011) Ethnobotanical survey of plants used by Gaddi tribe of Bharmour area in Himachal Pradesh. Himalayan Ecol 19:22–27

eFI (2022). https://sites.google.com/site/efloraofindia/. Accessed 12 Jan 2022

Eggersdorfer M, Wyss A (2018) Carotenoids in human nutrition and health. Arch Biochem Biophys 652:18–26

Ercisli S (2005) Rose (*Rosa* spp.) germplasm resources of Turkey. Genet Res Crop Evol 52:787–795

Ercisli S (2007) Chemical composition of fruits in some rose (*Rosa* spp.) species. Food Chem 104: 1379–1384

Fan C, Pacier C, Martirosyan DM (2014) Rose hip (*Rosa canina* L): a functional food perspective. J Funct Food Health Dis 4:493–509

Fang JM, Wang KC, Cheng YS (1991) Steroids and triterpenoids from *Rosa laevigata*. Phytochemistry 30:3383–3387

Fascella G, D'Angiolillo F, Mammano MM, Amenta M, Romeo FV, Rapisarda P, Ballistreri G (2019) Bioactive compounds and antioxidant activity of four rose hip species from spontaneous Sicilian flora. Food Chem 289:56–64

Fecka I (2009) Qualitative and quantitative determination of hydrolysable tannins and other polyphenols in herbal products from meadowsweet and dog rose. Phytochem Anal 20:177–190

Ferenczyova K, Kalocayova B, Bartekova M (2020) Potential implications of quercetin and its derivatives in cardioprotection. Int J Mol Sci 21:1585

Ferry DR, Smith A, Malkhandi J, Fyfe DW, deTakats PG, Anderson D, Baker J, Kerr DJ (1996) Phase I clinical trial of the flavonoid quercetin: pharmacokinetics and evidence for in vivo tyrosine kinase inhibition. Clin Cancer Res 2:659–668

Festa F, Aglitti T, Duranti G, Ricordy R, Perticone P, Cozzi R (2001) Strong antioxidant activity of ellagic acid in mammalian cells in vitro revealed by the comet assay. Anticancer Res 21:3903–3908

Franco D, Pinelo M, Sineiro J, Núñez MJ (2007) Processing of *Rosa rubiginosa*: extraction of oil and antioxidant substances. Bioresour Technol 98:3506–3512

Fujii T, Saito M (2009) Inhibitory effect of quercetin isolated from rose hip (*Rosa canina* L.) against melanogenesis by mouse melanoma cells. Biosci Biotechnol Biochem 73:1989–1993

Gao X, Uggla M, Rumpunen K (2005) Antioxidant activity of dried and boiled rose hips. Acta Hortic 690:239–249

Genc GE, Ozhatay N (2006) An ethnobotanical study in Çatalca (European part of Istanbul) II. Turk J Pharm Sci 3:73–89

Ghimire SK, Aumeeruddy-Thomas Y (2009) Ethnobotanical classification and plant nomenclature system of high altitude agro-pastoralists in Dolpo, Nepal. Bot Orient J Plant Sci 6:56–68

Grajzer M, Prescha A, Korzonek K, Wojakowska A, Dziadas M, Kulma A, Grajeta H (2015) Characteristics of rose hip (*Rosa canina* L.) cold-pressed oil and its oxidative stability studied by the differential scanning calorimetry method. Food Chem 188:459–466

Grochowski W (1990) Uboczna produkcja leśna. PWN, Warszawa, pp 379–383

Guo XD, Zhang DY, Gao XJ, Parry J, Liu K, Liu BL, Wang M (2013) Quercetin and quercetin-3-O-glucuronide are equally effective in ameliorating endothelial insulin resistance through inhibition of reactive oxygen species-associated inflammation. Mol Nutr Food Res 57:1037–1045

Hodisan T, Socaciu C, Ropan I, Neamtu G (1997) Carotenoid composition of *Rosa canina* fruits determined by thin-layer chromatography and high-performance liquid chromatography. J Pharm Biomed Anal 16:521–528

Hong L, Guo Z, Huang K, Wei S, Liu B, Meng S, Long C (2015) Ethnobotanical study on medicinal plants used by Maonan people in China. J Ethnobiol Ethnomed 11:32

Hornero-Méndez D, Mínguez-Mosquera MI (2000) Carotenoid pigments in *Rosa mosqueta* hips, an alternative carotenoid source for foods. J Agric Food Chem 48:825–882

Ilyasoğlu H (2014) Characterization of rosehip (*Rosa canina* L.) seed and seed oil. Int J Food Prop 17:1591–1598

Jacobsen SE, Sørensen M, Pedersen SM, Weiner J (2013) Feeding the world: genetically modified crops versus agricultural biodiversity. Agron Sustain Dev 33:651–662

Javanmard M, Asadi-Gharneh HA, Nikneshan P (2018) Characterization of biochemical traits of dog rose (*Rosa canina* L.) ecotypes in the central part of Iran. Nat Prod Res 32:1738–1743

Jimenez-Lopez J, Ruiz-Medina A, Ortega-Barrales P, Llorent-Martínez EJ (2017) *Rosa rubiginosa* and *Fraxinus oxycarpa* herbal teas: characterization of phytochemical profiles by liquid chromatography-mass spectrometry, and evaluation of the antioxidant activity. New J Chem 41:7681–7688

Johra FT, Bepari AK, Bristy AT, Reza HM (2020) A mechanistic review of β-carotene, lutein, and zeaxanthin in eye health and disease. Antioxidants 9:1046

Karhagomba IB, Mushagalusa TB, Nabino VB, Koh K, Kim HS (2013) The cultivation of wild food and medicinal plants for improving community livelihood: the case of the Buhozi site, DR Congo. Nutr Res Pract 7:510–518

Kazaz S, BaydaR H, Erbas S (2009) Variations in chemical compositions of *Rosa damascena* Mill. and *Rosa canina* L. fruits. Czech J Food Sci 27:178–184

Kerasioti E, Apostolou A, Kafantaris I, Chronis K, Kokka E, Dimitriadou C, Tzanetou EN, Priftis A, Koulocheri SD, Haroutounian SA, Kouretas D (2019) Polyphenolic composition of *Rosa canina*, *Rosa sempervivens* and *Pyrocantha coccinea* extracts and assessment of their antioxidant activity in human endothelial cells. Antioxidants 8:92

Kharazmi A, Winther K (1999) Rose hip inhibits chemotaxis and chemiluminescence of human peripheral blood neutrophils in vitro and reduces certain inflammatory parameters in vivo. Inflammopharmacology 7:377–386

Kim MH, Choi MK (2013) Seven dietary minerals (Ca, P, Mg, Fe, Zn, Cu, and Mn) and their relationship with blood pressure and blood lipids in healthy adults with self-selected diet. Biol Trace Elem Res 153:69–75

Kovacs S, Facsar G, Laszlo U, Toth M (2004) Phenological, morphological characteristics of some rose species found in Hungary. Acta Hortic 690:71–76

Kumar R, Bhagat N (2012) Ethnomedicinal plants of district Kathua (J&K). Int J Med Aromat Plants 2:603–611

Kumar K, Sharma YP, Manhas RK, Bhatia H (2015) Ethnomedicinal plants of Shankaracharya Hill, Srinagar, J&K, India. J Ethnopharmacol 170:255–274

Kumari P, Singh BK, Joshi GC, Tewari LM (2009) Veterinary ethnomedicinal plants in Uttarakhand Himalayan region, India. Ethnobot Leafl 13:1312–1327

Landete JM (2011) Ellagiţannins, ellagic acid and their derived metabolites: a review about source, metabolism, functions and health. Food Res Int 44:1150–1160

Larsen E, Kharazmi A, Christensen LP, Christensen SB (2003) An antiinflammatory galactolipid from rose Hip (*Rosa canina*) that inhibits chemotaxis of human peripheral blood neutrophils in vitro. J Nat Prod 66:994–995

Leaf A, Weber PC (1988) Cardiovascular effects of n-3 fatty acids. N Engl J Med 318:549–557

Li JR, Liu J, He DH, Xu HX, Ding LS, Bao WK, Zhou ZQ, Zhou Y (2013) Three new phenolic compounds from the leaves of *Rosa sericea*. Fitoterapia 84:332–337

Li BL, Yuan J, Wu JW (2021) A review on the phytochemical and pharmacological properties of *Rosa laevigata*: a medicinal and edible plant. Chem Pharm Bull 69:421–431

Lozada M, Ladio A, Weigandt M (2006) Cultural transmission of ethnobotanical knowledge in a rural community of northwestern Patagonia, Argentina. Econ Bot 60:374–385

Ma X, Zhang Y, Wang Z, Shen Y, Zhang M, Nie Q, Hou Y, Bai G (2017) Ursolic acid, a natural nutraceutical agent, targets caspase3 and alleviates inflammation-associated downstream signal transduction. Mol Nutr Food Res 61:1700332

Maxia A, Lancioni MC, Balia AN, Alborghetti R, Pieroni A, Loi MC (2008) Medical ethnobotany of the Tabarkins, a Northern Italian (Ligurian) minority in south-western Sardinia. Genet Resour Crop Evol 55:911–924

McRorie JW Jr, McKeown NM (2017) Understanding the physics of functional fibers in the gastrointestinal tract: an evidence-based approach to resolving enduring misconceptions about insoluble and soluble fiber. J Acad Nutr Diet 117:251–264

Moghanloo L, Ghahremani Nezhad F, Vafadar M (2019) Ethnobotanical study of medicinal plants in the central district of the Zanjan county, Zanjan province. Iran J Herb Drugs 9:121–131

Moure A, Franco D, Sineiro J, Domínguez H, Núñez MJ, Lema JM (2001) Antioxidant activity of extracts from *Gevuina avellana* and *Rosa rubiginosa* defatted seeds. Food Res Int 34:103–109

Nađpal JD, Lesjak MM, Mrkonjić ZO, Majkić TM, Četojević-Simin DD, Mimica-Dukić NM, Beara IN (2018) Phytochemical composition and in vitro functional properties of three wild rose hips and their traditional preserves. Food Chem 241:290–300

Nagatomo A, Nishida N, Fukuhara I, Noro A, Kozai Y, Sato H, Matsuura Y (2015) Daily intake of rosehip extract decreases abdominal visceral fat in preobese subjects: a randomized, double-blind, placebo-controlled clinical trial. Diabet Metab Syndr Obes 8:147

Najda A, Buczkowska H (2013) Morphological and chemical characteristics of fruits of selected *Rosa* sp. Modern Phytomorphol 3:99–103

Ninomiya K, Matsuda H, Kubo M, Morikawa T, Nishida N, Yoshikawa M (2007) Potent anti-obese principle from *Rosa canina*: structural requirements and mode of action of trans-tiliroside. Bioorg Med Chem Lett 17:3059–3064

Nowak R (2005) Chemical composition of hips essential oils of some *Rosa* L. species. Z Naturforsch C 60:369–378

Ouerghemmi S, Sebei H, Siracusa L, Ruberto G, Saija A, Cimino F, Cristani M (2016) Comparative study of phenolic composition and antioxidant activity of leaf extracts from three wild *Rosa* species grown in different Tunisia regions: *Rosa canina* L., *Rosa moschata* Herrm. and *Rosa sempervirens* L. Ind Crop Prod 94:167–177

Özdemir N, Pashazadeh H, Zannou O, Koca I (2022) Phytochemical content, and antioxidant activity, and volatile compounds associated with the aromatic property, of the vinegar produced from rosehip fruit (*Rosa canina* L.). LWT 154:112716

Padulosi S, Heywood V, Hunter D, Jarvis A (2011) Underutilized species and climate change: current status and outlook. In: Yadav SS, Redden RJ, Hatfield JL, Lotze-Campen H, Hall AE (eds) Crop adaptation to climate change, pp 507–521

Pang X, Liu H, Yu H, Zhang M, Bai L, Yan H (2019) A metal organic framework polymer monolithic column as a novel adsorbent for on-line solid phase extraction and determination of ursolic acid in Chinese herbal medicine. J Chromatogr B 1125:121715

Patel S (2013) Rose hips as complementary and alternative medicine: overview of the present status and prospects. Med J Nutr Metab 6:89–97

Patel S (2015) Emerging bioresources with nutraceutical and pharmaceutical prospects. Springer. https://doi.org/10.1007/978-3-319-12847-4

Patel S (2017) Rose hip as an underutilized functional food: evidence-based review. Trends Food Sci Technol 63:29–38

Polat R, Cakilcioglu U, Satıl F (2013) Traditional uses of medicinal plants in Solhan (Bingöl—Turkey). J Ethnopharmacol 148:951–963

Rao YK, Geethangili M, Fang SH, Tzeng YM (2007) Antioxidant and cytotoxic activities of naturally occurring phenolic and related compounds: a comparative study. Food Chem Toxicol 45:1770–1776

Rein E, Kharazmi A, Winther K (2004) A herbal remedy, Hyben Vital (stand. powder of a subspecies of *Rosa canina* fruits), reduces pain and improves general wellbeing in patients with osteoarthritis—a double-blind, placebo-controlled, randomised trial. Phytomedicine 11: 383–391

Rizeq B, Gupta I, Ilesanmi J, AlSafran M, Rahman MM, Ouhtit A (2020) The power of phytochemicals combination in cancer chemoprevention. J Cancer 11:4521

Roman I, Stănilă A, Stănilă S (2013) Bioactive compounds and antioxidant activity of *Rosa canina* L. biotypes from spontaneous flora of Transylvania. Chem Cent J 7:1–10

Rosu CM, Manzu C, Olteanu Z, Oprica L, Oprea A, Ciornea E, Zamfirache MM (2011) Several fruit characteristics of *Rosa* sp. genotypes from the Northeastern region of Romania. Not Bot Hortic Agrobot Cluj Napoca 39:203–208

Saaby L, Jäger AK, Moesby L, Hansen EW, Christensen SB (2011) Isolation of immunomodulatory triterpene acids from a standardized rosehip powder (*Rosa canina* L.). Phytother Res 25: 195–201

Sadia H (2020) Nutritional and antioxidant analysis of selected wild edible fruits in Pakistan Doctoral dissertation. Quaid-i-Azam University, Islamabad

Shakoor H, Feehan J, Mikkelsen K, Al Dhaheri AS, Ali HI, Platat C, Ismail LC, Stojanovska L, Apostolopoulos V (2020) Be well: a potential role for vitamin B in COVID-19. Maturitas 144: 108–111. https://doi.org/10.1016/j.maturitas.2020.08.007

Shameh S, Alirezalu A, Hosseini B, Maleki R (2019) Fruit phytochemical composition and color parameters of 21 accessions of five *Rosa* species grown in North West Iran. J Sci Food Agric 99: 574–5751

Sharma M, Shakya A, Sharma N, Shrivastava S, Shukla S (2012) Therapeutic efficacy of *Rosa damascena* Mill. on acetaminophen-induced oxidative stress in albino rats. J Environ Pathol Toxicol Oncol 31:193–201

Sharma P, Agnihotry A, Sharma PP (2015) An ethnobotanical study of medicinal plants in Murari Devi and surrounding areas (Mandi district, Himachal Pradesh), India. Indian For 141:68–78

Singh KN, Lal B (2008) Ethnomedicines used against four common ailments by the tribal communities of Lahaul-Spiti in western Himalaya. J Ethnopharmacol 115:147–159

Singh G, Rawat GS (2011) Ethnomedicinal survey of Kedarnath wildlife sanctuary in Western Himalaya, India. Indian J Fundam Appl Life Sci 1:35–46

Singh K, Sharma YP, Gairola S (2020) Morphological characterization of wild *Rosa* L. germplasm from the Western Himalaya, India. Euphytica 216:1–24

Singh K, Singh D, Lone JF, Bhat S, Sharma YP, Gairola S (2021) Nutraceutical potential of rose hips of three wild *Rosa* species from Western Himalaya, India. Not Bot Hortic Agrobot Cluj Napoca 49:12471–12471

Soelberg J, Jäger AK (2016) Comparative ethnobotany of the Wakhi agropastoralist and the Kyrgyz nomads of Afghanistan. J Ethnobiol Ethnomed 12:1–24

Speer H, D'Cunha NM, Alexopoulos NI, McKune AJ, Naumovski N (2020) Anthocyanins and human health—a focus on oxidative stress, inflammation and disease. Antioxidants 9:366

Sunderland TC (2011) Food security: why is biodiversity important? Int For Rev 13:265–274

Tang SM, Deng XT, Zhou J, Li QP, Ge XX, Miao L (2020) Pharmacological basis and new insights of quercetin action in respect to its anti-cancer effects. Biomed Pharmacother 121:109604

Thimmappa R, Geisler K, Louveau T, O'Maille P, Osbourn A (2014) Triterpene biosynthesis in plants. Annu Rev Plant Biol 65:225–257

Türkben C, Uylaser V, Incedayi B, Çelikkol I (2010) Effects of different maturity periods and processes on nutritional components of rose hip (*Rosa canina* L.). J Food Agric Environ 8:26–30

Ugulu I, Baslar S, Yorek N, Dogan Y (2009) The investigation and quantitative ethnobotanical evaluation of medicinal plants used around Izmir province, Turkey. J Med Plants Res 3:345–367

Ulian T, Diazgranados M, Pironon S, Padulosi S, Liu U, Davies L, Howes MJ, Borrell JS, Ondo I, Pérez-Escobar OA, Sharrock S (2020) Unlocking plant resources to support food security and promote sustainable agriculture. Plants People Planet 2:421–445

Upreti K, Tewari LM, Pangtey YP, Jalal JS (2010) Diversity and distribution of wild edible fruit plants of Uttarakhand. In: Lalit LM, Pangtey YPS, Tewari G (eds) Biodiversity potentials of the Himalaya. Gyanodaya Prakashan, Nainital, pp 157–196

Uzun M, Kaya A (2016) Ethnobotanical research of medicinal plants in Mihalgazi (Eskişehir, Turkey). Pharm Biol 54:2922–2932

Valdés L, Cuervo A, Salazar N, Ruas-Madiedo P, Gueimonde M, González S (2015) The relationship between phenolic compounds from diet and microbiota: impact on human health. Food Funct 6:2424–2439

Vitalini S, Iriti M, Puricelli C, Ciuchi D, Segale A, Fico G (2013) Traditional knowledge on medicinal and food plants used in Val San Giacomo (Sondrio, Italy)—an alpine ethnobotanical study. J Ethnopharmacol 145:517–529

Wan L, Tan HL, Thomas-Ahner JM, Pearl DK, Erdman JW, Moran NE, Clinton SK (2014) Dietary tomato and lycopene impact androgen signaling- and carcinogenesis-related gene expression during early TRAMP prostate carcinogenesis tomato and lycopene impact early prostate carcinogenesis. Cancer Prev Res 7:1228–1239

Wissemann V, Ritz CM (2005) The genus *Rosa* (Rosoideae, Rosaceae) revisited: molecular analysis of nrITS-1 and atpB-rbcL intergenic spacer (IGS) versus conventional taxonomy. Bot J Linn Soc 147:275–290

Xie G, Wang J, Xu X, Wang R, Zhou X, Liu Z (2016) Effect of different ripening stages on bioactive compounds and antioxidant capacity of wild *Rosa laevigata* Michx. Food Sci Technol 36:396–400

Yan G, Li S, Hu J, Zhai X, Ma W, Li N, Wang K (2014) Phenolic constituents from the roots of *Rosa laevigata* (Rosaceae). Biochem Syst Ecol 52:23–26

Zheng Y, Tian J, Yang W, Chen S, Liu D, Fang H, Zhang H, Ye X (2020) Inhibition mechanism of ferulic acid against α-amylase and α-glucosidase. Food Chem 317:126346

Žuna Pfeiffer T, Krstin L, Špoljarić Maronić D, Hmura M, Eržić I, Bek N, Stević F (2020) An ethnobotanical survey of useful wild plants in the north-eastern part of Croatia (Pannonian region). Plant Biosyst 154:463–473

Chapter 8
Ethnic Mountain Foods of Western and Eastern Himalayas, India

Malini Bhattacharyya, Anju Thattantavide, and Ajay Kumar

Abstract The Indian Himalayas is rich in flora and fauna. Plants growing in the Himalayan region are greatly valued by Himalayan dwellers for food and curing diseases. Due to a lack of sufficient transport, geographical and weather difficulties, local people depend on regional wild food plants. This chapter discusses eight local plants that are used in ethnic food preparation i.e. *Macrotyloma uniflorum* (Lam.) Verdc. (Gahat), *Echinochloa frumentacea* Link (Jhangora), *Cleome viscosa* L. (Jakhiya), *Urtica dioica* L. (Bichu), *Rhododendron campanulatum* D. Don (Buransh), *Eleusine coracana* Gaertn. (Mandwa), *Bambusa vulgaris* Nees (Bash), *Musa acuminata* Colla (Kach kola). Among these wild food plants viz. *M. uniflorum* (Gahat), *E. frumentacea* (Jhangora), *C. viscosa* (Jakhiya), *U. dioica* (Bichu), *R. campanulatum* (Buransh), and *E. coracana* (Mandwa) are selected from Western Himalayas, whereas *B. vulgaris* (Bash), and *M. acuminata* (Kach kola) are selected from Eastern Himalayan region. The traditional foods prepared from these wild food plants are not available in the mainstream Indian menu but are common in the daily cuisines of western and eastern Himalayan tribal communities. The importance of this article is to incorporate those plants in the Indian food basket and to popularize the cultivation and storage processes of these plants. In this article, we have further discuss some famous dishes prepared from the said crops and the importance of the culinary dishes for the mountain communities. This article also discusses the phytochemical constituents, nutritional values, and medicinal significance of the selected wild food plants.

Keywords Wild food plants · Eastern Himalayas · Western Himalayas · Mountain foods · Local foods

---

M. Bhattacharyya · A. Thattantavide
Department of Plant Science, School of Biological Sciences, Central University of Kerala, Kasaragod, Kerala, India

A. Kumar (✉)
Department of Plant Science, Central University of Kerala, Kasaragod, Kerala, India

A. Kumar et al. (eds.), *Wild Food Plants for Zero Hunger and Resilient Agriculture*, Plant Life and Environment Dynamics,
https://doi.org/10.1007/978-981-19-6502-9_8

8.1 Introduction

Ending global hunger or achieving zero hunger is still an unfinished agenda. The FAO statistics revealed that around 720–811 million people are suffering from hunger globally (FAO et al. 2021). The majority of the calorie intake by humans is contributed by 30 staple crops (Siddique et al. 2021). Increasing population, monoculture of high-yielding uniform varieties, global climate changes, increased susceptibility of cultivated crops to abiotic and biotic stress, inadequate distribution of food, the reduced nutritional value of staple crops, reduced availability of arable land, improper management of nutritional programs are the major challenges to global food security (Rosegrant and Cline 2003). Diversifying the food basket of consumers with nutritionally superior and abiotic and biotic stress-resilient crops can be a sustainable solution to the issue (Kumar et al. 2021a). Exploration of wild food plants to diversify the diet of humans can contribute significantly to food security (Parry et al. 2005). Various ethnic communities and rural populations depend on wild food plants to meet their major calorie requirement (Ghorbani et al. 2012). The Indian Himalayas have several tribal communities that rely on wild food plants to date. People living in the Himalayan region are uniquely dependent on nature (Uniyal et al. 2006; Malik et al. 2015). The Himalayas cover 18% of the Indian subcontinent. It contains more than 8000 plant species (Gairola et al. 2009). The Indian Himalayas is rich in biodiversity. It has almost 64 species of gymnosperms, 17,000 species of angiosperms, 2850 types of bryophytes, 1200 pteridophytes, and 2021 lichens richness (Joshi and Pant 2012). In India, 7500 species have been reported by scientists with various medicinal properties (Joshi and Pant 2012).

Indian Himalayas have a wide range, including tropical, subtropical, temperate, and alpine areas. Indian Himalayas is divided into two geographical regions: Western Himalayas and Eastern Himalayas. The Western Himalayan region consists of three Himalayan states i.e., Jammu and Kashmir, Himachal Pradesh, and Uttarakhand. Eastern Himalayas consist of many Indian states such as Sikkim, Assam, West Bengal, Mizoram, Tripura, Manipur, Meghalaya, and Nagaland. West Bengal is situated in the sub-Himalayan regions (Sati 2016). Figure 8.1 shows the satellite view of the Indian Himalayas. Western Himalayas is a dry area with less vegetation, while Eastern Himalayas is a more rain-prone area with dense vegetation. The states of India are originally present in the Himalayan region. The Himalayan region covers 12.84% global forest area (Negi 2009). But besides this rich species diversity and cultural variation, all Himalayan states suffer from different climatic extremities (Negi 2009). Therefore the biodiversity of the area is also very diverse. Figure 8.2a depicts the Western Himalayas and Fig. 8.2b depicts the Eastern Himalayas. Tribal and rural dwellers of Himalayan states are still dependent on plants as a source of food and as medicine (Kumar et al. 2011). From ancient times, plants have been used for food, shelter, and medicines. In recent times, civilization progress resulted in the decline of natural product use (Kumar et al. 2021a, b). People living in the Himalayan region depend on plants for their healthy well-being. This chapter explores the multipurpose uses of the eight wild food plants

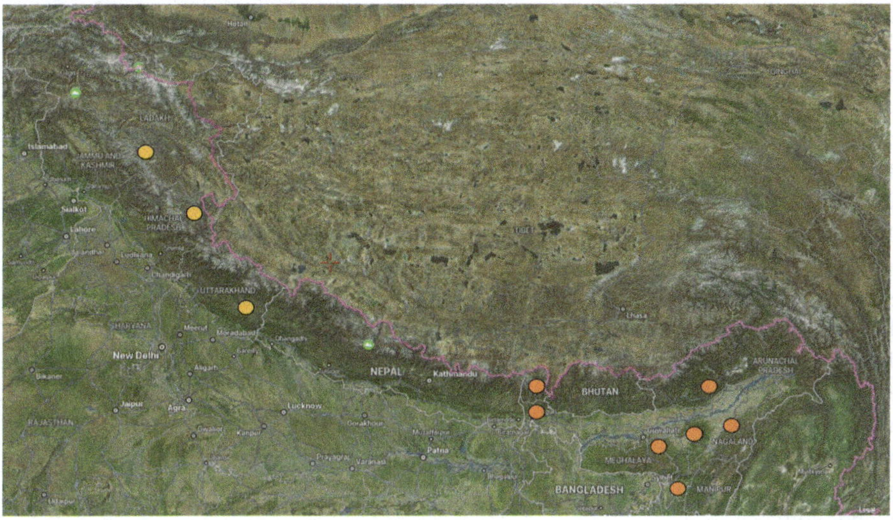

Fig. 8.1 Indian Himalayan states photograph—satellite view (https://satellites.pro/India_map#30.944573, 79.628906,6); Western and Eastern Himalayan states: Western Himalayan states (Jammu and Kashmir, Himachal Pradesh, and Uttarakhand) were marked by yellow and Eastern Himalayan states (Sikkim, West Bengal, Assam, Tripura, Meghalaya, Nagaland, Arunachal Pradesh, and Mizoram) were marked by orange colour

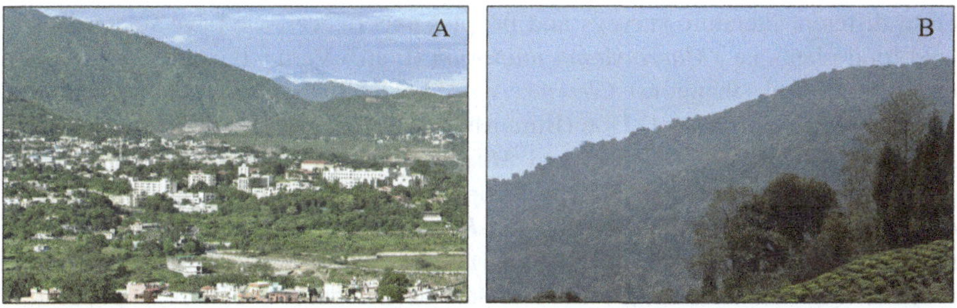

Fig. 8.2 (**a**) Dry Western Himalayas (Photograph collected from Nautiyal village; Pauri Garhwal district, Uttarakhand; 1765 m from sea level) (**b**) Monsoon-prone Eastern Himalayas (Photograph collected from Ravangla city, Sikkim; 2400 m from sea level)

for the tribal communities. The preparation of some important dishes from these plants is also detailed in this chapter. Brief discussion about the medicinal and ethnoveterinary uses of these eight WFPs is also given besides providing their nutritional and phytochemical composition.

The traditional knowledge regarding the wild food plants from the Himalayan region is significantly contributing toward the hunger satisfaction, poverty eradication, medicinal benefits and wiping out the economic insecurity of the consumers (Sharma et al. 1979; Singh et al. 1987). It was found that the younger generation is also depending on the plants to generate a small amount of income. Several other studies found that rural people are generating income and achieving food security during famine time with the help of wild plants. Several tribal communities residing in the Indian Himalayas still consume wild food plants. This chapter aims to explore the wild food plants that are used by the local communities of the Himalayan region. The majority of wild food plants are nutrient-rich and they can be introduced into modern-day agriculture (Goron and Raizada 2015). People from Africa and Asia consume plants that are less popular and neglected by the scientific community. Sometimes neglected, wild food plants are very hardy and stress-tolerant. So their cultivation can lead to reduction of hunger in drought-prone/harsh environmental conditions and poor-income areas (Goron and Raizada 2015). Due to geographical barriers and fewer transport facilities, Himalayan cuisines are not explored well. The purpose of this article is to explore some less-known underutilized wild food plants from the Himalayan region and popularize them in the mainstream food market. Here we have reviewed the medicinal benefits, phytoconstituents, and nutritional values of the eight selected wild food plants elaborately.

8.2 Selected Wild Food Plants Found in the Eastern and Western Himalayas

From different literature surveys and peoples experiences, we have selected eight wild food plants i.e., *Macrotyloma uniflorum* (Lam.) Verdc. (Gahat), *Echinochloa frumentacea* Link (Jhangora), *Cleome viscosa* L. (Jakhiya), *Urtica dioica* L. (Bichu), *Rhododendron arboreum* D.Don (Buransh), *Eleusine coracana* Gaertn. (Mandwa), *Bambusa vulgaris* Nees (Bash) and *Musa acuminata* Colla (Kach kola) in the present study. Western Himalayan people use *M. uniflorum*, *E. frumentacea*, *C. viscosa*, *U. dioica* as food plants and *R. arboreum* for the preparation of juice. Eastern Himalayan people consume *B. vulgaris* and *M. acuminata* for their food purpose (Fig. 8.3). *M. uniflorum* is a less-used legume plant. It is popular in pharmaceutical industries due to its medicinal potential (Siddhuraju and Manian 2007; Patel and Acharya 2020). *E. frumentacea* (Jhangora) is a weed plant and its cultivated variety is known as Barnyard millet. In India its cultivated variety is known as "Ragi". In some places in Africa, people consume *E. frumentacea* as food (Singh et al. 2010; Renganathan et al. 2020). *C. viscosa* (Jakhiya) is a highly useful plant in terms of its medicinal properties and available phytoconstituents (Saxena et al. 2000; Sengottuvelu et al. 2007; Singh and West 1991). But the use of this plant is restricted to some regional areas "Garhwal subdivision" of the western Himalayas. A secondary literature survey revealed the commercialization of this plant can not only reflect the culture and customs of a particular Himalayan region but can benefit other people in many ways a source of income generation, herbal medicine, and introduction of new cuisines. Moreover, the commercialization of its nonedible seed

Fig. 8.3 Habit photographs of the selected mountain food plants. (**a**) *Macrotyloma uniflorum* (Gahat), (**b**) *Echinochloa frumentacea* (Jhangora), (**c**) *Cleome viscosa* (Jakhiya), (**d**) *Urtica dioica* (Bichu), (**e**) *Rhododendron campanulatum* (Buransh), (**f**) *Eleusine coracana* (Mandwa), (**g**) *Bambusa vulgaris* (Bash), (**h**) *Musa acuminata* (Kach kola) (Photograph sources: (**a**) R Bisht, (**b**) R Bisht, (**c**) R Bisht, (**d**) R Bisht, (**e**) M Bhattacharyya, (**f**) R Bisht, (**g**) R Sarkar, and (**h**) R Bhattacharyya)

oil can act as a feedstock for biodiesel production and it can address the issues regarding the exploitation of non-renewable energy (Upadhyay 2011).

Planting *Cleome* plants in the agricultural fields can act as a natural insecticide, which can help farmers from crop loss and save money on chemical insecticides (Upadhyay 2011). *U. dioica* plant is commonly known as Stinging nettle as it has terrible stings with it (Sudhakar and Rolla 1985). These stings are the uniqueness of the plant. This plant is also highly neglected as a food source and restricted to some portions of the Himalayas but Badola (2011) reported that the plant can act as a source for the socio-economic upliftment of the Himalayan people (Badola 2011). *R. arboreum* reflects the agrobiodiversity of many Himalayan states such as Uttarakhand and Sikkim. It is well known for its large charming flowers. The flowers have created a source of income by exporting healthy flower juices (Som et al. 2019). It became a cash plant by developing different business strategies; from organic farming to small industries. The new generation is interested in selling *R. arboreum* (Buransh) flower juices (Bhatt et al. 2007). The flowers are notable for their health benefits also. *B. vulgaris* consumption was restricted to tribal regions of Tripura Himalaya, but now they have spread to urban areas also. It was a less-known food plant before and used for household purposes only. It has several medicinal properties (Jun 2015). The ethnomedicinal uses of the plant was documented by Srivastava and Shukla (2011). *M. acuminata* is available in the tropical and subtropical regions of India. It is also reported that this plant has migrated to the temperate lining of mountains. It has many medical properties (Mathew and Negi 2017). Other than that, it is highly useful for Indian societies (Yadav 2021) and it is traditionally used in preparation of several food preparations (Thorat and Bobade 2018). All Indian communities use this plant for the food purpose and ritual beliefs. The cultivated variety is the most popular and also it is a staple food for some communities, but the wild variety is closer to rural peoples' daily life. Its pseudostem and flowers are not consumed by many communities due to less knowledge regarding their consumption (Panigrahi et al. 2017). Fruit is only consumed by a large number of people. So the literature survey regarding these eight plants embodies the usefulness of the plants, their medicinal values, nutritional profile, and reported uses as ethnomedicines.

8.2.1 Traditional Ethnic Foods Prepared from the Wild Food Plants

Local communities prepare traditional dishes from the selected wild food plants. There are different methods for their preparation. Table 8.1 provides the details of the eight wild food plants and their preparation/consumption details. Table 8.2 provides the popular dishes and food applications of the eight plants. Figure 8.4 represents some dishes prepared by local communities. These dishes are linked with the culture and rituals of some Himalayan ethnic communities (Pant 2019). Figure 8.5 shows the mandatory cooking techniques and processing of the plants

Table 8.1 The details of selected eight wild food plants from the Eastern and Western Himalayan region

Sl. No.	Scientific names	Local name	Family	Used in Himalayan states	Plant Parts used	Method of preparation/ processing technique
1	*Macrotyloma uniflorum*	Gahat dal	Fabaceae	Uttarakhand	Seeds	Seeds are ground and curry is prepared
2	*Echinochloa frumentaceae*	Jhangora	Poaceae	Uttarakhand	Seeds	Khir is prepared by boiling seeds with milk
3	*Cleome viscosa*	Jakhiya	Cleomaceae	Uttarakhand	Seeds	Seeds are mixed with curry
4	*Urtica dioica*	Kandali	Urticaceae	Uttarakhand	Leaves	Leaves are boiled and curry is prepared
5	*Rhododendron campanulatum*	Burans	Ericaceae	Uttarakhand, Himachal Pradesh	Flowers	Juice is prepared
6	*Eleusine coracana*	Mandwa	Poaceae	Uttarakhand	Grains	Curry is prepared with ground grains
7	*Bambusa vulgaris*	Baash Kurul	Poaceae	Tripura	Young shoots	Young shoots are chopped and soaked with tap water for 30 min prior to curry preparation
8	*Musa acuminata*	Thor	Musaceae	West Bengal	Stems	Chopped and curry is prepared
		Mocha			Flower	Curry is prepared after eliminating style and stigma

Table 8.2 Some popular dish names of the selected wild food plant

Sl. No.	Plant name	Famous culinary dishes
1	*Macrotyloma uniflorum*	Gahat Phaanu; Gahat Paratha; Daal; Curry
2	*Echinochloa frumentacea*	Jhangora bhaat; Khir; Sweets
3	*Cleome viscosa*	Chatni; Spices; Tastemaker
4	*Urtica dioica*	Kandali ka saag; Curry
5	*Rhododendron arboreum*	Juice
6	*Eleusine coracana*	Halwa; Roti; Sweet dish
7	*Bambusa vulgaris*	Bash kurul; Curry preparation with dry fish, pork and red meat, Fry preparation
8	*Musa acuminata*	Mochar ghonto (from flower), Thor Chochchori (from stem), Thor chechki; Curry preparation with coconut, Snacks preparation with prawns

during dish preparation. Among eight food-producing plants, five plants were highlighted as they have unique food preparation techniques. For example, the preparation of curry from bichu ghas involves the grinding of leaves after boiling. The preparation of "Paratha" from gahat dal requires the crushing of grains for the paste preparation. The preparation of "jhangora vaat" also needs the selection of grains. In a similar way to preparation strategies for bamboo shoots curry needs chopped bamboo shoots rinsed with water. Preparation of "Mocha" dish requires boiling banana flowers without stigma and bract. Similarly, the preparation of "Thor chechki" from chopped pseudostems of *M. acuminata* needs to be boiled for 30 min with a little amount of turmeric to reduce bacterial infections.

8.2.2 Processing, Marketing, and Supply Chain Management of Wild Mountain Food Plants

The eight plants' edible parts are under-explored. Villagers lack proper cultivation knowledge and training. They lack proper irrigation techniques also. The knowledge regarding agriculture is passed from generation to generation. They store food grains/pulses/seeds in polyethene bags and jute bags. This study revealed that the above-mentioned plants belong to the daily needs of the village dwellers of the Himalayas. Besides their food value, literature surveys exposed that these plants have pharmacological activities and they are medicinally beneficial also. In recent times, mountain food products have received increasing attention (Cruz-Martins and Ferreira 2017). From the studies, it was reported that gahat dal is good for controlling kidney stones (Ravishankar and Priya 2012). Villagers also believe that Jhangora is a

Fig. 8.4 Some popular dishes of Western and Eastern Himalayas that are based on wild food plants. (**a**) Gahat dal (**b**) Jhangora ki kheer (**c**) Gahath ka paratha with Jakhiya chatni (**d**) Kandali ka saag (**e**) Mandwa ka halwa (**f**) Jhangora ka bhaat (**g**) Dry Bamboo shoot curry with red meat (**h**) Bamboo shoot mutton curry (**i**) "Mocha" (**j**) Bamboo shoot curry with dry fishes (**k**) *Rhododendron arboreum* juice and (**l**) Thor chechki. (Photograph sources: (**a**) R Bisht, (**b**) R Bisht, (**c**) R Bisht, (**d**) R Bisht, (**e**) R Bisht, (**f**) R Bisht, (**g**) R Sarkar, (**h**) R Sarkar, (**i**) R Bhattacharyya, (**j**) R Sarkar, (**k**) S Shah, (**l**) R Bhattacharyya)

good source of protein, but it should be verified with scientific experiments. Therefore selling these plants in the markets and exposing them to agricultural exhibitions can increase their popularity. It was found that from every household, 89–545 USD net income was computed by commercializing *R. arboreum* flower juice (Chauhan et al. 2021). Therefore, these plants are an ideal source of income generation for the villagers. But proper marketing and commercialization strategies should be applied for the successful trading and selling of the plants. Their medicinal values can also be explored to generate income. Figure 8.6 represents the marketing and exporting strategy of the wild food plants.

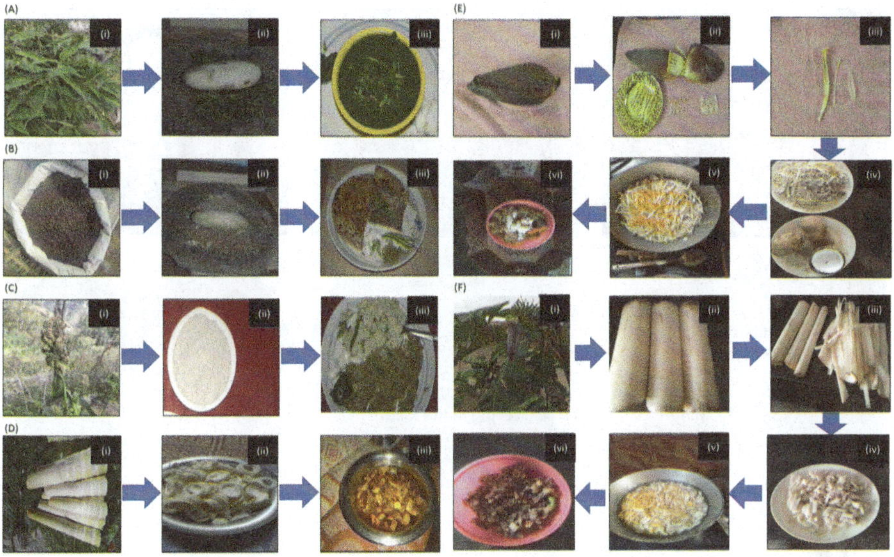

Fig. 8.5 Preparation of local dishes from the wild food plants discussed in this chapter. (**a**) Preparation of curry from bichu ghas. (**i**. Habit plant bichu ghas with leaves, **ii**. Leaves grinded after boiling, and **iii**. Photograph of curry served with onion, and green chillies) (**b**) Preparation of Paratha from daal (**i**. Dal grains, **ii**. Grains crushed and paste preparation, and **iii**. Paratha prepared and served with garlic—jakhiya chutney) (**c**) Preparation of jhangora vaat (Rice) (**i**. Habit of jhangora with grains, **ii**. Grains, and **iii**. Rice served with chausa dal (crushed *Vigna mungo* grains curry)) (**d**) Preparation strategies of Bamboo shoot curry (**i**. Chopped and peeled off young bamboo shoots, **ii**. Chopped bamboo shoots, and **iii**. Mutton curry preparation with shoots) (**e**) Preparation of Mocha dish from banana flowers (**i**. Banana inflorescence, **ii**. Flowers collected from inflorescence, **iii**. Stigma and bract removed from the flower before cooking, **iv**. Preparation ingredients, **v**. Cooking strategy needs large amounts of turmeric powder, and **vi**. The picture depicts the famous dish "Mocha") and (**f**) Preparation of Thor chechki from banana pseudostem (**i**. Banana pseudostem, **ii**. Chopped pseudostems, **iii**. Chopped pseudostems boiled for 30 min with a little amount of turmeric to reduce bacterial infections from food and **iv**. Cooked "Thor chechki") (Photograph sources: (**a**) R Bisht, (**b**) R Bisht, (**c**) R Bisht, (**d**) R Sarkar (**e**), and (**f**) R Bhattacharyya)

8.2.3 Ethnoveterinary Uses of the Selected Plants in Western and Eastern Himalayas

In developing countries, farmers raise cattle for food (meat and milk) demand. Therefore the use of ethnoveterinary medicine for the well-being of livestock is also common in such regions (Eiki et al. 2021). Plant sources are hazardless, cheap, and easy to obtain for ethnoveterinary purposes. Livestock is one of the strengths of agriculture-based countries. People living in rural areas depend on livestock for their economy and food (Mandal and Rahaman 2022). Due to the lack of doctors and ethnoveterinary hospitals in rural areas, villagers depend on plants for this purpose. The knowledge regarding ethnoveterinary plants is passed from ancestors

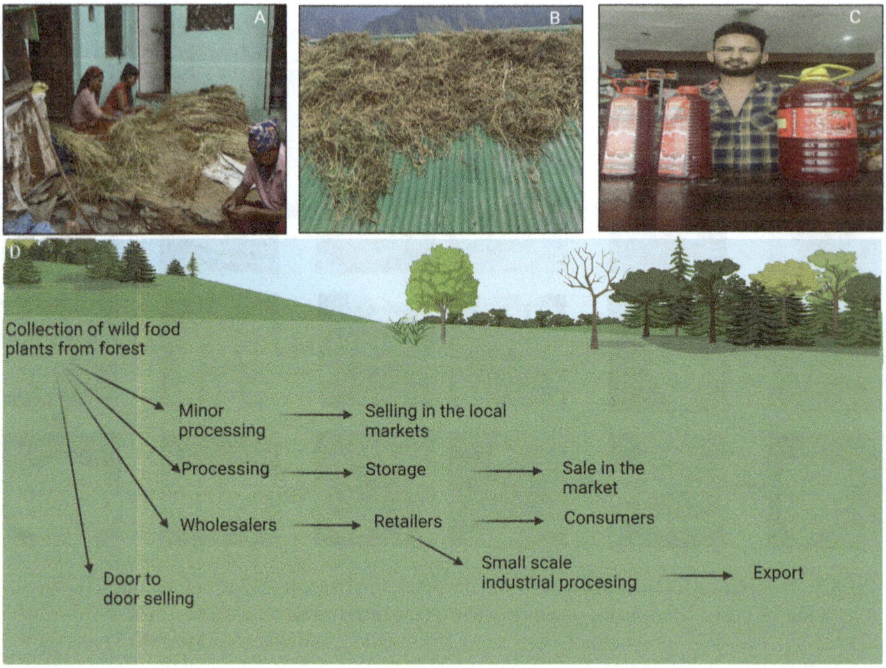

Fig. 8.6 Processing and marketing of ethnic foods. (**a**) Mandwa grains selection after harvesting in the high altitude village of Uttarkashi district of Uttarakhand state, (**b**) Harvested plants kept for drying, (**c**) *Rhododendron arboreum* juice prepared by organic farming groups from Chamoli district, Uttarakhand and sold by the departmental stores of Dehradun (Capital city): photographs show the departmental store owner (retailer) with some *Rhododendron* flower juice bottles and the owner has the connection with village people who are involved in farming and (**d**) Schematic representation of marketing channels of wild food plants. (Photograph sources: (**a**) R Bisht, (**b**) R Bisht, (**c**) S Shah; Figure created with biorender.com)

(Mandal and Rahaman 2022). Phytotherapy can protect rural people from losing livestock. This article can enlighten the possible ways to use the selected plants as animals' medicine. Table 8.3 presents ethnoveterinary uses and the reasons behind using the eight wild food plants.

8.3 Nutritional Value

Nutritional value refers to the presence of macro and micronutrients in a food sample. From the available literature, we have collected data related to the nutritional profile of the selected wild food plants. The high nutritional content in a food sample indicates the potential of the food to contribute to a healthy diet for the consumer. Nutritional richness in food is very important for the growth and development of the

Table 8.3 Ethnoveterinary uses of the selected wild food plants of the Himalayas

Sl. No.	WFP Name	Ethnoveterinary uses	Reasons behind	References
1	*M. uniflorum* (Gahat)	It is used in Himalayan areas as cattle food for the proper lactation of cattle	Seeds and pulp of the fruit are beneficial for the lactation of cattle	Sikarwar and Tiwari (2020)
2	*E. frumentacea* (Jhangora)	It is used as food for cattle	For maintaining a healthy diet for cattle	Renganathan et al. (2020)
3	*C. viscosa* (Jakhiya)	Survey analysis revealed the *C. viscosa* plant has a beneficial role in controlling diseases in cattle. In a case study on the forest-covered Pali district of Rajasthan tribal communities are using *C. viscosa* plants for controlling diseases in buffaloes	It helps the digestion process of cattle. Seeds with water can reduce fever and diarrhea	Meena et al. (2021)
		Useful for controlling epilepsy in animals	Leaves have healing properties	Patil and Deshmukh (2015)
4	*R. arboreum* (Buransh)	Good for cattle health	It can control external parasites	Som et al. (2019)
5	*E. coracana* (Mandwa)	Useful for controlling the mouth diseases of cows	It has antimicrobial activities	Gakuya et al. (2011)
6	*B. vulgaris* (Bash)	It is used to reduce labor pains in pigs. Sometimes it is fed as a postpartum cleanser	Not reported	Lans et al. (2007)
		It is used to control indigestion and diarrhea	Anti-inflammatory activity	Nair et al. (2021); Lodhi et al. (2016)
7	*M. acuminata* (Kach kola)	Peels are good for digestion	Not reported	Mandal and Rahaman (2022)

human body (Kader et al. 2012). Amino acids, proteins, carbohydrates, fat, fiber, minerals, and vitamins are the major nutrients found in a plant-based diet (Aberoumand 2009). *M. uniflorum* is rich in calcium (Ca), iron (Fe), magnesium (Mg), potassium (K), zinc (Zn), phosphorus (P), and manganese (Mn) as minerals (Aditya et al. 2019). *E. frumentacea* (Jhangora) also possess Ca, Fe Mg, K, Zn, P, and Mn as minerals (Kaur and Sharma 2020; Pandey et al. 2021). *C. viscosa* (Jakhiya) contains Ca, Fe, Mg, K, Zn, P, Cu, and Mn (Uraku and Uraku 2018). *R. arboreum* flower has Fe, Zn, Na, and Cu as mineral content (Kregiel et al. 2018). The plant *U. dioica* is rich in vitamins, minerals and fibers (Rutto et al. 2013; Maietti et al. 2021). Studies by Sharma et al. (2017) and Devi et al. (2014) reported a significant quantity of Ca, Fe, Mg, K, P, and Mn in *E. coracana*. *B. vulgaris* shoots are rich in Ca, Mg, K, P, and Na (Chongtham et al. 2011; Nongdam and Tikendra 2014). Studies revealed that both inflorescence and pseudostem of *M. acuminata* are rich in minerals. Pseudostem has Ca, Mg, K, P, Na, Zn, and Copper (Cu) in it (Islam et al. 2022). Therefore it is clear that the consumption of these wild food plants can help to eradicate hidden hunger or micronutrient deficiency from the consumers. Replacement of nutritional-less food with wild food plants can solve the issue to a certain extent at a cheap cost. The nutritional quantity of the selected wild food plants is displayed in Table 8.4 which is collected from other literature.

8.4 Medicinal Importance of the Selected Plants

The medicinal properties of the eight plants were reported by various other studies. *M. uniflorum* daal is well known for its anti-obesity activity. Obesity is the excessive weight gain phenomenon that finally leads to coronary heart disease, glucose intolerance, hypertension, and diabetes (Vadivelu et al. 2019). Other than that *M. uniflorum* is highly notable for its potential to reduce kidney stones (Patel and Acharya 2020). Therefore *M. uniflorum* daal is capable of protecting consumers from several diseases (Singh et al. 2012; Patel and Acharya 2020). *E. frumentacea* can also be used as a medicinal food crop. The plant contains less glucose, hence low glycemic index biscuits can be produced using *E. frumentacea* seeds (Anju and Sarita 2010). Studies revealed that *C. viscosa* can be called a "green pharmacy" since it has the potentiality to control several health issues (Choudhury et al. 2015; Upadhyay 2011). Leaves of *U. dioica* are considered herbal medicine for their capability to defend against various diseases (Dhouibi et al. 2021). *R. arboreum* is also a significant Himalayan plant with medicinal properties. Flowers have several health-protecting activities and can be consumed directly (Chauhan et al. 2021). The latest in-vitro study by Lingwan et al. (2021) reported that *R. arboreum* petals can inhibit SARS–CoV2 infection. The dermal wound healing property of *E. coracana* was proved by Hegde et al. (2005) on rats. It is also good for the health of pregnant women and increases bone strength. The presence of antioxidants points toward the anti-aging property of the plant (Luitel et al. 2020). Bamboo shoots are useful for the treatment of lung infection (Shukla et al. 2012). Wild *Musa* flower can control allergies, infections, Bronchitis, Tuberculosis, and other respiratory diseases (Mathew and Negi 2017; Camacho-Corona et al. 2008) *Musa* flower can reduce hypertension, cough, asthma, bronchitis, and joint pain (Di Staci et al. 2002; Subbaraya 2006; De Wet et al. 2016; Srinivas et al. 2021). Table 8.5 documented the disease remedial properties of the selected wild food plants.

Table 8.4 Nutritional composition of the wild food plants

WFP Name	Moisture	Energy	Protein	Fiber	Fat	Carbohydrate	Sugars	Ca	Fe	Mg	P	K	Na
M. uniflorum (Gahat dal)	9.28 ± 0.57 g/100 g	1379 ± 9 Kcal/100 g	21.73 ± 0.29 g/100 g	7.88 ± 0.02 g/100 g	0.62 ± 0.04 g/100 g	57.24 mg/100 g	–	269 ± 34.9 mg/100 g	8.76 ± 1.16 mg/100 g	152 ± 18.1 mg/100 g	296 ± 22.5 mg/100 g	1065 ± 42.4 mg/100 g	12.14 ± 0.17 mg/100 g
E. frumentacea (Jhangora)	8.74%	12.81%	10.10%	6.70%	4%	68.80%	–	19 mg/100 gms	5 mg/100 gms	83 mg/100 gms	281 mg/100 gms	408 mg/100 g	11 mg/100 g
C. viscosa (Jakhiya)	11.94%	7.60%	6%	0.50%	4.29 ± 0.07 mg/100 g	53.18%	5.3 mg/g	0.38%	0.04%	0.32%	0.33%	0.07%	0.02%
U. dioica (Kandali, bichu ghas)	7.04 ± 0.77%	307.24 ± 0.13 kcal/100 g	33.77 ± 0.35%	9.08 ± 0.14%	–	37.39 ± 0.72%	–	168.77 ± 1.47 mg/100 g	227.89 ± 0.21 mg/100 g	–	–	–	5.7 ± 0.1 mg/100 g
R. arboreum flower (Buransh)	79.40%	–	1.68%	2.9	1.52%	12.20%	–		405 ppm	–	–	–	385 ppm
E. coracana (Mandwa)	–	–	3.70%	–	–	54 mg/gm	–	0.01%	46.90%	0.13%	0.33%	0.17%	0.01%
B. vulgaris Bamboo shoot (Bash)	90.60 ± 0.82 g/100 g		3.64 ± 0.03 g/100 g	4.24 g/100 g	0.50 g/100 g	6.51 ± 0.05 g/100 g	–	320 mg/100 g		100 mg/100 g	220 mg/100 g	920 mg/100 g	400 mg/100 g
M. acuminata (Kach kola) pseudostem	6.2 ± 0.15%		3.6 ± 0.01%		1.8 ± 0.02%	62.8 ± 0.25		6.00 ± 0.19 mg/L		83.39 ± 1.91 mg/L	0.92 ± 0.03 g/L	2.03 ± 0.06 g/L	2.30 ± 0.08 g/L
M. acuminata (Kach kola) flower[a]	91 ± 0.54%	–	1.79 ± 0.23%	49.83%	0.43 ± 0.1	58.82%		377.63 mg/100 g	3.69 mg/100 g	250 mg/100 g	365.86 mg/100 g	365.9 mg/100 g	39.7 mg/100 g

(continued)

Table 8.4 (continued)

WFP Name	Zn	Cu	Mn	Si	Vit C	B1	B2	B3	Folate/ B9	Vit A	Vit E	Vit K	Carotenoids	References
M. uniflorum (Gahat dal)	2.71 ± 0.21 mg/ 100 g		3.13 ± 0.41 mg/100 g	–	–	0.32 ± 0.002 mg/ 100 g	0.24 ± 0.033 mg/ 100 g	1.82 ± 0.26 mg/ 100 g	163 ± 5.3 μg/ 100 g	–	0.27 ± 0.02 mg/ 100 g	–	–	Aditya et al. (2019)
E. frumentacea (Jhangora)	2.3 mg/ 100 g		5.94 mg/ 100 g	–	–	0.42 mg/ 100 g	0.19 mg/ 100 g	1.1 mg/ 100 g	18.3 μg/ 100 g	–	22 mg/ 100 g	–	–	Kaur and Sharma (2020); Pandey et al. (2021)
C. viscosa (Jakhiya)	0.00%	0.00%	0.01%	2.74%	2.15 (mg/g)	–	–	–	–	–	0.318 (mg/g)	–	–	Uraku and Uraku (2018)
U. dioica (Kandali, bichu ghas)	–	–	–	–	–	–	–	–	–	–	–	–	3496.67 ± 0.56 (μg/ g, db)	Kregiel et al. (2018), Rutto et al. (2013)
R. arboreum flower (Buransh)	32 ppm	26 ppm	50.2 ppm		11.5 (mg/100 mL)	–	–	–	–	–	–	–	–	Kumar et al. (2019)
E. coracana (Mandwa)	–	–	–	–	–	0.32%	0.05%	0.70%	–	–	–	–	–	Sharma et al. (2017), Devi et al. (2014)
B. vulgaris Bamboo shoot (Bash)	–	–	–	–	0.52 ± 0.10 (mg/100 g fresh weight)	–	–	–	–	–	–	0.52 mg/ 100 g		Chongtham et al. (2011); Nongdam and Tikendra (2014)
M. acuminata (Kach kola) pseudostem	1.00 ± 0.02 g/L	2.50 ± 0.09 g/ L	–	–	–	–	–	–	–	–	–	–	–	Islam et al. (2022), Maskey et al. (2020)
M. acuminata (Kach kola) flower[a]	4.01 mg/ 100 g	1.36 mg/ 100 g	8.77 mg/ 100 g	–	–	–	–	–	–	–	–	–	–	Fingolo et al. (2012)

[a]Nutritional composition of the whole inflorescence is reported

Table 8.5 Medicinal properties of the eight wild food plants eaten as a part of traditional foods

Medicinal properties of the wild food plants	References
1. *M. uniflorum* (Gahat)	
It can reduce the kidney stone	Gupta and Kanwar (2018); Patel and Acharya (2020)
Used in controlling bacterial and fungal infections	Gupta and Kanwar (2018)
Antilithiatic activity and antioxidant activity	Gupta and Kanwar (2018)
It can control obesity and digestion problems after high-fat diet	Vadivelu et al. (2019)
A rat model study revealed this species is useful in decreasing crystalluria	Patel and Acharya (2020)
Seeds can reduce cholesterol gallstones, serum total cholesterol, LDL, triglyceride, and very low-density lipoprotein	Bigoniya et al. (2014)
It has analgesic activity	Ashraf et al. (2018)
It has antifungal and antimicrobial activities	Mohan and Elyas (2019)
It has the potential to control liver injury. A study performed on obese rats showed *M. uniflorum* (Gahat) can reduce inflammation and oxidative stress developed in liver injury	Bharathi et al. (2018)
Studies in rats explored *M. uniflorum* phenolic acids can control heart problems	Panda et al. (2016)
Application of *M. uniflorum* seeds with fish scale collagen sponge act as wound dressing material	Muthukumar et al. (2014)
2. *E. frumentacea* (Jhangora)	
It has the potential to reduce diabetes	Saleh et al. (2013)
It has antioxidant and radical scavenging activity	Anis and Sreerama (2020)
3. *C. viscosa* (Jakhiya)	
Antipyretic, analgesic, anthelmintic, antimalarial, antimicrobial, gastroprotective, immunomodulatory activity	Mali et al. (2007); Mali (2010); Saxena et al. (2000); Tiwari et al. (2004); Sudhakar et al. (2006); Williams et al. (2003)
Essential oil of *C. viscosa* has antifungal and antimicrobial activities	Krishnamoorthy et al. (2021)
Antidiarrheal activity was reported	Parimala Devi et al. (2002); Devi et al. (2014)
It helps to recover digestive system disorder	Choudhury et al. (2015)
It has hepatoprotective activity	Dixit and Gupta (2009); Yadav et al. (2010); Nguyen et al. (2017); Sengottuvelu et al. (2007)
It has analgesic activity	Parimaladevi et al. (2003)
It has mosquitocidal properties against *Aedes aegypti* (Linn.) (Diptera: Culicidae)	Krishnappa and Elumalai (2013)
4. *U. dioica* (Bichu)	
It is used as tea or fresh leaves to treat disorders of the kidneys and urinary tract, gastrointestinal tract, locomotor system, skin, cardiovascular system, hemorrhage, influenza, rheumatism, and gout. In-vivo studies revealed that it helps to lower blood pressure and diabetic problems	El Haouari and Rosado (2018)

(continued)

Table 8.5 (continued)

Medicinal properties of the wild food plants	References
It controls hyperglycemia	Chehri et al. (2022)
This neglected crop has wound-healing activities	Kasouni et al. (2021)
Leaf extracts involved in glucose homeostasis in-vivo condition	Dhouibi et al. (2021)
Useful in controlling menopausal hot flashes	Kargozar et al. (2019); Taylor (2015)
It inhibits cell line proliferation and metastasis	Kardan et al. (2020)
Neurotoxicity, hepatotoxicity, and neurobehavioral changes after potassium bromate chronic exposure	Dhouibi et al. (2021)
It has wound-healing properties	Zouari Bouassida et al. (2017)
5. *R. arboreum* flower	
Useful for the treatment of diarrhea and dysentery, the corolla is helpful in controlling the fish bone stuck in the gullet, anti-inflammatory, antibacterial, and anti-nociceptive activity	Swamidasan et al. (2020)
Flower anthocyanin has radical scavenging properties	Bhatt et al. (2022)
Flowers have antidiarrheal activities	Verma et al. (2011)
6. *E. coracana* (Mandwa)	
Good for pregnant women's health, good for bone strength, antioxidant activity, antibacterial activity, and delays ageing	Luitel et al. (2020)
Studies on rats revealed this crop has the potential to heal dermal wounds	Hegde et al. (2005)
7. *B. vulgaris* (Bash) shoot	
Useful for the treatment of lung infection	Shukla et al. (2012)
8. *M. acuminata* (Kach kola) pseudo stem	
It can control tuberculosis and other respiratory diseases	Camacho-Corona et al. (2008)
It can control allergies, infections, bronchitis	Mathew and Negi (2017)
***M. acuminata* (Kach kola) flower**	
Can control hypertension	De Wet et al. (2016)
Helpful in curing cough, asthma, bronchitis	Di Staci et al. (2002)
Good for controlling joint pain, better blood circulation	Subbaraya (2006)
Lectin has anticancer activities	Srinivas et al. (2021)

8.5 Phytoconstituents of the Selected Mountain Wild Food Plants

Phytochemicals or phytoconstituents are the chemicals synthesized by the plant through primary or secondary metabolism. They play a major role in plant growth, development and defense mechanisms (Divya et al. 2017). They are non-nutritive in nature. They usually help plants by reducing the redox potential developed during stress exposure. They act as the reservoir of antioxidants and provide health benefits to human beings (Duduku et al. 2007). Table 8.6 represents the phytoconstituents and the secondary metabolites present in the eight plants.

Table 8.6 Phytoconstituents and secondary metabolites of the selected ethnic mountain wild food plants

Sl. No.	Plant name	Phytoconstituents/secondary metabolites	References
1	*M. uniflorum* (Gahat)	Kaempherol glycoside, Quercetin, Apigenin derivative, Choline (Ch), Rutin, Gallic acid, Caffeic acid, Malic acid, Stigmasterol	Gautam et al. (2020)
2	*E. frumentacea* (Jhangora)	Gallic acid, 3,4-Dihydroxy benzoic, Vanillic, Caffeic, Chlorogenic, Ferulic, *p*-Coumaric, Myricetin, Daidzein, Luteolin, Naringenin, Apigenin, Kaempferol, L-malic acid, trans-aconitic acid, (+)-isocitric acid, 5-*O*-caffeoylquinic acid, 4-*O*-caffeoylquinic acid, Isocarlinoside, 22″-*O*-rhamnosylisoorientin, and 7-*O*-(2″-*O*-glucuronosyl)glucuronosyltricin	Anis and Sreerama (2020); Kim et al. (2008)
3	*C. viscosa* (Jakhiya)	Tannins, Saponins, Flavonoids, Terpenoids, Steroids, Cleomeolide, Cleomaldic acid Cleosandrin, Cleomiscosin A, Cleomiscosin B, Cleomiscosin C, Cleomiscosin D, Dihydrokaempferide-3-glucuronide, Docosa noic acid, Naringenin-4-galactoside, Dihydrokaempferol-4′-xyloside, Naringenin-4-(xylosyl--(1,4)-glucoside), Naringenin Glycoside, Kaempferide-3-glucuronide, 3′,4′, 5′--trihydroxyflavanone-7-*O*-α-L-rhamnopyranoside, β-amyrin, Lupeol, Glyco flavanone, 3′,4′-dihydroxy-5-methoxyflavanone-7-*O*-α-L-rhamnopyranoside, stigmasta-5,24(28)-diene-3β-*O*-α-L-rhamnoside, Ergast-5-ene-3-*O*-α-L-rhamnopyranoside, 5,4′-di-*O*-methyleriodictyol-7-*O*-β-D-glucopyranoside, Gluocapparin, Glucocleomin, Cleomaldeic acid, 20-oxocembra-3,7,11,15-tetraen-19-oic acid, Visconoside A, Visconoside B, Vincetoxicoside A, Kaempferitrin, Kaempferide 3-*O*-*b*-D-glucopyranoside 7-*O*-a-L-rhamnopyranoside, Isorhamnetin 3-*O*-*b*-D-glucopyrano-side, Nevirapine, Lactam Nonaic acid, linoleic acid, palmitic acid, stearic acid, oleic acid, linolinic acid, Viscocic acid, Viscosin	Mali (2010)
4	*U. dioica* (Kandali, bichu ghas)	Flavonoids, Tannins, Coumarins, Lignans, sterols, fatty acids, polysaccharides, Isolectins, Ferulic acid, Naringin, Ellagic acid, Myricetin, Rutin, Secoisolariciresinol, 9,90-bisacetyl-neo-olivil, glucosides, β-sitosterol, Scopoletin, *p*-hydroxy-benzaldehyde, triterpenoic acids, monoterpendiols, Rutinosyl flavonoids, Quinic acid, mallic acid, Isorhamnetin-3-*O*-rutinoside, rutin, Quercetin-3-*O*-glucoside, Kaempferol-3-*O*-glucoside	Esposito et al. (2019), Dhouibi et al. (2021)
5	*R. arboreum* flower	Diterpenes, triterpenes, flavonoids, steroids, tannin, phenolics, saponin, glycosides, and alkaloids	Roy et al. (2014); Rawat et al. (2018)

(continued)

Table 8.6 (continued)

Sl. No.	Plant name	Phytoconstituents/secondary metabolites	References
6	*E. coracana* (Mandwa)	Saponins, carbohydrates, terpenoids, resins and triterpenoids	Bwai et al. (2014)
7	*B. vulgaris* (Bash) shoot	Carbohydrates, glycosides, saponins, alkaloids, flavonoids, phenolics and tannins, phytosterols, and triterpenoids, oils, and fats	Jun (2015); Fitri et al. (2020)
8	*M. acuminata* pseudostem	Phenolic compounds, flavonoids, terpenoids, alkaloids, protein, tannins, saponins, and cardiac glycosides	Onyema et al. (2016)

8.6 Conclusions and Future Prospects

The eight plants discussed in this chapter are highly useful for the people of the Eastern and Western Himalayas. The plants described in the study can be targeted for phytochemical and pharmacological studies to unravel their health-beneficial properties. This will help to understand the nutritional and therapeutic potential of non-popular wild food plants. The plants can be popularised by integrating them into agricultural and horticultural practices. Nursery management techniques can enhance the cultivation rate of these plants. The supply chain management procedures in the local and urban markets can subsequently raise income among rural dwellers. But more exploration in ethnobotanical and pharmacological studies is essential to understand the complete potential of the plants. The villagers are unaware of the conservation methods and sustainable production of the plants. Therefore elaborate research activities are required to conserve and popularize the plants. To sustain the usage of these local plants, the involvement of the local population is essential. Scientific studies about the long-term storage, processing, and export of those wild food plants are required for enhancing the income of the village people. Phytochemicals, essential oils, and secondary metabolites identification and characterization are also part of further research. The disease-curing potential of the plants should be evaluated in detail with more in-vitro and in-vivo studies. The Himalayan region is naturally resource rich but less explored and focused. Various government and nongovernmental organizations initiatives such as the formation of botanical gardens, seed banks, and various ex-situ preservation techniques can protect the plants from disappearance. The inclusion of food recipes in cultural activities, exploration of foods in front of travelers, and commercialization of foods can enhance the popularity of plants.

Acknowledgments MB acknowledges Mr. Rakesh Bisht, resident of Chamoli district, Uttarakhand for providing the photographs. MB thanks Mr. Sourav Shah from Dehradun, Uttarakhand for photographs and his help regarding processing and marketing of juice of *Rhododendron arboreum*. MB also thanks her parents for providing photographs of two Bengali dish preparations. MB further acknowledges Mr. Rupam Sarkar from Tripura and his family members for providing photographs of the famous traditional Tripura dish "Bamboo shoots".

References

Aberoumand A (2009) Preliminary assessment of nutritional value of plant-based diets in relation to human nutrients. Int J Food Sci Nutr 60:155–162

Aditya JP, Bhartiya A, Chahota RK, Joshi D, Chandra N, Kant L, Pattanayak A (2019) Ancient orphan legume horse gram: a potential food and forage crop of future. Planta 250:891–909. https://doi.org/10.1007/s00425-019-03184-5

Anis MA, Sreerama YN (2020) Inhibition of protein glycoxidation and advanced glycation end-product formation by barnyard millet (*Echinochloa frumentacea*) phenolics. Food Chem 315:126265. https://doi.org/10.1016/j.foodchem.2020.126265

Anju T Jr, Sarita S (2010) Suitability of foxtail millet (*Setaria italica*) and barnyard millet (*Echinochloa frumentacea*) for development of low glycemic index biscuits. Malays J Nutr 16(3):361–368

Ashraf J, Baig SG, Ahmed S, Hasan MM (2018) Analgesic, anti-inflammatory and diuretic activities of *Macrotyloma uniflorum* (Lam.) Verdc. Pak J Pharm Sci 31:59–1863

Badola H (2011) Nettle (*Urtica dioica* Linn.)—a potential wild resource for socio-economic upliftment in the Indian Himalayas. Nonwood News 23:29–30

Bharathi V, Rengarajan RL, Radhakrishnan R, Hashem A, Abd-Allah EF, Alqarawi AA, Anand AV (2018) Effects of a medicinal plant *Macrotyloma uniflorum* (Lam.) Verdc. formulation (MUF) on obesity-associated oxidative stress-induced liver injury. Saudi J Biol Sci 25:1115–1121. https://doi.org/10.1016/j.sjbs.2018.03.010

Bhatt D, Bisht K, Kulshreshtha K (2007) Agrobiodiversity of Uttarakhand: *Rhododendron arboreum* (Buransh) a potential additive in product development. J Eco-friendly Agric 2(2): 134–135

Bhatt V, Sendri N, Swati K, Devidas SB, Bhandari P (2022) Identification and quantification of anthocyanins, flavonoids, and phenolic acids in flowers of *Rhododendron arboreum* and evaluation of their antioxidant potential. J Sep Sci 45:2555. https://doi.org/10.1002/jssc.202200145

Bigoniya P, Bais S, Sirohi B (2014) The effect of *Macrotyloma uniflorum* seed on bile lithogenicity against diet induced cholelithiasis on mice. Anc Sci Life 33:242. https://doi.org/10.4103/0257-7941.147433

Bwai M, Afolayan M, Odukomaiya D, Abayomi O (2014) Proximate composition, mineral and phytochemical constituents of *Eleusine coracana* (finger millet). Int J Adv Chem 2:171. https://doi.org/10.14419/ijac.v2i2.3496

Camacho-Corona MR, Ramírez-Cabrera MA, González-Santiago O, Garza-González E, Palacios IP, Luna-Herrera J (2008) Activity against drug resistant-tuberculosis strains of plants used in Mexican traditional medicine to treat tuberculosis and other respiratory diseases. Phytother Res 22:82–85

Chauhan DS, Lal P, Shrama AK (2021) Extraction of *Rhododendron arboreum* Smith flowers from the forest for the livelihood and rural income in Garhwal Himalaya, India. Sci Rep 11:20844. https://doi.org/10.1038/s41598-021-00257-y

Chehri A, Yarani R, Yousefi Z, Novin Bahador T, Shakouri SK, Ostadrahimi A, Mobasseri M, Pociot F, Araj-Khodaei M (2022) Anti-diabetic potential of *Urtica dioica*: current knowledge and future direction. J Diabetes Metab Disord 21:931–940. https://doi.org/10.1007/s40200-021-00942-9

Chongtham N, Bisht MS, Haorongbam S (2011) Nutritional properties of bamboo shoots: potential and prospects for utilization as a health food. Compr Rev Food Sci Food Saf 10:153–168. https://doi.org/10.1111/j.1541-4337.2011.00147.x

Choudhury PR, Choudhury MD, Ningthoujam SS, Mitra A, Nath D, Talukdar AD (2015) Plant utilization against digestive system disorder in Southern Assam, India. J Ethnopharmacol 175: 192–197. https://doi.org/10.1016/j.jep.2015.09.020

Cruz-Martins N, Ferreira I (2017) Mountain Food Products: a broad spectrum of market potential to be exploited. Trends Food Sci Technol 67:12–18. https://doi.org/10.1016/j.tifs.2017.06.013

De Wet H, Ramulondi M, Ngcobo ZN (2016) The use of indigenous medicine for the treatment of hypertension by a rural community in northern Maputaland, South Africa. South Afr J Bot 103: 78–88

Devi PB, Vijayabharathi R, Sathyabama S, Malleshi NG, Priyadarisini VB (2014) Health benefits of finger millet (*Eleusine coracana* L.) polyphenols and dietary fiber: a review. J Food Sci Technol 51:1021–1040. https://doi.org/10.1007/s13197-011-0584-9

Dhouibi R, Affes H, Ben Salem M, Charfi S, Marekchi R, Hammami S, Zeghal K, Ksouda K (2021) Protective effect of *Urtica dioica* in induced neurobehavioral changes, nephrotoxicity and hepatotoxicity after chronic exposure to potassium bromate in rats. Environ Pollut 287: 117657. https://doi.org/10.1016/j.envpol.2021.117657

Di Staci LC, Oliveira GP, Carvalhaesa MA, Queiroz-Juniora M, Tiena OS, Kakinamia SH, Reisb MS (2002) Medicinal plants popularly used in the Brazilian Tropical Atlantic Forest. Fitoterapia 73:69–91

Divya BJ, Suman B, Venkataswamy M, Thyagaraju K (2017) A study on phytochemicals, functional groups and mineral composition of *Allium sativum* (garlic) cloves. Int J Curr Pharm Res 9:42–45

Dixit V, Gupta N (2009) Evaluation of hepatoprotective activity of *Cleome viscosa* Linn. extract. Indian J Pharmacol 41:36. https://doi.org/10.4103/0253-7613.48892

Duduku K, Rosalam S, Awang B (2007) Phytochemical antioxidants for health and medicine—A move towards nature. Biotechnol Mol Biol Rev 2:97–104

Eiki N, Sebola N, Sakong B, Mabelebele M (2021) Review on ethnoveterinary practices in Sub-Saharan Africa. Vet Sci 8:99. https://doi.org/10.3390/vetsci8060099

El Haouari M, Rosado JA (2018) Phytochemical, anti-diabetic and cardiovascular properties of *Urtica dioica* L. (Urticaceae): a review. Mini Rev Med Chem 19:63–71. https://doi.org/10.2174/1389557518666180924121528

Esposito S, Bianco A, Russo R, Di Maro A, Isernia C, Pedone P (2019) Therapeutic perspectives of molecules from *Urtica dioica* extracts for cancer treatment. Molecules 24:2753. https://doi.org/10.3390/molecules24152753

FAO, IFAD, UNICEF, WFP, WHO (2021) The State of Food Security and Nutrition in the World 2021. Transforming food systems for food security, improved nutrition and affordable healthy diets for all. FAO, Rome. https://doi.org/10.4060/cb4474en

Fingolo CE, Braga JMA, Vieira ACM, Moura MRL, Kaplan MAC (2012) The natural impact of banana inflorescences (*Musa acuminata*) on human nutrition. An Acad Bras Ciênc 84:891–898. https://doi.org/10.1590/S0001-37652012005000067

Fitri A, Asra R, Rivai H (2020) Overview of the traditional, phytochemical, and pharmacological uses of gold bamboo (*Bambusa vulgaris*). World J Pharm Pharm Sci 9:299–318. https://doi.org/10.20959/wjpps20208-16877

Gairola S, Sharma CM, Ghildiyal SK, Suyal S, Rana CS, Butola DS (2009) Biodiversity conservation and sustainable rural development in the Garhwal Himalaya. Report Opin 1:6–12

Gakuya DW, Mulei CM, Wekesa S (2011) Use of ethnoveterinary remedies in the management of foot and mouth disease lesions in a diary herd. Afr J Tradit Complement Altern Med 8(2): 165–169

Gautam M, Katoch S, Chahota RK (2020) Comprehensive nutritional profiling and activity directed identification of lead antioxidant, antilithiatic agent from *Macrotyloma uniflorum* (Lam.) Verdc. Food Res Int 137:109600. https://doi.org/10.1016/j.foodres.2020.109600

Ghorbani A, Langenberger G, Sauerborn J (2012) A comparison of the wild food plant use knowledge of ethnic minorities in Naban River Watershed National Nature Reserve, Yunnan, SW China. J Ethnobiol Ethnomed 8:17. https://doi.org/10.1186/1746-4269-8-17

Goron TL, Raizada MN (2015) Genetic diversity and genomic resources available for the small millet crops to accelerate a New Green Revolution. Front Plant Sci 6:157. https://doi.org/10.3389/fpls.2015.00157

Gupta S, Kanwar SS (2018) Phyto-molecules for kidney stones treatment and management. Biochem Anal Biochem 7:362. https://doi.org/10.4172/2161-1009.1000362

Hegde P, Anitha B, Chandra T (2005) In vivo effect of whole grain flour of finger millet (*Eleusine coracana*) and kodo millet (*Paspalum scrobiculatum*) on rat dermal wound healing. Indian J Exp Biol 43:254–258

Islam MS, Kasim S, Amin AM, Hun TG, Alam MK, Haque MA (2022) Banana-pseudostem sap growing media as a novel source of phytochemicals and mineral nutrients: influence on seedling growth of sweet corn. Chil J Agric Res 82:135–145. https://doi.org/10.4067/S0718-58392022000100135

Joshi B, Pant SC (2012) Ethnobotanical study of some common plants used among the tribal communities of Kashipur, Uttarakhand. Indian J Nat Prod Resour 3(2):262–266

Jun P (2015) Potential medicinal application and toxicity evaluation of extracts from bamboo plants. J Med Plants Res 9:681–692. https://doi.org/10.5897/JMPR2014.5657

Kardan M, Rafiei A, Golpour M, Ebrahimzadeh MA, Akhavan-Niaki H, Fattahi S (2020) *Urtica dioica* extract inhibits cell proliferation and induces apoptosis in HepG2 and HTC116 as gastrointestinal cancer cell lines. ACAMC 20:963–969. https://doi.org/10.2174/1871520620666200311095836

Kader AA, Perkins-Veazie P, Lester GE (2012) Nutritional quality and its importance to human health. The Commercial storage of fruits, vegetables, and florist and nursery stocks. p 166

Kargozar R, Salari R, Jarahi L, Yousefi M, Pourhoseini SA, Sahebkar-Khorasani M, Azizi H (2019) *Urtica dioica* in comparison with placebo and acupuncture: a new possibility for menopausal hot flashes: a randomized clinical trial. Complement Ther Med 44:166–173. https://doi.org/10.1016/j.ctim.2019.04.003

Kasouni AI, Chatzimitakos TG, Stalikas CD, Trangas T, Papoudou-Bai A, Troganis AN (2021) The unexplored wound healing activity of *Urtica dioica* L. extract: an in vitro and in vivo study. Molecules 26:6248. https://doi.org/10.3390/molecules26206248

Kaur H, Sharma S (2020) An overview of Barnyard millet (*Echinochloa frumentacea*). J Pharmacogn Phytochem 9:819–822

Kim C-S, Alamgir KM, Matsumoto S, Tebayashi S, Koh H-S (2008) Antifeedants of Indian Barnyard Millet, *Echinochloa frumentacea* Link, against Brown Planthopper, *Nilaparvata lugens* (Stål). Z Naturforsch C 63:755–760. https://doi.org/10.1515/znc-2008-9-1022

Kregiel D, Pawlikowska E, Antolak H (2018) *Urtica* spp.: ordinary plants with extraordinary properties. Molecules 23:1664. https://doi.org/10.3390/molecules23071664

Krishnamoorthy R, Gassem MA, Athinarayanan J, Periyasamy VS, Prasad S, Alshatwi AA (2021) Antifungal activity of nanoemulsion from *Cleome viscosa* essential oil against food-borne pathogenic *Candida albicans*. Saudi J Biol Sci 28:286–293. https://doi.org/10.1016/j.sjbs.2020.10.001

Krishnappa K, Elumalai K (2013) Mosquitocidal properties of *Basella rubra* and *Cleome viscosa* against *Aedes aegypti* (Linn.) (Diptera: Culicidae). Eur Rev Med Pharmacol Sci 17(9):1273–1277

Kumar M, Sheikh MA, Bussmann RW (2011) Ethnomedicinal and ecological status of plants in Garhwal Himalaya, India. J Ethnobiol Ethnomed 7:32. https://doi.org/10.1186/1746-4269-7-32

Kumar V, Suri S, Prasad R, Gat Y, Sangma C, Jakhu H, Sharma M (2019) Bioactive compounds, health benefits and utilization of *Rhododendron*: a comprehensive review. Agric Food Secur 8:1–7. https://doi.org/10.1186/s40066-019-0251-3

Kumar A, Anju T, Kumar S, Chhapekar SS, Sreedharan S, Singh S, Choi SR, Ramchiary N, Lim YP (2021a) Integrating omics and gene editing tools for rapid improvement of traditional food plants for diversified and sustainable food security. Int J Mol Sci 22:8093

Kumar M, Rawat S, Nagar B, Kumar A, Pala NA, Bhat JA, Bussmann RW, Cabral-Pinto M, Kunwar R (2021b) Implementation of the use of ethnomedicinal plants for curing diseases in the Indian Himalayas and its role in sustainability of livelihoods and socioeconomic development. Int J Environ Res Public Health 18:1509. https://doi.org/10.3390/ijerph18041509

Lans C, Georges K, Brown G (2007) Non-experimental validation of ethnoveterinary plants and indigenous knowledge used for backyard pigs and chickens in Trinidad and Tobago. Trop Anim Health Prod 39:375–385. https://doi.org/10.1007/s11250-007-9026-0

Lingwan M, Shagun S, Pahwa F, Kumar A, Verma DK, Pant Y, Kamatam LVK, Kumari B, Nanda RK, Sunil S, Masakapalli SK (2021) Phytochemical rich Himalayan *Rhododendron arboreum* petals inhibit SARS-CoV-2 infection *in vitro*. J Biomol Struct Dyn 1–11. https://doi.org/10.1080/07391102.2021.2021287

Lodhi S, Jain AP, Rai G, Yadav AK (2016) Preliminary investigation for wound healing and anti-inflammatory effects of *Bambusa vulgaris* leaves in rats. J Ayur Int Med 7:14–22

Luitel NP, Jordans MJD, Subba P, Komproe IH (2020) Perception of service users and their caregivers on primary care-based mental health services: a qualitative study in Nepal. BMC Fam Pract 21:202. https://doi.org/10.1186/s12875-020-01266-y

Maietti A, Tedeschi P, Catani M, Stevanin C, Pasti L, Cavazzini A, Marchetti N (2021) Nutrient composition and antioxidant performances of bread-making products enriched with stinging nettle (*Urtica dioica*) leaves. Foods 10:938. https://doi.org/10.3390/foods10050938

Mali RG (2010) *Cleome viscosa* (wild mustard): a review on ethnobotany, phytochemistry, and pharmacology. Pharm Biol 48:105–112. https://doi.org/10.3109/13880200903114209

Mali RG, Mahajan SG, Mehta AA (2007) In vitro screening of *Cleome viscosa* extract for anthelmintic activity. Pharm Biol 45:766–768

Malik ZA, Bhat JA, Ballabha R, Bussmann RW, Bhatt AB (2015) Ethnomedicinal plants traditionally used in health care practices by inhabitants of Western Himalaya. J Ethnopharmocol 172:133–144

Mandal SK, Rahaman CH (2022) Inventorization and consensus analysis of ethnoveterinary medicinal knowledge among the local people in Eastern India: perception, cultural significance, and resilience. Front Pharmacol 13:861577. https://doi.org/10.3389/fphar.2022.861577

Maskey B, Sangroula P, Shrestha NK (2020) Utilization of banana (*Musa acuminata*) pseudostem for biscuit making. Himalayan J Sci Technol 74–80. https://doi.org/10.3126/hijost.v4i0.33873

Mathew NS, Negi PS (2017) Traditional uses, phytochemistry and pharmacology of wild banana (*Musa acuminata* Colla): a review. J Ethnopharmacol 196:124–140. https://doi.org/10.1016/j.jep.2016.12.009

Meena ML, Dudi A, Singh D, Prasad R (2021) Traditional veterinary medicines for buffalo followed by tribal folks in Pali district of Rajasthan, India. Buffalo Bull 40(2):267–281

Mohan M, Elyas KK (2019) Anti-hyperglycemic properties of a purified proteinaceous protease inhibitor from *Macrotyloma uniflorum* seeds. Curr Top Med Chem 18:2502–2510. https://doi.org/10.2174/1568026619666181220110038

Muthukumar T, Prabu P, Ghosh K, Sastry TP (2014) Fish scale collagen sponge incorporated with *Macrotyloma uniflorum* plant extract as a possible wound/burn dressing material. Colloids Surf B: Biointerfaces 113:207–212. https://doi.org/10.1016/j.colsurfb.2013.09.019

Nair MNB, Punniamurthy N, Kumar SK (2021) Ethno-veterinary practices for animal health management and the associated medicinal plants from 24 locations in 10 Indian states. Curr Res Agric Vet Sci 3:86–102. https://doi.org/10.9734/bpi/cravs/v3/1691C

Negi S (2009) Forest cover in Indian Himalayan states—an overview. Indian J For 32:1–5

Nguyen TP, Tran CL, Vuong CH, Do THT, Le TD, Mai DT, Phan NM (2017) Flavonoids with hepatoprotective activity from the leaves of *Cleome viscosa* L. Nat Prod Res 31:2587–2592. https://doi.org/10.1080/14786419.2017.1283497

Nongdam P, Tikendra L (2014) The nutritional facts of bamboo shoots and their usage as important traditional foods of Northeast India. Int Sch Res Notices 2014:1–17. https://doi.org/10.1155/2014/679073

Onyema CT, Ofor CE, Okudo VC, Ogbuagu AS (2016) Phytochemical and antimicrobial analysis of banana pseudo stem (*Musa acuminata*). J Pharm Res Int 10:1–9. https://doi.org/10.9734/BJPR/2016/22593

Panda V, Laddha A, Nandave M, Srinath S (2016) Dietary phenolic acids of *Macrotyloma uniflorum* (horse gram) protect the rat heart against isoproterenol-induced myocardial infarction: cardioprotective activity of *Macrotyloma uniflorum* seed. Phytother Res 30:1146–1155. https://doi.org/10.1002/ptr.5620

Pandey S, Joshi N, Kumar M, Nautiyal P, Papnai G, Bhaskar R (2021) Nutritional profile & health benefits of Jhangora: a mini review. Pharm Innov 10:379–381. https://doi.org/10.22271/tpi.2021.v10.i3f.5799

Panigrahi PN, Dey S, Sahoo M, Dan A (2017) Antiurolithiatic and antioxidant efficacy of *Musa paradisiaca* pseudostem on ethylene glycol-induced nephrolithiasis in rat. Indian J Pharmacol 49:77–83. https://doi.org/10.4103/0253-7613.201026

Pant V (2019) Himalayan stinging nettle: rich source of protein and minerals. J Ethnobiol Tradit Med 130:1532–1548. https://doi.org/10.13140/RG.2.2.29576.44801

Parimala Devi B, Boominathan R, Mandal SC (2002) Evaluation of anti-diarrheal activity of *Cleome viscosa* L. extract in rats. Phytomedicine 9:739–742. https://doi.org/10.1078/094471102321621368

Parimaladevi B, Boominathan R, Mandal SC (2003) Studies on analgesic activity of *Cleome viscosa* in mice. Fitoterapia 74:262–266. https://doi.org/10.1016/S0367-326X(03)00020-0

Parry M, Rosenzweig C, Livermore M (2005) Climate change, global food supply and risk of hunger. Philos Trans R Soc B 360:2125–2138. https://doi.org/10.1098/rstb.2005.1751

Patel VB, Acharya N (2020) Effect of *Macrotyloma uniflorum* in ethylene glycol induced urolithiasis in rats. Heliyon 6:e04253. https://doi.org/10.1016/j.heliyon.2020.e04253

Patil US, Deshmukh OS (2015) Plants used in ethnoveterinary medicines by tribal people in Betul district, Madhya Pradesh, India. J Biosci 4(8):3049–3054

Ravishankar K, Priya PS (2012) Evaluation of diuretic effect of ethanolic seed extracts of *Macrotyloma uniflorum* and *C. melo* in rats. Int J Pharm Bio Sci 3(3):251–255

Rawat P, Bachheti RK, Kumar N, Rai N (2018) Phytochemical analysis and evaluation of in vitro immunomodulatory activity of *Rhododendron arboretum* leaves. Asian J Pharm Clin Res 11:123. https://doi.org/10.22159/ajpcr.2018.v11i8.25372

Renganathan VG, Vanniarajan C, Karthikeyan A, Ramalingam J (2020) Barnyard millet for food and nutritional security: current status and future research direction. Front Genet 11:500. https://doi.org/10.3389/fgene.2020.00500

Rosegrant MW, Cline SA (2003) Global food security: challenges and policies. Science 302:1917–1919

Roy JD, Handique AK, Barua CC, Talukdar A, Ahmed FA, Barua IC (2014) Evaluation of phytoconstituents and assessment of adaptogenic activity in vivo in various extracts of *Rhododendron arboreum* (leaves). Indian J Pharm Biol Res 2:49–56. https://doi.org/10.30750/ijpbr.2.2.9

Rutto LK, Xu Y, Ramirez E, Brandt M (2013) Mineral properties and dietary value of raw and processed stinging nettle (*Urtica dioica* L.). Int J Food Sci 2013:1–9. https://doi.org/10.1155/2013/857120

Saleh ASM, Zhang Q, Chen J, Shen Q (2013) Millet grains: nutritional quality, processing, and potential health benefits: millet grains. Compr Rev Food Sci Food Saf 12:281–295. https://doi.org/10.1111/1541-4337.12012

Sati V (2016) Geography of Himalaya. Progress of Indian geography: a country report (2012–2016) to 33rd IGU

Saxena BR, Koli MC, Saxena RC (2000) Preliminary ethnomedical and phytochemical study of *Cleome viscosa* L. Ethnobotany 12:47–50

Sengottuvelu S, Duraisamy R, Nandhakumar J, Sivakumar T (2007) Hepatoprotective activity of *Cleome viscosa* against carbon tetrachloride induced hepatotoxicity in rats. Pharmacogn Mag 3:121–124

Sharma PK, Dhyani SK, Shankar V (1979) Some useful medicinal plants of the district Dehradun and Siwalik. J Sci Res Plant Med 1:17–43

Sharma S, Sharma N, Handa S, Pathania S (2017) Evaluation of health potential of nutritionally enriched Kodo millet (*Eleusine coracana*) grown in Himachal Pradesh, India. Food Chem 214:162–168. https://doi.org/10.1016/j.foodchem.2016.07.086

Shukla R, Sumit G, Sajal S, Dwivedi P, Mishra A (2012) Medicinal importance of Bamboo. Int J Biopharm Phytochem Res 1:9–15

Siddhuraju P, Manian S (2007) The antioxidant activity and free radical-scavenging capacity of dietary phenolic extracts from horse gram (*Macrotyloma uniflorum* (Lam.) Verdc.) seeds. Food Chem 105(3):950–958

Siddique KHM, Li X, Gruber K (2021) Rediscovering Asia's forgotten crops to fight chronic and hidden hunger. Nat Plants 7:116–122. https://doi.org/10.1038/s41477-021-00850-z

Sikarwar RL, Tiwari A (2020) A review of plants used in ethnoveterinary medicine in Central India. Indian J Tradit Knowl 19:617–634

Singh PD, West ME (1991) Pharmacological investigations of sticky viscome extract (*Cleome viscosa* Linn) in rats, mice and guinea pigs. Phytother Res 5:82–84

Singh AK, Singh RN, Singh SK (1987) Some ethnobotanical plants of Terai region of Gorakhpur district—I. J Econ Tax Bot 9:407–410

Singh KP, Mishra HN, Saha S (2010) Moisture-dependent properties of barnyard millet grain and kernel. J Food Eng 96:598–606. https://doi.org/10.1016/j.jfoodeng.2009.09.007

Singh R, Singh MK, Chandra LR, Bhat D, Arora MS, Nailwal T, Pande V (2012) In-vitro antioxidant and free radical scavenging activity of *Macrotyloma uniflorum* (Gahat dal) from Kumauni region. Int J Fundam Appl Sci 1(1):7–10

Som K, Madhvi MS, Javaid I, Mohd Y (2019) Phytochemistry, traditional uses and pharmacology of *Rhododendron arboreum*: a review. Res J Pharm Technol 12(9):4565–4574. https://doi.org/10.5958/0974-360X.2019.00785.6

Srinivas BK, Shivamadhu MC, Jayarama S (2021) *Musa acuminata* lectin exerts anti-cancer effects on HeLa and EAC cells via activation of caspase and inhibitions of Akt, Erk, and Jnk pathway expression and suppresses the neoangiogenesis in in-vivo models. Int J Biol Macromol 166: 1173–1187. https://doi.org/10.1016/j.ijbiomac.2020.10.272

Srivastava S, Shukla R (2011) Medicinal importance of bamboo. Int J Biopharm Phytochem Res 1: 9–15

Subbaraya U (2006) Farmers knowledge of wild Musa in India. Food and Agriculture Organization of the United States, Rome

Sudhakar S, Rolla RS (1985) Medicinal plants of Upper East Godavari District (Andhra Pradesh) and need for establishment of Medicinal Farm. J Econ Tax Bot 7:399–406

Sudhakar M, Rao CV, Rao PM, Raju DB (2006) Evaluation of antimicrobial activity of *Cleome viscosa* and *Gmelina asiatica*. Fitoterapia 77:47–49. https://doi.org/10.1016/j.fitote.2005.08.003

Swamidasan R, Sanil R, Manasa R (2020) Medicinal values of *Rhododendron arboretum*: a comprehensive review. Int J Sci Res 9:1768–1771. https://doi.org/10.21275/SR20507184145

Taylor M (2015) Complementary and alternative approaches to menopause. Endocrinol Metab Clin North Am 44:619–648. https://doi.org/10.1016/j.ecl.2015.05.008

Thorat RL, Bobade HP (2018) Utilization of banana pseudo-stem in food applications. Int J Agric Eng 11:86–89

Tiwari U, Rastogi B, Thakur S, Jain S, Jain NK (2004) Studies on the immunomodulatory effects of *Cleome viscosa*. Indian J Pharm Sci 66:171–176

Uniyal SK, Singh K, Jamwal P, Lal B (2006) Traditional use of medicinal plants among the tribal communities of Chhota Bhangal, Western Himalaya. J Ethnobiol Ethnomed 2:14. https://doi.org/10.1186/1746-4269-2-14

Upadhyay R (2011) Plant latex: a natural source of pharmaceuticals and pesticides. Int J Green Pharm 5:169. https://doi.org/10.4103/0973-8258.91222

Uraku AJ, Uraku OH (2018) Quantitative assessment of phytochemicals and nutritional potential of leaves and seeds of *Cleome viscosa* from Abakaliki, Nigeria. J Biosci Biotechnol Discov 3:25–29. https://doi.org/10.31248/JBBD2017.036

Vadivelu B, Arumugam VA, Subbarayan S, Alshatwi AA, Krishnamoorthy R (2019) Effect of *Macrotyloma uniflorum* on antiobesity in rats fed with a high fat diet. Saudi J Biol Sci 26:1772–1778. https://doi.org/10.1016/j.sjbs.2018.05.003

Verma N, Singh AP, Gupta A, Sahu PK, Rao C (2011) Antidiarrheal potential of standardized extract of *Rhododendron arboretum* Smith flowers in experimental animals. Indian J Pharmacol 43(6):689–693. https://doi.org/10.4103/0253-7613.89827

Williams LAD, Vasques E, Reid W, Porter R, Kraus W (2003) Biological activities of an extract from *Cleome viscosa* L. (Capparaceae). Naturwissenschaften 90:4s68–4s4472. https://doi.org/10.1007/s00114-003-0460-1

Yadav A (2021) Banana (*Musa acuminata*): most popular and common Indian plant with multiple pharmacological potentials. World J Biol Pharm Health Sci 7:036–044. https://doi.org/10.30574/wjbphs.2021.7.1.0073

Yadav N, Chattopadhyay S, Pal A, Chanda D, Gupta A (2010) Hepatoprotective effects and safety evaluation of coumarinolignoids isolated from *Cleome viscosa* seeds. Indian J Pharm Sci 72:759. https://doi.org/10.4103/0250-474X.84589

Zouari Bouassida K, Bardaa S, Khimiri M, Rebaii T, Tounsi S, Jlaiel L, Trigui M (2017) Exploring the *Urtica dioica* leaves hemostatic and wound-healing potential. Biomed Res Int 2017:1–10. https://doi.org/10.1155/2017/1047523

Chapter 9
Re-emergence of Pseudocereals as Superfoods for Food Security and Human Health: Current Progress and Future Prospects

Adya Ram, Anju Thattantavide, and Ajay Kumar

Abstract Unlike cereals which are monocotyledons, pseudocereals are dicotyledonous plants and are known as pseudocereals because of the similarity of their grains with cereals. They were traditionally used in different civilizations and played a very prominent role in their food, medicine, and rituals. These grains are highly nutritious and are also medicinal since they have abundant nutrients and health-promoting phytochemicals. They are rich in high-quality proteins and essential amino acids. Unlike cereals, pseudocereals ensure the supply of various micro and macronutrients as well. They are considered as "superfood for the future" due to their high nutritional value. Their consumption had undergone massive decline due to negligence by the modern generations. But over the last few years, people have shown greater interest in the consumption of these nutritionally rich grains, and there is an increased acceptance of pseudocereals in the present times. The advancements in nutritional profiling technologies has shown their nutritional superiority over cereals. The major pseudocereals are amaranth, buckwheat, chia, and quinoa. These grains are gluten-free which makes their consumption safer for gluten-intolerant people. Genetic works on pseudocereals confirm that these plants are highly beneficial in improving health and are genetically very diverse which makes them capable to survive in a wide geographical range and tolerate various stresses. This chapter discusses the origin, distribution and taxonomy of the main pseudocerals with their nutritional composition. It further discusses the traditional, medicinal and nutritional importance of the pseudocereals in the context of climate change. A comparison of its nutritional components with the cereals provides understanding of the nutritional richness of the pseudocereals. The last two sections are devoted to the genetics and genomics of pseudocereals and their role in climate resilience.

A. Ram · A. Thattantavide
Department of Plant Science, School of Biological Sciences, Central University of Kerala, Kasaragod, Kerala, India

A. Kumar (✉)
Department of Plant Science, Central University of Kerala, Kasaragod, Kerala, India

A. Kumar et al. (eds.), *Wild Food Plants for Zero Hunger and Resilient Agriculture*, Plant Life and Environment Dynamics,
https://doi.org/10.1007/978-981-19-6502-9_9

Keywords Pseudocereals · Food security · Gluten-free diets · Climate-resilient crops · Bioactive compounds · Prebiotics · Probiotics

9.1 Introduction

Pseudocereals are a group of plants that are similar to cereals in their seed, grain structure, palatability, starch content and cooking characteristics but differ in the sense that they are dicotyledons (Morales et al. 2020). Most of the starchy grains of pseudocereals are smaller than the size of cereals but both are consumed in similar manner (Henrion et al. 2020). Pseudocereals are considered as the crop of twenty-first century due to their excellent nutritional quality (Pirzadah and Malik 2020). They have a higher lipid content than cereals and contain 75% unsaturated fatty acids, which can be used in the human diet in place of saturated and *trans* fat (Haros and Schonlechner 2017). These food crops which were once popular in ancient times as part of the diet of different populations and became less noticed due to the domination of cereals in the food pattern of the world population during twentieth century, hence they are called "sub-exploited food" (Morales et al. 2020). Pseudocereal nutrition is either comparable or superior to cereals, which makes them preferable to be used as a substitute for cereals. Nowadays, the the global attention towards pseudocereals have increased due to their nutritional qualities, agronomic properties, potential to diversify the natural resources and environmental adaptabiity (Carrasco and Soto 2010). The Latin American countries depend on the pseudocereals as their staple food crop and countries such as Russia and China are great producers of pseudocereal such as buckwheat (Morales et al. 2020). Amaranth and Quinoa were selected as the best vegetable food for people by The National Academy of Sciences (NAS) research. Also, they were selected as food in astronaut diet during space travels by the National Aeronautics and Space Administration (NASA; Morales et al. 2020). Another important fact about pseudocereals is that they are gluten-free, hence they are safe to be used by celiac-diseased patients who are reactant to the gluten in food items like wheat. The intake of gluten by the celiac patients causes damage to the mucosal wall of the small intestine, that results in conditions such as reduced absorption of nutrients and indigestion. Therefore gluten-free pseudocereals will be a relief for the celiac patients (Thakur et al. 2021). Besides that, various bioactive compounds present in pseudocereals provide a wide range of health benefits for the consumers apart from nutrition, hence they are also called "functional food" (Morales et al. 2020). The pseudocereals are enriched with various phenolic and flavonoid compounds such as vanillic acid, ferulic acid, quercetin and kaempferol which provides protection from chronic diseases such as cardiovascular disease, cancer, diabetes and hypertension (Thakur et al. 2021). Therefore, the inclusion of pseudocereals in people's diets is beneficial to both celiac patients and consumers suffering from various lifestyle related disorders.

The current global conditions also demand to diversify the diet of consumers with climate resilient, nutritionally superior and health beneficial food crops (Cheng et al. 2015). In this fast growing world, where the population is estimated to reach about

9 billion by 2050, depending only upon major cereals such as rice, maize and wheat alone for the availability of nutrients may not be good. Due to the dominance of limited number of food crops in the diet, the world is not able to mitigate the challenges of climate change and malnutrition (Cheng et al. 2015). The presence of high carbohydrate content in the grains, cereals are used for the 60% of calorie intake by the developing countries and 30% by the developed countries (Olugbire et al. 2021). At the same time, it was recorded that, the 11% of total global population is severely affected with malnutrition and 17% are suffering from hidden hunger or micro nutrient deficiency (Grote et al. 2021) which indicate the inability of major cereals to ensure food security. Additionally, produtivity of cereals has reduced significantly due to climate change and increased incidence of pathogens. Abbas (2022) documented that, the yeild of major cereals like rice, wheat, maize, barley and jowar were declined in a period of 2009-2019 due to the rise in temperature. This kind of research point towards the necessity to diversify the food systems with climate resilient and nutritionally superior crops. The popularization of inclusion of pseudocereals in the people's diets can solve the problem to a certain extent, and currently various countries are following this as discussed above. The tolerance of pseudocereals against abiotic stresses like drought, temperature, salinity, and heavy metal stress has already been proved which suggest their climate resilience (Pirzadah and Malik 2020). Besides that, they are capable of surviving and being productive even in marginal lands (Das 2016a). This also makes pseudocereals more relevant since urbanization has produced many marginal lands. The cereals do not have potential to grow in marginal lands due to their higher nutrient requirement for growth, but the capability of pseudocereals to survive under minimum nutrients can help them grow efficiently (Rodríguez et al. 2020). This may be another reason why pseudocereals are considered as "past food for future people" (Morales et al. 2020). The major pseudocereals that can be used to diversify our food menu with the assurance of availability of macro and micronutrients are grain amaranth, buckwheat, chia, and quinoa (Arslan-Tontul et al. 2022). All these crops are significant in their characteristic nutrient composition and presence of bioactive compounds. Table 9.1 represents the information about the taxonomy, origin, and distribution of these pseudocereals.

These pseudocereals are also useful in industrial and nutraceutical fields. Chia is very popular as an ingredient in skin care products (Huber et al. 2020). The high omega-3 fatty acid content in chia is a potential ingredient for maintaining and improving skin hydration (Ullah et al. 2016). Thus re-emergence of pseudocereals is a relevant topic to be discussed to ensure food security for the future generation and as a potential raw material in various nutraceutical industries.

Table 9.1 Taxonomy, origin, and distribution of pseudocereals

Pseudocereal	Family	Common name	Origin	Distribution
Amaranthus cruentus L.	Amaranthaceae	African spinach, Red amaranth, Purple amaranth (Ogwu 2020)	Southern Guatemala, South of Mexico, and Central America (Singh 2017)	Africa, America, Mesoamerica, North-western Argentina, East, Southeast and South Asia (Singh 2017)
Amaranthus hypochondriacus L.	Amaranthaceae	Prince's feather and amaranth (Ogwu 2020)	Northern Mexico (Singh 2017)	Africa, America, Mesoamerica, north-western Argentina, East, Southeast and South Asia (Singh 2017)
Amaranthus caudatus L.	Amaranthaceae	Red amaranth, Flower amaranth, Pendant amaranth (Ogwu 2020)	Andes (Singh 2017)	Africa, America, Mesoamerica, north-western Argentina, East, Southeast and South Asia (Singh 2017)
Chenopodium quinoa Willd.	Amaranthaceae	Quinoa	Andean region of South America (Jaikishun et al. 2019)	Bolivia, Peru, Ecuador, Colombia, Argentina, Chile (Jaikishun et al. 2019)
Salvia mexicana L.	Lamiaceae	Chia, Chian, Salba (Sosa-Baldivia et al. 2018)	México (Sosa-Baldivia et al. 2018)	Asia, Africa, North America, South America, Europe (Bhargava et al. 2006)
Fagopyrum esculentum Moench and *Fagopyrum tataricum* Gaertn.	Polygonaceae	Common Buckwheat (Rana et al. 2012), Sweet buckwheat (Norbu and Roder 2003), Tartary Buckwheat (Rana et al. 2012), bitter buckwheat (Norbu and Roder 2003)	Southern China (Zhou et al. 2018)	China, Korea, Japan, Tibet, Bhutan, Nepal, India, Poland, Europe, England, Ukraine, Russia (Suvorova and Zhou 2018)

9.2 History and Naming of the Pseudocereals

9.2.1 *Amaranth*

Amaranth, being popular as a leafy vegetable, has only gained very less significance in its grain use. These grains are highly nutritious and are abundant in various amino acids. The mostly used grain amaranth species are *Amaranthus caudatus*, *Amaranthus cruentus*, and *Amaranthus hypochondriacus* (Ogwu 2020). The name amaranth means immortal, everlasting, or non-wilting according to Greeks (Mlakar et al. 2009). Amaranth is known by various common names in different parts of the world. Some of them are listed below (Table 9.2):

9.2.2 *Quinoa*

Quinoa has been known by various local names in Andean regions. In Ecuador and Peru, they were known as kiuna and parca. In Bolivia, they were popular as supha, jopa, jupha, jiura, aara, ccallapi, and vocali. In Chile, it is "quinhua" and in Colombia, it is "suba" and "pasca" (Tahmasebi and Firuzkoohi 2017). In the Inca tradition, it is considered as "chisaya mama" (Tanwar et al. 2019) or "the mother grain", with holy status gifted to God (Vega-Gálvez et al. 2010).

Table 9.2 Common name of Amaranth in various countries

Country	Common name	Reference
Aztec civilization (present Mexico)	Huahtli	Myers (1996)
India	Rajgira ("king seed"), Ramdana ("seed sent by God"), Keerai	Singh (2017)
Nepal	Marcha, Nana, Pilim, Latav	Singh (2017)
China	Een choy, Yin choy, In-tsai, Hsien tsai, Xian cai	Rastogi and Shukla (2013)
Sri Lanka	Thampala	Rastogi and Shukla (2013)
Vietnam	Yan yang	Rastogi and Shukla (2013)
Bolivia	Coimi, Millmi	Rastogi and Shukla (2013)
Ecuador	Sangoracha, Alaco	Rastogi and Shukla (2013)

9.2.3 Chia

Many inconsistencies have occurred in the scientific and common nomenclature of chia. These inconsistencies in its name was one of the reasons for the decline in its production. Initially, chia was known as Chian in Nahua, a Mexican language which meant “oily” due to their high oil content (Sosa-Baldivia et al. 2018) and as “strength” in the Mayan ancestral language since they were used as a high-energy food in Yucatan (Akinfenwa et al. 2020). Mexican people used this term as a collective term to address all species belonging to the genus *Salvia*.

Carl Linnaeus assigned *Salvia hispanica* as the standardized name for the chia plants in 1753, which means “a Spanish plant to cure or save” in English, that lead to the assumption of origin of the plant in Spain and Italy. After 1832, the Spanish Language Academy finalized to switch over the common name from Chian to Chia. According to Spanish practice, they considered a single name for crops or plants which can be grouped with similar characteristics and functions, which intensified the misconception about naming as other plants also got the name chia by the Spanish people. During the mid-18s, Pablo De la Llave proposed to change the binomial name of chia to *Salvia chian*; thus highlighting its Mexican origin (De la Llave 1833). In 2006, through the marketing strategy of a Peru based company named Agrisalba SA, the chia production was increased rapidly by the claim that white chia (Salba) seeds are nutritionally superior than black mexican cultivar chia seeds. Even though it was later found that there was no such difference in the nutrient composition of the white Salba and black Chia, the name had gained wide popularity by that time through their marketing strategies. Chia seeds are also popular as ‘Salba’ (Salbachia (n.d.); Anacleto et al. 2018). But later on, to restore its Mexican identity, a major event in the history of the nomenclature of Chia took place on March 23, 2012 by naming the plant as *Salvia mexicana* as an acknowledgment of its geographical origin in Mexico (IPNI, 2012). This binomial naming corrected the English meaning of Chia from “a Spanish plant to cure or save” to “a Mexican plant to cure or save” (Sosa-Baldivia et al. 2018).

9.2.4 Buckwheat

Fagopyrum belongs to the family Polygonaceae. The genus name *Fagopyrum* has undergone many modifications by various botanists. In 1753, Linnaeus established *Polygonum* Linn by including *Fagopyrum* in *Polygonum*; without separating it as a genus under *Polygonum*. Finally, based on the morphological, palynological, and cytological studies, Graham and Wood (1965) and Hedberg (1946) came to the conclusion that *Fagopyrum* showed differences from *Polygonum* in their morphology, palynology, and cytology; and hence should be considered as a separate genus. Thus the genus name accepted now is *Fagopyrum* Miller. The term buckwheat can be considered as a collective term for the two widely cultivated and palatable

Table 9.3 Various vernacular names of buckwheat (Zhou et al. 2018)

Country	Vernacular names
India	Ogal
Nepal	Mite phapar
Bhutan	Jare
Russia	Grecicha kul'turnaja
Ukraine	Grechka
Poland	Gryka, Tatarka gryka, or Poganka
Czech Republic and Slovakia	Pohanka
Sweden	Bovete
Denmark	Boghvede
Finland	Tattari
Slovenia	Ajda, Hajdina, or Idina
Bosnia	Heljda
Serbia	Heljda
Montenegro	Heljda
Croatia	Heljda
French	Sarrasin, Blé noir, Renoue'e, Bouquette
Breton	Gwinizh-du
Italy	Fagopiro, Grano saraceno, Sarasin, Faggina
Germany	Buchweizen or Heidekorn
Korea	Maemil
Japan	Soba
Mandarin	Tian qiao mai
Mandarin (tartary buckwheat)	Ku qiao mai

variants of *Fagopyrum* i.e., common buckwheat (*F. esculentum* Moench) and tartary buckwheat (*F. tataricum* Gaertn) (Luthar et al. 2021; Sytar et al. 2016). Buckwheat have plenty of vernacular names in various parts of the world (Zhou et al. 2018). Table 9.3 lists out the various vernacular names of buckwheat.

9.3 The Traditional Importance of Pseudocereals

Ancient people in various parts of the world used these pseudocereals for several food and medicinal preparations (Rana et al. 2012; Adhikary et al. 2020). These small grains were also very important religiously in different countries (Adhikary et al. 2020). The religious, medicinal and popular culinary preparations from pseudocereals indicate the strong link of the grains with the day-to-day life of ancient people. These grains thus indicate the culture and history of many civilizations like the Inca and the Aztec. In the ancient Aztec and Inca civilizations, grain amaranth was an important crop with religious importance (Sauer 1950). They mixed amaranth seeds with honey which is consumed during some religious ceremonies (Sauer

1950, 1955). They are consumed by soaking in milk after fasting (Sauer 1955) and are used to prepare laddoos which upholds their importance in Hindu culture (Adhikary et al. 2020). Badagas people of Nilgiri hills used to offer puffed amaranth seeds on the Badagas funeral pyre (Adhikary et al. 2020). In Nigeria, amaranth was used in multiple ways traditionally and superstitiously (Ogwu 2020). Hot amaranth soup with fish is a popular food served as a post-delivery care to improve the health of the mother (Ogwu 2020). They were used as both animal and human food. Various local medicines were made from dried amaranth seeds (Ogwu 2020).

Quinoa grains had sacred importance in ancient Inca culture (Bastidas et al. 2016). Since they considered quinoa as the "mother of grains", these small seeds were very significant in religious, economic, and medicinal aspects (Bastidas et al. 2016). Archeological remains from hearths, burials, storage structures, and human digestive tracts confirm this fact (López et al. 2011). The residue of threshed quinoa seeds is burned and the ash along with cocoa leaves was chewed (Bonifacio 2003). A traditional South American beer called *Chicha* is made by adding quinoa and water, along with sugar and cinnamon if needed (Cutler and Cardenas 1947). Various salads and soups were prepared and consumed by South American people (Ludena Urquizo et al. 2017). Chia seeds were considered as a holy ingredient in Aztec civilization (Cahill 2003). In the ancient Mayan civilization, chia seeds were the staple diet for warriors since these small seeds provided a lot of energy (Scapin et al. 2016). In Mexico, a drink with chia seeds was consumed by runners known as Iskiates which helped them to cover a longer distance with the aid of a high energy supply from chia seeds (Scapin et al. 2016). Thus chia seeds were very important in the past and due to their medicinal importance, they were used in meals, drinks, and as pressed oil (Pal and Raj 2020). Chia seed oil was also a component of varnishes, cosmetics, and paints. The glossy finish of hand-made Mexican vessels was due to the use of chia seed oil in the paint (Cahill 2003).

Buckwheat is a major ingredient in various Bhutanese recipes (Norbu and Roder 2003). They prepare pancake-like *khuli*, noodles-like *puta*, fire-roasted pancakes called *Teyzey*, a form of unleavened circular bread called *Keptang* etc. and in Eastern Bhutan, a special dish called *Kontongs* was served during festivals (Norbu and Roder 2003). Sweet buckwheat dough sprinkled with a spoon of chilli powder and a drink called *Roth chang* was served on the 21st day of death as a symbol of mourning (Norbu and Roder 2003). In India, when cereals and pulses are not consumed during fasting, buckwheat and dishes prepared from buckwheat flour are consumed by people (Rana et al. 2012). Table 9.4 give details about various traditional uses of pseudocereals in different countries.

Even though, they had been widely used in the past, their popularity considerably decreased over the time due to various reasons. The monocropping techniques introduced by the agribusiness companies caused the production and consumption of only limited food items, especially cereals like rice, wheat, and maize. This not only homogenized the cultivation pattern but also our food habits. A decrease in amaranth cultivation can be viewed under this development of the monocropping style (De Shield 2015). A high degree of seed shattering, long maturity time, and

Table 9.4 Traditional uses of pseudocereals in different countries

Pseudocereal	Country/ Region	Traditional use/preparation	Reference(s)
Amaranth	Mexico	Zoale, a ceremonial paste made from grains was fed to the slaves before they were sacrificed to the God	Adhikary et al. (2020)
	Alegria	Used in the preparation of a confectionary	Adhikary et al. (2020)
	Nigeria	Red inflorescence is used for preparing a traditional drink for stomach aches	Ogwu (2020)
	Nepal, India and Pakistan	Popped seeds are used to make laddoos which are of religious importance in Hinduism	Adhikary et al. (2020)
Quinoa	South American Countries	Used to prepare salads, soups, porridges and fried patties	Ludena Urquizo et al. (2017)
		Seeds are used to make Chicha, seed flour is used to make tortillas and bread	Cutler and Cardenas (1947) and Nelson (1968)
		Used in the preparation of a sacred drink-*Mudai*	Bastidas et al. (2016)
Chia	Aztec empire	Flour from the seeds known as Chianpinolli was used in various Aztec beverages known as Chianatoles	Valdivia-López and Tecante (2015)
	Mexico	Beverage preparation such as "Agua de Chia" or "Chia fresca"	Cahill (2003)
		Chia seeds along with citrus fruits and water was consumed by Iskiates	Pal and Raj (2020)
Buckwheat	Bhutan	Used for preparation of *khuli, putta, teyzeye, Keptang* and *kongtong* For making a distilled alcohol-*ara*	Norbu and Roder (2003)
	Tang region of Bhutan	Cooked sweet buckwheat dough served on the 21st day after death followed by serving a drink called *roth chang*	Norbu and Roder (2003)
	Chumey region of Bhutan	Phob–biscuit-like preparation served during rituals	Norbu and Roder (2003)
	India	For preparing chillare (a kind of unleavened bread) and ghanti (a kind of local wine) Flour, namely "kuttu ka atta" is used as food during fasting	Rana et al. (2012)

lower yield when compared to other cereals added to the decline in the production of amaranth grain (Das 2016b). A principle cause for the decline in chia cultivation in Mexico was the Spanish colonization; Spaniards banned chia cultivation due to its wide use in Mexican rituals and the crop was replaced by other crops like wheat, barley, and sugarcane (Sosa-Baldivia et al. 2018). New food crops then became staple foods which neglected the traditional crops like chia and chia cultivation was thereafter limited only to the mountain ranges of Mexico where Spanish rule was limited (Sosa-Baldivia et al. 2018). The significance of buckwheat production mainly diminished because of the fact that their production per acre was considerably low than other cereals since they showed less response to the fertilizers and difficulty in developing hybrid varieties due to a self-incompatibility issue (Léder 2009). Another potential reason for the decline in buckwheat production might be its allergy-causing effects (Norbäck and Wieslander 2021). Exposure to buckwheat as food, during processing, and using buckwheat post-production materials like its husk in pillows lead to IgE-mediated allergy which causes severe allergic responses and anaphylaxis (Norbäck and Wieslander 2021). Even in the case of the decrease in quinoa cultivation and consumption, Spanish colonies had their negative impacts (Valencia-Chamorro 2003). They considered quinoa as a "Non-Christian" grain and hence prohibited it and introduced other cereals and thus cultivation of quinoa got confined only to communal lands called "aynokas" (Angeli et al. 2020). From these indications, it is clear that multiple factors affected the decrease in the cultivation and thereby the consumption of these pseudocereals.

9.4 Re-emergence of Pseudocereals as Superfoods

9.4.1 Nutritional Importance (Compared with Cereals)

Being rich in nutritional contents, pseudocereals are known as the "grains of twenty-first century" (Martínez-Villaluenga et al. 2020). Compared to cereals, pseudocereals are equal or rather superior in nutritional composition. These superfoods ensure to provide higher energy when consumed than rice, wheat or maize. Among the pseudocereals, chia provides the highest amount of energy (486 kcal; USDA 2018), which is far above the most popular cereals that we use.

Carbohydrate constitute the major nutrient fraction in pseudocereals, making up about 60–80% of their dry weight (Martínez-Villaluenga et al. 2020). It was found that these pseudocereals also ensure to provide a high amount of resistant starch which regulates the blood glucose and lipid level, reduce obesity, and improves intestinal biota (Skrabanja et al. 1998; Zhou et al. 2019). Dietary fiber is another important component that helps in proper digestion and bowel health (Heredia et al. 2002). Pseudocereals are rich in dietary fiber and satisfy the recommended level of daily dietary fibre intake (Jones 2014).

In the case of protein content, pseudocereals are the best source of high-quality protein which is comparable to cow milk (Bekkering and Tian 2019). They are made of 2S albumin, 11S and 7S globulin, which is superior to the protein provided by cereals (Pirzadah and Malik 2020). Since plant-based protein is gaining more acceptance in the present time due to its environmental sustainability and ensuring food security in a more cost-effective way, pseudocereals are on top of the list to meet the increasing demand for plant-based protein (Alonso-Miravalles and

O'Mahony 2018). An abundance of essential amino acids like methionine, phenylalanine, lysine, and isoleucine also make them preferable than cereals to be included in the day-to-day diet (Motta et al. 2019). This is not only applicable to the essential amino acid content but also to the total content of non-essential amino acids. Thus pseudocereals are an excellent source of protein in our diet which ensures proper growth and development as well as healing and repair of damaged cells in our body.

Total lipid content in pseudocereals is another reason why we should switch over to pseudocereals rather than solely relying upon conventional cereals. It should be noted that pseudocereals are safer to use since the major fraction of this lipid are unsaturated fatty acids and only a minor fraction is made up of saturated fatty acids (Martínez-Villaluenga et al. 2020). Therefore consumption of pseudocereals does not or only minimally cause health risks like atherosclerosis, cholesterol, obesity, diabetes or blood pressure (Kromhout and de Goede 2014; Simopoulos 2008).

Vitamins and minerals in the pseudocereals are also quite surprising due to their high content. This helps consumers to develop good health and well-being. Vitamin B content in various pseudocereals is either higher or comparable to that of the most consuming cereals (USDA 2005). They are also a promising source of vitamin E (tocopherol) which acts as an antioxidant in our body (Pirzadah and Malik 2020). The common cereals that we mostly depend upon only supply a minimal amount of tocopherol. In the case of mineral content, the whole pseudocereal grain is a rich source of various minerals, since they are primarily located in the bran (Martínez-Villaluenga et al. 2020). Therefore, they ensure a potential supply of minerals like magnesium, potassium, calcium, phosphorus, and iron. Table 9.5 provides nutritional composition of the pseudocerals in comparison with the cereals such as maize, wheat and rice. It is clear that pseudocerals are enriched with many important nutrients when compared with the cereals.

9.4.2 Medicinal Importance

9.4.2.1 Gluten-Free Foods

Gluten is a form of protein, made up of prolamin and glutelin fraction (El Khoury et al. 2018) which is mostly seen in cereals such as wheat, barley, and rye. The gluten content in the food improves the overall viscoelastic properties which are highly preferred during baking and other processes in the food industry (Giménez-Bastida et al. 2015). In some individuals, gluten generates some immune responses, causing inflammation in the intestine, this disease is termed as celiac disease (CD; Olivares et al. 2011). This is a genetic disorder and also overlaps with the other autoimmune diseases (Lundin and Wijmenga 2015) and the patients cannot consume gluten rich diets (Catassi and Fasano 2008). When ingested, the celiac disease susceptible people develop symptoms like abdominal pain, osteoporosis, neurological problems, anemia and irritable bowel syndrome (Green 2005). Gluten content in food may also cause health issues like dermatitis herpetiformis (DH), non-coeliac gluten

Table 9.5 The nutritional composition of cereals and pseudocereals

Nutritional content	Maize	Wheat	Rice	Amaranth	Buckwheat	Chia	Quinoa
Energy (kcal)	365	346	345	378	343	486	368
Carbohydrate (g)	74.3	71.2	78.2	63.1–70	63.1–82.1	42.12	48.5–77
Dietary fiber (g)	7.5	12.5	4.5	2.7–17.3	17.8	34.4	7–26.5
Total protein (%)	9.4	6.8	11.8	13.1–21.5	5.7–14.2	16.54	9.1–16.7
Essential amino acids (g/100 g)							
Isoleucine	3.6	3	4.1	2.7–4.2	1.1–4.1	0.801	0.8–7.4
Leucine	12.4	6.3	8.6	4.2–6.9	2.2–7.6	1.371	2.3–9.4
Lysine	2.7	2.3	4.1	4.8–8	4.2–8.6	0.97	2.4–7.8
Methionine	1.9	1.2	2.4	1.6–4.6	0.5–2.5	0.588	0.3–9.1
Phenylalanine	4.8	4.6	5.2	3.7–4.7	1.3–7.2	1.016	3–4.7
Threonine	3.7	2.4	4	3.3–5	3.9–4	0.709	2.1–8.9
Tryptophan	0.9	2.4	1.4	0.9–1.8	1.83	0.436	0.6–1.9
Valine	5.6	3.6	5.8	3.9–5	2.3–6.1	0.95	0.8–6.1
Non-essential amino acids (g/100 g)							
Alanine	7.81	3	0.57	3.5–6.2	4.6–9.6	1.044	3.2–5.7
Arginine	4.99	5.1	0.88	8.7–15.6	10.5–11.3	2.143	6.9–13.6
Aspartic acid	7.87	4.4	0.87	7.3–10.7	7.6–16.6	1.689	8
Cysteine	1.55	2.3	0.19	2.1–3.6	0.8–3.5	0.407	0.1–2.7
Glutamic acid	19.23	33	1.81	14.4–17.7	23.2–24.4	3.5	13.2
Glycine	4.39	3.8	0.46	6.7–15.2	6.2–13.2	0.943	2.2–6.1
Histidine	2.9	2.1	0.23	1.9–3.8	1.8–4.9	0.531	1.4–5.4
Proline	10.03	8.6	0.48	2.82–4.6	2.6–8.8	0.776	2.3–5.5
Serine	5.38	4.3	0.49	4.9–9.3	3.2–8.6	1.049	3.4–5.7
Tyrosine	3.76	3.5	0.53	3.3–3.7	0.6–4.9	0.536	2.5–3.7
Total lipid (g)	4.7	2.5	1.5	6.7	3.4	30.7	6.07
Vitamins (mg/100 g)							
Thiamine	0.4	0.5	0.06	0.01–0.1	0.1–3.3	0.62	0.3–0.4
Riboflavin	0.2	0.2	0.06	0.04–0.41	0.06–10.6	0.17	0.3–0.4
Niacin	3.6	5.5	1.9	<0.01–8.04	2.1–18	8.83	1.1–1.5
Tocopherol	0.49	0.06	0.11	15.4	9.5–16.4	0.5	24.7
Minerals (mg/100 g)							
Magnesium	127	138	65	254–266	390	335	207–502
Calcium	7	30	10	175–206	46.5–50.4	631	27.5–148.7
Phosphorus	210	298	160	441–455	330–395.3	860	140–530
Iron	2.7	3.5	0.7	12.17.4	11.8–14.9	7.72	1.1–16.7
Manganese	1.9	2.3	0.5	4	1.2–1.8	26.92	Not reported
Potassium	287	284	268	290–434	450	407	656–1475
Zinc	2.3	2.7	1.3	3.7–5.2	2.1–2.4	4.58	0.8–4.8
Sodium	15.9	0.6	–	0.6	-	16	11–31
Copper	0.14	7	2	0.77	0.9–1.6	13.88	1–9.5

$References for this table are given below for the respective crop

Maize	Pirzadah and Malik (2020), McKevith (2004), Bekkering and Tian (2019), Jood et al. (1995), Murdia et al. (2016)

(continued)

Table 9.5 (continued)

$References for this table are given below for the respective crop	
Wheat	Pirzadah and Malik (2020), McKevith (2004), Jancurová et al. (2009), Bekkering and Tian (2019), Ranhotra et al. (1996)
Rice	Pirzadah and Malik (2020), Jancurová et al. (2009), Bekkering and Tian (2019), Santos et al. (2013)
Amaranth	Martínez-Villaluenga et al. (2020), Boukid et al. (2018)
Buckwheat	Martínez-Villaluenga et al. (2020), Boukid et al. (2018)
Chia	Suri et al. (2016), Kulkarni et al. (2020)
Quinoa	Martínez-Villaluenga et al. (2020), Ranhotra et al. (1993)

sensitivity (NCGS), and gluten ataxia (GA) (El Khoury et al. 2018). The most recommended way to prevent the risks of these diseases is to switch over to gluten-free products (Green et al. 2015).

Even though gluten-free foods such as whole grain bread, cornflakes, bun, and whole grain cookie provide sufficient energy (Missbach et al. 2015), the protein, vitamin, mineral, and dietary fiber (El Khoury et al. 2018) supplied by these food items are limited (Missbach et al. 2015). The improper absorption due to the immune response activated by the CD patients leads to the deficiency of iron, folic acid, minerals like copper and zinc, and fat-soluble vitamins like vitamins A, D, E, and K (Caeiro et al. 2022). This causes nutritional deficiency to some extent in the affected ones.

One of the best way to ensure gluten-free food rich in protein and dietary fibres is to consume pseudocereals. It is also perhaps the best way to maintain a gluten-free diet in a more economical way. Normally used gluten-free food items like whole grain bread and cornflakes are more expensive than their gluten-containing forms (Panagiotou and Kontogianni 2017). Thus pseudocereals can be used in the daily diet of CD and other gluten-intolerant patients to lead a healthy life. The comparatively cheaper price of pseudocereals with no compromise in nutritional availability greatly helps to reduce the financial burden for CD patients. Since the protein in amaranth and quinoa as well as fibre in buckwheat are easily digestible and readily absorbed, they ensure proper absorption of nutrients contained in it (Caeiro et al. 2022). Amaranth flour was found to be used efficiently in making gluten-free bread in the study conducted by Gambus et al. (2002). The formulation of amaranth flour fortified with iron was a success, elevating the nutritional composition. Using flour of amaranth, buckwheat, and quinoa as a replacement for potato starch in baking bread not only increases protein, fibre, mineral, and tocopherol content but also improves the bread volume and produced softer crumbs (Alvarez-Jubete et al. 2010). Tarhana is a traditional Turkish fermented food normally prepared from cereals (Ozdemir et al. 2007). Use of quinoa instead of conventional rice flour or potato starch in the production of tarhana resulted in elevation of the crude protein and fat content in the tarhana and the mineral content was also high in tarhana made of quinoa flour with a notable increase in the concentration of K, Ca, Mg, and Fe (Demir 2014). Moreover, use of quinoa flour in tarhana preparation also improved the taste, odour, consistency, and acceptability of soups made from it (Demir 2014). Bread made of buckwheat flour showed inferior baking quality; but was high in protein, minerals like Cu and Mg, fatty acids like oleic and linoleic acids, folate, and antioxidant content (Giménez-Bastida et al. 2015). The bread made from buckwheat flour showed higher volume and softer crumbs. To overcome the poor baking quality of buckwheat flour, Renzetti et al. (2008) used microbial transglutaminase during bread preparation which resulted in increasing its baking qualities. Chia, when used as seed rather than flour, was found to be more acceptable by consumers and using 15% chia in bread baking does not change the overall acceptability of the bread and was also better in nutrient availability (Steffolani et al. 2014).

9.4.2.2 Prebiotics/Probiotics in Pseudocereals for Maintaining Gut Health

Prebiotics are non-digestible compounds that modulate the activity of the gut microbiota, through its metabolism by microorganisms in the gut, thereby providing beneficial physiological effects on the host (Bindels et al. 2015). These compounds are not fully metabolized, but induce the target metabolic processes in the host and thus confer health benefits (Markowiak and Śliżewska 2018). The term probiotics was originated from the Greek words "pro" and "bios" which means "for life" (Markowiak and Śliżewska 2018). According to the definition provided by FAO and WHO in 2002, probiotics are "live strains of strictly selected microorganisms which when administered in adequate amounts, confer a health benefit on the host" (FAO 2002). It may include one or more beneficial microbial strains which once reaching the intestine maintain the overall balance there. *Lactobacillus* strains, *Saccharomyces cerevisiae*, *Bacillus subtilis*, and *Aspergillus oryzae* are commonly used as probiotics (Choudhari et al. 2008). *L. rhamnosus* GR-1, a probiotic bacteria, was found to have the ability to support the immune system (Hemsworth et al. 2012). *L. rhamnosus* GR-1 can be trustfully used to improve overall immunity and quality of the life of HIV patients as they cause no issues related to the gastrointestinal tract (Hemsworth et al. 2012). Pseudocereals are now being evaluated as a very good source of prebiotics and probiotics (Ugural and Akyol 2022). When fermented, the nutritional values of pseudocereals are enhanced and this fermentation also benefits the supply of prebiotics or probiotics. Upon lactic acid fermentation, protein availability is enhanced and the growth of pathogenic bacteria is inhibited (Ugural and Akyol 2022). The richness of various nutrients in amaranth act as a key factor for the nourishment of various essential bacteria and the fermentation process benefits this bacterial growth in the gut (Ugural and Akyol 2022). In-vitro studies show that the fermentation of grain amaranth improved their antioxidant, metal chelating capacity, and protein digestibility (Olawoye and Gbadamosi 2017). Amaranth flour when fermented with *Lactobacillus rhamnosus* GG maintained its stability for 8 hours at 37°C and remained for 2 weeks in a range of 10^8–10^9 CFU/mL after storage and no fall in the level of starch, protein, and lipid was found after fermentation (Matejčeková et al. 2015). This suggests that fermented amaranth products can be developed which are beneficial for the consumers. The prebiotic activity may be present in amaranth, and studies are being progressed in this field (Ugural and Akyol 2022). The fermentation process in quinoa not only enhanced the beneficial riboflavin and folate level but also reduced the level of anti-nutritional compounds in it (Carrizo et al. 2016). The prebiotic and probiotic activities in quinoa are authenticated by another study by Ludena Urquizo et al. (2017). Of the various formulations tested, fructooligosaccharide (prebiotic), *Lactobacillus casei* Lc-01 (probiotic), fructooligosaccharide and *L. casei* Lc-01 (synbiotic), and formulations prepared from the extract of soy (70%) and quinoa (30%), the synbiotic association was found to be more beneficial for the gut microbiota (Ugural and Akyol 2022). The most researched probiotic bacteria to be present in fermented buckwheat is *L. rhamnosus* GG since buckwheat was a very suitable substrate for their growth and activity. They were able to survive in the acidic environment for even 3 days after consumption in the stomach and intestine (Lam et al. 2007). They play a vital role in improving the natural immune system and maintaining a healthy digestive system (Collado et al. 2007). *L. rhamnosus* GG also protects the body from pathogenic bacteria in the stomach and also helps in calming down various symptoms of atopic dermatitis, gastrointestinal disorders, and cow's milk allergy (Collado et al. 2007). Protein from tartary buckwheat was found to be beneficial in the growth and activity of prebiotic bacteria like *L. casei* LC2W and *Bifidobacterium longum* BD400 (Zhou et al. 2015). Thus the consumption of these pseudocereals not only ensure an ample supply of nutrients but also the complete health of a person including proper digestion and improved immunity. In the modern scenario where the consumers depend on fast food items, the presence of prebiotic and probiotic bacteria in our body will be at their minimum population; which will put our state of health in danger. Fibre-rich foods like pseudocereals may help us to overcome the issues that are probable to be developed due to frequent fast food consumption.

9.4.2.3 Bioactive Compounds

Bioactive compounds are non-nutrient compounds that have a profound role in improving human health (Martínez-Villaluenga et al. 2020). They may be either essential compounds like vitamins or non-essential like polyphenols (Biesalski et al. 2009). The presence of these bioactive compounds in pseudocereals makes them a functional food, which improves health more than the mere availability of nutrients to survive (Morales et al. 2020). Various pseudocereals are rich in different bioactive compounds. The major classes of bioactive compounds found in pseudocereals are saponins, phenolic compounds, polysaccharides, phytoecdysteroids, phytosterols, betalains, bioactive proteins, and peptides (Martínez-Villaluenga et al. 2020).

Saponin is the major bioactive compound in quinoa, which is the reason for its bitter taste (Gómez-Caravaca et al. 2011). Bioactive phenolic compounds include phenolic acids like vanillic acid, ferulic acid, gallic acid, salicylic acid, caffeic acid, and protocatechualdehyde; flavonoids like flavanols, flavones, flavanones, anthocyanins, isoflavonoids, and flavonols. Buckwheat is the richest source of phenolic compounds among the pseudocereals (Liu et al. 2019). The polysaccharides from quinoa show immune enhancing properties, thus they have the potential to be used in pharmaceutical preparations (Fan et al. 2019). Phytosterols are a plant homologous compounds of cholesterol in animals and are present in a relatively higher quantity in amaranth (Martínez-Villaluenga et al. 2020). The most important fact about phytoecdysteroids is that its presence in pseudocereals is exclusive to the bran of quinoa (Kumpun et al. 2011). In general, phytoecdysteriods are secondary metabolites that protect the plant from attack by nematodes and insects (Martínez-Villaluenga et al. 2020). Another interesting bioactive compound in quinoa is betalains, which is a nitrogen-containing aromatic indole derivative of tyrosine. Betalain extract is an approved natural colorant by the European Union (additive E-162) and the Food and Drug Administration (FDA) in food and pharmaceutical products (Martínez-Villaluenga et al. 2020). Bioactive peptides are the amino acid sequences that are inactive in their precursor protein structure but get biologically active when they are hydrolyzed in-vivo or in-vitro (Morales et al. 2020). Rutin is an important bioactive compound belonging to the class of flavonols in buckwheat and it is responsible for various therapeutic effects like anti-inflammatory, protection from UV radiation, anti-diabetic etc. (Koja et al. 2018). Table 9.6 shows the various bioactive compounds in different pseudocereals and their health benefits.

9.5 Pseudocereals as Climate-Smart Crops

Yet another interesting fact about these pseudocereals is that they can withstand climatic variations and stress conditions under which the major staple cereals may not be able to survive. Pseudocereals are hence able to grow in marginal lands with minimal nutrient availability with no compromise in their nutritional value (Pirzadah and Malik 2020). Quinoa can withstand a wide temperature range from −8 °C to 35 °C (Jacobsen et al. 2005). Zhou et al. (2018) revealed the genes responsible for ion sequestration, ABA homeostasis and signaling which make quinoa tolerant to abiotic stress. This team of researchers also proposed a model for salt accumulation in the salt bladder (Rodríguez et al. 2020). The pseudocereals show high photosynthetic diversity, which makes them capable of growing in various geographic and climatic conditions (Pirzadah and Malik 2020). Thus pseudocereal cultivation is effortless since they are climate-resilient and can be grown on any marginal land. In the present condition of increasing population and urbanization, the cultivation of pseudocereals is the best way to tackle poverty and malnutrition among the people in the near generation.

Table 9.6 Bioactive compounds in pseudocereals and their health benefits

Pseudocereal	Bioactive compounds	Health benefits	Reference
Amaranth	Squalene	Hypocholesterolemic, anti-cancerous, cardioprotective	Adhikary et al. (2020)
	Phytosterols	Hypocholesterolemic	Ogrodowska et al. (2014)
Buckwheat	Flavanols—rutin and quercetin	Anti-tumour, anti-oxidant, anti-inflammatory, anti-bacterial, anti-fungal, anti-diabetic, immunoregulatory, neuroregulatory, anti-atherosclerosis, hypotensive, gastroprotective	Huda et al. (2021), Kwon et al. (2018)
		Anti-tumour, anti-oxidant, anti-inflammatory, anti-bacterial, anti-fungal, anti-diabetic, immunoregulatory, neuroregulatory, anti-atherosclerosis	Huda et al. (2021)
	Flavones—orientin, vitexin, homoorientin, and isovitexin	Anti-inflammatory, anti-neoplastic	Huda et al. (2021)
	Flavanones—hesperidin	Anti-inflammatory, anti-viral	Huda et al. (2021)
	Flavanols—catechins and epicatechins	Anti-oxidant, anti-bacterial, anti-neoplastic, anti-ageing	Huda et al. (2021)
	Anthocyanin	Cardioprotective	Huda et al. (2021) and Kwon et al. (2018)
	Proanthocyanidin	Anti-tumour, anti-cancer	Huda et al. (2021)
	Stilbenes—resveratrol	Anti-cancer	Huda et al. (2021)
	Phenolic acids—Protocatechaldehyde, vanillic acid, caffeic acids, salicylic acid, gallic acid	Anti-oxidant, cardioprotective, anti-neoplastic	Martínez-Villaluenga et al. (2020) and Huda et al. (2021)
	Tannins	Anti-tumour, anti-oxidant	Huda et al. (2021)
	Polysaccharides	Anti-tumour, anti-oxidant, anti-inflammatory, hepatoprotective, hypolipidemic, immuno-regulatory, anti-diabetic, neuroprotective	Huda et al. (2021)
Chia	Sterols	Hypocholesterolemic	Ashura et al. (2021)
	Quercetin	Antioxidant, anti-inflammatory, antibacterial, antiviral, anti-hepatotoxic, hypocholesterolemic, vasodilation, blood thinning	Ashura et al. (2021)

(continued)

Table 9.6 (continued)

Pseudocereal	Bioactive compounds	Health benefits	Reference
	Caffeic acid	Hypoglycemic, activity, anti-oxidant activity, anti-cancer, memory protective, anti-hypertensive, neuroprotective	Ashura et al. (2021)
	Gallic acid	Antileukemic, anti-oxidant, anti-cancer, anti-neoplastic, anti-inflammatory, anti-diabetic	Ashura et al. (2021)
	Myricetin	Anti-bacterial, anti-gonadotrophic activity	Ashura et al. (2021)
	Kaempferol	Antioxidant, anti-diabetic, anti-cancer, cardioprotective, anti-inflammatory, anxiolytic, neuroprotective, analgesic, anti-allergic, anti-diabetic	Ashura et al. (2021)
Quinoa	Saponins	Anti-microbial, analgesic antioxidant, anti-inflammatory, cytotoxic, diuretic, hypocholesterolemic, hypoglycemic, anti-thrombotic, neuroprotective	Hernández-Ledesma (2019)
	Phytosterols	Anti-inflammatory, anti-cancer, anti-oxidant, hypocholesterolemic	Hernández-Ledesma (2019)
	Phytoecdysteroids	Anti-obesity, anti-osteoporotic, wound healing properties, anti-diabetic	Hernández-Ledesma (2019) and Vilcacundo and Hernández-Ledesma (2017)
	Phenolic compounds—phenolic acids, flavonoids, coumarins, lignans, quinine, phenols, phenylpropanoids, stilbenoids, xanthones	Anti-oxidant, anti-cancer, anti-inflammatory, anti-obesity, anti-diabetic, hypoglycemic, anti-diabetic	Hernández-Ledesma (2019)
	Polysaccharides	Anti-inflammatory, anti-cancer, lipid oxidation inhibitory agents, interleukins activators	Hernández-Ledesma (2019)
	Betalains	Anti-cancer, anti-lipidemic, anti-oxidant, antimicrobial	Hernández-Ledesma (2019)
	Bioactive proteins and peptides	Hypocholesterolemic, anti-diabetic, chemopreventive, anti-oxidant, anti-hypertensive	Hernández-Ledesma (2019)

9.6 Genetics and Genomics of Pseudocereals

One of the reasons for the negligence towards pseudocereals is the presence of anti-nutritional compounds which either cause low digestion of nutrients or make consumption difficult due to the characteristic taste and smell. Modern molecular methods can be used to find a way to overcome these limitations of pseudocereals (Yabe and Iwata 2020). Even though much research works has been done to know the evolutionary history and population structure of these plants, only very little genetic research work has been done on the genetic improvement of pseudocereal plants. This is due to the small and complex flowers of these plants which makes their handling during work difficult (Rodríguez et al. 2020). The wide scope of this topic opens the way to the development of a novel food habit in the world that ensures food security and proper nutrition. Genetic studies on the genes that are responsible for the climate resilience of pseudocereals also will benefit the development of new plant varieties which give the best yield with least requirements. The high content of essential amino acid lysine in amaranth was inferred to be due to the elevated expression of the *dihydrodipicolinate synthase* (*DHDPS*) gene in seeds and the presence of the *aspartate kinase 1 gene* (*AK1*) (Rodríguez et al. 2020). Stress-responsive transcriptome analysis of *Amaranthus hypochondriacus* by Délano-Frier et al. (2011) led to the understanding of the genes and their tolerance mechanism against various stress conditions. Genes involved in betalain biosynthesis, *cytochrome P450* (*CYP76AD1*) and *4,5-DOPA dioxygenase extradiol 1* (*DODA1*) were found using the physical and genetic maps by Lightfoot et al. (2017). A study on starch biosynthesis at a molecular level was done in *A. cruetens* by Park et al. (2017). It was already known that the amylose part of starch was produced by the granule-bound starch synthase (GBSS) encoded by the *Waxy* gene. Two isoforms of this gene viz. *CrGBSSI* in the perisperm and *CrGBSSIb* in the pericarp tissues of amaranth grains were found. The *CrGBSSI* gene is mainly expressed during the mid-late developmental stages of endosperm development whereas the *CrGBSSIb* gene expression was higher during initial developmental stages which implies that this enzyme may have a prominent role in the metabolism of amylose in pericarp starch (Park et al. 2017). The amylopectin part was coded by *soluble starch synthase* (*SSS*), starch branching enzyme, and starch debranching enzyme (Martin and Smith 1995). Park et al. (2012) revealed that the expression of *soluble starch synthase I* (*SSSI*) gene was high at the seed development stage. This expression pattern in amaranth indicates that the protein was essential for proper seed development (Park et al. 2012). This gene, even though expressed in a variety of tissues, showed a higher expression during the time of leaf development. The SSSI protein in grain amaranth was found to have three conserved regions common to all known starch synthases and *E. coli* glycogen synthase, which ultimately led to the conclusion that SSSI was detected in this study encoded a functional starch synthase enzyme. The SSSI gene showed a high degree of nucleotide polymorphism which may be beneficial in studying allele genealogies (Park et al. 2012). The transcriptome, gemone and physical map assembly of grain amaranth was performed by Clouse et al. (2016). A genome size of 377 Mb, a de novo transcriptome assembly with 66370 contigs from abiotic stress library and different tissues and a physical map of size 340 Mb generated in the BioNano Genomics platform by the study group. The molecular mechanism behind lysine richness of amaranth grain was unravelled by Sunil et al. (2014) through their genomics and transcriptomics study. They have annotated the genes involved in the lysine biosynthesis pathway of the plant through the comparative genomics and gene expression studies.

The shattering character of buckwheat was found using an AFLP marker and a linkage map was thus constructed around the *Shattering1* (*sht1)* gene locus by Matsui et al. (2004). Many research works were done for knowing the gene regulatory mechanism on the floral structure, salt tolerance, and aluminum toxicity in buckwheat. Whole-genome sequencing technologies like Illumina HiSeq 2000 were beneficial in deriving information about the genes involved in flavonoid biosynthesis, 2S albumin-type allergens biosynthesis, granule-bound starch synthases (GBSSs) etc. Identification of genes involved in the biosynthesis of rutin, one of the potent bioactive compounds in buckwheat, will definitely be helpful in the production of novel antioxidant compounds (Rodríguez et al. 2020). Two genes, *FtMYB16* and *FtMATE1* regulate the rutin biosynthesis with the former interacting with *Ftimportin-α1* mediates rutin biosynthesis and the later coding for the transporter of isoquercetin for the accumulation of rutin (Li et al. 2019). In another study by Koja et al. (2018), it was found that a recombinant *flavonol 3-O-glucoside 6″-O-rhamnosyltransferase* (*FeF3G6″ RhaT*) catalyzes the biosynthesis of rutin in the concerned rutin accumulating organs. *Flavonol 3-O-glucosyltransferase* (*F3GT*) first converts quercetin into isoquercetin which is then converted to rutin by *F3G6″ RhaT* (Koja et al. 2018). *MADS* genes, that control tissue and organ development, reproduction, and seed cracking were found to be relevant in the case of buckwheat since the fruit is very difficult to dehull. The study of cell walls showed that the presence of lignin, hemicellulose, cellulose, and pectin were the factors affecting fruit cracking. The easily cracking tartary buckwheat, rice-tartary had lower lignin, hemicellulose, and cellulase level but lower pectin level compared to the common tartary buckwheat. This trend in the rice-tartary favours easy dehulling and fruit cracking. Of the three potential targets of *SHP* genes in tartary buckwheat *MADS* gene family, expression of *FtpinG0009028100.01* and *FtpinG0009028000.01* was higher in easy cracking variety. *FtpinG0009028000.01* is the potential target for studying various aspects of dehulling characters of tartary buckwheat, which in the future may benefit developing novel varieties of consumer-friendly tartary buckwheat (Liu et al. 2019).

The chromosome number of quinoa was found to be $2n = 4x = 36$ (Stevens et al. 2006). The drought and salinity resistance of quinoa were studied by many researchers through transcriptome analysis. High lysine content was found due to the high expression of the gene encoding aspartokinase that converts aspartate to lysine. A high copy number of genes encoding the enzymes for vitamin B6 and dihydrofolate biosynthesis was found to be the reason for the higher availability of vitamin B and vitamin E in quinoa. *Triterpene saponin biosynthesis activating regulator-like* 2 (*TSARL2*) genes were found to be responsible for the saponin biosynthesis in quinoa seeds, which is an important bioactive compound. Also, its higher expression may cause a higher accumulation of saponin which imparts a bitter taste and certain anti-nutritional effects (Rodríguez et al. 2020). In a study done by Xiao-lin et al. (2022), 13 *Sucrose non-fermenting 1 (SNF1)-associated protein kinase 2* (*SnRK*$_2$) genes that encode proteins with complete serine/threonine protein kinase catalytic domains were found and named *CqSnRK*$_{2.1}$-*CqSnRK*$_{2.13}$. The *SnRK*$_2$ protein consists of protein kinase which helps a plant to withstand abiotic stress conditions like drought, salinity, and low temperature. The increased expression of *CqSnR*K$_{2.2}$, *CqSnRK*$_{2.11}$, and *CqSnRK*$_{2.12}$ in quinoa under drought treatment indicates the ability of quinoa to withstand drought conditions. Elevated expression of these genes was found to cause stomatal closure through the ABA signaling pathway which makes quinoa drought-resistant (Xiao-lin et al. 2022). The transcriptome sequencing of quinoa seedlings after salinity stress treatment revealed the molecular regulation involved in salt stress through the differential expression analysis. They documented that the genes coding for glucan endonuclease, glutathione S-transferase, phosphate transporter, beta-galactosidase, trichome birefringence-like protein, cytochrome P450 etc. were differently expressed (Ma et al. 2021). Many years before, the role of *Salt Overly Sensitive 1* (*SOS1*) in salinity tolerance mechanism of quinoa was analysed by Maughan et al. (2009) via genomic sequencing, expression analysis, fluorescent in-situ hybridisation and phylogenetic investigation, which was the first report on the salinity tolerance genes of Amaranthaceae family.

By using Illumina HiSeq 2500 and PacBio RSII, a draft genome of quinoa was assembled with a size of 1.1 Gb by Yasui et al. (2016a) and accordingly, Quinoa Genome DataBase was developed (QGDB; http://quinoa.kazusa.or.jp/). They also sequenced the genome of buckwheat using Illumina HiSeq 2000 and drafted data of 1.17 Gb size. Buckwheat Genome DataBase (BGDB; http://buckwheat.kazusa.or.jp) thus has about 35,816 annotated genes (Yasui et al. 2016b). This shows that, very limited genetic and genomic information of the pseudocereals is available. This suggests the dire need to do genetic and genomic studies on these pseudocereals to dissect their beneficial traits and to further improve them for greater food security and greater resilience. Further studies must focus on the genetics and genomics of the pseudocereals.

9.7 Challenges in Using Pseudocereals

Some compounds in pseudocereals were found to be anti-nutritional and harmful to human health. Excessive intake of them may cause serious health issues. Saponins in quinoa are anti-nutrient when consumed above the recommended limit, they cause hemolytic properties causing the release of hemoglobin due to the increased permeability of erythrocytes (Martínez-Villaluenga et al. 2020). They were also found to show membranolytic and fungi toxic activities (Melini and Melini 2021). Tannins present in pseudocereals were found to cause damage to the intestinal mucosa and thereby hinder mineral absorption. Their characteristic astringent smell also reduces their consumption (Melini and Melini 2021). Phytates also negatively affect mineral bioavailability by forming phytate complexes with Fe^{2+}, Zn^{2+}, Ca^{2+}, etc. (Melini and Melini 2021). The presence of trypsin inhibitors is also considered as anti-nutritional factor since they interfere with the action of trypsin in the intestine and thereby affect protein digestion (Satheesh and Fanta 2018). Oxalate present in amaranth, buckwheat, and quinoa cause susceptibility to secondary hyperoxaluria that ultimately leads to calcium oxalate stone formation (Jancurová et al. 2009). The presence of such anti-nutritional compounds thus becomes a reason for the reduced consumption and production of these pseudocereals. However, modern genetic tools can be used to silence the genes that govern the biosynthesis and accumulation of such antinutrients in the plants. These technologies can be used to reduce the contents of harmful compounds in the pseudocereals.

9.8 Conclusions and Future Prospects

Pseudocereals open a wide opportunity in the agricultural, pharmaceutical, and research field. Instead of limiting ourselves to hand-countable cereals, it is better to diversify our food menu by including these pseudocereals. This will not only help us improve our health and ensure the sufficient availability of nutrients but also increase the biodiversity of our nature. A new era with widespread cultivation of pseudocereals even in local lands should be a goal of the nations who fear food crisis in the future. Their importance ranging from the supply of various nutrients to unique bioactive compounds that improve the overall health of a person indicates how they can be effectively brought into the diet of the modern generation where the concept of leading a healthy life is a progressing trend. Cultivation of these grains will also be profitable since they demand only very less nutrients and maintenance. They are also a promising crop in producing gluten-free food products with the supply of all essential nutrients. Hence, they are important for CD patients in leading healthier lives in a more financially supportive way. Since only very limited works has been done in the field of genetic improvement of these pseudocereals, wide opportunities are being opened up before us to explore. Genetic improvement may be also helpful in overcoming the limitations that are being faced during its cultivation and also to reduce the anti-nutritional compounds in it. Maximum utilization of these grains will help to overcome the nutritional deficiency problems. Proper utilization of marginal lands for the cultivation of pseudocereals is thus economically beneficial. Pseudocereals thus are relevant in economic, social, nutrition, ecological, and functional aspects. A proper action plan for its public acceptance through awareness programs on its nutritional importance and scopes in its cultivation may help to diversify our food habits in a more healthier and economical way.

Being highly rich in nutrients and climate-smart crops, there is a wide scope for a large-scale acceptance of pseudocereals. Until now, only very little work has been done in the genetic improvement of pseudocereals. One of the major limitations of these pseudocereals is their small size and shattering of grains, which makes them less preferred by farmers to cultivate. Hence studies and work which would help to overcome these issues will benefit us to ensure a world without food scarcity and malnutrition. Knowledge about these nutritious pseudocereals is very limited since these grains were not a member of our food habits for nearly two or three generations. In India, where the population is increasing exponentially and the cultivable land is converted for non agricultural purposes, there would be a need soon where even marginal land should be converted into productive land to meet the food needs of people. Pseudocereals are thus important since they can be grown in marginal land with very limited nutrient availability and labour effort. As an important functional food, genetic improvement of pseudocereals plays a very important role which can enhance the bioactive compounds present in them. Using modern genome sequencing, gene modification and gene editing methods, many beneficial traits can be incorporated into the pseudocereals. Genes responsible for the biosynthesis and accumulation of of bioactive compounds and nutrients can be transformed into suitable vectors and their culturing thus helps in the large-scale production of improved crops. More studies are needed to improve the pseudocereals but focus should also be placed on the awareness programs to popularise the consumption of pseudocereals to tackle poverty, malnutrition, and hidden hunger.

References

Abbas S (2022) Climate change and major crop production: evidence from Pakistan. Environ Sci Pollut Res 29:5406–5414. https://doi.org/10.1007/s11356-021-16041-4

Adhikary D, Khatri-Chhetri U, Slaski J (2020) Amaranth: an ancient and high-quality wholesome crop. In: Waisundara VY (ed) Nutritional value of Amaranth. IntechOpen, pp 57–60

Akinfenwa AO, Cheikhyoussef A, Cheikhyoussef N, Hussein AA (2020) Cold pressed chia (*Salvia hispanica* L.) seed oil. In: Cold pressed oils. Elsevier, pp 181–190

Alonso-Miravalles L, O'Mahony J (2018) Composition, protein profile and rheological properties of pseudocereal-based protein-rich ingredients. Foods 7(5):73. https://doi.org/10.3390/foods7050073

Alvarez-Jubete L, Arendt EK, Gallagher E (2010) Nutritive value of pseudocereals and their increasing use as functional gluten-free ingredients. Trends Food Sci Technol 21(2):106–113. https://doi.org/10.1016/j.tifs.2009.10.014

Anacleto S, Ruiz-Ibarra G, Johnson F, Torre RR, Reyna R, Sharma M, Liu X (2018) A historical review of the scientific and common nomenclature associated with Chia: from *Salvia hispanica* to *Salvia mexicana* and Chian to Salba

Angeli V, Miguel Silva P, Crispim Massuela D, Khan MW, Hamar A, Khajehei F, Graeff-Hönninger S, Piatti C (2020) Quinoa (*Chenopodium quinoa* Willd.): an overview of the potentials of the "golden grain" and socio-economic and environmental aspects of its cultivation and marketization. Foods 9(2):216. https://doi.org/10.3390/foods9020216

Arslan-Tontul S, Candal Uslu C, Mutlu C (2022) Expected glycemic impact and probiotic stimulating effects of whole grain flours of buckwheat, quinoa, amaranth and chia. J Food Sci Technol 59:1460–1467. https://doi.org/10.1007/s13197-021-05156-8

Ashura K-K, Lillian DK, Oscar K, Leonard MPR (2021) Nutritional, health benefits and usage of chia seeds (*Salvia hispanica*): a review. Afr J Food Sci 15(2):48–59. https://doi.org/10.5897/AJFS2020.2015

Bastidas EG, Roura R, Rizzolo DAD, Massanés T, Gomis R (2016) Quinoa (*Chenopodium quinoa* Willd), from nutritional value to potential health benefits: an integrative review. Nutr Food Sci 6(3):497. https://doi.org/10.4172/2155-9600.1000497

Bekkering CS, Tian L (2019) Thinking outside of the cereal box: breeding underutilized (pseudo) cereals for improved human nutrition. Front Genet 10. https://doi.org/10.3389/fgene.2019.01289

Bhargava A, Shukla S, Ohri D (2006) *Chenopodium quinoa*—an Indian perspective. Ind Crop Prod 23(1):73–87. https://doi.org/10.1016/j.indcrop.2005.04.002

Biesalski H-K, Dragsted LO, Elmadfa I, Grossklaus R, Müller M, Schrenk D, Walter P, Weber P (2009) Bioactive compounds: definition and assessment of activity. Nutrition 25(11–12): 1202–1205. https://doi.org/10.1016/j.nut.2009.04.023

Bindels LB, Delzenne NM, Cani PD (2015) Towards a more comprehensive concept for prebiotics. Nat Rev Gastroenterol Hepatol 12:303–310

Bonifacio A (2003) *Chenopodium* sp.: genetic resources, ethnobotany, and geographic distribution. In: Food reviews international. Taylor and Francis Ltd., pp 1–7

Boukid F, Folloni S, Sforza S, Vittadini E, Prandi B (2018) Current trends in ancient grains-based foodstuffs: insights into nutritional aspects and technological applications. Compr Rev Food Sci Food Saf 7(1):123–136. https://doi.org/10.1111/1541-4337.12315

Caeiro C, Pragosa C, Cruz MC, Pereira CD, Pereira SG (2022) The role of pseudocereals in celiac disease: reducing nutritional deficiencies to improve well-being and health. Nutr Metab 2022:1–8. https://doi.org/10.1155/2022/8502169

Cahill JP (2003) Ethnobotany of chia, *Salvia hispanica* L. (Lamiaceae). Econ Bot 57(4):604–618

Carrasco E, Soto JL (2010) Importance of Andean grains. In: Bravo R, Valdivia R, Andrade K, Paludosi S (eds) Andean grains: advances, achievements and experiences developed in Quinoa, Cañihua and Kiwicha in Peru. Biodiversity International, Rome, pp 6–10

Carrizo SL, Montes de Oca CE, Laiño JE, Suarez NE, Vignolo G, LeBlanc JG, Rollán G (2016) Ancestral Andean grain quinoa as source of lactic acid bacteria capable to degrade phytate and produce B-group vitamins. Food Res Int 89:488–494. https://doi.org/10.1016/j.foodres.2016.08.013

Catassi C, Fasano A (2008) Celiac disease. In: Arendt EK, Dal Bello F (eds) Gluten-free cereal products and beverages. Academic Press, London

Cheng A, Mayes A, Dalle G, Demissew S, Massawe F (2015) Diversifying crops for food and nutrition security – a case of teff. Biol Rev 92(1):188–198. https://doi.org/10.1111/brv.12225

Choudhari A, Shinde S, Ramteke BN (2008) Prebiotics and probiotics as health promoter. Vet World 1(2):59–61

Clouse JW, Adhikary D, Page JT, Ramaraj T, Deyholos MK, Udall JA, Fairbanks DJ, Jellen EN, Maughan PJ (2016) The Amaranth genome: genome, transcriptome, and physical map assembly. Plant Genome 9:1. https://doi.org/10.3835/plantgenome2015.07.0062

Collado MC, Meriluoto J, Salminen S (2007) In vitro analysis of probiotic strain combinations to inhibit pathogen adhesion to human intestinal mucus. Food Res Int 40(5):629–636. https://doi.org/10.1016/j.foodres.2006.11.007

Cutler HC, Cardenas M (1947) Chicha, a native south American beer. Bot Mus Leafl Harv Univ 13(3):33–60

Das (2016a) Pseudocereals: an efficient food supplement. In: Das S (ed) Amaranthus: a promising crop of future. Springer, Singapore, pp 5–11. https://doi.org/10.1007/978-981-10-1469-7_2

Das S (2016b) Amaranths: the crop of great prospect. In: Amaranthus: a promising crop of future. Springer, Singapore, pp 13–48. https://doi.org/10.1007/978-981-10-1469-7_3

Délano-Frier JP, Avilés-Arnaut H, Casarrubias-Castillo K, Casique-Arroyo G, Castrillón-Arbeláez PA, Herrera-Estrella L, Massange-Sánchez J, Martínez-Gallardo NA, Parra-Cota FI, Vargas-Ortiz E, Estrada-Hernández MG (2011) Transcriptomic analysis of grain amaranth (*Amaranthus hypochondriacus*) using 454 pyrosequencing: comparison with *A. tuberculatus*, expression profiling in stems and in response to biotic and abiotic stress. BMC Genomics 12(1):363. https://doi.org/10.1186/1471-2164-12-363

De la Llave P (1833) Sobre una nueva especie diferente de Salvia. Registro Trimestre, pp 441–448

De Shield CL (2015) The cosmopolitan amaranth: a postcolonial ecology. Postcolon Text 10(1)

Demir MK (2014) Use of quinoa flour in the production of gluten-free Tarhana. FSTR 20(5): 1087–1092. https://doi.org/10.3136/fstr.20.1087

El Khoury D, Balfour-Ducharme S, Joye IJ (2018) A review on the gluten-free diet: technological and nutritional challenges. Nutrients 10(10):1410. https://doi.org/10.3390/nu10101410

Fan S, Li J, Bai B (2019) Purification, structural elucidation and in vivo immunity-enhancing activity of polysaccharides from quinoa (*Chenopodium quinoa* Willd.) seeds. Biosci Biotechnol Biochem 83(12):2334–2344. https://doi.org/10.1080/09168451.2019.1650635

FAO (2002) Guidelines for the evaluation of probiotics in food. Report of a Joint FAO/WHO Working Group on Drafting Guidelines for the evaluation of probiotics in food. 30.04–01.05.2002, London, Ontario, Canada

Gambus H, Gambus F, Sabat R (2002) The research on quality improvement of gluten-free bread by amaranthus flour addition. Zywnosc 9:99–112

Giménez-Bastida JA, Piskuła M, Zieliński H (2015) Recent advances in development of gluten-free buckwheat products. Trends Food Sci Technol 44(1):58–65. https://doi.org/10.1016/j.tifs.2015.02.013

Gómez-Caravaca AM, Segura-Carretero A, Fernández-Gutiérrez A, Caboni MF (2011) Simultaneous determination of phenolic compounds and saponins in quinoa (*Chenopodium quinoa* Willd.) by a liquid chromatography-diode array detection-electrospray ionization-time of flight mass spectrometry methodology. J Agric Food Chem 59:10815–10825. https://doi.org/10.1021/jf202224

Graham SA, Wood CE (1965) The genera of Polygonaceae in the southeastern United States. J Arnold Arboretum 46(2):91–121. https://doi.org/10.5962/p.37372

Green PHR (2005) The many faces of celiac disease: clinical presentation of celiac disease in the adult population. Gastroenterology 128(4):S74–S78. https://doi.org/10.1053/j.gastro.2005.02.016

Green PHR, Lebwohl B, Greywoode R (2015) Celiac disease. J Allergy Clin Immunol 135(5): 1099–1106. https://doi.org/10.1016/j.jaci.2015.01.044

Grote U, Fasse A, Nguyen TT, Erenstein O (2021) Food security and the dynamics of wheat and maize value chains in Africa and Asia. Front Sustain Food Syst 4:617009. https://doi.org/10.3389/fsufs.2020.617009

Haros CM, Schonlechner R (2017) Food uses of whole pseudocereals. In: Haros CM, Schonlechner (eds) Pseudocereals: chemistry and technology. John Wiley & Sons, Ltd., Hoboken, New Jersey, pp 163–192

Hedberg O (1946) Pollen morphology in the genus *Polygonum* L.(S.lat.) and its taxonomical significance. Svensk Bot Tidskr 40:371–404

Hemsworth JC, Hekmat S, Reid G (2012) Micronutrient supplemented probiotic yogurt for HIV-infected adults taking HAART in London, Canada. Gut Microbes 3(5):414–419

Henrion M, Labat E, Lamothe L (2020) Pseudocereals as healthy grains. In: Pojić M, Tiwari U (eds) Innovative processing technologies for healthy grains. John Wiley & Sons Ltd., Hoboken, New Jersey, pp 37–59

Heredia A, Jimenez A, Fernandez-Bolanos J, Guillen R, Rodriguez R (2002) Fibra Alimentaria. Biblioteca de Ciencias, Madrid, pp 1–117

Hernández-Ledesma B (2019) Quinoa (*Chenopodium quinoa* Willd.) as source of bioactive compounds: a review. BCHD 2(3):27. https://doi.org/10.31989/bchd.v2i3.556

Huber KL, Fernández JR, Webb C, Rouzard K, Healy J, Tamura M, Voronkov M, Stock JB, Stock M, Pérez E (2020) HYVIATM: a novel, topical chia seed extract that improves skin hydration. J Cosmet Dermatol 19(9):2386–2393. https://doi.org/10.1111/jocd.13469

Huda MN, Lu S, Jahan T, Ding M, Jha R, Zhang K, Zhang W, Georgiev MI, Park SU, Zhou M (2021) Treasure from garden: bioactive compounds of buckwheat. Food Chem 335:127653. https://doi.org/10.1016/j.foodchem.2020.127653

IPNI (International Plant Name Index) (2012) Lamiaceae, *Salvia mexicana* L. https://www.ipni.org/n/30001508-2

Jacobsen S-E, Monteros C, Christiansen JL, Bravo LA, Corcuera LJ, Mujica A (2005) Plant responses of quinoa (*Chenopodium quinoa* Willd.) to frost at various phenological stages. Eur J Agron 22(2):131–139. https://doi.org/10.1016/j.eja.2004.01.003

Jaikishun S, Li W, Yang Z, Song S (2019) Quinoa: in perspective of global challenges. Agronomy 9(4):176. https://doi.org/10.3390/agronomy9040176

Jancurová M, Minarovicova L, Dandar A (2009) Quinoa—a review. Czech J Food Sci 27(2):71–79

Jones JM (2014) CODEX-aligned dietary fiber definitions help to bridge the 'fiber gap'. Nutr J 13(1):34. https://doi.org/10.1186/1475-2891-13-34

Jood S, Kapoor AC, Singh R (1995) Amino acid composition and chemical evaluation of protein quality of cereals as affected by insect infestation. Plant Food Hum Nutr 48(2):159–167. https://doi.org/10.1007/BF01088312

Koja E, Ohata S, Maruyama Y, Suzuki H, Shimosaka M, Taguchi G (2018) Identification and characterization of a rhamnosyltransferase involved in rutin biosynthesis in *Fagopyrum esculentum* (common buckwheat). Biosci Biotechnol Biochem 82(10):1790–1802. https://doi.org/10.1080/09168451.2018.1491286

Kromhout D, de Goede J (2014) Update on cardiometabolic health effects of ω-3 fatty acids. Curr Opin Lipidol 25(1):85–90. https://doi.org/10.1097/MOL.0000000000000041

Kulkarni AT, Agarkar BS, Sawate AR, Kshirsagar RB (2020) Determination of physicochemical properties of chia seeds (*Salvia hispanica* L.). J Pharmacogn Phytochem 9(2):1858–1861

Kumpun S, Maria A, Crouzet S, Evrard-Todeschi N, Girault J-P, Lafont R (2011) Ecdysteroids from *Chenopodium quinoa* Willd., an ancient Andean crop of high nutritional value. Food Chem 125(4):1226–1234. https://doi.org/10.1016/j.foodchem.2010.10.039

Kwon SJ, Roy SK, Choi J-Y, Park J-H, Cho S-W, Sarker K, Woo S-H (2018) Recent research updates on functional components in buckwheat. J Agric Sci Chungbuk Nat'l Univ 34(1):1–8

Lam EKY, Tai EKK, Koo MWL, Wong HPS, Wu WKK, Yu L, So WHL, Woo PCY, Cho CH (2007) Enhancement of gastric mucosal integrity by *Lactobacillus rhamnosus* GG. Life Sci 80(23):2128–2136. https://doi.org/10.1016/j.lfs.2007.03.018

Léder I (2009) Buckwheat, amaranth, and other pseudocereal plants. In: Encyclopedia of life support systems, 1st edn. EOLSS Publishers Co Ltd, pp 1–17

Li J, Zhang K, Meng Y, Li Q, Ding M, Zhou M (2019) Ft MYB 16 interacts with Ftimportin-α1 to regulate rutin biosynthesis in tartary buckwheat. Plant Biotechnol J 17(8):1479–1481. https://doi.org/10.1111/pbi.13121

Lightfoot DJ, Jarvis DE, Ramaraj T, Lee R, Jellen EN, Maughan PJ (2017) Single-molecule sequencing and Hi-C-based proximity-guided assembly of amaranth (*Amaranthus hypochondriacus*) chromosomes provide insights into genome evolution. BMC Biol 15(1):74. https://doi.org/10.1186/s12915-017-0412-4

Liu M, Fu Q, Ma Z, Sun W, Huang L, Wu Q, Tang Z, Bu T, Li C, Chen H (2019) Genome-wide investigation of the MADS gene family and dehulling genes in tartary buckwheat (*Fagopyrum tataricum*). Planta 249(5):1301–1318. https://doi.org/10.1007/s00425-019-03089-3

López LM, Capparelli A, Nielsen AE (2011) Traditional post-harvest processing to make quinoa grains (*Chenopodium quinoa* var. quinoa) apt for consumption in Northern Lipez (Potosí, Bolivia): ethnoarchaeological and archaeobotanical analyses. Archaeol Anthropol Sci 3(1):49–70. https://doi.org/10.1007/s12520-011-0060-5

Ludena Urquizo FE, García Torres SM, Tolonen T, Jaakkola M, Pena-Niebuhr MG, von Wright A, Repo-Carrasco-Valencia R, Korhonen H, Plumed-Ferrer C (2017) Development of a fermented quinoa-based beverage. Food Sci Nutr 5(3):602–608

Lundin KEA, Wijmenga C (2015) Coeliac disease and autoimmune disease—genetic overlap and screening. Nat Rev Gastroenterol Hepatol 12(9):507–515. https://doi.org/10.1038/nrgastro.2015.136

Luthar Z, Fabjan P, Mlinarič K (2021) Biotechnological methods for buckwheat breeding. Plants 10(8):1547. https://doi.org/10.3390/plants10081547

Ma Q, Su C, Dong C (2021) Genome-wide transcriptomic and proteomic exploration of molecular regulations in Quinoa responses to ethylene and salt stress. Plants 10(11):2281. https://doi.org/10.3390/plants10112281

Markowiak P, Śliżewska K (2018) The role of probiotics, prebiotics and synbiotics in animal nutrition. Gut Pathog 10(1):21. https://doi.org/10.1186/s13099-018-0250-0

Martin C, Smith AM (1995) Starch biosynthesis. Plant Cell 7(7):971–985. https://doi.org/10.1105/tpc.7.7.971

Martínez-Villaluenga C, Peñas E, Hernández-Ledesma B (2020) Pseudocereal grains: nutritional value, health benefits and current applications for the development of gluten-free foods. Food Chem Toxicol 137:111178. https://doi.org/10.1016/j.fct.2020.111178

Matejčeková Z, Liptáková D, Valík Ľ (2015) Fermentation of milk- and water-based amaranth mashes. Acta Chim Slov 8(2):140–145. https://doi.org/10.1515/acs-2015-0024

Matsui K, Kiryu Y, Komatsuda T, Kurauchi N, Ohtani T, Tetsuka T (2004) Identification of AFLP makers linked to non-seed shattering locus (sht1) in buckwheat and conversion to STS markers for marker-assisted selection. Genome 47(3):469–474. https://doi.org/10.1139/g04-007

Maughan PJ, Turner TB, Coleman CE, Elzinga DB, Jellen EN, Morales JA, Udall JA, Fairbanks DJ, Bonifacio A (2009) Characterization of *Salt Overly Sensitive 1* (*SOS1*) gene homoeologs in quinoa (*Chenopodium quinoa* Willd.). Genome 52:647–657. https://doi.org/10.1139/G09-041

McKevith B (2004) Nutritional aspects of cereals. Nutr Bull 29(2):111–142

Melini V, Melini F (2021) Functional components and anti-nutritional factors in gluten-free grains: a focus on quinoa seeds. Foods 10(2):351. https://doi.org/10.3390/foods10020351

Missbach B, Schwingshackl L, Billmann A, Mystek A, Hickelsberger M, Bauer G, König J (2015) Gluten-free food database: the nutritional quality and cost of packaged gluten-free foods. PeerJ 3:e1337. https://doi.org/10.7717/peerj.1337

Mlakar SG, Turinek M, Jakop M, Bavec M, Bavec F (2009) Grain amaranth as an alternative and perspective crop in temperate climate. J Geogr 5(1):135–145

Morales D, Miguel M, Garcés-Rimón M (2020) Pseudocereals: a novel source of biologically active peptides. Crit Rev Food Sci Nutr 61(9):1537–1544. https://doi.org/10.1080/10408398.2020.1761774

Motta C, Castanheira I, Gonzales GB, Delgado I, Torres D, Santos M, Matos AS (2019) Impact of cooking methods and malting on amino acids content in amaranth, buckwheat and quinoa. J Food Compos Anal 76:58–65. https://doi.org/10.1016/j.jfca.2018.10.001

Murdia LK, Wadhwani R, Wadhawan N, Bajpai P, Shekhawat S (2016) Maize utilization in India: an overview. AJFSN 4(6):169–176

Myers RL (1996) Amaranth: new crop opportunity. In: Janick J (ed) Progress in new crops. ASHS Press, Alexandria, pp 207–220

Nelson DC (1968) Taxonomy and origins of *Chenopodium quinoa* and *Chenopodium nuttalliae*. Indiana University

Norbäck D, Wieslander G (2021) A review on epidemiological and clinical studies on buckwheat allergy. Plants 10(3):607. https://doi.org/10.3390/plants10030607

Norbu S, Roder W (2003) Traditional uses of buckwheat in Bhutan. In: Ethnobotany of buckwheat. Jinsol Publishing Co., pp 34–38

Ogrodowska D, Zadernowski R, Czaplicki S, Derewiaka D, Wronowska B (2014) Amaranth seeds and products—the source of bioactive compounds. Pol J Food Nutr Sci 64(3):165–170. https://doi.org/10.2478/v10222-012-0095-z

Ogwu MC (2020) Value of *Amaranthus* [L.] species in Nigeria. In: Waisundara VY (ed) Nutritional value of Amaranth. IntechOpen, pp 37–40

Olawoye BT, Gbadamosi SO (2017) Effect of different treatments on *in vitro* protein digestibility, antinutrients, antioxidant properties and mineral composition of *Amaranthus viridis* seed. Cogent Food Agric 3(1):1296402. https://doi.org/10.1080/23311932.2017.1296402

Olivares M, Laparra M, Sanz Y (2011) Influence of *Bifidobacterium longum* CECT 7347 and Gliadin Peptides on intestinal epithelial cell proteome. J Agric Food Chem 59(14):7666–7671. https://doi.org/10.1021/jf201212m

Olugbire OO, Olorunfemi S, Oke DO (2021) Global utilisation of cereals: sustainability and environmental issues. J Trop Agri Food Environ Exten 20(1):9–14. https://doi.org/10.4314/as.v20i1.2

Ozdemir S, Gocmen D, Yildirim Kumral A (2007) A traditional Turkish fermented cereal food: Tarhana. Food Rev Int 23(2):107–121. https://doi.org/10.1080/87559120701224923

Pal D, Raj K (2020) Chia seed in health and disease prevention: present usage and future perspectives. IJPSR 11(9):11. https://doi.org/10.13040/IJPSR.0975-8232.11(9).4123-33

Panagiotou S, Kontogianni MD (2017) The economic burden of gluten-free products and gluten-free diet: a cost estimation analysis in Greece. J Hum Nutr Diet 30(6):746–752. https://doi.org/10.1111/jhn.12477

Park Y-J, Nishikawa T, Tomooka N, Nemoto K (2012) Molecular cloning and expression analysis of a gene encoding soluble starch synthase I from grain amaranth (*Amaranthus cruentus* L.). Mol Breed 30(2):1065–1076. https://doi.org/10.1007/s11032-011-9692-5

Park Y-J, Nishikawa T, Matsushima K, Nemoto K (2017) Characterization of a new granule-bound starch synthase gene found in amaranth grains (*Amaranthus cruentus* L.). Mol Breed 37(9):111. https://doi.org/10.1007/s11032-017-0712-y

Pirzadah TB, Malik B (2020) Pseudocereals as super foods of 21st century: recent technological interventions. J Agric Res 2:100052. https://doi.org/10.1016/j.jafr.2020.100052

Rana JC, Chauhan RC, Sharma TR, Gupta N (2012) Analyzing problems and prospects of buckwheat cultivation in India. Eur J Plant Sci Biotechnol 6(2):50–56

Ranhotra GS, Gelroth JA, Glazer BK, Lorenz KJ, Johnson DL (1993) Composition and protein nutritional quality of quinoa. Cereal Chem 70(3):303–305

Ranhotra GS, Gelroth JA, Glaser BK, Lorenz KJ (1996) Nutrient composition of spelt wheat. J Food Compos Anal 9(1):81–84. https://doi.org/10.1006/jfca.1996.0009

Rastogi A, Shukla S (2013) Amaranth: a new millennium crop of nutraceutical values. Crit Rev Food Sci Nutr 53(2):109–125. https://doi.org/10.1080/10408398.2010.517876

Renzetti S, Dal Bello F, Arendt EK (2008) Microstructure, fundamental rheology and baking characteristics of batters and breads from different gluten-free flours treated with a microbial transglutaminase. J Cereal Sci 48:33–45. https://doi.org/10.1016/j.jcs.2007.07.011

Rodríguez JP, Rahman H, Thushar S, Singh RK (2020) Healthy and resilient cereals and pseudo-cereals for marginal agriculture: molecular advances for improving nutrient bioavailability. Front Genet 11:49. https://doi.org/10.3389/fgene.2020.00049

Salbachia (n.d.). http://salbasmart.com/manufacturers/

Santos KFDN, Silveira RDD, Martin-Didonet CCG, Brondani C (2013) Storage protein profile and amino acid content in wild rice *Oryza glumaepatula*. Pesq Agropec Bras 48(1):66–72. https://doi.org/10.1590/S0100-204X2013000100009

Satheesh N, Fanta SW (2018) Review on structural, nutritional and anti-nutritional composition of Teff (*Eragrostis tef*) in comparison with Quinoa (*Chenopodium quinoa* Willd.). Cogent Food Agric 4(1):1546942. https://doi.org/10.1080/23311932.2018.1546942

Sauer JD (1950) The grain amaranths: a survey of their history and classification. Ann Mo Bot Gard 37(4):561. https://doi.org/10.2307/2394403

Sauer JD (1955) Revision of the dioecious amaranths. Madrono 13:5–46. https://www.jstor.org/stable/41422838

Scapin G, Schmidt MM, Prestes RC, Rosa CS (2016) Phenolics compounds, flavonoids and antioxidant activity of chia seed extracts (*Salvia hispanica*) obtained by different extraction conditions. Int Food Res J 23:2341–2346

Simopoulos AP (2008) The importance of the Omega-6/Omega-3 fatty acid ratio in cardiovascular disease and other chronic diseases. Exp Biol Med (Maywood) 233(6):674–688. https://doi.org/10.3181/0711-MR-311

Singh AK (2017) Early history of crop introductions into India: II. *Amaranthus* (L.) spp. Asian Agri-History 21(4):319–324

Skrabanja V, Laerke HN, Kreft I (1998) Effects of hydrothermal processing of buckwheat (*Fagopyrum esculentum* Moench) groats on starch enzymatic availability in vitro and in vivo in rats. J Cereal Sci 28(2):209–214. https://doi.org/10.1006/jcrs.1998.0200

Sosa-Baldivia A, Ibarra GR, de la Torre RRR, López RR, López AM (2018) The chia (*Salvia hispanica*): past, present and future of an ancient Mexican crop. Aust J Crop Sci 12(10): 1626–1632. https://doi.org/10.21475/ajcs.18.12.10.p1202

Steffolani E, de la Hera E, Pérez G, Gómez M (2014) Effect of chia (*Salvia hispanica L*) addition on the quality of gluten-free bread: effect of chia on gluten-free bread quality. J Food Qual 37 (5):309–317. https://doi.org/10.1111/jfq.12098

Stevens MR, Coleman CE, Parkinson SE, Maughan PJ, Zhang H-B, Balzotti MR, Kooyman DL, Arumuganathan K, Bonifacio A, Fairbanks DJ, Jellen EN, Stevens JJ (2006) Construction of a quinoa (*Chenopodium quinoa* Willd.) BAC library and its use in identifying genes encoding seed storage proteins. Theor Appl Genet 112(8):1593–1600. https://doi.org/10.1007/s00122-006-0266-6

Sunil M, Hariharan AK, Nayak S, Gupta S, Nambisan SR, Gupta RP, Panda B, Choudhary B, Srinivasan S (2014) The draft genome and transcriptome of *Amaranthus hypochondriacus*: a C4 dicot producing high-lysine edible pseudo-cereal. DNA Res 21(6):585–602. https://doi.org/10.1093/dnares/dsu021

Suri S, Passi SJ, Goyat J (2016) Chia seed (*Salvia hispanica* L.)—a new age functional food. IJATES 4(3):286–299

Suvorova G, Zhou M (2018) Distribution of cultivated buckwheat resources in the world. In: Buckwheat germplasm in the world. Academic, pp 21–35

Sytar O, Brestic M, Zivcak M, Phan Tran L-S (2016) The contribution of buckwheat genetic resources to health and dietary diversity. Curr Genomics 17(3):193–206. https://doi.org/10.2174/1389202917666160202215425

Tahmasebi M, Firuzkoohi M (2017) Discribation of quinoa plant. Agric Sci Lett 1(1):29–52

Tanwar B, Goyal A, Irshaan S, Kumar V, Sihag MK, Patel A, Kaur I (2019) Quinoa. In: Johnson J, Wallace T (eds) Whole grains and their bioactives. Wiley, New York, pp 269–305. https://doi.org/10.1002/9781119129486.ch10

Thakur P, Kumar K, Dhaliwal HS (2021) Nutritional facts, bio-active components and processing aspects of pseudocereals: a comprehensive review. Food Biosci 42:101170. https://doi.org/10.1016/j.fbio.2021.101170

Ugural A, Akyol A (2022) Can pseudocereals modulate microbiota by functioning as probiotics or prebiotics? Crit Rev Food Sci Nutr 62(7):1725–1739. https://doi.org/10.1080/10408398.2020.1846493

Ullah R, Nadeem M, Khalique A, Imran M, Mehmood S, Javid A, Hussain J (2016) Nutritional and therapeutic perspectives of Chia (*Salvia hispanica* L.): a review. J Food Sci Technol 53(4): 1750–1758. https://doi.org/10.1007/s13197-015-1967-0

USDA (2005) USDA National Nutrient Database for Standard Reference, Release 18. Nutrient Data Laboratory Home Page. http://www.nal.usda.gov/fnic/foodcomp

USDA (2018) National Nutrient Database for Standard Reference Release 28. Basic report 12006, seeds, chia seeds, dried. https://fdc.nal.usda.gov/fdc-app.html#/food-details/170554/nutrients

Valdivia-López MÁ, Tecante A (2015) Chia (*Salvia hispanica*): a review of native Mexican seed and its nutritional and functional properties. In: Henry J (ed) Advances in food and nutrition research. Academic, pp 53–75

Valencia-Chamorro SA (2003) Quinoa. In: Caballero B (ed) Encyclopedia of food science and nutrition, vol 8, pp 4895–4902

Vega-Gálvez A, Miranda M, Vergara J, Uribe E, Puente L, Martínez EA (2010) Nutrition facts and functional potential of quinoa (*Chenopodium quinoa* Willd.), an ancient Andean grain: a review. J Sci Food Agric 90(15):2541–2547. https://doi.org/10.1002/jsfa.4158

Vilcacundo R, Hernández-Ledesma B (2017) Nutritional and biological value of quinoa (*Chenopodium quinoa* Willd.). Curr Opin Food Sci 14:1–6. https://doi.org/10.1016/j.cofs.2016.11.007

Xiao-lin Z, Bao-qiang W, Xiao-hong W (2022) Identification and expression analysis of the CqSnRK2 gene family and a functional study of the CqSnRK2.12 gene in quinoa (*Chenopodium quinoa* Willd.). BMC Genomics 23(1):397. https://doi.org/10.1186/s12864-022-08626-1

Yasui Y, Hirakawa H, Oikawa T, Toyoshima M, Matsuzaki C, Ueno M, Muzuno N, Nagatoshi Y, Imamura T, Miyago M, Tanaka M, Mise T, Tanaka T, Muzukoshi H, Mori M, Fujita Y (2016a) Draft genome sequence of an inbred line of *Chenopodium quinoa*, an allotetraploid crop with great environmental adaptability and outstanding nutritional properties. DNA Res 23:535–546. https://doi.org/10.1093/dnares/dsw037

Yasui Y, Hirakawa H, Ueno M, Matsui K, Katsube-Tanaka T, Yang SJ, Aii J, Sato S, Mori M (2016b) Assembly of the draft genome of buckwheat and its applications in identifying agronomically useful genes. DNA Res 23(3):215–224. https://doi.org/10.1093/dnares/dsw012

Yabe S, Iwata H (2020) Genomics-assisted breeding in minor and pseudo-cereals. Breed Sci 70(1): 19–31. https://doi.org/10.1270/jsbbs.19100

Zhou X, Lu Y, Zhou Y, Xiao Y, He L (2015) The effect of tartary buckwheat protein on the growth of probiotics. In: Proceedings of the 2015 2nd International Workshop on Materials Engineering and Computer Sciences. Atlantis Press, Jinan

Zhou M, Tang Y, Deng X, Ruan C, Tang Y, Wu Y (2018) Classification and nomenclature of buckwheat plants. In: Buckwheat germplasm in the world. Academic, pp 9–20

Zhou Y, Zhao S, Jiang Y, Wei Y, Zhou X (2019) Regulatory function of buckwheat-resistant starch supplementation on lipid profile and gut microbiota in mice fed with a high-fat diet. J Food Sci 84(9):2674–2681. https://doi.org/10.1111/1750-3841.14747

Chapter 10
Nutraceutical Potential of Tropical Wild Edible Plants of India

K. Abha Manohar, Vangapandu Thriveni, Gopal Shukla, and Sumit Chakravarty

Abstract Humankind nowadays embraces living naturally and healthily. Wild edible plants are medicinally and nutritionally rich. Wild food plants are valued like never before in the current scenario. These plants harbour many different species which are less known to modern science. Wild plants can be effectively used for nutrition and medicinal purposes. The global nutraceutical market is growing at a high pace. In India, it is evolving and gaining momentum. Underutilized wild plants possess huge potential for the country to gain foreign exchange. Incorporation of wild edible plants into the diet is indeed essential for the well-being of humankind. It is essential to raise scientific backup and popularization of this idea. The value of a useful wild edible plant species should account for its nutraceutical value and contribution towards biodiversity. Most of these species are less known and there is no proper documentation on their population estimation and conservation needs. Thus, there is a great scope for this newly emerging science along with incorporating traditional knowledge in research and quality improvement by considering the global climatic events which harms the persistence of wild edible plant species. This chapter discusses the nutraceutical potential of some of the wild edible plants of India.

Keywords Nutrition · Health · Assessment · Popularization · Sustainable utilization · Preservation

K. Abha Manohar (✉) · V. Thriveni
Department of Agronomy and Agroforestry, Centurion University of Technology and Management, R. Sitapur, Odisha, India
e-mail: abha.manohar@cutm.ac.in

G. Shukla · S. Chakravarty
Department of Forestry, Uttar Banga Krishi Viswavidyalaya, Pundibari, West Bengal, India

A. Kumar et al. (eds.), *Wild Food Plants for Zero Hunger and Resilient Agriculture*, Plant Life and Environment Dynamics,
https://doi.org/10.1007/978-981-19-6502-9_10

10.1 Introduction

Recent years have witnessed sudden progress in the use of plant resources in healthcare. The age-old relationship of human beings with plants is inevitable. Plant diversity is very vital for India's ecological, environmental, and healthcare sectors as well (Rai et al. 2021). Indigenous people play a key role in the usage of traditional ways of using plants as a healing remedy for various diseases, culinary purposes, and other construction and craft works too. The ancient relationship between man and plants is ethnobotany or as aboriginal botany (Gras et al. 2021). Wild plants are consumed by humans for health and nutrition in a sustainable manner. The sustainable development goals (SDGs) of the United Nations (UN) aim to achieve a better and sustainable future for all. The third sustainable development goal (SDG3) of UN's sustainable development agenda aims to ensure healthy lives and promote well-being for all at all ages which means health is a key goal in the sustainable development agenda of the UN. Globalization has contributed to food homogenization throughout the world (De Medeiros et al. 2021). Wild food plants do not need any agricultural management practice or use of pesticides and fertilizers (Soares and Porto 2009). Tropical forests are very rich terrestrial ecosystems, with a variety of natural resources to sustain the livelihood of indigenous communities (Kumar et al. 2006). Unfortunately, they are depleting at an alarming rate and the flora is most affected (Dierick and Hölscher 2009). India is one of the 17 mega-diverse countries and therefore needs special attention concerning wild plant conservation (Kumari et al. 2011). The conservation status of wild edible plants is a challenge as correct knowledge of population and usage is lacking (Moyo et al. 2015). Harvesting, consumption, and trade of wild edible plants are essential for attaining food security of the rural community (Cordain et al. 2005; Albert 2012). India has rich genetic diversity of tropical root and tuberous plants such as yams, colocasias, taros, ginger, arrowroots, zedoary, ginger lily, wild turmeric, and orchids (Misra et al. 2013; Kumar et al. 2013). Not only these tubers, but India is also rich in diverse wild food plants, and tribal communities in the tropics still rely on a number of wild food plants for their health and nutritional needs. They consume the wild food plants by collecting from the wild. They are either eaten fresh, cooked, processed or stored for longer usage. This chapter provides a brief update on the usage of the wild food plants in the tropics and their nutraceutical potential.

10.2 Relevance of Plant-Based Nutraceuticals in the Present Era

The linkage between biodiversity and health is inevitable. It is very relevant in present times as globalization and economic development have damaged our eating habits thus questioning healthy living. Foods with high calories, carbohydrates, and unsaturated fats have made human health terrible in the present times. Back to nature is the only way to combat most of the lifestyle-related diseases/disorders. In ancient time, people were using only natural products for nutrition and medical purposes as well. But modernization has destroyed most of the plant resources and thus such traditional practices are on the decline. Still, in some places of the world, people are relying on wild plant based ethnic medicines mostly because of their poor economic status (Alebie et al. 2017; Mohammed et al. 2009). The sudden surge in the usage of medicinal and aromatic plants in healthcare is mainly due to the various side effects created by unsafe synthetic pharmacological agents. The ever-increasing world population exerts great pressure on agricultural land and other resources (Abberton et al. 2016). Since wild food plants are in plenty and are found in diverse agroecological regions, their usage can improve the resiliency of local food systems. In the current times, people are more concerned about maintaining their health with safer and more effective natural products. The majority of present-day nutraceuticals belong to terrestrial origin. But marine-based nutraceuticals are also slowly gaining momentum as they are introduced into the markets (Apostolidis and Lee 2016).

Wild edible plant (WEP) biodiversity underpins forest contributions to living beings and provides ecosystem products and services that are necessary to human health and prosperity. Local and indigenous varieties of food may have higher nutrient content than common varieties. Good nutrition is good for health. If we can commercialize wild edible plants, we can promote the production and consumption of fresh fruits and vegetables. WEPs offer vital solutions to some of the world's most existential challenges of nutritional demands. Many WEPs contains trace elements, vitamins, and minerals essential for human growth and development (Dogan et al. 2013). Some of the WEPs are nutritionally very rich, therefore, they can help in improving the food security of the households in the low-income countries. They may improve resilience, nutrition, and taste. WEP possesses an array of secondary metabolites in them which may be carefully explored for diverse applications like antibiotics, bio repellents, veterinary nutrition, food additives, and cosmetics (Hussein and El-Anssary 2018).

10.3 People and Wild Food Plants

Wild edible plants play a significant role in tribal diet and healthy well-being but very less authentic information is documented (Patil et al. 2014). There are 50,000–80,000 plants belonging to angiosperms that are used for medicinal uses and out of which many plant species are threatened with extinction (Kala and Sajwan 2007). 60% of the world and 80% of the population of developing countries depend on traditional medicine (Khan and Ahmad 2019). People were using natural products for nutrition as well as for healthcare since ancient time. The ethnic people are using medicinal plants for their medicinal purposes firstly because they are easily available and secondly because of the poor economic conditions prevailing in such communities. There are so many unknown sources and uses of plant species among tribal people. The knowledge source is depleting as the younger generation does not care about such things and they are moving towards cities. It has been estimated that 5–10% of the existing plant species in India have been surveyed for biologically active compounds (Singh et al. 2012). Interestingly, 25% of medical drugs used in developed countries are based on plant sources (Cragg et al. 1997).

Various studies report the isolation and characterization of nutritional components and secondary metabolites of underutilized tropical plants (Onweluzo et al. 1995; Sena et al. 1998; Kala and Mohan 2010; Dolezal et al. 2001) which included tropical legumes (*Afzelia africana, Brachystegia eurycoma, Detarium microcarpum and Mucuna flagellipes*), tree leaves (*Ziziphus mauritiana, Ceratotheca sesamoides, Moringa oleifera, Leptadenia hastata, Hibiscus sabdariffa, Amaranthus viridis and Adansonia digitata*), leafy vegetables (*Moringa oleifera, Adansonia digitata, Colocasia esculenta, Corchorus tridens, Cassia tora, Amaranthus spinosus*), and lesser-known wild fruits (*Aronia melanocarpa, Cornus mas, Berberis vulgaris, Pyracantha coccinea, Elaeagnus angustifolia, Prunus speciosa*). A study envisages nutritional composition of lesser-known fruits used by the ethnic communities from Indian forests and reported 218 species of wild edible fruit plants (Nazarudeen 2010). Out of them, ten fruits were analyzed for their nutritional value and compared with ten common fruit crops (Nazarudeen 2010). An underutilized fruit plant *Elaeagnus kologa* from Western Ghats was studied by Paulsamy et al. (2010). Another study reports nutritional properties of the less exploited wild legume plant *Rhynchosia bracteata,* rich in magnesium, iron, potassium, and phosphorous which contained higher amounts of crude protein, crude fat, ash, and nitrogen-free extractive constitutes 20.18, 6.16, 6.12, and 61.31%, respectively.

The current research shows that underutilized wild food plants possess high nutritional values. These species were used in various cuisines throughout the globe to flavour and garnish other food products. The wild edible plant species were a good source of vitamins and minerals, but have now become less relevant as the bulk of knowledge on wild plants is orally passed on from generation to generation and not written or documented, thus depleting. In order to arrest the decline of the traditional knowledge of wild edible plants, we need to fasten the attempts to gather traditional knowledge and disseminate the same to save the precious information for future usage. Considering the importance of the wild edible plants for food and nutritional purposes, it is essential to collaborate with the ethnic people and take them into confidence for safeguarding and conservation of the knowledge base before it vanishess.

10.4 What Are Nutraceuticals?

Dr. Stefen DeFelise coined the term Nutraceutical and it is composed of two words i.e. “Nutrition” and “Pharmaceutical” in 1989. Nutraceuticals are such foods or medicines that provide health benefits through nutritional and medicinal value (Wildman 2007; Jain and Ramawat 2013; Yadav and Dugaya 2013). These include

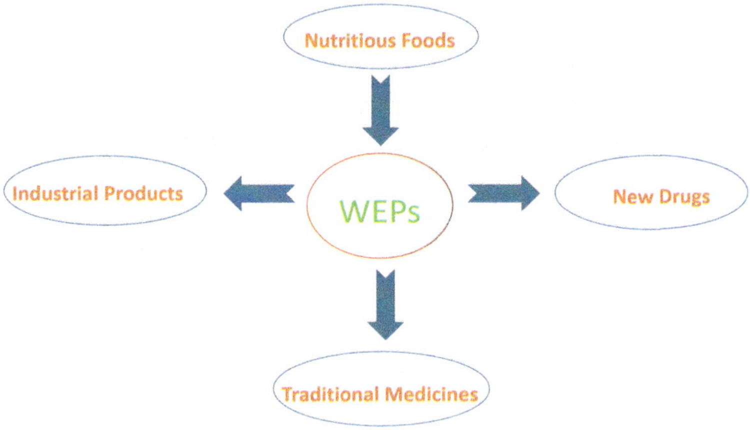

Fig. 10.1 Uses of wild edible plants

dietary supplements, herbal products, engineered or designed foods that are health-promoting, disease-curative or preventive, functional foods, and beverage items that are rich in vitamins, carbohydrates, proteins, and minerals. A conceptual diagram of nutraceuticals is given in Fig. 10.1. They are any kind of non-toxic food component which have been scientifically proven to benefit health or prevent diseases (Kathleen and Stephen 2013). Abbet et al. (2014) report that in the Swiss Alps, tea is most frequently used followed by wild vegetables. The nutraceutical industry is booming under the prevailing COVID-19 pandemic situation as people tend to prefer natural products over chemical derivatives. Nutraceutical wild plants belong to a diverse groups, starting from grass, herbs, shrubs, woody lianas, and tall trees from varied habitats. Mostly roots or tubers are used as a major food as these contain starch and other secondary metabolites (Kesavan and Swaminathan 2008). Some plants used for consumption with their uses and consumable parts are given in Table 10.1.

Besides their nutritional and medicinal value, many of the species lack detailed analysis of their bioactive ingredients. However, research on these aspects has been picking pace in the recent years (Motamed and Naghibi 2010; Egea-Gilabert et al. 2013; García-Herrera et al. 2014) as the edible plants from the wild have been appreciated to have immense nutritional value and play a vital role to rural and tribal communities in the form of nutrition and medicine (Seal et al. 2017). For example, *Cucumis hardwickii* widely consumed in many parts of India grows naturally in the household and is put to use for the treatment of various ailments (Chatterjee and Maitra 2014). Similarly *Rubus acuminatus* is rich in vitamin C. This can be used as antioxidant source. Figure 10.2 provides a conceptual diagram of the meaning and uses of nutraceuticals.

Table 10.1 Wild edible plants with their therapeutic value

Scientific name	Family	IUCN status	Potential uses	Parts used for food
Aloe barbadensis	Asphodelaceae	NL	Dilates capillaries, anti-inflammatory, emollient, wound-healing properties	Leaf
Ephedra sinica	Ephedraceae	LC	Bronchodilator, vasoconstrictor, reduces bronchial Edema	Ariel parts
Allium sativum	Amarillidaceae	NL	Antibacterial, antifungal, antithrombotic, hypotensive anti-inflammatory	Bulb
Glycyrrhiza glabra	Fabaceae	LC	Expectorant, secretolytic, treatment of peptic ulcer	Rhizome
Zingiber officinale	Zingiberaceae	DD	Carminative, antiemetic, cholagogue, positive inotropic	Rhizome
Delonix elata	Fabaceae	LC	Inflammatory joint disorders, colon cancer	Leaves
Erythrina variegeta	Fabaceae	LC	Antioxidant, antibacterial, and anti-inflammatory effects	Leaves, pods
Eclipta alba	Asteraceae	LC	Prevention of ageing and rejuvenates hair, teeth, bone, memory, eyesight, and hearing	Leaves
Ginkgo biloba	Ginkoaceae	EN	Antagonist, improves memory	Fruit, leaves
Echinacea purpurea	Asteraceae	NT	Antiviral, anti-inflammatory, Immuno modular	Flower, leaves
Panax ginseng	Araliaceae	NL	Stimulates immune and nervous system	Tuber
Hypericum perforatum	Hypericaceae	LC	Against hepatitis-c, HIV, antidepressant	Leaves, flowers
Curcuma longa	Zingiberaceae	DD	Anticancer, anti-inflammatory, antiseptic	Rhizome
Valeriana officinalis	Caprifoliaceae	LC	Menstrual cramps, abdominal pain, tranquilizer	Rhizome
Hydrastis canadensis	Ranunculaceae	VU	Antihaemorragic, antimicrobial, astringent	Rhizome
Cassia angustifolia	Fabaceae	NL	Purgative	Flowers and leaves
Ferula asafoetida	Apiaceae	NL	Expectorant, carminative, purgative	Seed
Aegle marmelos	Rutaceae	NT	Good health of digestive system, appetizer	Bulb, leaves
Centella asiatica	Apiaceae	LC	Antianxiety, nervine tonic	Whole plant
Abelmoschus esculentus	Malvaceae	NL	Antidiabetic	Fruit
Citrullus colocynthis	Cucurbitaceae	NL	Blood sugar level moderation	Fruit, seed
Allium hookery	Amarylliadaceae	NL	Reduction in cholesterol	Shoot
Alocasia macrorrhiza	Araceae	NL	Antimicrobial, anticancer	Stem and corn
Amaranthus spinosus	Amaranthaceae	NL	Antidiabetic, antipyretic, antigonorrheal	Shoot, leaves

Ammomum subulatam	Zingiberaceae	NL	Tooth care, aphrodisiac, stimulant	Fruit
Artocarpus heterophyllus	Moraceae	NL	Antibacterial, antifungal, antidiabetic, anti-inflammatory	Fruit
Bambusa tulda	Poaceae	NL	Antioxidantal property	Tender shoot
Bischofia javanica	Phyllanthaceae	LC	Anti-inflammatory	Fruit
Clerodendrum colebrookianum	Verbenaceae	NL	Anti-nociceptive, anti-oxidant, anti-hypertensive, anticancer, antimicrobial	Tender leaves
Dendrocalamus strictus	Poaceae	NL	Memory enhancer, antioxidant	Tender shoot
Rubus ellipticus	Rosaceae	LC	Nephroprotective, antioxidant, anti-diabetic, antimicrobial, antiproliferative	Berry, root
Morus alba	Moraceae	LC	Antioxidant, anti-tumour, anti-cancer, neuro-protective, hypolipidemic, antidiabetic, immunomodulator	Stem
Pyracantha crenulata	Rosaceae	NL	Antiurolithogenic, antimicrobial, antioxidant, anti-inflammatory, diuretic, anti-elastase, anti-collagnase, anti-tyrosinase	Fruit
Myrica esculenta	Myricaceae	NL	Antioxidant, antifungal	Fruits, leaves
Berberis asiatica	Berberidaceae	NL	Antioxidant, anti-diabetic, anti-tyrosinase, anti-collagnase, anti-elastase	Shoots and roots
Haematocarpus validus	Menispermaceae	NL	Jaundice, cancer, hypertension	Fruit
Xylopia brasiliensis	Annonaceae	NL	Carminative, febrifuge, and stomachic	Seed, bark
Araucaria angustifolia	Araucariaceae	CR	Rheumatism, respiratory infections, fatigue, anemia	Seeds
Bactris gasipaes	Arecaceae	NL	Vermifuge, seasoning foods	Fruit
Protium spruceanum	Burseraceae	LC	Hyperglycemia, glucosidase inhibition, protection against oxidative stress	Fruit
Cecropia pachistachya	Cecropiaceae	NL	Pneumonia; oedema, diabetes, Parkinson's disease	Bark, timber
Maytenus aquifolium	Celesteraceae	NL	Antiulcerogenic effect	Leaves
Eugenia uniflora	Myrtaceae	LC	Antimicrobial, antioxidant, insecticide and protective	Fruit
Piper hispidinervum	Piperaceae	NL	Astringent, diuretic, stimulant	Whole plant
Esenbeckia leiocarpa	Rutaceae	VU	Alzheimer's disease	Fruit
Theobroma grandiflorum	Sterculiaceae	LC	Abdominal pains, angina, high blood pressure	Fruit, seed
Trema micrantha	Ulmaceae	LC	Coughs, sore throats, asthma, bronchitis, gonorrhea, yellow fever, toothache	Leaves, wood

(Abbreviations: *NL* Not Listed, *LC* Least Concern, *DD* Data Deficient, *EN* Endemic, *NT* Near Threatened) (Jhansi and Manjula 2013; IUCN 2021; Payum et al. 2014; Momin et al. 2018)

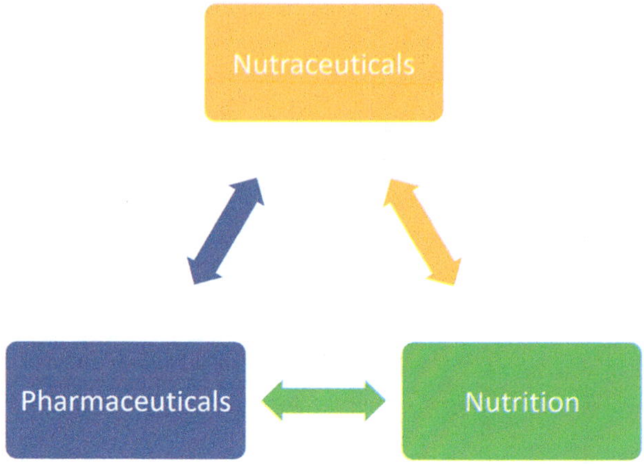

Fig. 10.2 Conceptual diagram of nutraceuticals

10.5 Nutraceutical Market

Globally the nutraceutical market is on the rise day-by-day, USA being the number one leading country with the largest market (Chauhan et al. 2013). In the Asia-Pacific, Japan is leading (The 2nd National Symposium on "Nutraceuticals, Herbals & Functional Foods", 2016). In the international market, nutraceuticals have become a multi-billion dollar industry with about USD 117 billion estimated cost of investment (Rishi 2006). India has rich genetic diversity of tropical wild food plants such as yams, arrowroot, ginger, zedoary, ginger lily, wild turmeric, and orchids (Misra et al. 2013; Kumar et al. 2013). With the vast biological resource potential, India needs more investments as a potential country for leading the nutraceutical market. In India, the market is growing at a rate of 21% per annum. In 2014, the nutraceutical market was INR 44 billion but could be worth more than INR 95 billion in the future (Nwosu and Kingsley 2020). Several countries especially developing including India and under-developed countries face famines, food shortages, malnutrition, endemic lethal diseases, antibiotic resistance, and other severe nutrition-related disorders (Bohin et al. 2010). One major conccern in nutraceutical industry is the lack of regulations and proper laws. There is a misconception that botanical drugs are safe to consume. But this may cause severe consequences because of adulterations and toxicity thereby increasing threats to the consumer. Every nutraceutical product in the market should undergo a safety check, efficacy check, and quality testing before reaching the market (Pandey et al. 2010). In India FSSAI (Food Safety and Standards Authority of India) deals with the regulations in nutraceuticals (Nwosu and Kingsley 2020). Therefore, the nutraceutical industry is in its growing phase and wild food plant-based nutraceuticals will be in much demand in the future. India should tap this potential. It should create nutraceuticals from the wild edible plants and market them. It will also need safety checks, therefore, the safety and quality of the plant-based nutraceuticals should be robust.

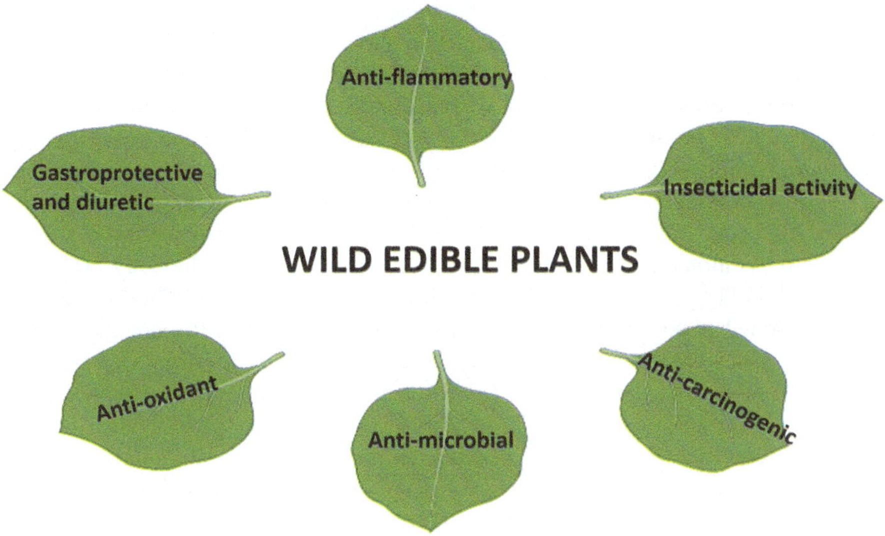

Fig. 10.3 Pharmaceutical values of WEP

10.6 Wild Edible Plants with Medicinal Attributes

Medicinal plants are getting increased attention across the world not only for their therapeutic but also their ecological, social, economic, and nutraceutical benefits (Arnold and Perez 2001; Larsen et al. 2000; Olsen 2005; Pradhan and Badola 2008; Raj et al. 2018; Lepcha et al. 2022). Wild edible plant bioactives holds an important category of nutraceuticals commonly used by people who opt for traditional health care, mainly forest dwellers or tribal people. These include various active phytochemicals viz. flavonoids, uramines, terpenoids, sulfides, polyphenols, carotenoids, lignans, coumarins, saponins, sterols, and phthalates (Cragg et al. 1997). Figure 10.3 provides an overview of the important bioactivities of the wild edible plants.

Wild edible plants have been used for food and medicinal purposes since time immemorial. Their health properties are attached to their chemical constituents (vitamins, flavonoids, terpenoids, carotenoids, phytoestrogens, and minerals) (Calucci et al. 2003; Suhaj et al. 2006). *Terminalia esculanta, Myrica esculanta,* and *Morus alba, Berberis asiatica* were found rich in phenolics and anthocyanins respectively (Bhatt et al. 2017). Wild edible plants viz. *Cichorium intybus, Picris hieracioides,* and *Sanguisorba minor* are reportedly found suitable for domestication and cultivation in Italy (Ceccanti et al. 2020). Sundriyal et al. (1998) reports wild plant species viz. *Spondias axillaris, Machilus edulis, Baccaurea sapida, Eriolobus indica, Elaeocarpus sikkimensis, Bassia butyracea* and a variety of edible medicinal plants like *Aconitum heterophyllum, Nardostychys jatamansi, Picrorhiza scrophulariiflora, Podophyllum hexandrum, Heracleum wallichii* and *Swertia chirata* from Sikkim Himalaya are exploited on commercial scale and their regeneration in the natural habitats is threatened because of over-exploitation. In an inventory of wild food plants of Indian Himalaya ranges, 675 wild plant species, under 149 diverse families, are used as food and are either consumed raw, roasted, boiled, fried, cooked or as oil, spice, seasonings, jams, pickles, jelly, etc. West Himalaya shows the highest diversity (50.96%) of edible plants and East Himalaya has the maximum number of endemics (18 taxa) and wild relatives of crop plants (39 taxa). Many of these species have cultural value and some of them are worshiped (Sundriyal et al. 1998; Akhtar 2001). From the Indian sub-Himalaya, a study claimed 67 ethnomedicinal plant species belonging to 44 families and 60 genera (22 trees, eight shrubs, 34 herbs, and three climber species). The documented ethnomedicinal plant species were used for folk therapeutics to cure 39 diseases including cancer. Many communities use seeds of *Corchorus capsularis, Millettia pinnata, Abelmoschus esculentus, Datura metel* and *Momordica charantia* and *Piper longum* to cure cough, cold and fever. The whole of *Ampelocissus barbata, Andrographis paniculata, Acacia nilotica, Ailanthes excelsa, Bridelia tomentosa, Morinda angustifolia, Ricinus communis and Semecarpus anacardium, Calotropis procera, Hibiscus rosa-sinensis, and Leucas cephalotes* and *Blumea lacera* were mostly used by the communities in many folk medicines (Rajendran et al. 2002; Mahishi et al. 2005; Ignacimuthu et al. 2008; Malla et al. 2012; Suresh et al. 2013; Lepcha et al. 2022). Table 10.1 shows the list of some wild edible plants used to cure diseases. In the pharmaceutical industry, many wild edible plants are used. The secondary metabolites derived from wild plants are used in many pharmaceutical preparations in various medicinal systems including Ayurveda, Unani, Allopathy, Homoeopathy, etc. (Vineeta et al. 2018). Some of the wild edible plants and their phytochemicals which are utilized for medicinal purposes are enlisted in Table 10.2.

10.7 Conservation Aspects and Challenges

10.7.1 Conservation

Wild food plant resources are nutritionally rich and inexpensive (Momin et al. 2018) with health-promoting constituents. Available literature reports 1532 wild edible food species used in India alone (Reddy et al. 2007). In these nutraceuticals, root and tubers are widely consumed after grains (Edison et al. 2006). The collection and consumption of wild edible plants help to enhance the livelihood of rural poor and forest dwellers (Kumar and Jena 2016). As these wild resources are deliberately collected from the forests, their population is depleting as their distribution is mostly restricted. Many of the uses of the food plants in the forests are unknown to this present generation owing to the erosion of indigenous knowledge (Nazarudeen 2010). Most of the uses of these plants known to indigenous populations are mostly passed on from one generation to another orally and these days as globalization overtook the livelihoods of people, such knowledge transfer has also slowly

Table 10.2 Phytochemical components of some important wild edible plant species of India

Scientific name	Phytochemicals	Utilized part	Reference
Adenanthera pavonina	Tannin, protein, alkaloid, glycoside, lignin, cellulose	Seed	Panda (2014)
Abrus precatorius	Alkaloid, glycoside, flavonoid, phenol	Seed	Panda (2014)
Abutilon indicum	Steroid, sapogenin, carbohydrates, flavonoids	Seed	Panda (2014)
Aegle marmelos	Alkaloid, phenylpropanoid, terpenoid	Fruit	Mahapatra and Panda (2012)
Aerva lanata	Alkaloids, flavonoids, methylgrevillate, lupeol, lupeol acetatebenzoic acid, β-sitosteryl acetate, tannic acid	Leaf	Panda (2014)
Agave americana	Carbohydrates, reducing sugar, alkaloid, volatile oil, tannin, steroid, glycoside, flavonoid	Leaf	Namrata et al. (2011)
Angelica glauca	Coumarin, saponin, alkaloid, tannin, carbohydrate, protein, fat, steroid glycoside, flavonoid	Leaf	Namrata et al. (2011)
Artocarpus heterophyllus	Flavonoid, sterol, prenylflavone	Fruit	Majumdar and Datta (2009)
Asparagus filicinus	Asparagine, arginine, tyrosine, flavonoid, resin, tannin	Root	Namrata et al. (2011)
Averrohoea carambola	Saponin, alkaloid, flavonoid, tannin	Fruit	Brahma et al. (2013)
Baccaurea sapida	Phenol, flavonoid	Fruit	
Bambusa bambos	Tannins, steroids, phenols, glycosides, flavanoids, carbohydrates and proteinsannins, steroids, phenols, glycosides, flavanoids, carbohydrates and proteins Tannin, steroid, phenol, glycoside, flavonoid, carbohydrate, protein	Culm	Panda (2014)
Bauhinia purpurea	Carbohydrate, protein, alkaloid, saponin, phenol, flavonoid, fat, anthocyanin, terpenoid, steroid	Flower	Namrata et al. (2011)
Bauhinia variegata	Steroid, glucoside, tannin, carbohydrate, amide, reducing sugar, vitamin C, crude protein, fiber, calcium, phosphorus, quercetin, rutin, quercitrin	Flower	Namrata et al. (2011)
Bauhinia retusa	Steroid, glucoside, tannin, carbohydrate, amide, reducing sugar, vitamin C, crude protein, fibre, calcium, phosphorus, quercetin, rutin, quercitrin	Flower	Panda (2014)
Boerhavia diffusa	Flavonoid, glycoside, alkaloid, rotenoid, steroid, triterpenoid, lipid, lignin, carbohydrate, protein, glycoprotein	Leaf	Panda (2014)

(continued)

Table 10.2 (continued)

Scientific name	Phytochemicals	Utilized part	Reference
Capparis zeylanica	Fatty acid, flavonoid, tannin, E-octadec-7-en-5-ynoic acid, alkaloid, saponin glycoside, terpenoid, *p*-hydroxybenzoic, saponin, syringic, vanillic, ferulic, *p*-coumanic acid	Fruit	Mahapatra and Panda (2012)
Carissa species	Tannin,2-phenly ethanol, linalool, β-caryophylline, isoamyl alcohol, benzyl acetate	Fruit	Majumdar and Datta (2009)
Cassia tora	Saponin, glycoside, protein, tannin, carbohydrate	Leaf	Panda (2014)
Centella asiatica	Carbohydrates, tannin, steroids, terpenoids, alkaloids, flavonoids, cardiac glycoside, saponin	Leaf	Panda (2014)
Chenopodium album	Flavonoid as phenolic amide, saponin, cinnamic acid amide, alkaloid chinoalbicin, apocortinoid, xyloside, phenol, lignin	Leaf	Panda (2014)
Colocasia esculanta	Alkaloid, glycoside, flavonoids, terpenoid, saponin, phenol	Leaf, tuber	Panda (2014)
Cordia myxa	Carbohydrates, tannin, protein, glycoside, alkaloid, flavonoid, phenolic compound, saponin	Fruit	Vineeta et al. (2018)
Dillenia indica	Alkaloid, glycoside, steroid, flavonoid, tannin, saponin, phenol	Fruit	Thakur and Rukhsaar (2014)
Dioscorea bulbifera	Flavonoid, terpenoid, saponin, steroid, cardiac glycoside, carbohydrate	Tuber	Namrata et al. (2011)
Diospyros melanoxylon	Alkaloid, flavonoid, tannin, terpenoid, essential oil	Fruit, seed	Panda (2014)
Diplazium esculentum	Alkaloid, reducing sugar, anthraquinones, anthranolglycoside, cyanidin, phenol, saponin, protein	Leaf	Namrata et al. (2011)
Elaeocarpus floribundus	Saponin, tannin, cardiacglycoside, flavonoid, steroid	Seed	Brahma et al. (2013)
Embilica officinalis	Fixed oils phosphatide, essential oil, tannin, mineral, vitamin, amino acid, fatty acid, glycoside	Fruit, seed	Sharma et al. (2017)
Eriobotrya japonica	Essential oil, triterpene, flavonoid	Fruit	Mogole et al. (2020)
Fagopyrum cymosum	Flavonoid, steroid, organic acid	Seed	Namrata et al. (2011)
Ficus hispida	Alkaloids, carbohydrates, proteins, amino acids, sterol, phenol, flavonoids, gum, mucilage, glycoside, saponin, terpene	Fruit	Mahapatra and Panda (2012)
Ficus racemosa	Flavonoid, triterpenoid alkaloid, glycoside, sterol, tannin	Fruit	Mahapatra and Panda (2012)
Flacourtia indica	Phenol, tannin, carbohydrate, saponin, alkaloid, protein, flavonoid, phytosterol, furanoid, coumarin, quinine	Fruit	Mahapatra and Panda (2012)

(continued)

Table 10.2 (continued)

Scientific name	Phytochemicals	Utilized part	Reference
Garcinia cowa	Flavonoid, phenol	Fruit	Panda (2014)
Gmelina arborea	Steroid, triterpenoid, saponin, protein, phenol, flavonoid, carbohydrate	Fruit	Agrahar-Murugkar and Subbulakshmi (2005b)
Grewia asiatica	Mucilage, flavonoid, alkaloid	Fruit	Mahapatra and Panda (2012)
Grewia subaequalis	Polyphenol, anthocyanin, flavonoid, tannin	Fruit	Burlingame (2000)
Indigofera pulchella	Terpenoid, flavonoid, steroid	Flower	Namrata et al. (2011)
Lepidium sativum	Carbohydrates, phenol, flavonoid, alkaloid, protein, saponin, lipid	Leaf	Namrata et al. (2011)
Madhuca indica	Alkaloid, tannin, protein, carbohydrate	Flower, seed	Mahapatra and Panda (2012)
Melia azedarach	Alkaloid, tannin, saponin, phenol, glycoside, steroid, terpenoid, flavonoid	Leaf	Panda (2014)
Mesua ferrea	Alkaloid, phenol	Fruit	Panda (2014)
Mimusopus elengi	Mannitol, flavonoids, alkaloids	Fruit	Gami et al. (2012)
Musa balbisiana	Saponin, glycoside, fatty acid	Fruit, flower	Majumdar and Datta (2009)
Nasturtium officinale	Phenol, flavonoids	Leaf	Namrata et al. (2011)
Nipa fruticans	Phenol, flavonoids	Leaf	Prasad et al. 2013
Oxalis corniculata	Flavonoid, tannin, phytosterol, phenol, glycoside, fatty acid, galacto-glycerolipid, volatile oil	Leaf	Panda (2014)
Paspalum scrobiculatum	Tannin, phenol, saponin, protein, carbohydrate	Seed	Namrata et al. (2011)
Phytolacca acinosa	Saponin	Root	Namrata et al. (2011)
Passiflora foetida	Carbohydrate, glycoside, flavonoid, resin, balsam, alkaloid, phenol	Fruit	Panda (2014)
Rhododendron arboreum	Polyphenol, quinone, sterol, triterpene, fatty acid	Flower	Namrata et al. (2011)
Schleichera oleosa	Phenol, flavonoid, tannin, fatty acid, sterol	Fruit	Agrahar-Murugkar and Subbulakshmi (2005a)
Shorea robusta	Phenol, tannin, saponin, glycoside, alkaloid	Fruit	Namrata et al. (2011)
Sida cordifolia	Alkaloid, flavonoid, tannin	Leaf	Namrata et al. (2011)
Simarouba glauca	Carbohydrate, protein, alkaloid, phenol, flavonoid, saponin, tannin, phytosterol, glycoside	Seed	Majumdar and Datta (2009)

(continued)

Table 10.2 (continued)

Scientific name	Phytochemicals	Utilized part	Reference
Spondias cytherea	Phenol, sterol, triterpene, saponin, essential oil, amino acid, polysaccharide	Fruit	Brahma et al. (2013)
Strebulus asper	Alkaloid, glycoside, sterol	Flower	Sharma et al. (2017)
Terminalia bellerica	Flavonoid, alkaloid, terpenoid, tannin	Fruit	Sharma et al. (2017)
Urtica parviflora	Flavonoid, tannin, scopoletin, sterol, fatty acid, polysaccharide, isolectin, sterol	Leaf	Taheri et al. (2022)
Ziziphus mauritiana	Glycoside, saponin, phenol, lignin, tannin	Fruit	Majumdar and Datta (2009)

vanished. Documentation of such precious knowledge will help in saving it for future generations and it will increase the cultural pride of such communities (Bhutia et al. 2018). Many of the wild edible plant resources are yet be assessed for their threat status. There is no exact data on the assessment of a number of individuals, population distribution, ecology, threats, trade opportunities, conservation evidence. If put into proper use, domestication can be used as a sustainable option to popularize many of these species without harming the pristine forest populations (Ros-Tonen 2000; Lowore 2020) i.e. conservation by commercialization or domestication (Lowore 2020; Kusters et al. 2006). Conservation involves careful management of natural resources and their judicious preservation and cessastion of their destruction and over-exploitation (Okigbo et al. 2008; Aremu et al. 2012). Through in situ and ex situ conservation efforts, we can conserve these valuable resources to ensure their sustainable use by present and future generations. Sustainable use of the wild resources help to minimize the adverse affects on their population (Asfaw and Tadesse 2001). For the species showing population decline in the wild, ex situ conservation is an important option. Some times, the wild plant collectors lack awareness about the sustainable harvesting practices, so the collectors and gatherers need to get awareness and training about the sustainable harvesting techniques of wild edible plant resources (Delang 2006). Creating awareness about the sustainable harvesting practices will ensure sustainable development in the particular region.

Utilization should supplement conservation and vice versa (Fig. 10.4). Then only the system will be sustainably feasible. These wild edible plant supplements during times of food shortage and any other emergency situation (Edison et al. 2006; Behera et al. 2010). Some of the wild edible plant species which are in immediate need of conservation interventions with their IUCN status is given in Table 10.3.

10.7.2 Challenges

Threats of the wild edible plants are many viz. forest fires, agriculture expansion, indiscriminate harvesting, fuel wood collection from the forest, settlements inside

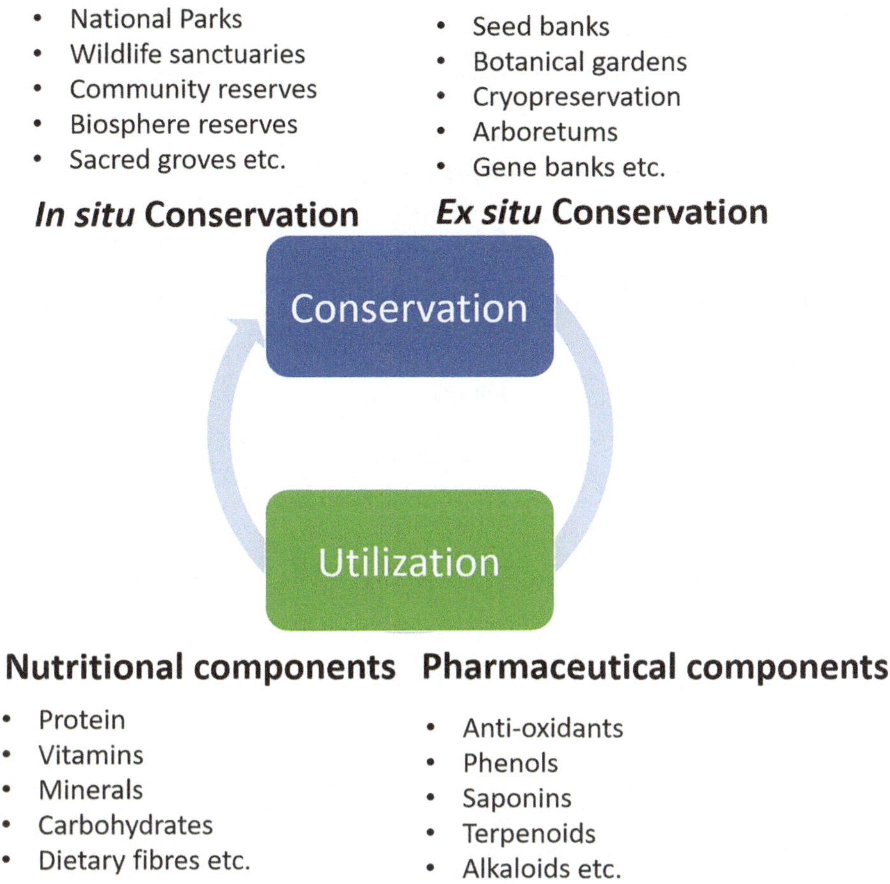

Fig. 10.4 Cycle of utilization and conservation of wild edible plants

the forest, overgrazing, and lack of awareness among collectors (Thakdeal et al. 2020; Duguma 2020). Because of the limited distribution of these resources, demand cannot meet supply, and thereby great pressure is there on forests. If we do not conserve the ecosystems where these wild food plants reside, these live sources will be lost and will not be available for future generations to use (Asfaw 2009). We should plan strategies including wild edible plants in the rehabilitation of degraded land or forest areas or integration of such plants in farms (Asfaw 2009). Future use of the wild plants must involve sustainable harvesting (Holm et al. 2008; Emanuel et al. 2005). Proper harvesting without endangering the survival of the wild plants is crucial for the sustainable consumption and production of the wild edible plant resources. Wild food plants provide a safety net function to the people relying on their consumption, but the climate poses a severe threat to this function, therefore strategies are needed to avoid the impacts of the climate change on the safety net function of the wild edible plants (Paumgarten et al. 2018).

Table 10.3 IUCN Red listed wild edible plants in India which needs conservation interventions (IUCN 2021; Rao and Rao 2014)

Scientific name	Family	IUCN status
Aegle marmelos	Rutaceae	Vulnerable
Angiopteris erecta	Angiopteridaceae	Endangered
Baliospermum montanum	Euphorbiaceae	Vulnerable
Cayratia pedata	Vitaceae	Vulnerable
Celastrus paniculatus	Celastraceae	Vulnerable
Costus spinosus	Costaceae	Vulnerable
Curcuma pseudomotana	Zingiberaceae	Vulnerable
Entada purseatha	Mimosaceae	Endangered
Litsea glutinosa	Lauraceae	Critically endangered
Gloriosa superba	Liliaceae	Vulnerable
Gnetum ula	Gnetaceae	Vulnerable
Gymnerama sylvestris	Asclepediaceae	Vulnerable
Lasia spinosa	Araceae	Endangered
Oroxylum indicum	Bignoniaceae	Vulnerable
Piper longicaudatum	Piperaceae	Endangered
Pterocarpus marsupium	Fabaceae	Near threatened

10.8 Conclusion and Future Perspectives

Wild food plant resources are inevitable for food and nutritional security. These underutilized resources have wide potential in their crude and processed form, which can earn additional income for the forest-dwelling and rural people. Nutraceutical potential of many wild food plants is yet to be discovered and it should be done on a high-priority basis for the betterment of society. Research should focus to bridge the gap between traditional knowledge and scientific interventions. From this perspective, more research should be undertaken to ensure the quality and safity of the wild plant based products. Policy formulations are much needed to control harvesting and quality aspects. The potential of wild edible plant species as nutraceuticals is a new area of research as many specific uses of these species are unknown. Species identification, population estimation, traditional uses, conservation, popularization, chemical constituents' extraction, identification of metabolites, cultural values, and marketing, many studies have to be done. Ethnobotanical studies can be used as a base for further research. Since traditional knowledge resources are depleting throughout the world as a result of modernization and globalization, it is high time that we preserve and use these resources wisely and sustainably for future generations as well. Inclusion of such amazing plants in the diet may also improve immune strength thereby having much resistance to diseases. Encouraging agroforestry to domesticate nutritionally and economically best species is one way as this will also help in income generation as per the upheaval in interest among the people.

References

Abberton M, Batley J, Bentley A, Bryant J, Cai H, Cockram J (2016) Global agricultural intensification during climate change: a role for genomics. Plant Biotechnol J 14:1095–1098

Abbet CR, Mayor D, Roguet R, Spichiger M, Hamburger, Potterat O (2014) Ethnobotanical survey on wild alpine food plants in Lower and Central Valais (Switzerland). J Ethnopharmacol 151: 624–634

Agrahar-Murugkar D, Subbulakshmi G (2005a) Nutritive values of wild edible and species consumed by the Khasi tribe of India. Ecol Food Nutr 44:207–223

Agrahar-Murugkar D, Subbulakshmi GJFC (2005b) Nutritional value of edible wild mushrooms collected from the Khasi hills of Meghalaya. Food Chem 89:599–603

Albert S (2012) Food security for Africa: an urgent global challenge. Agric Food Secur 1:2. https://doi.org/10.1186/2048-7010-1-2

Alebie G, Urga B, Worku A (2017) Systematic review on traditional medicinal plants used for the treatment of malaria in Ethiopia: trends and perspectives. Malar J 16:307–320

Apostolidis E, Lee CM (2016) Emerging sources for marine nutraceuticals. In: Vattem DA, Maitin V (eds) Functional foods, nutraceuticals and natural products. DEStech Publications Inc., p 797

Akhtar N, Jipson J, Callanan MA (2001) Learning words through overhearing. Child development 72:416–430

Aremu AO, Finnie JF, Van Staden J (2012) Potential of South African medicinal plants used as anthelmintics—their efficacy, safety concerns and reappraisal of current screening methods. S Afr J Bot 82:314–150

Arnold JEM, Perez MR (2001) Can Non Timber Forest Products match tropical forest conservation, development objectives? Ecol Econ 39:437–447

Asfaw Z (2009) The future of wild food plants in southern Ethiopia: ecosystem conservation coupled with enhancement of the roles of key social groups. Acta Hortic 806:701–708

Asfaw Z, Tadesse M (2001) Prospects for the sustainable use and development of wild food plants in Ethiopia. Econ Bot 55:47–62

Behera TK, Behera S, Bharathi LK, John KJ, Simon PW, Staub JE (2010) 2 bitter gourd: botany, horticulture, breeding. Hort rev 37:101

Bhatt ID, Sandeep R, Amit B, Ranbeer SR (2017) Nutraceutical potential of selected wild edible fruits of the Indian Himalayan region. Food Chem 215:84–91

Bhutia KD, Suresh CP, Nazir PA, Shukla G, Chakravrty S (2018) Nutraceutical potential of some wild edible fruits of Sikkim, Himalaya. India Ethno Med 12:106–112

Bohin L, Goransson U, Alsmark C et al (2010) Natural products in modern life sciences. Phytochem Rev 9(2):279–301

Brahma S, Narzary H, Basumatary S (2013) Wild edible fruits of Kokrajhar district of Assam, North-East India. Asian J Plant Sci Res 3(6):95–100

Burlingame B (2000) Wild nutrition. J Food Compost Anal 13:99–100

Calucci L, Pinzono C, Zandomeneghi M, Capocchi A (2003) Effects of gamma-irradiation on the free radical and antioxidant contents in nine aromatic herbs and spices. J Agric Food Chem 51: 927–934

Ceccanti C, Landi M, Incrocci L, Pardossi A, Venturi F, Taglieri I, Ferroni G, Guidi L (2020) Comparison of three domestications and wild-harvested plants for nutraceutical properties and sensory profiles in five wild edible herbs: is domestication possible? Foods 9:1065. https://doi.org/10.3390/foods9081065

Chatterjee R, Maitra S (2014) An inventory of the diversity and ethnomedicinal properties of cucurbitaceous vegetables in the homestead gardens of Sub Himalayan Districts of West Bengal, India. Res Rev J Bot Sci 3:40–45

Chauhan B, Kumar G, Kalam N, Ansari SH (2013) Current concepts and prospects of herbal nutraceutical: a review. J Adv Pharm Technol Res 4(1):4–8

Cordain L, Eaton SB, Sebastian A (2005) Origins and evolution of the western diet: health implications for the 21st century. Am J Clin Nutr 81:341–354

Cragg GM, Newman DJ, Snader KM (1997) Natural Products in drug discovery and development. J Nat Prod 60:52–60

De Medeiros PM, Barbosa DM, dos Santos GMC, da Silva RRV (2021) Wild food plant popularization and biocultural conservation: challenges and perspectives. In: Jacob MCM, Albuquerque UP (eds) Local food plants of Brazil. Ethnobiology. Springer, Cham. https://doi.org/10.1007/978-3-030-69139-4_16

Delang CO (2006) The role of wild food plants in poverty alleviation and biodiversity conservation in tropical countries. Prog Dev Stud 4:275–286

Dierick D, Hölscher D (2009) Species-specific tree water use characteristics in reforestation stands in the Philippines. Agr For Meteorol 149:1317–1326. https://doi.org/10.1016/j.agrformet.2009.03.003

Dogan Y, Ugulu U, Durkan N (2013) Wild edible plants sold in the local markets of Izmir, Turkey. Pak J Bot 45:177–184

Dolezal M, Velisek J, Famfulikova P (2001) Chemical composition of less-known wild fruits. Biologically—active phytochemicals in food: analysis, metabolism, bioavailability and function. In: Proceedings of the Eurofoodchem XI Meeting. pp 26–28

Duguma HT (2020) Wild edible plant nutritional contribution and consumer perception in Ethiopia. Int J Food Sci Technol 2020

Egea-Gilabert C, Niñirola D, Conesa E, Candela ME, Fernández JA (2013) Agronomical use as baby leaf salad of *Silene vulgaris* based on morphological, biochemical and molecular traits. Sci Hortic 152:35–43

Edison S, Unnikrishnan M, Vimala B (2006) Biodiversity of tropical tuber crops in India. National Biodiversity Authority, Chennai, pp 3–60

Emanuel PL, Shackleton CM, Baxter JS (2005) Modelling the sustainable harvest of *Sclerocarya birrea* subsp. caffra fruits in the South African lowveld. For Ecol Manag 214:91–103

Gami B, Pathak S, Parabia M (2012) Ethnobotanical, phytochemical and pharmacological review of *Mimusops elengi* Linn. Asian Pac J Trop Biomed 2:743–748

García-Herrera P, Sánchez-Mata MC, Cámara M, Fernández-Ruiz V, Díez-Marqués C, Molina M, Tardío J (2014) Nutrient composition of six wild edible Mediterranean Asteraceae plants of dietary interest. J Food Compost Anal 34:163–170

Gras A, Hidalgo O, D'Ambrosio U, Parada M, Garnatje T, Vallès J (2021) The role of botanical families in medicinal ethnobotany: a phylogenetic perspective. Plants 10:163

Holm JA, Miller CJ, Cropper WP (2008) Population dynamics of the dioecious Amazonian palm *Mauritia flexuosa*: simulation analysis of sustainable harvesting. Biotropica 40:550–558

Hussein RA, El-Anssary AA (2018) Plants secondary metabolites: the key drivers of the pharmacological actions of medicinal plants. In: Builders PF (ed) Herbal medicine. IntechOpen. https://doi.org/10.5772/intechopen.76139

Ignacimuthu S, Ayyanar M, Sankarasivaraman K (2008) Ethnobotanical study of medicinal plants used by Paliyar tribals in Theni district of Tamil Nadu, India. Fitoterapia 79:562–568

IUCN (2021) The IUCN red list of threatened species. Version 2014.3. http://www.iucnredlist.org/

Jain N, Ramawat KG (2013) Nutraceuticals and antioxidants in prevention of diseases. Nat prod 2559–2580

Jhansi D, Manjula K (2013) Functional and nutraceutical properties of herbals and its applications in food. Int J Sci Res 5:196–198

Kala CP, Sajwan BS (2007) Revitalizing Indian systems of herbal medicine by the national medicinal plants board through institutional networking and capacity building. Curr Sci 93: 797–806

Kala BK, Mohan VR (2010) Chemical composition and nutritional evaluation of lesser known pulses of the genus, Mucuna. Adv Biores 1:105–116

Kathleen C, Stephen D (2013) Nutraceuticals: what are they and do they work? vol 7. Kentucky Equine Research, Inc., Versailles, pp 1–50

Kesavan PC, Swaminathan MS (2008) Strategies and models for agricultural sustainability in developing Asian countries. Philos Trans R Soc Lond Ser B Biol Sci 363:877–891

Khan MSA, Ahmad I (2019) Herbal medicine: current trends and future prospects. In: Khan MSA, Ahmad I, Chattopadhyay D (eds) New look to phytomedicine advancements in herbal products as novel drug leads. Apple Publishers, pp 657–685. https://doi.org/10.1016/C2017-0-01165-5

Kumar VP, Chauhan NS, Padh H, Rajani M (2006) Search for antibacterial and antifungal agents from selected Indian medicinal plants. J Ethnopharmacol 107:182–188

Kumar S, Jena PK (2016) Tools from biodiversity: wild nutraceutical plants. Mathematical advances towards sustainable environmental systems. pp 181–213

Kumar S, Parida AK, Jena PK (2013) Ethno-medico-biology of ban aalu (*Dioscorea* species): a neglected tuber crops of Odisha, India. Int J Pharm Life Sci 4:3143–3150

Kumari P, Joshi GC, Tewari LM (2011) Diversity and status of ethno-medicinal plants of Almora district in Uttarakhand, India. Int J Biodivers Conserv 3:298–326

Kusters K, Achdiawan R, Belcher B, Pérez MR (2006) Balancing development and conservation? An assessment of livelihood and environmental outcomes of nontimber forest product trade in Asia, Africa, and Latin America. Ecol Soc 11:20

Larsen HO, Olsen CS, Boon TE (2000) The non-timber forest policy process in Nepal: actors, objectives and power. For Policy Econ 1:267–281

Lepcha LD, Shukla G, Moonis M, Bhat JA, Kumar M, Chakravarty S (2022) Seasonal relation of NTFPs and socio-economic indicators to the household income of the forest-fringe communities of Jaldapara National Park. Acta Ecol Sin 42:180–187

Lowore J (2020) Understanding the livelihood implications of reliable honey trade in the Miombo woodlands in Zambia. Front For Glob Chang 3:28

Mahapatra AK, Panda PC (2012) Wild edible fruit diversity and its significance in the livelihood of indigenous tribals: evidence from eastern India. Food Secur 4:219–234

Mahishi P, Srinivasa BH, Shivanna MB (2005) Medicinal plant wealth of local communities in some villages in Shimoga District of Karnataka, India. J Ethnopharmacol 98:307–312

Majumdar K, Datta BK (2009) Traditional wild edible fruits for the forest dwellers of Tripura, India. East Himalayan Society for Spermatophyte Taxonomy. Pleione 3:167–178

Malla S, Shukla G, Chakravarty S (2012) Utilization and conservation of wild plants by the tribal communities of Tripura. Indian For 138:1002–1007

Misra RC, Sahoo HK, Pani DR, Bhandari DC (2013) Genetic resources of wild tuberous food plants traditionally used in Similipal Biosphere Reserve, Odisha, India. Genet Resour Crop Evol 60:2033–2054. https://doi.org/10.1007/s10722-013-9971-6

Mogole L, Omwoyo W, Mtunzi F (2020) Phytochemical screening, anti-oxidant activity and α-amylase inhibition study using different extracts of loquat (*Eriobotrya japonica*) leaves. Heliyon 6:e04736

Mohammed R, Israt JM, Fahmidul HAKM, Ariful Md. HM, Parvin K, Rownak J, Majeedul HC, Taufiq R (2009) An ethnobotanical survey and pharmacological evaluation of medicinal plants used by the Garo Tribal Community living in Netrakona district, Bangladesh. Adv Nat Appl Sci 3:402–418

Momin KC, Sangma AN, Suresh CP, Singh YS, Rao SR (2018) Blood fruit [*Haematocarpus validus* (Miers) Bakh. f. ex Forman]—a potential nutraceutical and therapeutic fruit plant. Int J Min Fruits Med Aromat Plant 4:44–49

Motamed SM, Naghibi F (2010) Antioxidant activity of some edible plants of the Turkmen Sahra region in northern Iran. Food Chem 119:1637–1642

Moyo M, Adeyemi O, Aremu, Staden JV (2015) Medicinal plants: an invaluable, dwindling resource in sub-Saharan Africa. J Ethnopharmacol 174:595–606

Namrata, Kumar L, Ghosh D, Dwivedi SC, Singh B (2011) Wild edible plants of Uttarakhand Himalaya: a potential neutraceutical source. J Med Plant Res 5:670–684

Nazarudeen A (2010) Nutritional composition of some lesser-known fruits used by ethnic communities and local folks of Kerala. Indian J Tradit Knowl 9:398–402

Nwosu OK, Ubaoji KI (2020) Nutraceuticals: history, classification and market demand. In: Egbuna C, Dable Tupas G (eds) Functional foods and nutraceuticals. Springer International Publishing, Cham, pp 13–22

Olsen CS (2005) Valuation of commercial central Himalayan medicinal plants. Ambio 34:607–610
Onweluzo JC, Onuoha KC, Obanu ZA (1995) Isolation and characterization of proteins of some lesser known tropical legumes of Africa. J food Sci Technol 32:349–351
Okigbo RN, Eme UE, Ogbogu S (2008) Biodiversity and conservation of medicinal and aromatic plants in Africa. Biotechnol Mol Biol Rev 3:127–134
Panda T (2014) Traditional knowledge on wild plants as livelihood food in Odisha, India. J Biol Earth Sci 4:B144–B159
Pandey M, Verma RK, Saraf SA (2010) Nutraceuticals: new era of medicine and health. Asian J Pharm Clin Res 3:11–11
Patil P, Taware S, Kulkarni D (2014) Traditional knowledge of broom preparation from Bhor and Mahad region of western Maharashtra, India. Biosci Discov 5:218–220
Paulsamy S, Kumar PS, Kumar AMA, Kumar PS (2010) *Elaeagnus kologa* Schlecht.—an under *Elaeagnus kologa* Schlecht.—an under Nilgiris, the western ghats. Indian J Nat Prod Resour 1:258–260
Paumgarten F, Locatelli B, Witkowski ETF (2018) Wild foods: safety net or poverty trap? A south African case study. Hum Ecol 46:183–195
Payum T, Das AK, Shankar R (2014) Nutraceutical folk food plants used among indigenous people of east Siang district of Arunachal Pradesh, India. Am J Pharm Technol Res 4:697–704
Pradhan BK, Badola HK (2008) Ethnomedicinal plant use by Lepcha tribe of Dzonguvalley, bordering Khangchendzonga Biosphere Reserve, in North Sikkim, India. J Ethnobiol Ethnomed 4:22
Prasad N, Yang B, Kong KW, Khoo HE, Sun J, Azlan A, Ismail A, Romli ZB (2013) Phytochemicals and antioxidant capacity from *Nypa fruticans* Wurmb. Fruit. Evid Based Complement Alternat Med 2013:154606. https://doi.org/10.1155/2013/154606
Rai P, Vineeta, Shukla G, Manohar KA, Bhat JA, Kumar A, Kumar M, Cabral-Pinto M, Chakravarty S (2021) Carbon storage of single tree and mixed tree dominant species stands in a reserve forest—case study of the Eastern Sub-Himalayan Region of India. Land 10:435. https://doi.org/10.3390/land10040435
Raj AJ, Biswakarma S, Pala NA, Shukla G, Vineeta, Kumar M, Chakravarty S, Bussman RW (2018) Indigenous uses of ethnomedicinal plants among forest-dependent communities of northern Bengal, India. J Ethnobiol Ethnomed 14:8
Rajendran SM, Chandrasekhar K, Sundaresan V (2002) Ethnomedicinal lore of Valya tribe in Seithur hills of Virudhunagar district, Tamil Nadu, India. Indian J Tradit Knowl 1:59–71
Rao JP, Rao BV (2014) Need for conservation of red listed medicinal plants from tropical deciduous forests of Maredumilli, Andhra Pradesh, India. Int J Med Plants 106:468–480
Reddy KN, Pattanaik C, Reddy CS, Raju VS (2007) Traditional knowledge on wild food plants in Andhra Pradesh. Indian J Tradit Knowl 6:223–229
Rishi RK (2006) Nutraceutical: borderline between food and drug. Pharma review 2006. https://www.kppub.com/articles/herbal-safety-pharmareviwe-004/nutraceuticals-functional/article/13296428/the-global-market-for-making-for-nutraceuticals-set-for-robust-growth. Accessed 6 Jan 2020
Ros-Tonen MAF (2000) The role of non-timber forest products in sustainable tropical forest management. Holz Roh Werkst 58:196–201
Seal T, Chaudhuri K, Pillai B (2017) Nutraceutical and antioxidant properties of *Cucumis hardwickii* Royle: a potent wild edible fruit collected from Uttarakhand, India. J pharmacogn phytochem 6:1837–1847
Sena LP, Vanderjagt DJ, Rivera C, Tsin ATC, Muhamadu I, Mahamadou O (1998) Analysis of nutritional components of eight famine foods of the Republic of Niger. Plant Foods Hum Nutr 52:17–30
Sharma IP, Kanta C, Semwal SC, Goswami N (2017) Wild fruits of Uttarakhand (India): ethnobotanical and medicinal uses. Int J Complement Altern Med 8:1–8
Singh R, Singh S, Jeyabalan G, Ali A (2012) An overview on traditional medicinal plants as aphrodisiac agent. J Pharmacogn Phytochem 1:43–56

Soares WL, Porto MF d S (2009) Estimating the social cost of pesticide use: an assessment from acute poisoning in Brazil. Ecol Econ 68:2721–2728

Suhaj M, Rácová J, Polovka M, Brezová V (2006) Effect of gamma-irradiation on antioxidant activity of black pepper (*Piper nigrum* L.). Food Chem 97:696–704

Sundriyal M, Sundriyal RC, Sharma E, Purohit AN (1998) Wild edibles and other useful plants from the Sikkim Himalaya, India. Oecologia 7:43–54

Suresh CP, Bhutia KD, Shukla G, Pradhan K, Chakravarty S (2013) Free list of wild edible fruit species of Sikkim Himalayas, their uses. In: Proc Secur Intl Symp Min Fruits Med Plants. pp 17–37

Taheri Y, Quispe C, Herrera-Bravo J, Sharifi-Rad J, Ezzat SM, Merghany RM, Shaheen S, Azmi L, Mishra AP, Sener B, Kılıç M, Sen S, Acharya K, Nasiri A, Cruz-Martins N, Fokou PVT, Ydyrys A, Bassygarayev Z, Daştan SD, Alshehri MM, Calina D, Cho WC (2022) *Urtica dioica*-derived phytochemicals for pharmacological and therapeutic applications. Evid Based Complement Alternat Med 24:4024331

Thakdeal KL, Debela HF, Dereje BJ (2020) Management and utilization of wild nutraceutical plants. Ethiop J Sci Technol 7:83–97

Thakur M, Rukhsaar S (2014) Qualitative phytochemical screening, total phenolic content, in vitro antioxidant activity and antimicrobial activities in methanolic extracts of *Lactarius sanguifluus* (Paulet) Fr. J Pure Appl Microbiol 8:4735–4741

Vineeta, Shukla G, Pala NA, Dobhal S, Chakravarty S (2018) Influence of seed priming on germination and seedling growth of *Adeanthera pavonina* in subhumid region of West Bengal, India. Indian J Trop Biodiv 26:87–91

Wildman REC (2007) Nutraceuticals and functional foods. In: Wildman handbook of nutraceuticals and functional foods, 1st edn. CRC Press, New York, pp 1–9

Yadav M, Dugaya D (2013) Non-timber forest products certification in India: opportunities and challenges. Environ Dev Sustain 15:567–586

Chapter 11
Utilization of Wild Food Plants for Crop Improvement Programs

Anju Thattantavide and Ajay Kumar

Abstract Crop wild relatives or wild plants related to domesticated crops are potential resources for crop improvement. They can be landraces, crop progenitors, and plants closely related to the taxa and are not part of agricultural history. They are enriched with gene pools that are capable of revolutionizing current agriculture. As they possess allelic variations necessary for disease resistance, abiotic stress tolerance, nutritional value, and ecological adaptations, proper utilization of these better characteristics from wild crop relatives for crop improvement programs can change the agricultural landscape of the world. Presently, the majority of crops have less allelic diversity than their wild relatives and they are susceptible to diseases and other abiotic and biotic stresses, which is also known as the domestication bottleneck. The major cause of the domestication bottleneck was demand for high-yielding varieties and unintended neglectence of other characteristics. But in the wild crop relatives, all such genes were preserved and agricultural scientists started to create new improved lines of crops from crop wild relatives since the 1940s. Advancement in technologies should be reflected in the advancement of crop wild relative dependent crop improvement programs also. Here, we summarize the utilization of wild crops for crop improvement programs through various strategies such as genetic mapping, transgenic approaches, application of genomic tools, and gene editing.

Keywords Crop wild relatives · Crop improvement · Abiotic-stress tolerance · Biotic-stress tolerance · Crop diversity

A. Thattantavide
Department of Plant Science, School of Biological Sciences, Central University of Kerala, Kasaragod, Kerala, India

A. Kumar (✉)
Department of Plant Science, Central University of Kerala, Kasaragod, Kerala, India

A. Kumar et al. (eds.), *Wild Food Plants for Zero Hunger and Resilient Agriculture*, Plant Life and Environment Dynamics,
https://doi.org/10.1007/978-981-19-6502-9_11

11.1 Introduction

In the twentieth century, the world witnessed a drastic improvement in crop yield through the development of advanced crop breeding technologies and new agronomic practices (Miflin 2000). Malnutrition, food scarcity, and the price of food reduced during the period. The first half of crop improvement programs in the twentieth century was based on the increased use of chemical fertilizers, pesticides, and land. But later, the advancement in crop breeding and biotechnology developed effective methods such as hybridization and mutagenesis which accelerated the crop yield (Murphy 2011). But it resulted in various harmful changes in the environment as well. The major challenge to food security during the twentieth century was the poor distribution of food products, war, the political system, and poverty (Miflin 2000). But during the twenty-first century, new challenges such as the rise in the global population increased consumption of meat, water scarcity, climate change due to global warming and reduced land availability emerged (Rosegrant and Cline 2003). The unfavourable climatic conditions are leading to continuous crop failures (Raza et al. 2019). The Intergovernmental Panel on Climate Change (IPCC) estimated an average of 2% reduction in crop yield due to the variation in climatic conditions and its severity can increase by the year 2050 (Porter et al. 2014). Therefore, it is necessary to execute novel and sustainable crop improvement methods for the health and well-being of the people.

The expansion of agrobiodiversity is recommended as an efficient solution to overcome the multiple challenges of food security (FAO 2011). Since, globally humans depend on only 82 crops for consumption out of 12,650 edible plants, it is essential to explore other neglected plants also (Kahane et al. 2013). The rise of industrialization led to the practice of monoculture of a few commercially important crops such as rice, maize, wheat, sugarcane, and cotton (Salaheen and Biswas 2019). The dependency on a few crops can result in genetic homogenization and vulnerability to pathogens and pests. It reduces the fertility, and microbial diversity of the soil (Salaheen and Biswas 2019). Therefore, cultivation of multiple crop varieties is essential to meet the dimensions of food security. The diversification of food crops can help in the eradication of malnutrition and hidden hunger from the population effectively (Keatinge et al. 2011). The loss of crop diversity was driven by the process of domestication. The loss of crop diversity is linked with the loss of genes with agronomically important traits (Vigouroux et al. 2002). The excessive selection by targeting agronomically important traits at the time of domestication resulted in reduced genetic diversity of crops and isolation from crop wild relatives (CWR; Dempewolf et al. 2017). Domestication resulted in reduced allelic diversity in the crops when compared with their wild crop relatives, which is also known as the 'domestication bottleneck' that further leads to demand for excess production and cultivation of uniform crop varieties (Hammer 1984). The selection procedure at the time of domestication unintentionally eliminated the traits related to abiotic and biotic tolerance from the gene pool. But those alleles remained in their wild relatives (Hammer 1984).

The importance of CWRs in ensuring food security has been well recognised (Allen et al. 2019; Fonseka et al. 2020). CWRs are the relatives of crop plants with a diverse gene pool and the potential to be used in crop breeding. The gene pool concept can explain the relationship between CWR and the corresponding crop (Fielder et al. 2016). According to the gene pool concept (GP), for each and every crop, there is the availability of a diverse genetic pool, which is classified as primary, secondary, and tertiary gene pool. The primary gene pool encompasses the gene pool of cultivated forms and wild/weedy plants of the crops which are termed GP-1A and GP-1B respectively. The secondary gene pool (GP2) includes less related species of the crops, where the transfer of these genes to crops is difficult to execute with conventional breeding practices. The tertiary gene pool (GP3) is the crop, where gene transfer is more or less impossible/possible with the aid of advanced techniques such as genetic engineering, somatic fusion, and embryo rescue (Maxted et al. 2006). CWRs possess considerable potential to tackle the abiotic and biotic stress of crop plants because they are well adapted to marginal areas and possess tolerance against abiotic and biotic stresses (Redden et al. 2015). Global food insecurity and climate change are demanding the production of stress-resilient and nutritionally diverse crops to feed the future generations (Kumar et al. 2021). Therefore, the utilization of genes related to agronomically important traits for breeding and crop improvement from CWR can provide significant outcomes for the world. But the extinction of CWRs is reported in various parts of the world. For example, in Mesoamerica, where the origin of chili pepper, maize, squash, cotton, vanilla, and avocado occurred, the extinction of 35% of selected CWR was reported in the International Union of Conservation of Nature (IUCN) Red List due to various anthropogenic activities and climate change (Goettsch et al. 2021). The extinction of CWRs was also reported in Italy (Perrino and Wagensommer 2022). Similarly, the CWRs of the different parts of the globe are threatened and the conservation is limited in gene banks (Redden et al. 2015). Loss of CWRs pose a great threat to the sustainability of food security.

11.2 The Wild Crop Relatives' Diversity

CWRs are distributed in different habitats and adapted to various ecological conditions. They are resilient to changing climate conditions and show better adaptation. CWRs are distributed in different corners of the world (González-Orozco et al. 2020). But CWR richness is found in Vavilov's centers of diversity (Vincent et al. 2013). Understanding the diversity of CWRs help to develop conservation strategies and crop improvement programs (Maxted et al. 2006). There exist several organizations that function to document CWR diversity. For example, in Mexico, the National CWR inventory documented 310 CWR varieties for selective crops including important vegetables, fruits, cereals, pseudocereals, tubers, legumes, forages, nuts, and industrial/medicinal crops (Contreras-Toledo et al. 2019). The study reported 17 CWR varieties for *Amaranthus*, 12 for *Annona*, 19 for *Agave,* 11 for

Cucurbita, 21 for *Manihot,* 14 for *Phaseolus,* 15 for *Tripsacum* and 20 for *Solanum.* By understanding the threats to CWR diversity in the United Kingdom, The Convention on Biological Diversity (CBD) 2010, formed a National CWR inventory in the UK (Maxted et al. 2007). The inventory of CWRs in the UK had 250 varieties during that time. Khoury et al. (2020) developed a CWR inventory for the USA and documented 600 CWR taxa including the relatives of beans, barley, apples, peppers, berries, cotton, onion, grapes, plums, potatoes, sweet potatoes, and sunflowers (Khoury et al. 2020). Besides that, Fielder et al. (2015) reported CWR of 126 plant species and 22 subspecies in the English National inventory. A 76% percentage of the CWRs were related to food crops and rich CWR diversity was recorded for the genus *Chenopodium* L., *Vicia* L., and *Trifolium* L. The CWR inventory for the pseudocereals in Argentina possesses 42 species and 4 subspecies. *Amaranthus* documented 24 CWR plants and *Chenopodium* documented 18 CWRs (Curti et al. 2017). Besides that, the world vegetable Center—AVRDC has 61,000 vegetable accessions from 155 countries. The majority of these accessions are the landraces or wild relatives gathered from different regions of the world (Redden et al. 2015).

In the case of India, Pandey et al. (2008) reported 105 CWRs from Western Himalayas, 38 CWRs from Eastern Himalayas, 53 CWRs Northeastern region, 82 CWRs from the Gangetic plains, 42 CWRs from the Indus plain, 123 CWRs from the Western Ghats, 101 CWRs from the Eastern Ghats and 32 CWRs from Islands. China has a huge diversity of wild crop relatives including CWRs of rice, sorghum, banana, soybean, apple, millet, and grape. The native species of CWRs found in China include *Malus baccata* (L.) Borkh. (apple; *M. domestica* Borkh.), *Glycine soja* Siebold & Zucc. (soybean; *G. max* (L.) Merr.), *Musa balbisiana* Colla and *M. basjoo* Siebold & Zucc. ex Linuma (banana; *M. acuminata* Colla), *Oryza rufipogon* Griff. (rice; *O. sativa* L.), *Sorghum propinquum* (Kunth) Hitchc. (sorghum; *S. bicolor* (L.) Moench), and *Vitis amurensis* Rupr. (grape; *V. vinifera* L.) (Kell et al. 2015).

Through the species distribution model, González-Orozco et al. (2020) reported 22 CWRs for *Theobroma cacao* L. from Columbia including *T. glaucum, T. bernouilli,* and *Herrania* sp. In another study, Thomas et al. (2017) reported the CWRs of cultivated rice viz. *Oryza glumaepatula*, *O. grandiglumis*, *O. alta*, and *O. latifolia*. The Armenia has a collection of three wild ancestors of wheat out of four, *Triticum boeoticum* Boiss., *T. urartu* Thum. ex Gandil. and *T. araraticum* Jakubz (Avagyan 2008). The other CWR species found in Armenia include *Lens orientalis* (Boiss.) Hand. (lentil; *L. ervoides*), *Pisum elatius* M. Bieb., *P. sativum* L. *subsp. humile* (Holmb.) Greut (pea; *P. sativum*) and *Linum bienne* Mill. (flax; *Linum usitatissimum*). Similarly, the CWR of rye, barley, mountain ash, apple, grapevine, plum, grass pea, turnip and flax are also reported from Armenia. The Armenian State Agrarian University reported a total of 2834 CWRs for *Triticum* L., *Secale* L., *Hordeum* L., *Aegilops* L., *Beta* L., *Daucas* L., *Spinacia* L., *Coriandrum* L., *Lathyrus* L., *Lens* Mill, *Vicia* L., *Trifolium* L., *Medicago* L. and *Amygdalus* L. The center for vegetable-melon and industrial crops has 11 accessions of CWR including *Lycopersicon* L., *S. melongena* L., *Melo* Adans., and *Citrullus* Schrad. *ex*

Eckl. & Zeyh. The scientific centre for agronomy and plant protection has a total of 17 CWR accessions for *Triticum* L. and *Hordeum* L. (Avagyan 2008).

In South Africa, there are a number of in-situ conserved CWR taxa under different categories. Fifty-six endemic and restriction in distribution CWRs, seven endemic and relatively widespread CWRs, twenty-two least concerned endemic CWRs, thirty four indigenous regionally distributed South African CWRs, 118 widely distributed CWRs in the South African region, and twenty nine naturalized exotic plants (Holness et al. 2019). Perrino and Wagensommer (2021) reported 14 non-endemic threatened CWRs from Italy including *Ipomoea sagittata* Poir., *Brassica insularis* Moris, and *Lathyrus amphicarpos* L. Further, several studies reported the diversity of CWRs in different corners of the world such as Tunisia (Mezghani et al. 2019), Sweden (Weibull and Phillips 2020), North Africa (Lala et al. 2018), Iran (Magos Brehm et al. 2008), Sri Lanka (Ratnayake et al. 2021), Spain (Rubio Teso et al. 2018), North America (Heinitz et al. 2019), Mexico (De La Torre et al. 2018), and Indonesia (Rahman et al. 2019).

11.3 Importance of Crop Wild Relatives

CWRs are under-evaluated and underutilized plants with several important characters and huge importance. CWRs are either ancestors of cultivated crops or relatives of wild ancestors or evolutionarily closely or distantly associated plant (Zhang et al. 2017). The sequential rounds of selection of specific agriculturally important traits have resulted in reduced allelic diversity in domesticated crops than in their wild ancestors (Dempewolf et al. 2017). Since CWRs were not selected for domestication they are rich in diversity compared to cultivated crops and are distributed in a wide range of climatic, geographical, and ecological conditions also (Smýkal et al. 2018). Sorghum is a major staple crop in both semi-arid and arid regions of the developing world (Hariprasanna and Patil 2015). There exists controversy about the proper origin of the sorghum plant, but archaeological studies provided evidence regarding its origin in BC 3000 in eastern Sudan. At the same time, few investigations claim multiple domestication events of the sorghum crop. Additionally, CWRs have the potential to adapt to changing environment and climatic conditions. The genetic composition of the CWRs can contribute to the stress adaptation and tolerance capacity of the plants (Benlioğlu and Adak 2019). They act as a reservoir for alleles of these important traits. The importance of CWRs was proved during the green revolution period (the 1960s–1970s) since the genes for crop improvement were used from the landraces and crop relatives (Ford-Lloyd et al. 2011).

11.4 Crop Improvement: Importance and Methods Adopted

The current scenario of the world appeals for the establishment of efficient and sufficient crop improvement programs. There are several reasons that demand extensive and efficient crop improvement strategies. The major reasons are - reduction in arable land because of non-sustainable farming practices, soil degradation, soil erosion, unpredictable weather, and environmental change such as rise in sea level, drought, salinity, heat current, floods, and cyclones due to global climate change (Takeda and Matsuoka 2008). Availability of irrigation water is also reducing and all these factors are declining the yield and stability of major food crops such as maize, rice and wheat (Kang et al. 2015). The necessity for the cultivation of land for bio-energy sources is also leading to reduction in agricultural land. These major factors are the driving force for plant scientists to develop efficient crop improvement methods (Rosegrant and Cline 2003). Advanced technology is required to overcome the issues in agriculture that are prevailing globally. Particularly, it is urgent to develop climate-resilient crops that display tolerance against biotic and abiotic stress. A rich diversity in germplasm, alleles, and phenotypes in agricultural crops is the basic requirement for effective crop improvement (Byrne et al. 2018). Emerging tools in molecular biology and genetics are capable of boosting crop improvement programs which are discussed in the following sections. The diversity of alleles in wild crop relatives and the application of advanced molecular biology breeding tools can be exploited for crop improvement programs (Takeda and Matsuoka 2008). Various methods can be employed for crop improvement using CWRs which are described in the upcoming sections (Fig. 11.1).

11.4.1 Plant Introduction

The introduction of new varieties with desirable traits from outside or within the country into a local environment is known as plant introduction, which brings higher crop yield or targeted desirable traits to the new area from the introduced variety (Yadav and Chen 2007). Guo et al. (2009) reported that in the USA, approximately 4000 plant species were introduced there due to the Columbian exchange which included the CWR also. Among the introduced taxa, several plants had potential to be utilized for crop improvement with genetic diversity. For example the cold-adapted *Medicago sativa* L. subsp. *falcata* (L.) Arcang. from South Dakota is the CWR of alfalfa (*M. sativa* L.). During the Columbian exchange, it was introduced in the US and helped to develop intermountain-adapted varieties (Bossdorf et al. 2005). CWRs are not cosmopolitan in distribution. Since the majority of CWRs are found in the place of origin of a crop plants, the introduction of stress-tolerant, disease-resistant, high-yielding, and nutritionally superior CWRs from one location to another can increase the diversity of crops in a new area. Later, through intraspecific

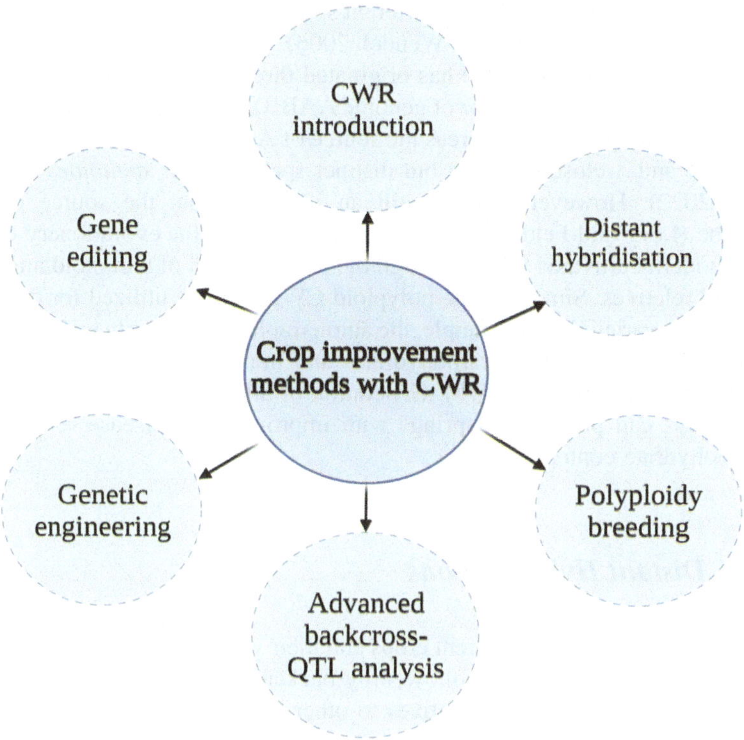

Fig. 11.1 Use crop wild relatives for crop improvement by employing various methods (Created with www.biorender.com)

hybridization and interspecific hybridization, the traits can be exchanged between the crops or a new crop variety can be created.

11.4.2 Polyploidy Breeding

Polyploidy is commonly found in plants, where there exist more than two sets of chromosomes (Barringer 2007). Polyploidy is the result of mutations inside the cells and it is generally heritable in nature (Barringer 2007). Through the process of selection, analysis, and recombination, polyploids can be exploited for obtaining genetic gain (Bharadwaj 2015). Hence polyploidy breeding is very important in crop improvement. The dosage effect during polyploidy breeding can produce additive effects of alleles and thereby increase the phenotypic number. Polyploidy breeds display more adaptive potential. It produces more diverse allele combinations and increases heterozygosity in the offspring. The genome interaction and expression can result in novelty of phenotypes (Bharadwaj 2015). The development of alloploids (polyploids derived from different species) can help to generate improved phenotypes of crops (Udall and Wendel 2006). The hexaploid wheat (genome AABBDD *Triticum aestivum* L.) has originated through the hybridisation between tetraploid *T. turgidum* (progenitor of genome AABB), and *Aegilops tauschii* (diploid donor of the D subgenome), whereas the source of A and B subgenome respectively are *T. uratu* and a closely related but distinct species of *A. speltoides* (Levy and Feldman 2022). However there is still an illusion about the source of the B subgenome (Levy and Feldman 2022). It points towards the evolutionary development of modern cultivated wheat through the hybridization of tetraploid and diploid wheat wild relatives. Similarly, the polyploid CWRs can be utilized for developing improved crop varieties. For example, the autotetraploid ryegrass (a relative of wheat and barley) possesses higher carbohydrates and improved disease resistance than diploid plants (Van Bogaert 1975). Generation of allopolyploids by utilizing tetraploid ryegrass can produce offsprings with improved pest-disease resistance and high carbohydrate content.

11.4.3 Distant Hybridizations

The hybridization between different crops and their wild relatives (both interspecific and intergeneric hybridization) followed by backcrossing is a method to segregate desirable traits from crop wild relatives to other crops (Chauhan et al. 2021). The origin of the crop, evolutionary pattern, and cytological pattern should be considered during distant hybridisation for obtaining successful results (Choudhary et al. 2017). Rarely distant hybridization results in the production of a novel crop also. Hybrid sterility, cross incompatibility, and generation of non-viable hybrids are the major hurdles in distant hybridization (Chauhan et al. 2021). There are successful reports of the production of hybrids from CWR and cultivated crops by overcoming those barriers. The transfer of traits for leaf roll virus, late blight virus, potato virus X, and powdery mildew in Okra was achieved through the distant hybridization between its wild relatives (Kalloo 1992). In the case of tomatoes also, genes for late blight, tomato leaf curl virus, and bacterial resistance were transferred from wild relatives (Kalloo 1991). Besides stress resistance, wild relatives were used to improve the nutritional quality of the crops. For example, the lycopene content of cultivated tomatoes was improved by distant hybridization with wild tomato *Solanum pimpinellifolium* (Sheesh et al. 2011).

11.4.4 Advanced Backcross: QTL Analysis

Advanced backcross—QTL analysis (AB-QTL) is used for the detection and transfer of important Quantitative Trait Locus (QTLs) from CWRs/uncultivated plants into crop plants. The method used for combining marker-based QTL with other crops for their improvement (Tanksley and Nelson 1996). Huang et al. (2003) conducted AB-QTL analysis in the German winter wheat variety (a relative of *T. aestivum*) and synthetic wheat line and developed a hybrid (Huang et al. 2003). With the aid of single marker regression analysis and interval mapping, they mapped the QTLs related to yield, yield component, ear emergence, and plant height. Huang et al. (2003) proved that it is possible to improve the traits in staple crops by the transfer of QTLs from wild relatives through the AB-QTL method. Similarly, Fulton et al. (1997) performed a backcross between wild relatives of tomato *L. peruvianum* and cultivated variety *L. esculentum*. 166 QTLs were discovered in the next generation related to 29 traits. At least one QTL was identified as associated with a wild allele out of half of these 29 agronomically beneficial traits (Fulton et al. 1997). Other research groups have also explored the potential of CWR and crop variety through the AB-QTL method. There are plenty of successful reports such as *L. pimpinellifolium* v/s elite tomato variety (Tanksley et al. 1996), *L. hirsutum* v/s cultivated tomato (Bernacchi et al. 1998), *Hordeum vulgare* ssp. *Vulgare* v/s *H. v.* ssp. *spontaneum* (Pillen et al. 2003) and *O. rufipogon* v/s *O. sativa* (Li et al. 2002).

11.4.5 Genetic Engineering

The discovery of restriction enzymes have revolutionized the field of crop improvement by allowing the modification of an organism through the recombination of genes between organisms (Dale 1992). The insertion of genes responsible for superior traits in one plant/organism can be transferred to other plants through genetic engineering and thereby the modified offspring display novel phenotypes (Käppeli and Auberson 1998) (Fig. 11.2). In the case of CWRs, gene modification or genetic engineering can be performed in two ways, i.e., transfer of the genes for any desirable trait from CWR into a crop plant and modification of CWR by introducing the foreign genes. Various factors such as sexual compatibility between the plants, same flowering time, mode of pollen transport, and short distance for cross-pollination influence the efficiency of genetic engineering (Dale 1992). The diverse gene pool of CWRs can be used to modify a wide variety of crops.

11.4.6 Gene Editing

The common method for the introgression of beneficial genes from CWR to crops is selection and backcrossing. The fine mapping of genes in CWR is required to unravel their superior qualities and it can significantly increase their role in crop improvement (Yates et al. 2004). Understanding the exact location of the target genes can increase the possibility of their usage in gene editing techniques such as clustered regularly interspaced short palindromic repeats and CRISPR-associated protein 9 (CRISPR-Cas9) (Monsur et al. 2020). Gene editing can help to delete or

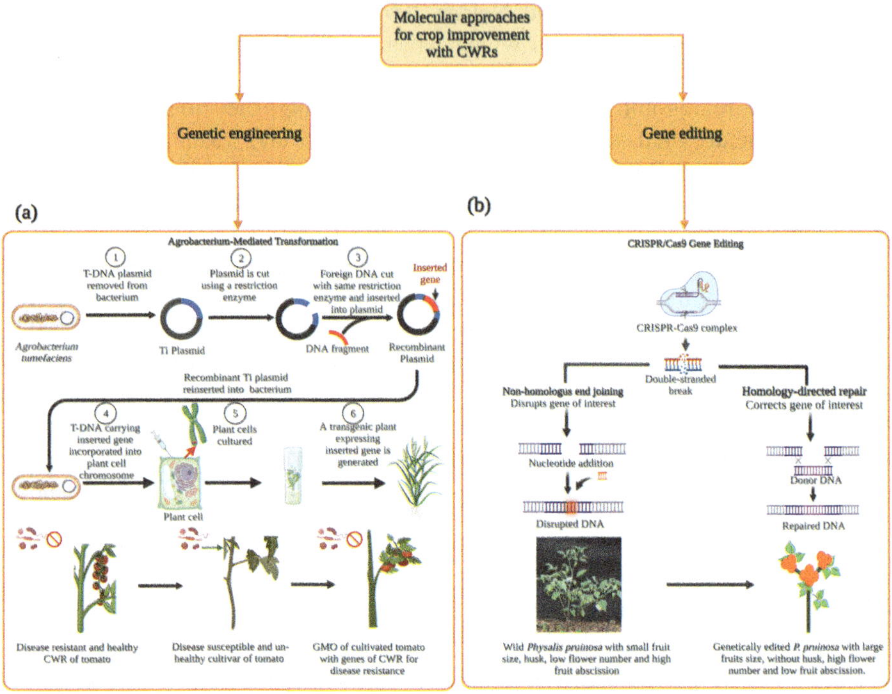

Fig. 11.2 The molecular mechanisms behind genetic engineering and gene editing for crop improvement. Both techniques can be applied to modify the CWR varieties for their improved performance (Created with www.biorender.com). The agronomically important traits from CWRs can be transferred to major crops through genetic engineering (**a**) and gene editing (**b**)

edit target genes and regulatory elements. Gene editing technology such as CRISPR-Cas9 does not produce genetically modified organisms but only creates genetic variation within the gene of that organism with the aid of guide RNA (Fig. 11.2). CWRs can be *de-novo* domesticated with the aid of genetic engineering (Zsögön et al. 2017). *De-novo* domestication is the incorporation of domesticated genes into non-domesticated plants, and in the case of CWR, it demands a model crop with complete genome sequence availability or a wild plant with germplasm characterized (Fernie and Yan 2019). *De-novo* domestication provides new food crops into the food basket with desirable traits. The unfavorable traits in the CWRs can be eliminated with gene editing and in similar ways novel traits can be implemented. The successful gene editing and *de-novo* domestication in the orphan Solanaceae plant, *Physalis pruinosa* were performed by Lemmon et al. (2018). Through the study, they obtained *P. pruinosa* plants with increased flower size, increased fruit locule and plant architecture. Zsögön et al. (2017) reviewed that, with the application of CRISPR-Cas9 gene editing the wild tomato *S. galapagense* can be improved and its *de-novo* domestication can be achieved. *S. galapagense* is a CWR of tomato with small, orange-colored fruit, indeterminate growth, single fruit per truss, and salt resistance. The CRISPR-Cas9 gene editing by targeting the genes *SELF PRUNING, FW2.2, FASCIATED, LOCULE NUMBER, LYCOPENE BETA CYCLASE AND COMPOUND INFLORESCENCE,* it is possible to manipulate *S. galapagense* plant with determinate growth, red fruits with large size, multiple fruits per truss and salt resistance. There are plenty of CWRs that can be *de-novo* domesticated and gene editing can be accelerated in CWRs through their complete germplasm characterisation. Besides the CRISPR-Cas9 technique, zinc-finger nucleases (ZFNs) and transcription activator-like effector nucleases (TALENs) are also popular gene editing tools. Both of these methods are protein engineering-based techniques, while CRISPR-Cas9 is an RNA-guided system. CRISPR-Cas9 is more popular because of its potential to create stable and heritable mutations similar to natural mutations without changing the natural variable traits (Renzi et al. 2022).

11.5 Crop Wild Relatives as a Promising Resource for Crop Improvement

The Russian botanist Nikolai Ivanovich Vavilov revealed the possibility of exploring CWR for crop improvement by pointing out the utilization of wild plants *Aegilops* L., *Dasypyrum* (Coss. & Durieu), *Sécale* L. and *Haynaldia villosa* (L.) Schur. in wheat breeding (Vavilov 1949). The capability of CWRs and landraces for crop breeding was recognized during the twentieth century itself. But the great number of crops and lack of ideas regarding exact CWR with valuable and precise genetic variation for breeding was the major limitation in their wider application (Coyne et al. 2020). But now with the aid of advanced molecular biology tools, bioinformatics, and phenotyping methods, the possibility of utilizing CWR in various agricultural applications have improved (Tanksley and McCouch 1997). The worldwide distribution of CWRs except in Antarctica points toward their potential to be used in various crop improvement programs (Larson et al. 2014). But out of approximately 60,000 crop relatives, 95% are not represented sufficiently in gene banks and research on CWR-based breeding has lagged for years (Zhang et al. 2017). At the same time, numerous efforts are also initiated to incorporate the potential of CWRs in the breeding programs. In the published research papers, a high amount of research was observed on the crops such as rice, wheat, barley, and tomato. The evaluation of the biotic and abiotic stress-tolerance potential of CWRs is the primary step. For example, the abiotic and biotic stress tolerance characteristics of wild relatives of maize, which are used for the production of improved lines are shown in Fig. 11.3. Majorly the CWR-based breeding is restricted to a few crops since the breeding program demands cross compatibilities, fertility of F1 progenies, the relationship between the cultivated plant and wild relatives, high availability of the CWR, sustainable conservation status of the CWR, proper utilization of CWR and extensive exploration of CWR, geographical distribution of CWR and good financial support (Zamir 2001).

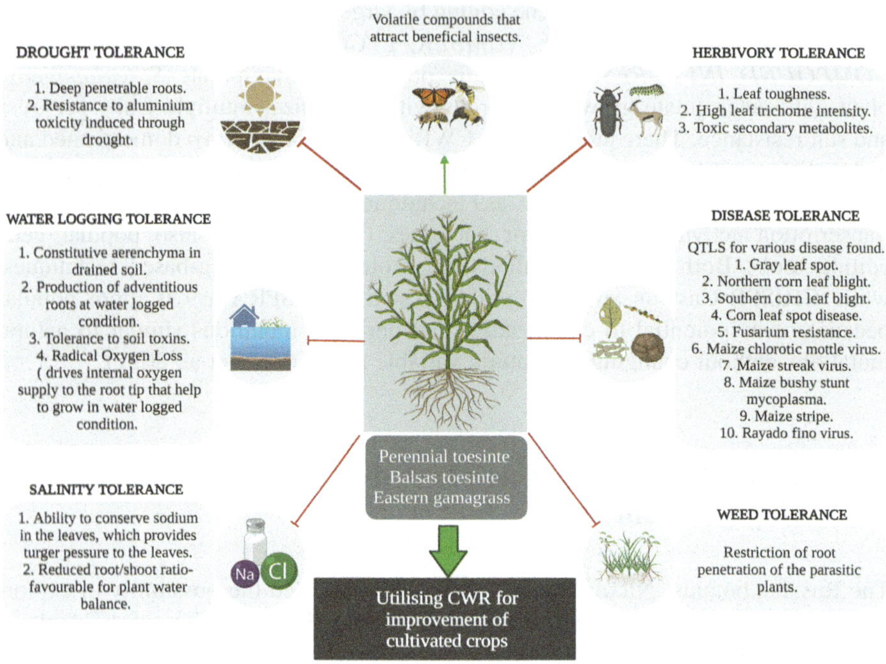

Fig. 11.3 The abiotic and biotic stress tolerant traits of maize wild relatives that can be successfully used for crop improvement (Mammadov et al. 2018) (Created with www.biorender.com)

Even though there exist a number of difficulties in achieving successful breeding results with CWR, there are several examples where the discovery and transfer of genes for agronomically important traits have occurred which are discussed in subsequent sections. During earlier decades, classical backcrossing, map-based cloning, and introgression were used as major tools for gene transfer (Qi et al. 2014). At present, advanced modern molecular biology approaches are used. For the discovery of agronomically important genes from CWR, next-generation sequencing and high-throughput phenotyping technologies are used. As a result, genes and quantitative trait locus (QTLs) related to the traits are recorded which can be applied for breeding programs (Zhang et al. 2017). In the case of gene transfer, modern technologies such as direct gene transfer, vector-mediated gene transfer, and gene editing are used for speed, reliability, and precise transfer (Jørgensen et al. 1996). The examples of successfully improved varieties of crops for various phenotypes and outcomes are discussed in the following subsections.

11.5.1 Biotic Stress Tolerance

The cultivated plants are susceptible to various kinds of biotic stresses such as pathogens, pests, and weeds and they are less tolerant to these stresses compared to their wild relatives (Mammadov et al. 2018). The negative correlation of the tolerance gene with the yield-related trait is the reason behind the negligence of these traits during artificial selection at the time of domestication (Wise 2007). But in 2009, Kaplan et al. (2009) proved that it is possible to breed crops for multiple stress tolerance without reducing the yield (Kaplan et al. 2009). Through conventional breeding and modern transgenic approaches, the introduction of biotic stress-tolerant genes from CWRs into stress-susceptible crops can be achieved (Mammadov et al. 2018). Hu et al. (2016) reviewed the development of molecular breeding in rice for planthopper resistance and documented approximately 29 brown planthoppers (BPH) resistant genes including *Bph14*, *Bph26*, *Bph17,* and *Bph29* in wild rice. They reported studies on the development of introgression lines, pyramided lines, and near-isogenic lines by marker-assisted backcross method from wild relatives of rice (Hu et al. 2016).

The CWRs are considered the stock house of plenty of beneficial traits, that can be used for crop improvement programs. For example, Mammadov et al. (2018) reviewed the biotic and abiotic stress tolerance traits of rice, maize, soybean, and cotton (Mammadov et al. 2018). They reported that the CWRs of maize viz. *Z. mays* subsp. *pulviglumis*, *Z. diploperennis*, *Z. mays* spp. *pulviglumis* are tolerant to fall armyworm. Maize spotted stalk borer and western corn rootworm tolerance are exhibited by *Z. mays* subsp. *mexicana*, *Z. mays* subsp. *Pulviglumis, Z. perennis* and teosinte and eastern gamagrass respectively. Mammadov et al. (2018) listed the pest tolerance, disease tolerance, and weed tolerance of CWR of the four plants in the review in detail. There are a lot of successful breeding efforts, where CWRs were explored to improve the phenotypic trait of major and minor crops. The introgression of tomato resistance genes from *S. pimpinellifolium* in tomato, leaf rust resistance genes from *Agropyron elongatum* in wheat, potato late blight resistance gene from *Solanum demissumm* in potato, *grassy stunt* virus genes from *O. nivara* in rice are few examples of this (Hajjar and Hodgkin 2007). The hybridization of *Sorghum bicolor* and *S. macrospermum* resulted in disease and pest resistance (Price et al. 2005). The list of various genes/QTLs for beneficial agronomic traits found in CWRs are presented in Table 11.1.

11.5.2 Abiotic Stress Tolerance

The abiotic stresses such as increased temperature, salinity, drought, heavy metals, and flooding are the barriers in the life span of high-yielding crops. Mantri et al. (2012) reported that, due to exposure to various kinds of abiotic stress, major crops are undergoing huge yield losses globally (Mantri et al. 2012). Further loss in

Table 11.1 Various genes/QTLs for beneficial agronomic traits found in CWRs

Domesticated crop	Agronomic trait in the CWR	Genes/QTLs for the trait	Crop wild relative	Reference
Oryza sativa L.	Bacterial blight resistance	*Xa21*	*O. longistaminata* A.Chev, & Roehr.	Jena (2010)
	Blast planthopper resistance	*Pi40*	*O. australiensis* Domin	Jena (2010)
	Brown planthopper resistance	*Bph18*	*O. australiensis*	Jena (2010)
	Xanthomonas oryzae pv. *oryzae* resistance	*Xa21*	*O. saiiva* ssp. *indica*	Song et al. (1995)
	Bacterial blight resistance	*Xa38*	*O. rufipogon* Griff.	Ellur et al. (2016)
	Rice tungro disease resistance		*O. longistaminata* and *O. rufipogon*	Khush et al. (2004)
	Grassy stunt resistance	*GS*	*O. rufipogon*	Sanchez et al. (2013)
	Heat and/or drought tolerance	QTLs	*O. glaberrima* Steud.	Sanchez et al. (2013)
		Cpn60, Hsp90, Hsp70, THI1	*O. meridionalis* Ng.	Scafaro et al. (2010)
	Aluminum toxicity tolerance	QTLs	*O. rufipogon*	Sanchez et al. (2013)
	Increase in yield	*yld1, yld2*	*O. rufipogon*	Sanchez et al. (2013)
	Salinity tolerance	*L-myo-inositol-1-phosphate synthase*	*Porteresia coarctata* (Roxb.) *Tateoka*	Majee et al. (2004)
		MYB, bHLH, AP2-EREBP, WRKY, bZIP and *NAC* transcription factors	*P. coarctata*	Garg et al. (2014)
	Cold resistance	QTLs	*O. japonica*	Zeng et al. (2009)
	Flood tolerance	*SUB1A-1*	*O. rhizomatis* and *O. eichingeri*	Niroula et al. (2012)
	Yield	QTLs	*O. rufipogon*	Xie et al. (2008)
		QTLs	*O. rufipogon*	Fu et al. (2010)
	Fragrance	*fgr* gene	*O. rufipogon*	Prathepha (2009)

(continued)

Table 11.1 (continued)

Domesticated crop	Agronomic trait in the CWR	Genes/QTLs for the trait	Crop wild relative	Reference
	Panicle length, primary branch number, secondary branch number spikelet number per panicle, spikelet density	QTLs (*QPl1b, QPbn8, QSd4, QSd11b, QSpp4*)	*O. rufipogon*	Luo et al. (2009)
Gossypium hirsutum var. *religiosum* (L.) Watt	Hairiness	*H6*	*G. raimondii* Ulbr.	Saunders (1965)
	Bacterial resistance gene	*B6*	*G. arboreum* L.	Knight (1948)
	Cotton rust resistance		*G. anomalum* G. Watt	Blank and Leathers (1963)
	Salinity tolerance	*WRKY* genes	*G. aridum* (Rose & Standl.) Skovst.	Fan et al. (2015)
		QTLs	*G. tomentosum* Nutt. ex Seem.	Oluoch et al. (2016)
	Drought tolerance	QTLs	*G. tomentosum*	Zheng et al. (2016)
Glycine max (L.) Merr.	Soybean cyst nematode resistance	QTLs (cqSCN-006 & cqSCN-007)	*G. soja* Siebold & Zucc.	Yu and Diers (2017)
	Asian soybean rust resistance	ASR resistance sources	*G. canescens* F.J. Herm., *G. clandestine* Wendl., *G. argyrea* Tindale, *G. microphylla* Tindale,	Hartman et al. (2016)
	Drought and salinity tolerance	*GsWRKY20*	*G. soja*	Tang et al. (2014)
	Alkali and salinity tolerance	*GsJAZ2*	*G. soja*	Zhu et al. (2012)
Zea mays L.	Gray leaf spot resistance	QTLs	*Z. mays* subsp. *parviglumis* Iltis & Doebley	Lennon et al. (2016)
	Rust resistance	$Rp1^{td}$	*Tripsacum dactyloides* (L.) L.	Bergquist (1981)
	Leaf blight resistance	*Ht3*	*T. dactyloides*	Hooker (1981)
Cicer arietinum L.	Root lesion nematodes & *Phytophthora* root rot resistance		*C. reticulatum* Ladizinsky & *C. echinospermum* P.H. Davis	Hajjar and Hodgkin (2007)

(continued)

Table 11.1 (continued)

Domesticated crop	Agronomic trait in the CWR	Genes/QTLs for the trait	Crop wild relative	Reference
Hordeum vulgare L.	Improved ß-amylase thermostability and fermentability	*Bmy*1-*Sd3*, -*Sd4*, and -*Sd5*	*H. spontaneum* K. Koch	Eglinton et al. (2001)
	Drought resistance	*Dehydrins*, *hsdr4*, and *eibi1*	*H. spontaneum*	Chen et al. (2008)
Triticum aestivum subsp. *aestivum*	Yield increases	*Lr19*	*Agropyron elongatum* var. *corsicum* (Hack.) Fiori	Reynolds et al. (2001)
	Drought tolerance	QTLs	*Triticum turgidum* ssp. *dicoccoides* (Körnicke ex Asch. & Graebn.) Thell.	Peleg et al. (2009)
		Chromosome segment (7DL)	*A. elongatum*	Placido et al. (2013)
	Salinity tolerance	*Nax2*	*T. monococcum* L.	Munns et al. (2012)
		TmHKT1;4-A2, *TmHKT1;5-A*	*T. monococcum*	James et al. (2012)
	Powdery mildew resistance	*Pm36*	*T. turgidum* var. *dicoccoides*	Blanco et al. (2008)
		pm42	*T. turgidum* var. *dicoccoides*	Hua et al. (2009)
	Leaf rust resistance	QTLs	*T. timopheevii* (Zhuk.) Zhuk./*T. tauschii* (Coss.) Schmalh.	Irina et al. (2007)
	Leaf rust resistance	*Lr59*	*A. peregrina* Mill.	Marais et al. (2008)
	Wheat stem rust resistance	*Sr35*	*T. monococcum*	Saintenac et al. (2013)
	Stripe rust resistance	*Yr5* and *Yr15*	*T. dicoccoides* Koern.	Murphy et al. (2009)
	High iron and zinc content	Ph^{I}	Wild *Triticum* and *Aegilops* species	Rawat et al. (2008)
Solanum lycopersicum L.	Accumulation of carotenoids	*Phytoene synthase-1* gene (*Psy-1*)	Wild accessions of tomato	Meléndez-Martínez et al. (2010)

(continued)

Table 11.1 (continued)

Domesticated crop	Agronomic trait in the CWR	Genes/QTLs for the trait	Crop wild relative	Reference
	Salt tolerance	*DREB1A* and *VP1.1*	*S. pimpinellifolium* L.	Rao et al. (2015)
	Powdery mildew resistance	QTLs	*L. parviflorum* C.M. Rick, E. Kesicki, J. F. Forbes, and M. Holle	Bai et al. (2003)
	Late blight resistance	QTLs	*Solanum habrochaites* Knapp and Spooner	Haggard and St. Clair (2015)
	Fruit yield increase	QTLs	*S. pennellii* (Correll) D'Arcy	Eshed and Zamir (1995)
	Fruit elongation	*OVATE, SUN*	*S. pimpinellifolium*	Wu et al. (2015)
Solanum tuberosum L.	Cold sweetening resistance	QTLs	*S. raphanifolium* Cárdenas and Hawkes	Hamernik et al. (2009)
	Late blight disease resistance	*Rpi-rzc1*	*S. pubescens* Ruiz and Pav.	Śliwka et al. (2012)
	Late blight disease resistance	R genes	*S. demissum* Lindl.	Hein et al. (2009)

crop yield is predicted due to the change in global climate conditions. Therefore, it is important to improve the abiotic stress tolerance of crop plants to stabilize global food security (Waqas et al. 2019). The genetic resistance of CWRs towards abiotic stresses can be explored to improve tolerance in crops (Hajjar and Hodgkin 2007). The capacity of CWRs to survive in extreme environmental conditions and its high adaptation potential to new environments is contributes to their abiotic stress tolerance (Henry 2014). Munns et al. (2012) performed a cross-breeding between salt tolerant CWR variety of wheat *T. monococcum* and cultivated wheat, and the progeny displayed a 25% of yield increase in high saline conditions (Munns et al. 2012). The inbred lines developed through introgression between wild barley *H. spontaneum* displayed drought tolerance in the field trialed progeny (Honsdorf et al. 2014). The drought tolerance of CWR of chickpea (i.e. *Cicer reticulatum), H. spontaneum* (CWR of barley), acid sulfate tolerance, and drought tolerance of *O. rufipogon* and *O. longistaminata* respectively (CWRs of rice) and salinity and drought tolerance of *L. chilense*, and *L. Pennellii* (CWRs of tomato) are utilized for producing abiotic stress tolerant crops by various research groups (Rick and Chetelat 1995; Baum et al. 2003; Nguyen et al. 2003; Yadav et al. 2004). However the conventional breeding approach of to interrogate drought tolerance in maize was unsuccessful due to the complexity of trait inheritance (Ray 1999). Therefore, it is clear that CWRs can significantly contribute to the development of abiotic stress-tolerant crops using different crop breeding programs.

11.5.3 Nutritional Improvement

Nearly 690 million people suffered from hunger in 2019, the reports indicate that the goal to achieve zero hunger by 2030 seems impossible (Furman et al. 2021). Improvement in the nutritional quality of food crops is a major concern for plant breeders. Malnutrition and micronutrient deficiency in the population are rising challenges to global food security (Khush et al. 2012). Micronutrient deficiency severely affects the growth and development of children, pregnant women, and others (Welch 2021). The cultivation of a uniform variety of crops by focusing only on their yield instead of regional landraces, changes in climate change, and prioritization of crop breeders only for high-yield varieties are the major reasons for the reduction of nutrients in the crops (FAO 2019; Furman et al. 2021; Onyeje 2022). In 2021, Medina-Lozano et al. (2021) compared the nutritional (vitamin C) value of traditional, wild and commercial lettuce and it was found that the wild varieties contain more vitamin C than the other two (Medina-Lozano et al. 2021). The elemental composition of cultivated lentil seeds (*Lens culinaris*) and wild relatives *L. lamottei, L. nigricans,* and *L. ervoides* was performed by Gupta et al. (2016) and found that there is no significant difference between the crops in the case of magnesium, iron, calcium, zinc and copper content (Gupta et al. 2016). Schier et al. (2019) also compared the elemental composition of wild and cultivated *Phaseolus vulgaris* with the aid of ionomics tools and found that the manganese, phosphorus, Fe, Ca, Cu and Mg content of wild plants are higher than cultivated plants (Schier et al. 2019). Characterization of wild chickpea varieties for unraveling their nutritional significance was performed by (Sharma et al. 2021) and found that *C. chorassanicum* is a highly promising crop for crop improvement with a high level of Ca and protein. Other wild varieties also reported high mineral content in the seeds. They suggested that these CWRs are crossable with cultivated chickpeas and can be used for crop improvement of cultivated varieties. Higher antioxidant activity and polyphenol content was observed in the wild relatives of a blackberry than in cultivated ones. Hence it indicates their potential to improve the cultivated variety for antioxidant activity (Deighton et al. 2000). Therefore it is evidenced that, through the breeding efforts such as hybridization, genetic engineering, and gene editing it is possible to utilize the nutritional superiority of CWRs to improve cultivated crops. *De-novo* domestication of nutritionally rich CWRs can also be performed to eliminate the undesirable traits to modify and make them suitable for cultivation. By exploring the nutritional potential of CWRs, it is possible to reduce hidden hunger and malnutrition among consumers.

11.5.4 Yield Improvement

The requirement to feed the burgeoning population demands high-yielding crop varieties (Hall et al. 2017). To satisfy the hunger of the global population and to acquire food security it is essential to improve the yielding potential of cultivated crops (Maletta 2014). Simultaneously introducing diverse crop varieties into the food basket of consumers can also be done (Adjimoti and Kwadzo 2018). But generally, CWRs are associated with low yield and high shattering rate. The occurrence of a high-yield variety of CWRs is very less. Hence, *de-novo* domestication can be implemented to improve the yield potential of CWR (Salamini et al. 2002; Gasparini et al. 2021). But the wild relative of *P. sativum, P. fulvum* displayed positive seed yield and seed yield components (Mikić et al. 2013). Hence proper evaluation of different varieties of CWRs should be performed for developing crop improvement strategy.

11.6 Challenges and Strategies to Overcome Barriers in Using CWR for Crop Breeding

Even though the CWRs have the potential to improve agronomic traits of the crops, crop breeding with CWRs is burdened with several challenges and barriers. But the advancement in research can be used to overcome these challenges. Interspecific crossability can restrict the transfer of gene(s) from the CWR to the cultivated crops. Fu et al. (2009) reported the failure of hybrid formation between the precious tetraploid variety of rice *Oryza meyeriana* Baill and *O. sativa* due to crossability issue. Tissue culture techniques such as embryo rescue (in-vitro culture of immature and weak embryos to prevent them from degradation), ovary and ovule culture, chromosome exchange induction, and chromosome doubling can be applied to overcome this (Bhanu 2018). Another major hindrance is the lack of information regarding the ploidy level, taxonomy and traits, gene-trait relation, and expression pattern of allelic combinations of CWRs, but extension of research on diversity, genomics, and epistasis of CWRs can contribute more information to these areas (Dempewolf et al. 2017). Misidentified or unidentified varieties and underrepresented germplasm collection also act as limiting factor in CWR-based breeding. Therefore the collection of CWRs should be increased with the assistance of eminent taxonomists by analyzing the gaps in the collection (García et al. 2017). Even after collection also, several CWRs are not characterized and evaluated from gene banks, therefore it should be performed systematically with proper coordination. It is essential to ensure the proper assessment of the germplasm from multi-locations. Bioinformatics tools can also be utilized for this purpose (Dempewolf et al. 2017). The precise transfer of genes from CWR is also a key bottleneck in CWR crop breeding. But marker-assisted selection, transgenic technologies, and modern embryo rescue technologies can be used for the precise transfer of genes (Zhang et al. 2017). Another major obstacle in CWR-based breeding is the lack of access to certain CWR material, the formation of national-level policies can facilitate access to CWR for different researchers around the globe. There is a requirement to improve the collaboration and coordination among researchers to develop a breeding continuum of CWRs. The governments should adopt new strategies to improve manpower and resources for CWR-based breeding to obtain results drastically (Dempewolf et al. 2017). Through these procedures, it is possible to overcome the challenges of the utilization of CWRs in crop improvement.

11.7 Conclusions

This chapter summarizes the potential of crop wild relatives in crop improvement programs, various breeding methods to utilize CWR for breeding, and the challenges of CWR during crop breeding. The advancement in molecular technologies, hybridization techniques and genetic characterization is required to increase the potential of CWR in crop improvement. Gene editing technologies can be applied to overcome the problems related to crossing incompatibility, and undesirable traits. The inclusion of CWRs in crop breeding can result in the production of crops possessing wild genes and an increased interest of breeders to grow CWRs. It can help to improve both ex-situ and in-situ conservation of CWRs and it will increase the crop diversity in the hands of breeders. Since the information regarding the genetic diversity of CWRs is very less, it is urgent to apply advanced molecular biology techniques to dissect the genetic characters of CWRs. The crop breeding programs based on CWRs ultimately depend on the availability of resources. The diversity, region, and phenotypic characters of each CWR should be analyzed in detail to execute crop breeding programs efficiently. Collaboration and coordination among researchers are required to develop a breeding continuum of CWRs. The governments should adopt new strategies to improve manpower and resources for CWR-based breeding to obtain better results. The exploration, crop improvement, effective cultivation practices, and popularization are the key steps for utilizing the potential of CWRs (Singh et al. 2019).

References

Adjimoti GO, Kwadzo GTM (2018) Crop diversification and household food security status: evidence from rural Benin. Agric Food Secur 7:82. https://doi.org/10.1186/s40066-018-0233-x

Allen E, Gaisberger H, Brehm JM, Maxted N, Thormann I, Lupupa T, Dulloo ME, Kell SP (2019) A crop wild relative inventory for Southern Africa: a first step in linking conservation and use of valuable wild populations for enhancing food security. Plant Genet Resour 17:128–139. https://doi.org/10.1017/S1479262118000515

Avagyan A (2008) Crop wild relatives in Armenia: diversity, legislation and conservation status. In: Maxted N, Ford-Lloyd BV, Kell SP, Iriondo JM, Dulloo ME, Turok J (eds) Crop wild relative conservation and use. CABI, Nosworthy Way

Bai Y, Huang C, van der Hulst R, Meijer-Dekens F, Bonnema G, Lindhout P (2003) QTLs for tomato powdery mildew resistance (*Oidium lycopersici*) in *Lycopersicon parviflorum* G1.1601 co-localize with two qualitative powdery mildew resistance genes. Mol Plant-Microbe Interact 16:169–176. https://doi.org/10.1094/MPMI.2003.16.2.169

Barringer BC (2007) Polyploidy and self-fertilization in flowering plants. Am J Bot 94:1527–1533. https://doi.org/10.3732/ajb.94.9.1527

Baum M, Grando S, Backes G, Jahoor A, Sabbagh A, Ceccarelli S (2003) QTLs for agronomic traits in the Mediterranean environment identified in recombinant inbred lines of the cross "Arta" x *H. spontaneum* 41-1. Theor Appl Genet 107:1215

Benlioğlu B, Adak MS (2019) Importance of crop wild relatives and landraces genetic resources in plant breeding programmes. J Exp Agric Int 37:1–8. https://doi.org/10.9734/jeai/2019/v37i330268

Bergquist RR (1981) Transfer from *Tripsacum dactyloides* to corn of a major gene locus conditioning resistance to *Puccinia sorghi*. Phytopathology 71:518–520. https://doi.org/10.1094/Phyto-71-518

Bernacchi D, Beck-Bunn T, Eshed Y, Lopez J, Petiard V, Uhlig J, Zamir D, Tanksley S (1998) Advanced backcross QTL analysis in tomato. I. Identification of QTLs for traits of agronomic importance from *Lycopersicon hirsutum*. Theor Appl Genet 97:381–397. https://doi.org/10.1007/s001220050908

Bhanu AN (2018) Crossability studies of interspecific hybridization among *Vigna* species. Biomed J Sci Tech Res 2:001–007

Bharadwaj D (2015) Polyploidy in crop improvement and evolution. In: Bahadur B (ed) Plant biology and biotechnology, vol I. Plant diversity, organization, function and improvement. Springer, India, pp 619–638

Blanco A, Gadaleta A, Cenci A, Carluccio AV, Abdelbacki AMM, Simeone R (2008) Molecular mapping of the novel powdery mildew resistance gene Pm36 introgressed from *Triticum turgidum* var. *dicoccoides* in durum wheat. Theor Appl Genet 117:135–142. https://doi.org/10.1007/s00122-008-0760-0

Blank LM, Leathers CR (1963) Environmental and other factors influencing development of southwestern cotton rust (*Puccinia stakmanii*). Phytopathology 53:921–928

Bossdorf O, Auge H, Lafuma L, Rogers WE, Siemann E, Prati D (2005) Phenotypic and genetic differentiation between native and introduced plant populations. Oecologia 144:1–11. https://doi.org/10.1007/s00442-005-0070-z

Byrne PF, Volk GM, Gardner C, Gore MA, Simon PW, Smith S (2018) Sustaining the future of plant breeding: the critical role of the USDA-ARS National Plant Germplasm System. Crop Sci 58:451–468. https://doi.org/10.2135/cropsci2017.05.0303

Chauhan A, Sharma D, Banyal SK, Madhvi, Divya (2021) Potential, challenges and strategies involved in gene introgression from wild relatives of vegetable crops: a review. Agric Rev. https://doi.org/10.18805/ag.R-2036

Chen G, Li C, Shi Y, Nevo Y (2008) Wild barley, *Hordeum spontaneum*, a genetic resource for crop improvement in cold and arid regions. Sci Cold Arid Reg 115–124

Choudhary M, Singh V, Muthusamy V, Wani S (2017) Harnessing crop wild relatives for crop improvement. Int J Life Sci 6:73. https://doi.org/10.5958/2319-1198.2017.00009.4

Contreras-Toledo AR, Cortés-Cruz M, Costich DE, Rico-Arce ML, Brehm JM, Maxted N (2019) Diversity and conservation priorities of crop wild relatives in Mexico. Plant Genet Resour 17: 140–150. https://doi.org/10.1017/S1479262118000540

Coyne CJ, Kumar S, von Wettberg EJB, Marques E, Berger JD, Redden RJ, Ellis THN, Brus J, Zablatzká L, Smýkal P (2020) Potential and limits of exploitation of crop wild relatives for pea, lentil, and chickpea improvement. Legume Sci 2:e36. https://doi.org/10.1002/leg3.36

Curti RN, Sajama J, Ortega-Baes P (2017) Setting conservation priorities for Argentina's pseudocereal crop wild relatives. Biol Conserv 209:349–355. https://doi.org/10.1016/j.biocon.2017.03.008

Dale PJ (1992) Spread of engineered genes to wild relatives. Plant Physiol 100:13–15. https://doi.org/10.1104/pp.100.1.13

De La Torre JFS, González S R, Cruz G EJ, Pichardo G JM, Quintana C M, Contreras T AR, Cadena I J (2018) Crop wild relatives in Mexico: an overview of richness, importance, and conservation status. In: Greene SL, Williams KA, Khoury CK, Kantar MB, Marek LF (eds) North American crop wild relatives, vol 1. Conservation strategies. Springer International Publishing, Cham, pp 63–96

Deighton N, Brennan R, Finn C, Davies HV (2000) Antioxidant properties of domesticated and wild *Rubus* species. J Sci Food Agric 80:1307–1313. https://doi.org/10.1002/1097-0010(200007)80:9<1307::AID-JSFA638>3.0.CO;2-P

Dempewolf H, Baute G, Anderson J, Kilian B, Smith C, Guarino L (2017) Past and future use of wild relatives in crop breeding. Crop Sci 57:1070–1082

Eglinton JK, Evans DE, Brown AHD, Langridge P, McDonald G, Jefferies SP, Barr AR (2001) The use of wild barley (*Hordeum vulgare* ssp. *spontaneum*) in breeding for quality and adaptation. In: Proceedings of the 10th Barley Technical Symposium. ACT, Canberra, pp 16–20

Ellur RK, Khanna A, Krishnan GS, Bhowmick PK, Vinod KK, Nagarajan M, Mondal KK, Singh NK, Singh K, Prabhu KV, Singh AK (2016) Marker-aided incorporation of Xa38, a novel bacterial blight resistance gene, in PB1121 and comparison of its resistance spectrum with xa13 + Xa21. Sci Rep 6:29188. https://doi.org/10.1038/srep29188

Eshed Y, Zamir D (1995) An introgression line population of *Lycopersicon pennellii* in the cultivated tomato enables the identification and fine mapping of yield-associated QTL. Genetics 141:1147–1162

Fan X, Guo Q, Xu P, Gong Y, Shu H, Yang Y, Ni W, Zhang X, Shen X (2015) Transcriptome-wide identification of salt-responsive members of the wrky gene family in *Gossypium aridum*. PLoS One 10:e0126148. https://doi.org/10.1371/journal.pone.0126148

FAO (2011) Food, agriculture and cities. Save and grow: a new paradigm of agriculture. A policymaker's guide to the sustainable intensification of smallholder crop production. Food and Agriculture Organization of the United Nations, Rome

FAO (2019) Voluntary guidelines for the conservation and sustainable use of farmers' varieties/landraces |policy support and governance| Food and Agriculture Organization of the United Nations. https://www.fao.org/policy-support/tools-and-publications/resources-details/en/c/1263074/. Accessed 15 Jul 2022

Fernie AR, Yan J (2019) De novo domestication: an alternative route toward new crops for the future. Mol Plant 12:615–631. https://doi.org/10.1016/j.molp.2019.03.016

Fielder H, Brotherton P, Hosking J, Hopkins JJ, Ford-Lloyd B, Maxted N (2015) Enhancing the conservation of crop wild relatives in England. PLoS One 10:e0130804. https://doi.org/10.1371/journal.pone.0130804

Fielder H, Smith C, Ford-Lloyd B, Maxted N (2016) Enhancing the conservation of crop wild relatives in Scotland. J Nat Conserv 29:51–61. https://doi.org/10.1016/j.jnc.2015.11.002

Fonseka RM, Fonseka HHD, Abhyapala KMRD (2020) Crop wild relatives: an underutilized genetic resource for improving agricultural productivity and food security. In: De Silva RP, Pushpakumara G, Prasada P, Weerahewa J (eds) Agricultural research for sustainable food systems in Sri Lanka, vol 2. A pursuit for advancements. Springer, Singapore, pp 11–38

Ford-Lloyd BV, Schmidt M, Armstrong SJ, Barazani O, Engels J, Hadas R, Hammer K, Kell SP, Kang D, Khoshbakht K, Li Y, Long C, Lu B, Ma K, Nguyen VT, Qiu L, Ge S, Wei W, Zhang Z, Maxted N (2011) Crop wild relatives—undervalued, underutilized and under Threat? Bioscience 61:559–565. https://doi.org/10.1525/bio.2011.61.7.10

Fu X, Lu Y, Liu X, Li J (2009) Crossability barriers in the interspecific hybridization between *Oryza sativa* and *O. meyeriana*. J Integr Plant Biol 51:21–28. https://doi.org/10.1111/j.1744-7909.2008.00728.x

Fu Q, Zhang P, Tan L, Zhu Z, Ma D, Fu Y, Zhan X, Cai H, Sun C (2010) Analysis of QTLs for yield-related traits in Yuanjiang common wild rice (*Oryza rufipogon* Griff.). J Genet Genomics 37:147–157. https://doi.org/10.1016/S1673-8527(09)60033-3

Fulton TM, Beck-Bunn T, Emmatty D, Eshed Y, Lopez J, Petiard V, Uhlig J, Zamir D, Tanksley SD (1997) QTL analysis of an advanced backcross of *Lycopersicon peruvianum* to the cultivated tomato and comparisons with QTLs found in other wild species. Theor Appl Genet 95: 881–894. https://doi.org/10.1007/s001220050639

Furman B, Noorani A, Mba C (2021) On-farm crop diversity for advancing food security and nutrition. IntechOpen

García RM, Parra-Quijano M, Iriondo JM (2017) A multispecies collecting strategy for crop wild relatives based on complementary areas with a high density of ecogeographical gaps. Crop Sci 57:1059–1069. https://doi.org/10.2135/cropsci2016.10.0860

Garg R, Verma M, Agrawal S, Shankar R, Majee M, Jain M (2014) Deep transcriptome sequencing of wild halophyte rice, *Porteresia coarctata*, provides novel insights into the salinity and submergence tolerance factors. DNA Res 21:69–84. https://doi.org/10.1093/dnares/dst042

Gasparini K, Moreira JR, Peres LEP, Zsögön A (2021) De novo domestication of wild species to create crops with increased resilience and nutritional value. Curr Opin Plant Biol 60:102006. https://doi.org/10.1016/j.pbi.2021.102006

Goettsch B, Urquiza-Haas T, Koleff P, Acevedo Gasman F, Aguilar-Meléndez A, Alavez V, Alejandre-Iturbide G, Aragón Cuevas F, Azurdia Pérez C, Carr JA, Castellanos-Morales G, Cerén G, Contreras-Toledo AR, Correa-Cano ME, De la Cruz Larios L, Debouck DG, Delgado-Salinas A, Gómez-Ruiz EP, González-Ledesma M, González-Pérez E, Hernández-Apolinar M, Herrera-Cabrera BE, Jefferson M, Kell S, Lira-Saade R, Lorea-Hernández F, Martínez M, Mastretta-Yanes A, Maxted N, Menjívar J, de los Ángeles Mérida Guzmán M, Morales Herrera AJ, Oliveros-Galindo O, Orjuela-R. MA, Pollock CM, Quintana-Camargo M, Rodríguez A, Ruiz Corral JA, Sánchez González J, Sánchez-de la Vega G, Superina M, Tobón Niedfeldt W, Tognelli MF, Vargas-Ponce O, Vega M, Wegier A, Zamora Tavares P, Jenkins RKB (2021) Extinction risk of Mesoamerican crop wild relatives. Plants People Planet 3:775–795. https://doi.org/10.1002/ppp3.10225

González-Orozco CE, Galán AS, Ramos PE, Yockteng R (2020) Exploring the diversity and distribution of crop wild relatives of cacao (*Theobroma cacao* L.) in Colombia. Genet Resour Crop Evol 67:2071–2085. https://doi.org/10.1007/s10722-020-00960-1

Guo Q, Falcone J, Brownsmith J (2009) Building the database for introduced plants in the United States. In: McManus KA, Gottschalk KW (eds) Proceedings of the 20th U.S. Department of Agriculture interagency research forum on invasive species. Department of Agriculture, Forest Service, Northern Research Station, Annapolis, p 73

Gupta SD, Thavarajah D, McGee RJ, Coyne CJ, Kumar S, Thavarajah P (2016) Genetic diversity among cultivated and wild lentils for iron, zinc, copper, calcium and magnesium concentrations. Aust J Crop Sci 10:1381–1387. https://doi.org/10.3316/informit.501057674329853

Haggard J, St. Clair D (2015) Combining ability for *Phytophthora infestans* quantitative resistance from wild tomato. Crop Sci 55:240. https://doi.org/10.2135/cropsci2014.04.0286

Hajjar R, Hodgkin T (2007) The use of wild relatives in crop improvement: a survey of developments over the last 20 years. Euphytica 156:1–13. https://doi.org/10.1007/s10681-007-9363-0

Hall C, Dawson TP, Macdiarmid JI, Matthews RB, Smith P (2017) The impact of population growth and climate change on food security in Africa: looking ahead to 2050. Int J Agric Sustain 15:124–135. https://doi.org/10.1080/14735903.2017.1293929

Hamernik AJ, Hanneman RE, Jansky SH (2009) Introgression of wild species germplasm with extreme resistance to cold sweetening into the cultivated potato. Crop Sci 49:529–542. https://doi.org/10.2135/cropsci2008.04.0209

Hammer K (1984) Das Domestikations syndrom. Kult 32:11–34. https://doi.org/10.1007/BF02098682

Hariprasanna K, Patil JV (2015) Sorghum: origin, classification, biology and improvement. In: Madhusudhana R, Rajendrakumar P, Patil JV (eds) Sorghum molecular breeding, New Delhi, pp 3–20
Hartman GL, Rupe JC, Sikora EJ, Domier LL, Davis JA, Steffey KL (2016) Compendium of soybean diseases and pests, 5th edn. The American Phytopathological Society
Hein I, Birch P, Danan S, Lefebvre V, Odeny D, Gebhardt C, Trognitz F, Bryan G (2009) Progress in mapping and cloning qualitative and quantitative resistance against phytophthora infestans in potato and its wild relatives. Potato Res 52:215–227. https://doi.org/10.1007/s11540-009-9129-2
Heinitz CC, Uretsky J, Dodson Peterson JC, Huerta-Acosta KG, Walker MA (2019) Crop wild relatives of grape (*Vitis vinifera* L.) throughout North America. In: Greene SL, Williams KA, Khoury CK, Kantar MB, Marek LF (eds) North American crop wild relatives, vol 2. Important species. Springer International Publishing, Cham, pp 329–351
Henry R (2014) Genomics strategies for germplasm characterization and the development of climate resilient crops. Front Plant Sci 5:68. https://doi.org/10.3389/fpls.2014.00068
Holness S, Hamer M, Brehm J, Raimondo D (2019) Priority areas for the in situ conservation of crop wild relatives in South Africa. Plant Genet Resour Charact Util 17:1–13. https://doi.org/10.1017/S1479262118000503
Honsdorf N, March TJ, Berger B, Tester M, Pillen K (2014) High-throughput phenotyping to detect drought tolerance QTL in wild barley introgression lines. PLoS One 9:e97047. https://doi.org/10.1371/journal.pone.0097047
Hooker AL (1981) Resistance to *Helminthosporium turcicum* from *Tripsacum floridanum* incorporated into corn. Maize Genet Coop News Lett 55:87–88
Hu J, Xiao C, He Y (2016) Recent progress on the genetics and molecular breeding of brown planthopper resistance in rice. Rice 9:30. https://doi.org/10.1186/s12284-016-0099-0
Hua W, Liu Z, Zhu J, Xie C, Yang T, Zhou Y, Duan X, Sun Q, Liu Z (2009) Identification and genetic mapping of pm42, a new recessive wheat powdery mildew resistance gene derived from wild emmer (*Triticum turgidum* var. *dicoccoides*). Theor Appl Genet 119:223–230. https://doi.org/10.1007/s00122-009-1031-4
Huang XQ, Cöster H, Ganal MW, Röder MS (2003) Advanced backcross QTL analysis for the identification of quantitative trait loci alleles from wild relatives of wheat (*Triticum aestivum* L.). Theor Appl Genet 106:1379–1389. https://doi.org/10.1007/s00122-002-1179-7
Irina L, Laikova L, Popova O, Unger O, Börner A, Röder M (2007) Detection of quantitative trait loci for leaf rust resistance in wheat—*T. timopheevii*/*T. tauschii* introgression lines. Euphytica 155:79–86. https://doi.org/10.1007/s10681-006-9303-4
James RA, Blake C, Zwart AB, Hare RA, Rathjen AJ, Munns R (2012) Impact of ancestral wheat sodium exclusion genes *Nax1* and *Nax2* on grain yield of durum wheat on saline soils. Funct Plant Biol 39:609–618. https://doi.org/10.1071/FP12121
Jena KK (2010) The species of the genus *Oryza* and transfer of useful genes from wild species into cultivated rice, *O. sativa*. Breed Sci 60:518–523. https://doi.org/10.1270/jsbbs.60.518
Jørgensen RB, Hauser T, Mikkelsen TR, Østergård H (1996) Transfer of engineered genes from crop to wild plants. Trends Plant Sci 1:356–358. https://doi.org/10.1016/S1360-1385(96)82597-0
Kahane R, Hodgkin T, Jaenicke H, Hoogendoorn C, Hermann M, Keatinge JDH, d'Arros Hughes J, Padulosi S, Looney N (2013) Agrobiodiversity for food security, health and income. Agron Sustain Dev 33:671–693. https://doi.org/10.1007/s13593-013-0147-8
Kalloo G (1991) Interspecific and intergeneric hybridization in tomato. In: Kalloo G (ed) Genetic improvement of tomato. Springer, Berlin, Heidelberg, pp 73–82
Kalloo G (1992) Utilization of wild species. In: Kalloo G, Chowdhury JB (eds) Distant hybridization of crop plants. Springer, Berlin, Heidelberg, pp 149–167
Kang YJ, Satyawan D, Shim S, Lee T, Lee J, Hwang WJ, Kim SK, Lestari P, Laosatit K, Kim KH, Ha TJ, Chitikineni A, Kim MY, Ko J, Gwag J, Moon J, Lee Y, Park B, Varshney RK, Lee S

(2015) Draft genome sequence of adzuki bean, *Vigna angularis*. Sci Rep 5:8069. https://doi.org/10.1038/srep08069

Kaplan I, Dively GP, Denno RF (2009) The costs of anti-herbivore defence traits in agricultural crop plants: a case study involving leafhoppers and trichomes. Ecol Appl 19:864–872. https://doi.org/10.1890/07-1566.1

Käppeli O, Auberson L (1998) How safe is safe enough in plant genetic engineering? Trends Plant Sci 3:276–281. https://doi.org/10.1016/S1360-1385(98)01251-5

Keatinge JDH, Yang R-Y, Hughes J d'A, Easdown WJ, Holmer R (2011) The importance of vegetables in ensuring both food and nutritional security in attainment of the millennium development goals. Food Secur 3:491–501. https://doi.org/10.1007/s12571-011-0150-3

Kell S, Qin H, Chen B, Ford-Lloyd B, Wei W, Kang D, Maxted H (2015) China's crop wild relatives: diversity for agriculture and food security. Agric Ecosyst Environ 209:138–154. https://doi.org/10.1016/j.agee.2015.02.012

Khoury CK, Carver D, Greene S, Frances A (2020) Crop wild relatives of the United States require urgent conservation action. Proc Natl Acad Sci U S A 117:33351–33357

Khush GS, Angeles E, Virk PS, Brar DS (2004) Breeding rice for resistance to Tungro virus at IRRI. J Breed Genet 6:101–106

Khush GS, Lee S, Cho J, Jeon JS (2012) Biofortification of crops for reducing malnutrition. Plant Biotechnol Rep 6:195–202. https://doi.org/10.1007/s11816-012-0216-5

Knight RL (1948) The genetics of black arm resistance; *Gossypium arboreum* L. J Genet 49:109–116. https://doi.org/10.1007/BF02986827

Kumar A, Anju T, Kumar S, Chhapekar SS, Sreedharan S, Singh S, Choi SR, Ramchiary N, Lim YP (2021) Integrating omics and gene editing tools for rapid improvement of traditional food plants for diversified and sustainable food security. Int J Mol Sci 22:8093. https://doi.org/10.3390/ijms22158093

Lala S, Amri A, Maxted N (2018) Towards the conservation of crop wild relative diversity in North Africa: checklist, prioritisation and inventory. Genet Resour Crop Evol 65:113–124. https://doi.org/10.1007/s10722-017-0513-5

Larson G, Piperno DR, Allaby RG, Purugganan MD, Andersson L, Arroyo-Kalin M, Barton L, Climer Vigueira C, Denham T, Dobney K, Doust AN, Gepts P, Gilbert MTP, Gremillion KJ, Lucas L, Lukens L, Marshall FB, Olsen KM, Pires JC, Richerson PJ, Rubio de Casas R, Sanjur OI, Thomas MG, Fuller DQ (2014) Current perspectives and the future of domestication studies. Proc Natl Acad Sci U S A 111:6139–6146. https://doi.org/10.1073/pnas.1323964111

Lemmon ZH, Reem NT, Dalrymple J, Soyk S, Swartwood KE, Rodriguez-Leal D, Van Eck J, Lippman ZB (2018) Rapid improvement of domestication traits in an orphan crop by genome editing. Nat Plants 4:766–770. https://doi.org/10.1038/s41477-018-0259-x

Lennon JR, Krakowsky M, Goodman M, Flint-Garcia S, Balint-Kurti PJ (2016) Identification of alleles conferring resistance to gray leaf spot in maize derived from its wild progenitor species teosinte. Crop Sci 56:209–218. https://doi.org/10.2135/cropsci2014.07.0468

Levy AA, Feldman M (2022) Evolution and origin of bread wheat. Plant Cell 34(7):2549–2567. https://doi.org/10.1093/plcell/koac130

Li D, Sun C, Fu Y, Li C, Zhu Z, Chen L, Cai H, Wang X (2002) Identification and mapping of genes for improving the yield from Chinese common wild rice (*O. rufipogon* Griff.) using advanced backcross QTL analysis. Chin Sci Bull 47:1533–1537. https://doi.org/10.1360/02tb9337

Luo X, Tian F, Fu Y, Yang J, Sun C (2009) Mapping quantitative trait loci influencing panicle-related traits from Chinese common wild rice (*Oryza rufipogon*) using introgression lines. Plant Breed 128:559–567. https://doi.org/10.1111/j.1439-0523.2008.01607.x

Magos Brehm J, Maxted N, Ford-Lloyd BV, Martins-Loução MA (2008) National inventories of crop wild relatives and wild harvested plants: case-study for Portugal. Genet Resour Crop Evol 55:779–796. https://doi.org/10.1007/s10722-007-9283-9

Majee M, Maitra S, Dastidar KG, Pattnaik S, Chatterjee A, Hait NC, Das KP, Majumder AL (2004) A novel salt-tolerant L-myo-inositol-1-phosphate synthase from *Porteresia coarctata* (Roxb.) Tateoka, a halophytic wild rice: molecular cloning, bacterial overexpression, characterization, and functional introgression into tobacco-conferring salt tolerance phenotype. J Biol Chem 279: 28539–28552. https://doi.org/10.1074/jbc.M310138200

Maletta HE (2014) From hunger to food security: a conceptual history. SSRN. https://ssrn.com/abstract=2484166. https://doi.org/10.2139/ssrn.2484166

Mammadov J, Buyyarapu R, Guttikonda SK, Parliament K, Abdurakhmonov IY, Kumpatla SP (2018) Wild relatives of maize, rice, cotton, and soybean: treasure troves for tolerance to biotic and abiotic stresses. Front Plant Sci 9:886

Mantri N, Patade V, Penna S, Ford R, Pang E (2012) Abiotic stress responses in plants: present and future. In: Ahmad P, Prasad MNV (eds) Abiotic stress responses in plants: metabolism, productivity and sustainability. Springer, New York, pp 1–19

Marais G, Mccallum B, Marais A (2008) Wheat leaf rust resistance gene Lr59 derived from *Aegilops peregrina*. Plant Breed 127:340–345. https://doi.org/10.1111/j.1439-0523.2008.01513.x

Maxted N, Ford-Lloyd BV, Jury S, Kell S, Scholten M (2006) Towards a definition of a crop wild relative. Biodivers Conserv 15:2673–2685. https://doi.org/10.1007/s10531-005-5409-6

Maxted N, Scholten M, Codd R, Ford-Lloyd B (2007) Creation and use of a national inventory of crop wild relatives. Biol Conserv 140:142–159. https://doi.org/10.1016/j.biocon.2007.08.006

Medina-Lozano I, Bertolín JR, Díaz A (2021) Nutritional value of commercial and traditional lettuce (*Lactuca sativa* L.) and wild relatives: vitamin C and anthocyanin content. Food Chem 359:129864. https://doi.org/10.1016/j.foodchem.2021.129864

Meléndez-Martínez AJ, Fraser PD, Bramley PM (2010) Accumulation of health promoting phytochemicals in wild relatives of tomato and their contribution to in vitro antioxidant activity. Phytochemistry 71:1104–1114. https://doi.org/10.1016/j.phytochem.2010.03.021

Mezghani N, Khoury CK, Carver D, Achicanoy HA, Simon P, Flores FM, Spooner D (2019) Distributions and conservation status of carrot wild relatives in Tunisia: a case study in the western Mediterranean Basin. Crop Sci 59:2317–2328. https://doi.org/10.2135/cropsci2019.05.0333

Miflin B (2000) Crop improvement in the 21st century. J Exp Bot 51:1–8. https://doi.org/10.1093/jexbot/51.342.1

Mikić A, Mihailović V, Dimitrijević M, Petrović S, Ćupina B, Đorđević V, Kosev V, Milošević B, Jovanović Z, Milovac Z (2013) Evaluation of seed yield and seed yield components in red–yellow (*Pisum fulvum*) and Ethiopian (*Pisum abyssinicum*) peas. Genet Resour Crop Evol 60: 629–638. https://doi.org/10.1007/s10722-012-9862-2

Monsur MB, Shao G, Lv Y, Ahmad S, Wei X, Hu P, Tang S (2020) Base Editing: the ever expanding clustered regularly interspaced short palindromic repeats (CRISPR) tool kit for precise genome editing in plants. Genes 11:466. https://doi.org/10.3390/genes11040466

Munns R, James RA, Xu B, Athman A, Conn SJ, Jordans C, Byrt CS, Hare RA, Tyerman SD, Tester M, Plett D, Gilliham M (2012) Wheat grain yield on saline soils is improved by an ancestral Na+ transporter gene. Nat Biotechnol 30:360–364. https://doi.org/10.1038/nbt.2120

Murphy D (2011) Crop improvement in the 20th century. In: Murphy D (ed) Plants, biotechnology and agriculture. CABI, Wallingford, pp 194–219

Murphy LR, Santra D, Kidwell K, Yan G, Chen X, Campbell KG (2009) Linkage maps of wheat stripe rust resistance genes Yr5 and Yr15 for use in marker-assisted selection. Crop Sci 49: 1786–1790. https://doi.org/10.2135/cropsci2008.10.0621

Nguyen BD, Brar DS, Bui BC, Nguyen TV, Pham LN, Nguyen HT (2003) Identification and mapping of the QTL for aluminium tolerance introgressed from the new source, *Oryza rufipogan* Griff., into indica rice (*Oryza sativa* L.). Theor Appl Genet 106:583–593. https://doi.org/10.1007/s00122-002-1072-4

Niroula RK, Pucciariello C, Ho VT, Novi G, Fukao T, Perata P (2012) SUB1A-dependent and -independent mechanisms are involved in the flooding tolerance of wild rice species. Plant J Cell Mol Biol 72:282–293. https://doi.org/10.1111/j.1365-313X.2012.05078.x

Oluoch G, Zheng J, Wang X, Khan MKR, Zhou Z, Cai X, Wang C, Wang Y, Li X, Wang H, Liu F, Wang K (2016) QTL mapping for salt tolerance at seedling stage in the interspecific cross of

Gossypium tomentosum with *Gossypium hirsutum*. Euphytica 209:223–235. https://doi.org/10.1007/s10681-016-1674-6

Onyeje BT (2022) Micronutrient deficiency among Nigerian population: the hidden hunger. Int Acad J Med Clin Pract 7:1–14

Pandey A, Tomer AJ, Bhandari DC, Pareek SK (2008) Towards collection of wild relatives of crop plants in India. Genet Resour Crop Evol 55:187–202. https://doi.org/10.1007/s10722-007-9227-4

Peleg Z, Fahima T, Krugman T, Abbo S, Yakir D, Korol AB, Saranga Y (2009) Genomic dissection of drought resistance in durum wheat x wild emmer wheat recombinant inbreed line population. Plant Cell Environ 32:758–779. https://doi.org/10.1111/j.1365-3040.2009.01956.x

Perrino EV, Wagensommer RP (2021) Crop Wild Relatives (CWR) priority in Italy: distribution, ecology, in situ and ex situ conservation and expected actions. Sustainability 13:1682. https://doi.org/10.3390/su13041682

Perrino EV, Wagensommer RP (2022) Crop Wild Relatives (CWRs) Threatened and endemic to Italy: urgent actions for protection and use. Biology 11:193. https://doi.org/10.3390/biology11020193

Pillen K, Zacharias A, Léon J (2003) Advanced backcross QTL analysis in barley (*Hordeum vulgare* L.). Theor Appl Genet 107:340–352. https://doi.org/10.1007/s00122-003-1253-9

Placido DF, Campbell MT, Folsom JJ, Cui X, Kruger GR, Baenziger PS, Walia H (2013) Introgression of novel traits from a wild wheat relative improves drought adaptation in wheat. Plant Physiol 161:1806–1819. https://doi.org/10.1104/pp.113.214262

Porter J, Xie L, Challinor AJ, Cochrane K, Howden SM, Iqbal MM, Lobell DB, Travasso MI (2014) Food security and food production systems. In: Field CB, Barros VR, Dokken DJ, Mach KJ, Mastrandrea MD, Bilir TE, Chatterjee M, Ebi KL, Estrada R, Genova B, Girma AS, Kisse ES, Levy AS, MacCracken S, Mastrandrea PR, White LL (eds) Climate change 2014: impacts, adaptation, and vulnerability. Part A: global and sectoral aspects. Contribution of Working Group II to the Fifth Assessment Report of the Intergovernmental Panel on Climate Change. Cambridge University Press, Cambridge and New York, pp 485–533

Prathepha P (2009) The fragrance (fgr) gene in natural populations of wild rice (*Oryza rufipogon* Griff.). Genet Resour Crop Evol 56:13–18. https://doi.org/10.1007/s10722-008-9337-7

Price HJ, Hodnett GL, Burson BL, Dillon SL, Rooney WL, Price HJ, Hodnett GL, Burson BL, Dillon SL, Rooney WL (2005) A *Sorghum bicolor* × *S. macrospermum* hybrid recovered by embryo rescue and culture. Aust J Bot 53:579–582. https://doi.org/10.1071/BT04213

Qi X, Li M, Xie M, Liu X, Ni M, Shao G, Song C, Kay-Yuen Yim A, Tao Y, Wong F, Isobe S, Wong C, Wong K, Xu C, Li C, Wang Y, Guan R, Sun F, Fan G, Xiao Z, Zhou F, Phang T, Liu X, Tong S, Chan T, Yiu S, Tabata S, Wang J, Xu X, Lam H (2014) Identification of a novel salt tolerance gene in wild soybean by whole-genome sequencing. Nat Commun 5:4340. https://doi.org/10.1038/ncomms5340

Rahman W, Magos Brehm J, Maxted N (2019) Setting conservation priorities for the wild relatives of food crops in Indonesia. Genet Resour Crop Evol 66:809–824. https://doi.org/10.1007/s10722-019-00761-1

Rao ES, Kadirvel P, Symonds RC, Geethanjali S, Thontadarya RN, Ebert AW (2015) Variations in *DREB1A* and *VP1.1* Genes show association with salt tolerance traits in wild tomato (*Solanum pimpinellifolium*). PLoS One 10:e0132535. https://doi.org/10.1371/journal.pone.0132535

Ratnayake SS, Kariyawasam CS, Kumar L, Hunter D, Liyanage ASU (2021) Potential distribution of crop wild relatives under climate change in Sri Lanka: implications for conservation of agricultural biodiversity. Curr Res Environ Sustain 3:100092. https://doi.org/10.1016/j.crsust.2021.100092

Rawat N, Tiwari VK, Singh N, Randhawa GS, Singh K, Chhuneja P, Dhaliwal HS (2008) Evaluation and utilization of Aegilops and wild *Triticum* species for enhancing iron and zinc content in wheat. Genet Resour Crop Evol 56:53. https://doi.org/10.1007/s10722-008-9344-8

Ray JD (1999) Introgressing root aerenchyma into maize. Maydica 44:113–117

Raza A, Razzaq A, Mehmood SS, Zou X, Zhang X, Lv Y, Xu J (2019) Impact of climate change on crops adaptation and strategies to tackle its outcome: a review. Plants 8:34. https://doi.org/10.3390/plants8020034

Redden RJ, Yadav SS, Maxted N, Dulloo ME, Guarino L, Smith P (2015) Crop wild relatives and climate change. Wiley

Renzi JP, Coyne CJ, Berger J, von Wettberg E, Nelson M, Ureta S, Hernández F, Smýkal P, Brus J (2022) How could the use of crop wild relatives in breeding increase the adaptation of crops to marginal environments? Front Plant Sci 13:886162

Reynolds MP, Calderini DF, Condon AG, Rajaram S (2001) Physiological basis of yield gains in wheat associated with the LR19 translocation from *Agropyron elongatum*. Euphytica 119:139–144. https://doi.org/10.1023/A:1017521800795

Rick CM, Chetelat RT (1995) Utilization of related wild species for tomato improvement. Acta Hortic 412:21–38. https://doi.org/10.17660/ActaHortic.1995.412.1

Rosegrant MW, Cline SA (2003) Global food security: challenges and policies. Science 302:1917–1919. https://doi.org/10.1126/science.1092958

Rubio Teso L, Torres Lamas E, Parra-Quijano M, de la Rosa L, Fajardo J, Iriondo JM (2018) National inventory and prioritization of crop wild relatives in Spain. Genet Resour Crop Evol 65:1237–1253. https://doi.org/10.1007/s10722-018-0610-0

Saintenac C, Zhang W, Salcedo A, Rouse MN, Trick HN, Akhunov E, Dubcovsky J (2013) Identification of wheat gene *Sr35* that confers resistance to *Ug99* stem rust race group. Science 341:783–786. https://doi.org/10.1126/science.1239022

Salaheen S, Biswas D (2019) Chapter 2—Organic farming practices: integrated culture versus monoculture. In: Biswas D, Micallef SA (eds) Safety and practice for organic food. Academic, Cambridge, pp 23–32

Salamini F, Özkan H, Brandolini A, Schäfer-Pregl R, Martin W (2002) Genetics and geography of wild cereal domestication in the near east. Nat Rev Genet 3:429–441. https://doi.org/10.1038/nrg817

Sanchez PL, Wing RA, Brar DS (2013) The wild relative of rice: genomes and genomics. In: Zhang Q, Wing RA (eds) Genetics and genomics of rice. Springer, New York, pp 9–25

Saunders JH (1965) Genetics of hairiness transferred from *Gossypium raimondii* to *G. hirsutum*. Euphytica 14:276–282. https://doi.org/10.1007/BF00149511

Scafaro AP, Haynes PA, Atwell BJ (2010) Physiological and molecular changes in *Oryza meridionalis* Ng., a heat-tolerant species of wild rice. J Exp Bot 61:191–202. https://doi.org/10.1093/jxb/erp294

Schier HE, Eliot KA, Herron SA, Landfried LK, Migicovsky Z, Rubin MJ, Miller AJ (2019) Comparative analysis of perennial and annual *Phaseolus* seed nutrient concentrations. Sustainability 11:2787. https://doi.org/10.3390/su11102787

Sharma S, Lavale SA, Nimje C, Singh S (2021) Characterization and identification of annual wild *Cicer* species for seed protein and mineral concentrations for chickpea improvement. Crop Sci 61:305–319. https://doi.org/10.1002/csc2.20413

Sheesh PS, Sheesh SP, Rohini, Sharma AK (2011) Distant hybridization in tomato for Lycopene improvement. J Indian Bot Soc 90:399–344

Singh A, Dubey PK, Chaurasia R, Dubey RK, Pandey KK, Singh GS, Abhilash PC (2019) Domesticating the undomesticated for global food and nutritional security: four steps. Agronomy 9:491. https://doi.org/10.3390/agronomy9090491

Śliwka J, Jakuczun H, Chmielarz M, Hara-Skrzypiec A, Tomczyńska I, Kilian A, Zimnoch-Guzowska E (2012) Late blight resistance gene from *Solanum ruiz-ceballosii* is located on potato chromosome X and linked to violet flower colour. BMC Genet 13:11. https://doi.org/10.1186/1471-2156-13-11

Smýkal P, Nelson MN, Berger JD, Von Wettberg EJB (2018) The impact of genetic changes during crop domestication. Agronomy 8:119. https://doi.org/10.3390/agronomy8070119

Song W, Wang G, Chen L-L, Kim H, Pi L, Holsten T, Gardner J, Wang B, Zhai W, Zhu L, Fauquet C, Ronald P (1995) A receptor kinase-like protein encoded by the rice disease resistance gene, Xa21. Science 270:1804–1806. https://doi.org/10.1126/science.270.5243.1804

Takeda S, Matsuoka M (2008) Genetic approaches to crop improvement: responding to environmental and population changes. Nat Rev Genet 9:444–457. https://doi.org/10.1038/nrg2342

Tang L, Cai H, Zhai H, Luo X, Wang Z, Cui L, Bai X (2014) Overexpression of *Glycine soja* WRKY20 enhances both drought and salt tolerance in transgenic alfalfa (*Medicago sativa* L.). Plant Cell Tissue Organ Cult 118:77–86. https://doi.org/10.1007/s11240-014-0463-y

Tanksley SD, McCouch SR (1997) Seed banks and molecular maps: unlocking genetic potential from the wild. Science 277:1063–1066. https://doi.org/10.1126/science.277.5329.1063

Tanksley SD, Nelson JC (1996) Advanced backcross QTL analysis: a method for the simultaneous discovery and transfer of valuable QTLs from unadapted germplasm into elite breeding lines. Theor Appl Genet 92:191–203. https://doi.org/10.1007/BF00223376

Tanksley SD, Grandillo S, Fulton TM, Zamir D, Eshed Y, Petiard V, Lopez J, Beck-Bunn T (1996) Advanced backcross QTL analysis in a cross between an elite processing line of tomato and its wild relative *L. pimpinellifolium*. Theor Appl Genet 92:213–224. https://doi.org/10.1007/BF00223378

Thomas E, Tovar E, Villafañe C, Bocanegra JL, Moreno R (2017) Distribution, genetic diversity and potential spatiotemporal scale of alien gene flow in crop wild relatives of rice (*Oryza* spp.) in Colombia. Rice 10:13. https://doi.org/10.1186/s12284-017-0150-9

Udall JA, Wendel JF (2006) Polyploidy and crop improvement. Crop Sci 46:S-3–S-14. https://doi.org/10.2135/cropsci2006.07.0489tpg

Van Bogaert G (1975) A comparison between colchicine induced tetraploid and diploid cultivars of *Lolium* species. Swiss Federal Research Station for Agronomy, Zurich-Reckenholz

Vavilov NI (1949) The origin, variation, immunity and breeding of cultivated plants. Chronicle Bot 13:1–54

Vigouroux Y, McMullen M, Hittinger CT, Houchins K, Schulz L, Kresovich S, Matsuoka Y, Doebley J (2002) Identifying genes of agronomic importance in maize by screening microsatellites for evidence of selection during domestication. Proc Natl Acad Sci 99:9650–9655. https://doi.org/10.1073/pnas.112324299

Vincent H, Wiersema J, Kell S, Fielder H, Dobbie S, Castañeda-Álvarez NP, Guarino L, Eastwood R, León B, Maxted N (2013) A prioritized crop wild relative inventory to help underpin global food security. Biol Conserv 167:265–275. https://doi.org/10.1016/j.biocon.2013.08.011

Waqas MA, Kaya C, Riaz A, Farooq M, Nawaz I, Wilkes A, Li Y (2019) Potential mechanisms of abiotic stress tolerance in crop plants induced by thiourea. Front Plant Sci 10:1336

Weibull J, Phillips J (2020) Swedish Crop Wild Relatives: towards a national strategy for in situ conservation of CWR. Genet Resour 1:17–23. https://doi.org/10.46265/genresj.2020.1.17-24

Welch A (2021) Micronutrient malnutrition across the life course, sarcopenia and frailty. Proc Nutr Soc 80:279–282. https://doi.org/10.1017/S0029665121001968

Wise MJ (2007) Evolutionary ecology of resistance to herbivory: an investigation of potential genetic constraints in the multiple-herbivore community of *Solanum carolinense*. New Phytol 175:773–784. https://doi.org/10.1111/j.1469-8137.2007.02143.x

Wu S, Clevenger JP, Sun L, Visa S, Kamiya Y, Jikumaru Y, Blakeslee J, van der Knaap E (2015) The control of tomato fruit elongation orchestrated by *SUN*, *OVATE* and *FS8.1* in a wild relative of tomato. Plant Sci 238:95–104. https://doi.org/10.1016/j.plantsci.2015.05.019

Xie X, Jin F, Song M, Suh J, Hwang H, Kim Y, McCouch SR, Ahn S (2008) Fine mapping of a yield-enhancing QTL cluster associated with transgressive variation in an *Oryza sativa* x *O. rufipogon* cross. Theor Appl Genet 116:613–622. https://doi.org/10.1007/s00122-007-0695-x

Yadav SS, Chen W (2007) Chickpea breeding and management. CABI, Oxfordshire

Yadav SS, Kumar J, Turner NC, Berger J, Redden R, McNeil D, Materne M, Knights EJ, Bahl PN (2004) Breeding for improved productivity, multiple resistance and wide adaptation in chickpea (*Cicer arietinum* L.). Plant Genet Resour 2:181–187. https://doi.org/10.1079/PGR200448

Yates HE, Frary A, Doganlar S, Frampton A, Eannetta NT, Uhlig J, Tanksley SD (2004) Comparative fine mapping of fruit quality QTLs on chromosome 4 introgressions derived from two wild tomato species. Euphytica 135:283–296. https://doi.org/10.1023/B:EUPH.0000013314.04488.87

Yu N, Diers BW (2017) Fine mapping of the SCN resistance QTL cqSCN-006 and cqSCN-007 from *Glycine soja* PI 468916. Euphytica 213:54. https://doi.org/10.1007/s10681-016-1791-2

Zamir D (2001) Improving plant breeding with exotic genetic libraries. Nat Rev Genet 2:983–989. https://doi.org/10.1038/35103590

Zeng Y, Yang S, Cui H, Yang X, Xu L, Du J, Pu X, Li Z, Cheng Z, Huang X (2009) QTLs of cold tolerance-related traits at the booting stage for NIL-RILs in rice revealed by SSR. Genes Genomics 31:143–154. https://doi.org/10.1007/BF03191147

Zhang H, Mittal N, Leamy LJ, Barazani O, Song B (2017) Back into the wild—apply untapped genetic diversity of wild relatives for crop improvement. Evol Appl 10:5–24. https://doi.org/10.1111/eva.12434

Zheng JY, Oluoch G, Riaz Khan MK, Wang XX, Cai XY, Zhou ZL, Wang CY, Wang YH, Li XY, Liu F, Wang KB (2016) Mapping QTLs for drought tolerance in an F2:3 population from an inter-specific cross between *Gossypium tomentosum* and *Gossypium hirsutum*. Genet Mol Res 15. https://doi.org/10.4238/gmr.15038477

Zhu D, Cai H, Luo X, Bai X, Deyholos MK, Chen Q, Chen C, Ji W, Zhu Y (2012) Over-expression of a novel JAZ family gene from *Glycine soja*, increases salt and alkali stress tolerance. Biochem Biophys Res Commun 426:273–279. https://doi.org/10.1016/j.bbrc.2012.08.086

Zsögön A, Cermak T, Voytas D, Peres LEP (2017) Genome editing as a tool to achieve the crop ideotype and de novo domestication of wild relatives: case study in tomato. Plant Sci 256:120–130. https://doi.org/10.1016/j.plantsci.2016.12.012

Chapter 12
Conservation of Wild Food Plants and Crop Wild Relatives: Planning, Strategies, Priorities, and Legal Frameworks

Anupama Sajeev, Kiran Vyshnav Eliyan, Anju Thattantavide, Sajana Sreedharan, and Ajay Kumar

Abstract Wild food plants (WFPs) are nutritionally rich and are consumed by the indigenous communities whereas crop wild relatives (CWRs) are the wild relatives of the domesticated crops with huge role in crop improvement. These plants are not only a source of macro and micronutrients, but also carry important traits which can be utilized for crop improvement programs. WFPs and CWRs have higher levels of environmental stress tolerance as compared to their cultivated relatives. CWRs can act as a donor of various abiotic and biotic stress tolerant genes and some of them are superior in micro and macro nutrients. The effective utilization of WFPs for crop improvement can revolutionize crop breeding research. WFPs are locally important as they provide nutritional security to consumers since they can grow in harsh ecological and environmental conditions. For these reasons, the potential of WFPs and CWRs is increasingly realised. At the same time, WFPs and CWRs face various threats due to anthropogenic and environmental stresses. Several studies have reported the loss of CWRs which is not good for the planet. Therefore, loss in their biodiversity should be arrested and their conservation deserves utmost importance. The habitats of WFPs are also endangered due to several anthropogenic and climate associated changes. This chapter mainly focuses on the importance of the protection of WFPs and CWRs at the global, national, local, and regional levels by effective planning, employing suitable strategies, and framing priorities based on multiple criteria. Priority-based conservation strategies and legal frameworks to ensure their protection are also briefly discussed in this chapter. The preference for the conservation of WFPs is based on their socio-economic value and potential uses. Protection is mainly through in-situ and ex-situ methods. Coordination between

Anupama Sajeev, Kiran Vyshnav Eliyan, and Anju Thattantavide contributed equally to this work.

A. Sajeev · K. V. Eliyan · A. Thattantavide · S. Sreedharan
Department of Plant Science, School of Biological Sciences, Central University of Kerala, Kasaragod, Kerala, India

A. Kumar (✉)
Department of Plant Science, Central University of Kerala, Kasaragod, Kerala, India

A. Kumar et al. (eds.), *Wild Food Plants for Zero Hunger and Resilient Agriculture*, Plant Life and Environment Dynamics,
https://doi.org/10.1007/978-981-19-6502-9_12

different in-situ and ex-situ conservation methods, and modifying legal frameworks based on regional requirements with proper utilization of WFPs would be the most sustainable way to meet global food demands.

Keywords Wild food plants · Crop wild relatives · In-situ and ex-situ conservation · Conservation planning · Conservation strategies · Botanic gardens

12.1 Introduction

Wild food plants (WFPs) are plant species that bear immense potential as a significant source of genetic diversity for crop improvement (von Wettberg et al. 2020). Crop wild relatives (CWRs) are the wild counterparts considered as the progenitors of domesticated crops (Ford-Lloyd et al. 2011). The importance of CWRs is well acknowledged for prospecting new genetic sources of variation for domesticated crops (Zhang et al. 2017). The Russian botanist N. I Vavilov was the first person to put forward the prospective use of CWR in plant breeding (Vavilov 1951). Later, Harlan and Wet (1971) conceptualized the importance of CWRs with the aid of their gene pool concept, where CWR can be categorized into primary and secondary gene pools (Choudhary et al. 2017). Gene flow occurs easily between the CWR and domesticated varieties, which increases the significance of CWRs in plant breeding, which denotes the genetic similarity between cultivated crops and CWRs (Maxted et al. 2006). Immense genetic diversity, adaptation to the local environment, and changing climatic conditions further signify their role in crop improvement (Mammadov et al. 2018). Moreover, WFPs are utilized by rural and tribal communities for their nutritional, medicinal, and economic services (Ahmad and Pieroni 2016; Mishra et al. 2021). Thus, the role of WFPs to ensure global food security is notable as they provide food and income for the rural populations while being a genetic resource for crop improvement. It means that wild relatives of the cultivated crops are crucial for the crop improvement programs. Some of the WFPs eaten by the indigenous communities are also wild relatives of some of the crops suggesting an overlap between the CWRs and WFPs. Therefore, strategies that aims to conserve the CWRs are equally applicable to the WFPs.

Modern food system practices that are based on the cultivation of a few selected crops, increase the vulnerability of monocultures to diseases, pest attacks and various abiotic stresses (Altieri 2019). As a measure to improve crop productivity, the availability of CWRs is highly crucial, considering the need for the identification and sourcing of genes related to nutritional traits, stress, and disease tolerance (Hanson 1952). For example, wild relatives of wheat such as emmer wheat are nutritionally rich and stress-tolerant (El Haddad et al. 2021). *Physalis*, a wild relative of tomato is highly nutritional and it has been widely used as a WFP in various parts of the world especially in South America (Zamora-Tavares et al. 2015). The plant may be used as a source of important genes that can be transformed into tomatoes or other related crops. WFP such as *Eleusine africana* Kenn.-O'Byrne is a CWR of finger millet and it is a endowed with some good traits that can be utilized for the improvement of cultivated finger millet (Dida and Devos 2006). The biotic stress tolerance of the CWRs can be effectively explored to improve pathogen-resistance in the cultivated crops. For example, *Brassica insularis* Moris, a wild variety native to the Mediterranean region, is tolerant to the fungus *Leptosphaeria maculans* (El Mokni et al. 2022). CWR of wheat, *Triticum turgidum* L. is notable for its resistance to fungal pathogens (Gnanesh et al. 2014). There are many more such examples of WFPs and CWRs that are superior and carry important traits crucial for their role in crop improvements. In fact, some of the CWRs are consumed by the indigenous communities as WFPs by collecting from the wild or through limited cultivation in their vicinity. There are many examples of WFPs which are consumed across the globe and have high nutritional value. However, recent studies have shown decline in the consumption of the WFPs by the current generation. This decline in the consumption of the WFPs has been attributed to several factors such as unavailability of the wild plants, change of life styles, and lack of knowledge of their value (Pawera et al. 2020). The marketing of commercially profitable crops on a large scale is also the major reason behind the negligence of WFPs. Several WFPs face conservation issues due to negligence and lack of documentation. It is important to conserve the WFPs and CWRs considering their potential use as a genetic resources. Moreover, climate change, along with issues such as habitat loss, industrialization, anthropogenic threats, and biopiracy, pose serious survival threats to the plants (Siddiqui and Shukla 2015). The present chapter briefly discusses about planning, strategies, and prioritization of important activities for the conservation of WFPs and CWRs.

12.2 Relevance of Crop Wild Relatives and Wild Food Plants

The CWRs are one of the best sources of plant genetic diversity. They represent an important resource for improvement of agricultural production around the world (Brozynska et al. 2016). CWRs are an important source of genetic diversity that can be harnessed for plant breeding and helps to improve agricultural production around the world (Bohra et al. 2022). They grow in a variety of habitats including harsh environmental conditions and are hence important for plant breeding initiatives (Crop Wild Relatives Global Portal 2021). CWRs are rich in genetic diversity as compared to their domesticated counterparts (Zhang et al. 2017). Our current food systems are highly homogenized. Diversification of the food crops is crucial to feed the burgeoning population, ensure their nutritional security and curb hunger (Sarkar et al. 2019). The current conditions demand the diversification of food systems according to the consumer perception and productivity. To fulfill the human calorie requirements, more stress-tolerant and nutritionally superior crops should be developed with CWRs (Brozynska et al. 2016). For example, consider the case of tomato plants, its wild relatives showed superiority in various attributes such as sugar content, yield and a higher proportion of soluble solids (Robertson and Labate 2006). The majority of the WFPs and CWRs are distributed in wild habitats, indigenous communities or forest dwellers depend on these plants as a major source of their diet (Borelli et al. 2020). It is also a boon to the non-indigenous population as an income generator and acts as a means of supplementary feeding to these people (Nyakoojo and Tugume 2020). Customary and traditional knowledge of the local communities about the WFPs pertaining to their use as food signifies the essential nutritional aspects and the feeding diversity of wild relatives in ensuring local food security (Ahmad and Pieroni 2016). Domesticated varieties demand the most optimum conditions in the context of abiotic and biotic factors and many of them have turned susceptible to various stresses which significantly affect their agronomic performance (Edmeades et al. 2001).

In contrast, WFPs can sustain adverse environmental conditions to a greater extent and thus ensure strong support to the global food systems in times of unfavorable conditions (Ladio and Lozada 2004). The presence of a high quantity of micronutrients suggests their potential for subsequent utilization in development of improved agronomic crop varieties (Sánchez-Mata et al. 2016). WFPs have gained paramount importance in providing a balanced diet to a seemingly increased health-concerned population and thus ensure healthy means of food consumption (Ju et al. 2013). Homogenized food systems and extensive competition in the global market create a greater probability for the underprivileged people of the society to face malnutrition and hunger (Schanbacher 2010). These traditional WFPs offer boundless possibilities in relieving the pressure of inequitable food access to several indigenous communities of West Sumatra, Indonesia (Pawera et al. 2020). WFPs with richness in bioactive compounds and micronutrients can sustain a larger vulnerable population by strengthening their bodily requirements and improving their immunity to combat various ailments (El-Ramady et al. 2022). A study by Mutie et al. (2020) in the drylands of Kitui county of Kenya shows that WFPs complement the food system of the population to thrive during unpredicted climatic alterations. The rural people of Kitui, market the products from WFPs such as gums and resins after processing them properly by applying their traditional knowledge (Mutie et al. 2020). The status of WFPs as an ecosystem service corresponding to their functional importance in the anthropological context was elucidated by Schulp et al. (2014). They provided information regarding spatial distribution, and interconnections between benefits, demand and supply of WFPs widely utilized in the European Union. The study reported that a large number of European citizens consume WFPs regularly.

The relevance of CWRs was better elucidated by Bohra et al. (2022), and stated that emerging nations should utilize their resources for crop improvement. The gene pool of CWRs is rich in certain disease resistance genes and abiotic stress tolerant genes, which can be utilized for improvement of the climatic change susceptible staple crops (Porch et al. 2013). CWRs have been considered paramount to crop improvement for humankind since the green revolution. A noteworthy example is the pivotal role of introgression breeding in conferring resistance to *Puccinia graminis* from the *Aegilops tauschii*, a type of wild wheat, to curb the stem rust and advancement in cereal production (Bohra et al. 2022). CWRs of industrially valuable crops utilized for fuel and aesthetic beauty also find their place in the need to be conserved for a sustainable healthy economy (Maxted et al. 2010). Judicious manipulation of the CWRs in plant breeding approaches directs more diversity in the genetic level of the cultivated crop varieties. Ziska (2021) reviewed the selection efforts by nature and breeders that are better adapted to increased CO_2 level in the atmosphere. They reported CWRs had better adaptation to increasing CO_2 conditions than the domesticated counterparts. In this context, attempts to identify the genetic loci conferring the adaptable traits were widely studied to produce elite cultivated varieties. In the study done by El Haddad et al. (2021), an evaluation of the performance of three critical dryland crops (durum wheat, barley, and lentil) developed by incorporating CWR in their pedigrees in the context of yield potential, yield stability across different environments were conducted. The results showed that the accessions derived from CWRs had better yields across all the varied conditions. For example, three accessions of wheat developed from *Triticum araraticum* and *T. urartu* were enriched with Zn content and barley accessions produced from *Hordeum spontaneum* were rich in proteins. The CWR-derived elites even had better food performances, further substantiating the utility of these wild alleles in crop improvement.

WFPs are important parts of local markets and are very important for associated communities, especially the women and young population. This remains valid in the current scenario of COVID-19 fractured world. As the COVID-19 disrupted the transport and cargo and also led to widespread disruptions in the food supply chains that involved long distance travels, people started relying on the local foods including the WFPs as they are locally obtained and involve short supply chains which are least affected due to COVID-19 (Borelli et al. 2020). This suggests the role and relevance of the WFPs in the future if we face a situation similar to COVID-19 or other natural disasters or wars. The relevance of CWRs in crop breeding and providing food security is required to be explored more and benefits of advanced molecular techniques such as genetic modification and gene editing should be reaped for improving the crops. Through the successful establishment of gene editing tools, rapid-domestication of CWRs can be achieved according to the breeders designs (Lemmon et al. 2018; Zhu and Zhu 2021; Yu et al. 2021). Figure. 12.1 represents various important aspects of these wild plants (CWRs and WFPs) that are beneficial for in plant breeding.

12.3 Status of Crop Wild Relatives and Wild Food Plants at the Global, National, Regional and Local Levels

For thousands of years, humans have relied on wild plants for their diets, and several people still rely on these species to satisfy their nutritional requirements (Turner et al. 2011). Wild edible plants are components of the cultural and genetic heritage of

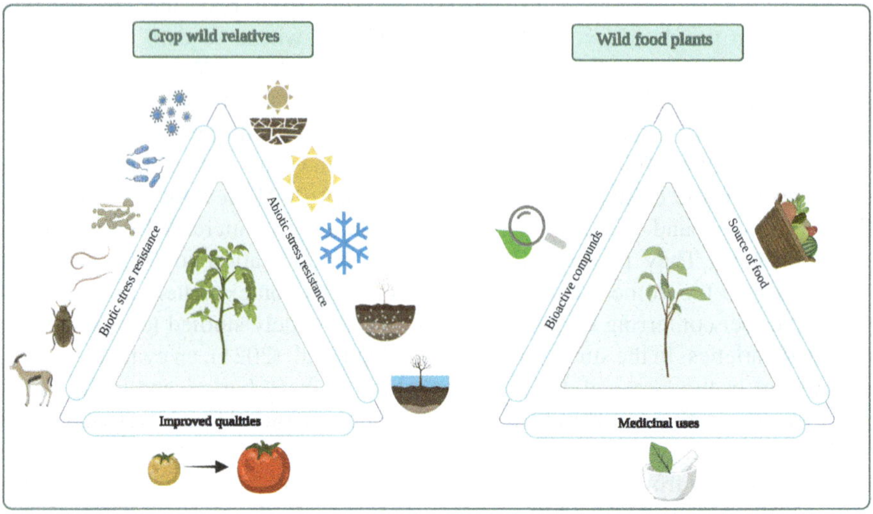

Fig. 12.1 Importance of CWRs and WFPs (Created with BioRender.com). CWRs possess various abiotic and biotic stress tolerant traits and the crop breeding methods can be applied for their improvement. WFPs are rich in bioactive compounds and hence they are used for various medicinal purposes besides their food value

different parts of the world and have aided rural and sub-urban communities during severe famine and scarcity (Pinela et al. 2017). An increase in the interest in WFPs is observed among modern societies that resulted in the accumulation of extensive ethnobotanical knowledge (Tardío et al. 2006). Research has shown decline in the wild species diversity due to several reasons such as socio-economic changes, industrialization of the diets, expansion of food markets, changes in land usage patterns and unsustainable harvesting practices (Bharucha and Pretty 2010). Kidane and Kejela (2021) have attributed this decline to agricultural expansion, poor management, overgrazing, and loss of indigenous knowledge. Only around 150 of the estimated 30,000 edible plant species are cultivated by humans, with only 30 species accounting for the majority of our dietary requirements (Shelef et al. 2017). Despite the fact that we produce optimum food to feed a significant proportion of the human population, nutrient-rich food is still out of reach for more than 1.5 billion people, and the food system needs to be transformed to ensure nutritional security for consumers (FAO 2020a). The relevance of agro-biodiversity and WFPs in risk management and establishment of resilient and sustainable food systems is increasingly recognized worldwide (Kahane et al. 2013; Boedecker et al. 2014). As a result, more research should be encouraged on documentation of WFP diversity, uses, and conservation strategies, as well as re-establishment of traditional knowledge for the sustainable usage of WFPs in a rapidly changing world. CWRs on the other hand, aided in crop domestication for thousands of years (Perrino and Perrino 2020) and are widely used in breeding experiments and crop improvement programs for enhancement of plant performance (Zhang et al. 2017; Bohra et al. 2022). The in-situ and ex-situ strategies are critical for the long-term conservation of forest genetic resources in the context of changing societal requirements and climate change (FAO 2020b). CWR conservation and utilization are becoming increasingly popular in international food security policy and research (Smýkal et al. 2017). However, there is a lack of coordination in the crop research community to convey developments, best practices, constraints, and opportunities for adoption of CWRs. Attempts aimed at systematically assessing the utility and diversity of CWR species are infrequent (Dempewolf et al. 2017). To avoid the extinction of CWRs and WFPs diversity and to optimize their availability, particularly for crop development, proper conservation, and sustainable usage must be ensured at the global, regional, national, and local levels.

Assessment of the existing state of CWRs and WFPs conservation and utilization can help uncover gaps in their conservation and aid in the planning priorities. In many situations, such an evaluation is already available, either as part of a country report, research program, or as a separate study (FAO 2017). An attempt to assess genetic resources of crop plants, livestock and forest trees on a global level was initiated by the Food and Agricultural Organization (FAO) of the United States through the program "The State of the World's Biodiversity for Food and Agriculture" and WFPs were also included in this report (FAO 2019). A total of 91 countries submitted a report and among them, only 69 countries provided a detailed report. A total of 1995 WFPs were recognized combining all the reports besides other groups of wild species of mammals, birds, fungi, insects, fish, crustacea, molluscs, reptiles and amphibians (FAO 2019). Ulian et al. (2020), reviewed the global status of edible plants by a combined data from major projects and research with a focus on neglected and underutilized species (NUS), dividing them into three categories; wild, cultivated, and wild and cultivated species as well as mentioned edible parts, uses and distribution of NUS. Regarding CWRs, Milla (2020) undertook a global assessment of food crops, examining 866 food crop species and 901 wild progenitors, as well as their cultivation and domestication histories, geographical distribution, and a time-calibrated genetic phylogeny of food crops. Castañeda-Álvarez et al. (2016) utilized gene bank databases, herbariums, and biodiversity data to simulate the global distribution of 1076 taxa connected to 81 crops. They discovered that CWRs are underrepresented on gene banks, with no germplasm accessions for 313 (29.1% of total) taxa connected with 63 crops, and a further 257 (23.9%) taxa represented by less than ten accessions. Overall, global trends in the abundance of WFPs and CWR were reported to be decreasing (FAO 2019). Botanic gardens also play an important role in plant genetic resource conservation (Kumar 2021). Meyer and Barton (2019) outlined the reserves of CWRs in botanic gardens, they reported 28.6% of global priority CWR taxa and 75.4% of US priority CWR taxa in botanic gardens. Botanic gardens maintain 22 global priority and 108 US priority CWR taxa not reported by crop gene banks, based on a comparison with priority CWR holdings in crop gene banks. Several ethnobotanical studies have reported the usage and diversity of WFPs from different parts of the world (Al-Fatimi 2021; Guarrera and Savo 2016; Baldi et al. 2022; de Medeiros et al.

2021; Xu et al. 2020; Winstead and Jacobson 2022; Ding et al. 2021; Menendez-Baceta et al. 2017).

At the national level, a wide range of collaborators are involved in the conservation and sustainable use of CWRs and WFPs. In many cases, national governments have introduced National Plant Genetic Resources for Food and Agriculture (PGRFA) programs to drive efforts towards global PGRFA protection and sustainable usage (FAO 2017). Many nations report a lack of efficient information-sharing and collaboration systems among stakeholders, notably between those working in the food and agricultural sector and those working on environmental and wildlife matters (Pilling et al. 2020). While most nations have routinely established threatened species conservation strategies for the past three decades, the establishment of conservation efforts for CWRs at the national level is still in its budding stages (Iriondo et al. 2016). The same goes with WFPs, though the number of publications on the relevance of WFPs is rising, mostly at the local level, there is a lack of data and information at the national level, and conservation assessments are still inadequate; due to this, the contribution of WFPs to the national economy and their overall importance is under-evaluated (Borelli et al. 2020). It is very concerning that biodiversity for food and agriculture, as well as the ecosystem services it provides, is reportedly declining in so many production systems in several nations (FAO 2019). WFPs provide food and financial security at the local level and are a source of traditional medicines too (Uprety et al. 2012; Hickey et al. 2016; Asprilla-Perea and Díaz-Puente 2019; Ray et al. 2020; Punchay et al. 2020). For local communities, WFPs remain a preferred alternative to commercial food crops because they are easily accessible and economically feasible (Delang 2006a). Many ethnobotanical researches and surveys are being conducted at regional and local levels to better understand how human societies collect, manage, and interpret the local plants which they utilize as food and medicine (Pardo-de-Santayana and Macía 2015). Most such recent studies reveal a gradual decrease in traditional knowledge about wild plant usage and management among local communities (Singh and Bedi 2017; Luo et al. 2019; Yeşil et al. 2019; Thorn et al. 2020; Punchay et al. 2020; Aziz et al. 2021; Pascual-Mendoza et al. 2021; Ghanimi et al. 2022). According to a review by Schunko et al. (2022) regarding the change in perceptions of WFPs and mushrooms among local people, 92% of all reported changes in wild edibles are related to their decreased abundance, and fruits and vegetables account for 76% of the wild edibles with decreasing abundance, whereas CWRs account for 23%.

12.4 Threats to Wild Food Plants and Crop Wild Relatives

The existing agrobiodiversity is threatened mainly by the combined effects of anthropogenic activities and natural calamities (Chaudhary et al. 2020). The rise in human population is contributing significantly to the devastation of agrobiodiversity (Khumalo et al. 2012; Montenegro de Wit 2016) in different ways, such as uncontrolled and unsustainable practices of natural resources exploitation, urbanization, (Ebert and angels 2020) introduction of exotic species, and forest land usage for expanding agricultural operations (FAO 2017; Hunter 2012). Apart from these human interventions, unavoidable climate change, desertification, the spread of invasive species and natural disasters also contribute to the deterioration of the diversity of WFPs and CWRs (Fatima et al. 2020; Norton et al. 2017; FAO 2017; Gupta et al. 2020). The extinction risk analysis of wild varieties of some vital crop species in Mesoamerica, known for its agricultural significance as the center of crop origins and highly diverse CWRs, revealed that among 224 CWRs analyzed, 35% of them are under the risk of extinction (Goettsch et al. 2021). Land conversion for agriculture, as well as current agricultural techniques such as herbicide use and urban expansion, were some major threats to CWR species in Mesoamerica (Goettsch et al. 2021). Climate change causes biotic and abiotic stresses to WFPs through varying rainfall and temperature patterns, heat waves, weeds and pest outbreaks, and changes in sea level and atmospheric CO_2 concentrations (Raza et al. 2019). According to van Treuren et al. (2020), climate change has a significant influence on the distribution of red-listed CWRs in the Netherlands. Their findings indicate that the distribution areas of CWRs that are categorized as being critically endangered have been declining owing to changing climatic conditions. A similar study by Vincent et al. (2019) also points out the impact of climate change on CWRs distribution. They analyzed the number of CWRs under different crop types expected to lose 50% or more of their current ranges by 2070 due to climate change and found that the root, bulb, or tuberous vegetable CWRs are facing distribution loss, followed by cereals and leguminous CWRs. Some CWRs such as *Zea perennis* and *Vicia hyaeniscyamus* are expected to lose 100% of their distribution area by 2070. Jarvis et al. (2008) studied the influence of climate change on wild relatives of three crop plants, potato (*Solanum tuberosum*), peanut (*Arachis hypogaea*), and cowpea (*Vigna unguiculata*), and decline and fragmentation of distribution of area were observed together with increased extinction risk of some species. South African WFPs distribution patterns tend to vary under uncertain temperatures and the geographic range of certain traditional WFPs is changing due to rising temperature (Wessels et al. 2021). The nonavailability of optimal climatic conditions reduces the WFP species richness (Wessels et al. 2021), conversely the increasing temperature is predicted to influence some species positively by providing optimal growth conditions and lengthened growing season (Phillips et al. 2017).

The Green Revolution of the 1960s and 1970s popularized modern cultivars, which were widely adopted by farmers. It quickly displaced the genetically diverse traditional and wild landraces that had existed for centuries, making traditional diversity only to be found in gene banks or other conservation initiatives (Ford-Lloyd et al. 2011). As a result of plant breeding techniques, the agricultural diversity was decreased through the introduction of mono-cropping system with genetically uniform cultivars (Dwivedi et al. 2016). The plant breeding studies should utilise the potential of CWR diversity, but CWRs have been underrepresented in gene bank-like conservation programs (Castañeda-Álvarez et al. 2016). Besides that, the genetic pollution or the gene flow from exotic and invasive species along with genetically modified and conventional crops causes a significant threat to agrobiodiversity (Gepts and Papa 2003) and its in-situ conservation efforts. Traditionally, protected areas have not been designed by considering CWRs, and legislation is needed to encourage CWRs and WFPs protection in the same way that rare breeds of domestic animals are safeguarded (Bettencourt et al. 2007). Genetic erosion is the common threat to WFPs diversity in biotically disturbed ecosystems (Fu 2017), and conservation efforts in combination with information gathering about the plant features are vital for maintaining diversity (Pandey et al. 2005).

Uncontrolled use of wild resources and modification of the wild landscape can drastically reduce agrobiodiversity (Upreti and Upreti 2002). For example, due to human population pressure and agricultural expansion, frequent deforestation reduces the forest land area. In the case of Northwest Ethiopia, wildfires, fuelwood collection, overgrazing, and overharvesting were reported as a threat to WFP diversity (Berihun and Molla 2017). A similar study suggests that introduction of exotic species such as *Eucalyptus* and agricultural land expansion forced the WFPs to be grown in farmlands, farm boundaries, and watershed areas instead of their natural habitat (Kidane and Kejela 2021). Even in protected regions, the lack of strict legislation for the conservation of wild flora leads to unrestricted use. However, even when rules are very strictly enforced, the illegal harvesting of traded medicinal plants and timber is obeserved in from different parts (Khakurel et al. 2021). This shows that rules are not always effective but a community consciousness and awareness about the loss of biodiversity among the locals is also crucial for conserving the WFP biodiversity.

The traditional knowledge about WFPs is also under threat, and the younger generation abandons it. Traditional knowledge is passed down from generation to next generation (Yuan et al. 2014). The risk of losing traditional knowledge is increasing among the younger generation due to the death of older adults, increased deforestation of natural forests, monoculture of economic plants, reduction in the availability of WFPs (Cao et al. 2020), and socio-economic changes (Ju et al. 2013). A study by Reyes-García et al. (2013) on changes in traditional knowledge about WFPs among Amazonian indigenous people found that the usage of wild plant and knowledge regarding them have decreased. The loss of traditional knowledge can undoubtedly result in biodiversity loss, particularly WFP species. Therefore, preservation of local traditional knowledge is essential for safeguarding agrobiodiversity (Luo et al. 2019).

12.5 Conservation Priorities and Strategies

12.5.1 Conservation Priorities

CWRs have always been at the forefront in the improvment of the agronomic traits of the modern varieties, enabling resistance to adapt and survive to a multitude of stresses in their natural environment. Therefore, conservation of the CWRs is a prominent concern in the context of its importance in various aspects of cultivated plant species (Heywood et al. 2007). In an anthropological context, CWRs serve as an important resources to strengthen global food security. Medicinal, food, and ornamental value are some of the significant prioritization criteria (Ciancaleoni et al. 2021). The evaluation of habitat, climate and associated threat, anthropological impacts on the CWRs is essenial to outline the priorities for conservation (Castañeda-Álvarez et al. 2016). Besides that, the extent of utilisation of the gene pool of the CWR in plant breeding, socio-economic status of the cultivated crop, degree of threat exposed and endemicity of the wild relative are also considered (Idohou et al. 2013). The taxonomic and geographical information related to CWRs can help to refine the priority strategies (Castañeda-Álvarez et al. 2016). The geographic distribution of CWRs is an essential criterion in the formulation of conservation priorities as some CWR plants are native to two or more countries or to just one country which calls for better coordination among nations (Lala et al. 2018). By understanding the importance of CWRs in global food supply chain, production network and food security, the CWRs of some important crops are prioritized for conservation at global, national, regional and local level (Rahman et al. 2019). For example in Indonesia, out of 234 prioritized CWRs, 99 were crucial at global and national level such as the CWRs of banana, rice, mango, sorghum, sweet potato, citrus and coconut. 69 CWRs were important at national and regional level such as sugarcane and tropical fruits and 70 taxa including fig and yams were important only at global level (Rahman et al. 2019). The prioritization strategies at regional level are more focused on the production value, location, energy supply, nutrient content and threats faced by the plant (Zair et al. 2018).

The development of prioritization strategies for conservation helps to narrow down the checklist of CWR to be conserved and broadens the conservation activities which results in the effective conservation (Labokas et al. 2018). Maxted et al. (2007) discussed the role of national inventories in generation of conservation action plans of CWRs in United Kingdom. National CWR inventory serves an important function by describing the significant CWR crops for conservation by analysing their threat status, diversity, present conservation status, recognition of conservation sites, pattern of distribution and development of action plans (Maxted et al. 2007). They suggested that, the development of National CWR inventory, can easily prioritize the CWR taxa for conservation, which is easy to identify their threats through the genetic and ecogeographical investigation. The recognition of threats for the CWR can identify the gaps in conservation action plans and thereby effective in-situ and ex-situ conservation programs can be established (Maxted et al. 2007). The process of inclusion of CWR members into the prioritization list follows a highly selective methodology where the major focus is on the endemicity of the wild variety, and its socioeconomic importance to the country's economic status (Rahman et al. 2019).

Prioritization should be aimed at wild relatives with restricted distribution rather than CWR population with wider global existence (Perrino and Wagensommer 2021). According to Brehm et al. (2010), there is no stringent method for setting the criteria for prioritization; it largely depends on the aim of our conservation and the available information. Brehm et al. (2010) formulated nine prioritization criteria: habitat status, economic worth, ethnobotanical significance, worldwide distribution, country-wise distribution, ex-situ conservation status, in-situ conservation status, legal norms, and threat status and they used this criteria on the Portuguese CWRs. Ciancaleoni et al. (2021) suggested that prioritization of CWRs for their role in the functioning of the ecosystem is noted but focus on the species level is more worthy. Prioritization also considers the phylogenetic distance between the CWR and the cultivated variety (Viruel et al. 2021). The closer the links between CWRs and the related crops, the more ease with which gene donation from the wild relatives can occur (Maxted et al. 2010). Consideration of all the priority criteria is important to find appropriate in-situ and ex-situ conservation modes (Brehm et al. 2010). While prioritising conservation strategies, aspects such as threat status, closeness to the domesticated crop, traits of interest, distribution, status of conservation, availability and legal frameworks are important (Engels and Thormann 2020). The establishment of a priority list of CWRs at different geographic locations following different criteria serves as an index to aid in constructing sustainable policies for the preservation of high-priority CWRs (Castañeda-Álvarez et al. 2016). Categorization of taxa based on a final priority score (FPS) ranging from zero to ten, sets a base for collecting the wild relatives and increasing their representation in the gene banks (Castañeda-Álvarez et al. 2016). Khoury et al. (2013) prioritised 21 taxa from 69 genera of USA which are related to major food crops. Within the major categories formulated as part of the criteria, there are subcriteria to accommodate all the variants of CWRs with proper positions signifying their value. IUCN Red List Categories and Criteria represent the baseline for grouping the CWRs including Extinct in the Wild (EW), Critically Endangered (CR), Vulnerable (VU), Near Threatened (NT), Least Concern (LC), Data Deficient (DE) and Not Evaluated (NE) (IUCN 2020). The commercialization potential and the current and future trade potential of the cultivated crops also serve as a baseline under economic status to prioritize the conservation of its CWR (Perrino and Perrino 2020). In this manner, prioritization baselines and strategies differ with regional, and international boundaries, where the value is based on the yield of CWR-related crops and the production quantity of the related food crops. A high priority is given for the CWRs that face threats due to the alteration of natural habitats within prescribed boundaries (Fitzgerald et al. 2013). It is also mandatory that assigning priority to conserve any taxon should necessarily correlate to its existing conservation initiatives (Brehm et al. 2010). After the establishment of conservation strategies, if the conserved CWR attains a significant population, they can be readily excluded from the current conservation priority list, and other CWRs can be considered for conservation (Nduche et al. 2021). In South Africa, Holness et al. (2019) reported that the identification and conservation of CWR endemics with narrower distribution were given priority. Those with broader distribution status were not targeted for conservation policies since they face a lesser rate of endangerment. As depicted in the study by Mponya et al. (2020), climate change adaptation is also a feature considered as a prioritization criterion. Maxted et al. (2010) provides an important 14 point strategy for the improvement of conservation of CWRs for the next 10 years. Some of these points include analysis of gaps, bioclimatic modelling, establishment of genetic reserves, monitoring and popularisation of the conserved species, development of

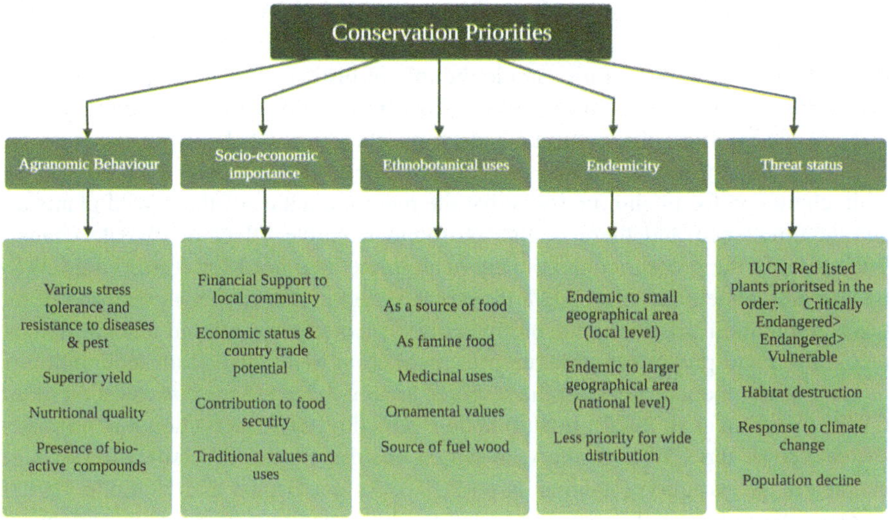

Fig. 12.2 Priorities for the conservation of WFPs and CWRs (Created with BioRender.com; see Sect. 12.5.1)

core collections and characterisation and evaluation of the CWRs for the beneficial traits (Maxted et al. 2010). But the utilisation of CWRs for breeding programs, consumption by consumers and popularity across the globe can provide actual results for the conservation policies. Ethnobotanical importance of the CWRs is also important for their conservation (Ciancaleoni et al. 2021). Ensuring the protection of the WFPs and CWRs should be prioritized for their sustainability and to strengthen the global food security (El Mokni et al. 2022). Figure 12.2 provides an outline of the priorities that act as a guide for the conservation of the CWRs and WFPs.

12.5.2 Conservation Strategies

For effecive conservation of the CWRs, estimation of the number of the CWR species is crucial (Maxted et al. 2012). Both ex-situ and in-situ conservation methods are important for the conservation of the CWRs, however most of the earlier conservation programs were focused on ex-situ conservation, which are found to be less effective (Maxted et al. 2012). Although ex-situ conservation methods were taken in the earlier instances, it is associated with several issues such as difficulty in collection, maintainence and propagation and costs associated with it (Meilleur and Hodgkin 2004). These issues including the research on natural introgressions between the crops and the CWRs led to increase in the interest in in-situ conservation (Meilleur and Hodgkin 2004). Meilleur and Hodgkin (2004) provides an extensive list of the examples of in-situ conserved CWRs from various countries. While focussing on in-situ conservation, it is also important to consider the other factors that influence the CWRs, for example the role of birds for the wild species of Chili (Tewksbury et al. 1999). Since CWRs grow in the wild, they carry adaptive traits that are absent from the domesticated crops, therefore CWRs from the wild are crucial for the introgressions and crop improvement (Warschefsky et al. 2014). The main challenge for in-situ modes is that the natural habitats of these wild plants are getting threatened and there is also profound negative effect of climate change. Therefore, in the present context, there is an immediate need for conserving CWRs by alternative methods as well (Meilleur and Hodgkin 2004). A vast majority of the CWR population existing in the biodiversity-protected areas remain unidentified, so recognition and proper representation of them along with significant moderation of in-situ management protocol of the reserved areas can lead to the conservation of the CWRs (Vincent et al. 2022). In-situ mode is a method of conservation that promotes sustenance of the CWRs in the wild which enables further diversification of wild characters by evolution (Rahman et al. 2019). As per Brown and Hodgkin (2015), incorporation of these diverse characters into the agriculture stream is important. As described by Heywood et al. (2007), various countries adopt diverse CWR prioritization factors with major emphasis on geographic surroundings for the adoption of a suitable mode of conservation method.

The ex-situ mode proceeds with efforts such as the collection of seeds of CWRs where educational societies and government collaborates are involved in the selection of seeds of significant CWRs (Zair et al. 2021). The role of botanical gardens in CWR conservation pertains to the protection of wild relatives of the staple crops. The botanical gardens also help in the dissemination of the information regarding CWRs to the general public which may trigger further conservation strategies (Kumar 2021). Seed banks store seeds of crops and wild relatives used for a variety of purposes and act as an important ex-situ conservation method, the seed banks act as important repositories of the crops (Hay and Probert 2013). However, seed banks depend on several important factors such as the seed germination, maturity and development (Hay and Probert 2013). It is important to note that conservation strategies are devised regionally and depend upon several factors including the availability of CWR diversity and threats associated with it.

A check list of the CWRs of Czech Republic was prepared and a conservation strategy was devised for their conservation (Taylor et al. 2017). Fielder et al. (2015) inventorised 148 priority CWRs of England and suggested in-situ and ex-situ conservation plants.

Cryopreservation can be used to conserve the seeds of CWR and WFPs which failed in ex-situ seed banks and this technique is particularly important for endangered species of CWRs (Pence et al. 2017). Plant tissue culture (PTC) has diverse applications including the storage of germplasm (Thorpe 2007). In-vitro culture is globally accepted as an efficient method for the germplasm storage and propagation, and for plants where seeds are not viable or seeds are not available, it is possible to apply in-vitro conservation techniques (Rajasekharan and Sahijram 2015). The PTC is widely applied as a tool for the conservation of endangered plants (Oseni et al.

2018). The major benefit of tissue culture is that it needs only a small space and a few plants for the initial culture (Oseni et al. 2018). Lauzer et al. (1992) developed clones of wild yams viz. *Dioscorea abyssinica* Hoch, and *D. mangenotiana* Miège through their nodal culture. Singh et al. (2019) developed an efficient protocol for the propagation of wild relative of chickpea (*Cicer arietinum* L.) viz. *Cicer microphyllum* Benth. from excised embryo. Hence, it is clear that in-vitro propagation can be successfully applied for the conservation of germplasm of CWRs and WFPs for long term and short term.

Conservation strategies including both ex-situ and in-situ modes should be analyzed so that the most efficient means of conservation can be targeted (Padulosi et al. 2011). An integrative strategy by utilising both modes of conservation can be applied to balance the difficulties and limitations of ex-situ and in-situ conservation (Meilleur and Hodgkin 2004). A major policy undertaken in Europe for the conservation of the CWR populations is to identify hotspots and their inclusion into Important CWR areas, which call for the immediate need for conserving this wild wealth (Maxted et al. 2015). Conservation of these valuable genetic resources from the wild is important. Protection of these plants ensure their sustainable presence even in the future. The involvement of various local communities in the conservation of the CWRs and WFPs is important to ensure their sustainable use (FAO 2017). International Treaty on Plant Genetic Resources for Food and Agriculture (ITPGRFA) mandates that CWRs should be made publically available, which is important for the conservation of and research on the CWRs (Tyack et al. 2020). Since CWR are rich in nutrient composition, by focusing on this aspect, its utility as food and fodder, the farmer community and the animal husbandry sector would promote its cultivation and thereby conservation (Perrino and Wagensommer 2021).

Assessing and addressing anthropogenic harm inflicted upon the CWRs is a major strategy to reduce unsustainable utilization and thereby ensure the proper management of CWR resources (Engels and Thormann 2020). Those communities staying near the forests collect the WFPs from the forest and rely on their existence for livelihoods and other aspects, therefore, these communities promote the conservation of the wild species as their surivival is linked to the availability of the WFPs (Delang 2006b). Therefore, there is an intimate relationship between the communities and the WFP resources and the communities take interest in the usage, management and conservation of the WFPs (Evans 1993). This strategy should also be therefore combined with the conservation of WFP resources and CWRs. Formulation and implementation of appropriate conservation priorities and strategies at different levels is a vital step in establishing a proper CWR conservation frame (Rahman et al. 2019). Collaborative efforts between developing and developed nations is also important for the future conservation of the CWRs and WFPs. Considering all this, it is clear that regional, national and global strategies are crucial for the conservation of the wild genetic resources of a sustainable present and future, it is also important to identify the crucial CWRs and priorities should be devised to conserve them (Ford-Lloyd et al. 2011). On the basis of this small discussion, it can be stated that strategies for the conservation of the WFPs and CWRs depend upon their use value, locality, availability, threat and conservation status. It also depends

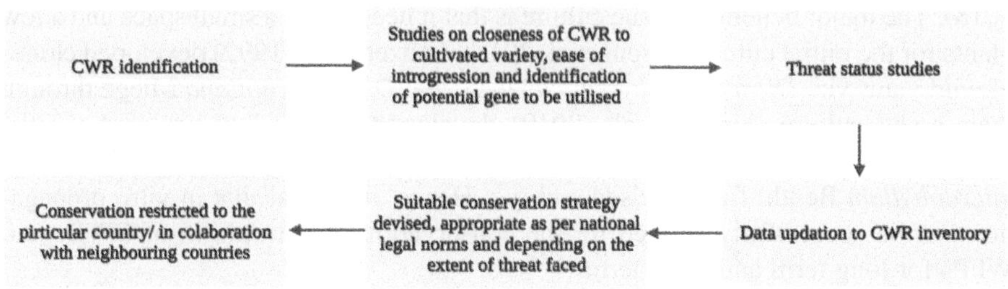

Fig. 12.3 An example of an appropriate and effective strategy for the conservation of WFPs and CWRs (Created with BioRender.com; see Sect. 12.5.2)

on the geographical location of the CWRs or WFPs and more importantly the local and national policies of a respective country are also crucial for the determination of the conservation strategy. Identification and inventorisation of the CWRs diversity and their hotspots are also very important to begin with any conservation strategy. Figure 12.3 provides an outline for the formulation of a conservation strategy for the conservation of the CWRs and WFPs.

12.6 Legal Frameworks for the Protection of CWRs and WFPs

Since CWRs and WFPs are collected from the wild in different countries, their conservation falls within the jurisdiction of a particular country (Montenegro de Wit 2017). The CWRs and WFPs from different countries are also different and plants may need different strategies for protection, therefore the legal frameworks of each country should be different. In the African countries, they focus on the multi-use regulatory framework by including access to the local people to utilize the WFPs for traditional uses, thereby ensuring coordination among the local population and the public institutions (Cunningham 2014). In 2010, Convention on Biological Diversity (CBD) formulated and presented a plan consisting of 20 targets known as the Aichi Biodiversity Targets which are spread across 5 strategic goals (CBD 2020; Garcia and Rice 2020). Target 13 emphasized the sustenance and conservation of the cultivated plants and animals including their wild relatives by 2020 (CBD 2020). It further aimed to ensure appropriate measures to protect the diverse gene pool of the wild relatives to arrest their genetic erosion (CBD 2020). With an aim to arrest the loss of plant biodiversity, CBD also established a Global Strategy for Plant Conservation (GSPC) program comprising of 16 targets (CBD 2011). The GPSC envisioned a symbiotic relationship between the plant genetic diversity and the humans with a major focus on the sustainable use of the plant resources for livelihoods and wellbeing while protecting the plant diversity including their habitats (CBD 2011). Target 9 of GPSC aimed to conserve 70% of crop diversity including their wild relatives with due respect towards indigenous knowledge and its protection. The International Treaty on Plant Genetic Resources for Food and Agriculture in Article 5 details about conservation, exploration, collection and evaluation of the plant genetic resources culminating in the maintenance and documentation of proper records of the plant genetic wealth (FAO 2001). Moreover, the treaty has a functional funding policy to extend support to developing countries to activate the efforts that fall under their idealized conservation protocol (Esquinas-Alcázar 2005). The Second Global Plan of Action for Plant Genetic Resources for Food and Agriculture, adopted by FAO member countries in 2011, signifies particular remarks on the broader aspects of conservation and suitable integrated approaches for its judicious utilization by the people (Mba et al. 2020). Legal frameworks are usually postulated within national borders, where effective strategies are formulated with the coordination of neighboring countries. As per CBD, the CWR protection falls under the country's jurisdiction. On a broader scale, developing countries have imposed national-level based restrictions on the unlimited access to plant genetic resources for food and agriculture (PGFRA; Esquinas-Alcázar 2005). Concerning in-situ conservation, the primary prerequisite is to have a solid national strategy, which gets further integrated with other countries, to widen the approach to a global base. The Convention on International Trade in Endangered Species of Wild Fauna and Flora (CITES) is an important agreement with the governments that ensures judicious global wildlife trade and ensures that there is no survival threat to the wild animals and plants (Fuchs 2008). In the present situation of increase in the illicit trade of wild plants, considering its demand and relevance in food security, this convention stands at the forefront to ensure the sustainable use and availability of wild plants in the market. It emphasizes trade, which should work as per the national legal norms (Lavorgna et al. 2018). ITPGRFA mandates that monetary benefits obtained through the commercialization of the seed banks should be used for the conservation and management of the crop genetic resources (Perrino and Perrino 2020). There is an illegal trade of some endangered plants that are highly important economically and medicinally. For the protection of WFPs, especially those which are endangered and critically endangered, laws of each country should be strengthened and international coordination should be established to tackle the illegal trade of the important wild plant resources (Reeve 2014). Pharmaceutical companies are also known to exploit wild plant resources without caring about their sustainability. There should be stringent laws globally to regulate the harvest of WFP resources for pharmaceutical purposes. National governments need to consider the international treaties and conventions to make their legislations that deal with the protection of the CWRs and WFPs. Several countries have evolved their own laws for the effective implementation of the international treaties and conventions. However, there may be certain locally important CWRs and WFPs and there may be local challenges to their conservation that needs local solutions. Therefore, while evolving a legislation, local needs, challenges and issues should also be considered.

12.7 Future Prospects of CWRs and WFPs

With an expected increase in population to nine billion by 2050, the future will revolve around finding answers to how this massive population can be fed (Godfray et al. 2010). Climate change is also posing great threat to the food security by reducing crop productions in various corners of the world. As the land resources are also declining and the current crops are susceptible to stresses, using CWRs for the improvement of the domesticated crops is an important area to enhance the food production and create resilient crops that can withstand vagaries of the environment. Plants are in a constant race to combat the pathogens which mutate faster, due to which cultivated varieties are always exposed to the biotic and stressful environment. The significant fact is that, CWRs possess resistance to these kinds of novel emerging pathogen strains, therefore, the future would constantly rely on the wild genetic resources (Hajjar and Hodgkin 2007). Plant breeding utilize the significant traits of the CWRs and incorporate them into the narrow genetic base of cultivated varieties (Tanksley and McCouch 1997). CWR accessions in Genbank are incomplete, which means this wild wealth's identity is still to be acknowledged. Therefore documentation of these wild resources would be a key in the future to utilize this genetic diversity in crop improvement programs and the medium that can be utilized would be well-formulated protocols and complementary taxonomic input (Dempewolf et al. 2017). CWR stands as an underutilized wild species whose worth increase in the future as the climate change threatens them and all other domesticated crops. The underutilised crops hold the key to the future, they will be helpful in the diversification of the crops and crop production systems (Padulosi et al. 2011). Considering all these insights on the importance of CWRs and on how things are going to be in the future, the core need is its sustainable utilization and a strong hand in its conservation.

12.8 Conclusions

This chapter focuses on the requirement for the conservation of CWR and WFPs detailing their significance. Taking the current trend of climate change and the subsequent consequences, the urge to protect wild genetic wealth is the need of the hour. Ensuring global food security needs multiple initiatives from different corners. Therefore, to feed the world with appropriate nutritional quality, increasing the yield of staple food crops by genetic enhancement need the existence of the wild progenitors of the crops with remarkable genetic superiority. All the conservation needs and appropriate priorities depend upon the proper action plans considering the threat status at global, national, regional, and local levels. The sustenance of wild genetic resources will be extremely vulnerable in the future if we do not make plans, strategies and take action appropriately. Anthropogenic activities are at the centre of the depletion of wild resources. In the context of the harm inflicted by man on natural resources, the possible way to counteract this is to value indigenous knowledge and support the local communities to earn economic benefits as well. The importance of a collaborative approach to club in-situ and ex-situ modes with local community participation is noteworthy. These underutilized wild relatives should be utilized sustainably and made available in the markets, vital scientific research should be complemented to monitor their population to avoid extinctions. Sustainable utilization along with conservation would be the most desirable framework to ensure that their diversity thrives in the wild. Considering all the facts, we conclude that, over time, things can't be changed, judiciously devised protocols for conservation and the urge to protect biodiversity and sustainable utilization would be the key.

References

Ahmad K, Pieroni A (2016) Folk knowledge of wild food plants among the tribal communities of Thakht-e-Sulaiman Hills, North-West Pakistan. J Ethnobiol Ethnomed 12:17. https://doi.org/10.1186/s13002-016-0090-2

Al-Fatimi M (2021) Wild edible plants traditionally collected and used in southern Yemen. J Ethnobiol Ethnomed 17(1):49. https://doi.org/10.1186/s13002-021-00475-8

Altieri MA (2019) Agroecology: principles and practices for diverse, resilient, and productive farming systems. In: Oxford research encyclopedia of environmental science. Oxford University Press

Asprilla-Perea J, Díaz-Puente JM (2019) Importance of wild foods to household food security in tropical forest areas. Food Secur 11(1):15–22. https://doi.org/10.1007/s12571-018-0846-8

Aziz MA, Ullah Z, Al-Fatimi M, De Chiara M, Sõukand R, Pieroni A (2021) On the trail of an ancient middle eastern ethnobotany: traditional wild food plants gathered by Ormuri speakers in Kaniguram, NW Pakistan. Biology 10(4):302. https://doi.org/10.3390/biology10040302

Baldi A, Bruschi P, Campeggi S, Egea T, Rivera D, Obón C, Lenzi A (2022) The renaissance of wild food plants: insights from Tuscany (Italy). Foods 11(3):300. https://doi.org/10.3390/foods11030300

Berihun T, Molla E (2017) Study on the diversity and use of wild edible plants in Bullen District Northwest Ethiopia. J Bot 2017:1–10. https://doi.org/10.1155/2017/8383468

Bettencourt E, Ford-Lloyd BV, Dias S (2007) Genetic erosion and genetic pollution of crop wild relatives: the PGR forum perspective and achievements. In: Maxted N, Ford-Lloyd BV, Kell SP, Iriondo JM, Dulloo ME, Turok J (eds) Crop wild relative conservation and use. CABI, Wallingford, pp 277–286

Bharucha Z, Pretty J (2010) The roles and values of wild foods in agricultural systems. Philos Trans R Soc B 365(1554):2913–2926. https://doi.org/10.1098/rstb.2010.0123

Boedecker J, Termote C, Assogbadjo AE et al (2014) Dietary contribution of Wild Edible Plants to women's diets in the buffer zone around the Lama forest, Benin—an underutilized potential. Food Secur 6:833–849. https://doi.org/10.1007/s12571-014-0396-7

Bohra A, Kilian B, Sivasankar S, Caccamo M, Mba C, McCouch SR, Varshney RK (2022) Reap the crop wild relatives for breeding future crops. Trends Biotechnol 40(4):412–431. https://doi.org/10.1016/j.tibtech.2021.08.009

Borelli T, Hunter D, Powell B, Ulian T, Mattana E, Termote C, Pawera L, Beltrame D, Penafiel D, Tan A, Taylor M, Engels J (2020) Born to eat wild: an integrated conservation approach to secure wild food plants for food security and nutrition. Plants 9(10):1299. https://doi.org/10.3390/plants9101299

Brehm JM, Maxted N, Martins-Loução MA, Ford-Lloyd BV (2010) New approaches for establishing conservation priorities for socio-economically important plant species. Biodivers Conserv 19(9):2715–2740. https://doi.org/10.1007/s10531-010-9871-4

Brown AHD, Hodgkin T (2015) Indicators of genetic diversity, genetic erosion, and genetic vulnerability for plant genetic resources. In: Ahuja MR, Jain SM (eds) Genetic diversity and erosion in plants: indicators and prevention. Springer International Publishing, Cham, pp 25–53

Brozynska M, Furtado A, Henry RJ (2016) Genomics of crop wild relatives: expanding the gene pool for crop improvement. Plant Biotechnol J 14(4):1070–1085. https://doi.org/10.1111/pbi.12454

Cao Y, Li R, Zhou S, Song L, Quan R, Hu H (2020) Ethnobotanical study on wild edible plants used by three trans-boundary ethnic groups in Jiangcheng County, Pu'er, Southwest China. J Ethnobiol Ethnomed 16(1):66. https://doi.org/10.1186/s13002-020-00420-1

Castañeda-Álvarez NP, Khoury CK, Achicanoy HA, Bernau V, Dempewolf H, Eastwood RJ, Guarino L, Harker RH, Jarvis A, Maxted N, Müller JV, Ramirez-Villegas J, Sosa CC, Struik PC, Vincent H, Toll J (2016) Global conservation priorities for crop wild relatives. Nat Plants 2(4):16022. https://doi.org/10.1038/nplants.2016.22

CBD (2020) Aichi biodiversity targets. https://www.cbd.int/sp/targets/. Accessed 11 Dec 2022

CBD (2011) Vision. https://www.cbd.int/gspc/vision.shtml. Accessed 11 Dec 2022

Chaudhary P, Bhatta S, Aryal K, Joshi BK, Gauchan D (2020) Threats, drivers, and conservation imperative of agrobiodiversity. J Agric Environ 21:44–61. https://doi.org/10.3126/aej.v21i0.38441

Choudhary M, Singh V, Muthusamy V, Wani SH (2017) Harnessing crop wild relatives for crop improvement. Int J Life Sci 6:73. https://doi.org/10.5958/2319-1198.2017.00009.4

Ciancaleoni S, Raggi L, Barone G, Donnini D, Gigante D, Domina G, Negri V (2021) A new list and prioritization of wild plants of socio economic interest in Italy: toward a conservation strategy. Agroecol Sustain Food Syst 45(9):1300–1326. https://doi.org/10.1080/21683565.2021.1917469

Crop Wild Relatives Global Portal (2021) Crop wild relatives: the importance of crop wild relatives. http://www.cropwildrelatives.org/resources/in-situ-conservation-manual/elearning-modules/elearning/the-importance-of-crop-wild-relatives/. Accessed 26 Jul 2022

Cunningham AB (ed) (2014) Applied ethnobotany, 1st edn. Routledge, London

de Medeiros PM, Figueiredo KF, Gonçalves PHS, Caetano RDA, Santos ÉMDC, dos Santos GMC, Barbosa DM, de Paula M, Mapeli AM (2021) Wild plants and the food-medicine continuum—an ethnobotanical survey in Chapada Diamantina (Northeastern Brazil). J Ethnobiol Ethnomed 17(1):37. https://doi.org/10.1186/s13002-021-00463-y

Delang CO (2006a) The role of wild food plants in poverty alleviation and biodiversity conservation in tropical countries. Prog Dev Stud 6(4):275–286. https://doi.org/10.1191/1464993406ps143oa

Delang CO (2006b) Not just minor forest products: the economic rationale for the consumption of wild food plants by subsistence farmers. Ecol Econ 59(1):64–73. https://doi.org/10.1016/j.ecolecon.2005.10.006

Dempewolf H, Baute G, Anderson J, Kilian B, Smith C, Guarino L (2017) Past and future use of wild relatives in crop breeding. Crop Sci 57(3):1070–1082. https://doi.org/10.2135/cropsci2016.10.0885

Dida MM, Devos KM (2006) Finger millet. In: Kole C (ed) Cereals and millets. Springer, Berlin, Heidelberg, pp 333–343

Ding X-Y, Zhang Y, Wang L, Zhuang H-F, Chen W-Y, Wang Y-H (2021) Collection calendar: the diversity and local knowledge of wild edible plants used by Chenthang Sherpa people to treat seasonal food shortages in Tibet, China. J Ethnobiol Ethnomed 17(1):40. https://doi.org/10.1186/s13002-021-00464-x

Dwivedi SL, Ceccarelli S, Blair MW, Upadhyaya HD, Are AK, Ortiz R (2016) Landrace germplasm for improving yield and abiotic stress adaptation. Trends Plant Sci 21(1):31–42. https://doi.org/10.1016/j.tplants.2015.10.012

Ebert AW, Engels JMM (2020) Plant biodiversity and genetic resources matter! Plants 9(12):1706. https://doi.org/10.3390/plants9121706

Edmeades GO, Cooper M, Lafitte R, Zinselmeier C, Ribaut JM, Habben JE, Löffler C, Bänziger M (2001) Abiotic stresses and staple crops. In: Nösberger J, Geiger HH, Struik PC (eds) Crop science: progress and prospects. Papers presented at the Third International Crop Science Congress, Hamburg, Germany, 17–22 August 2000. CABI, Wallingford, pp 137–154

El Haddad N, Sanchez-Garcia M, Visioni A, Jilal A, El Amil R, Sall AT, Lagesse W, Kumar S, Bassi FM (2021) Crop wild relatives crosses: multi-location assessment in durum wheat, barley, and lentil. Agronomy 11(11):2283. https://doi.org/10.3390/agronomy11112283

El Mokni R, Barone G, Maxted N, Kell S, Domina G (2022) A prioritised inventory of crop wild relatives and wild harvested plants of Tunisia. Genet Resour Crop Evol 69(5):1787–1816. https://doi.org/10.1007/s10722-021-01340-z

El-Ramady H, Hajdú P, Törős G, Badgar K, Llanaj X, Kiss A, Abdalla N, Omara AE-D, Elsakhawy T, Elbasiouny H, Elbehiry F, Amer M, El-Mahrouk ME, Prokisch J (2022) Plant nutrition for human health: a pictorial review on plant bioactive compounds for sustainable agriculture. Sustainability 14(14):8329. https://doi.org/10.3390/su14148329

Engels JMM, Thormann I (2020) Main challenges and actions needed to improve conservation and sustainable use of our crop wild relatives. Plants 9(8):968. https://doi.org/10.3390/plants9080968

Esquinas-Alcázar J (2005) Protecting crop genetic diversity for food security: political, ethical and technical challenges. Nat Rev Genet 6(12):946–953. https://doi.org/10.1038/nrg1729

Evans MI (1993) Conservation by commercialisation. In: Ladik CM, Hladik A, Linares OF, Pagezy H, Semple A, Hadley M (eds) Tropical forests, people and food: biocultural interactions and applications to development, MAB series, vol 13. UNESCO, Paris and Parthenon Publishing Group, Carnforth, pp 815–822

FAO (2017) Voluntary guidelines for the conservation and sustainable use of crop wild relatives and wild food plants. United Nations, New York

FAO (2019) The state of the world's biodiversity for food and agriculture. In: Bélanger J, Pilling D (eds) FAO Commission on Genetic Resources for Food and Agriculture Assessments. Rome. 572 pp. http://www.fao.org/3/CA3129EN/CA3129EN.pdf

FAO (2001) International treaty on plant genetic resources for food and agriculture. https://www.fao.org/plant-treaty/en/; https://www.fao.org/3/i0510e/i0510e.pdf. Accessed 11 Dec 2022

FAO, IFAD, UNICEF, WFP, WHO (2020a) The State of Food Security and Nutrition in the World 2020. Transforming food systems for affordable healthy diets. Food and Agriculture Organization of the United Nations, Rome. https://doi.org/10.4060/ca9692en

FAO, UNEP (2020b) The State of the World's Forests 2020. Forests, biodiversity and people. Food and Agriculture Organization of the United Nations, Rome

Fatima A, Farid M, Safdar K, Fayyaz A, Ali SM, Adnan S, Nawaz M, Munir H, Raza N, Zubair M (2020) Loss of agro-biodiversity and productivity due to climate change in continent Asia: a review. In: Hasanuzzaman M (ed) Plant ecophysiology and adaptation under climate change: mechanisms and perspectives I. Springer, Singapore, pp 51–71

Fielder H, Brotherton P, Hosking J, Hopkins JJ, Ford-Lloyd B, Maxted N (2015) Enhancing the conservation of crop wild relatives in England. PLoS One 10(6):e0130804. https://doi.org/10.1371/journal.pone.0130804

Fitzgerald H (2013) The national crop wild relative strategy report for Finland. MTT report. EU Seventh Framework Programme, Jokioinen

Ford-Lloyd BV, Schmidt M, Armstrong SJ, Barazani O, Engels J, Hadas R, Hammer K, Kell SP, Kang D, Khoshbakht K, Li Y, Long C, Lu B-R, Ma K, Nguyen VT, Qiu L, Ge S, Wei W, Zhang Z, Maxted N (2011) Crop wild relatives—undervalued, underutilized and under threat? Bioscience 61(7):559–565. https://doi.org/10.1525/bio.2011.61.7.10

Fu Y-B (2017) The vulnerability of plant genetic resources conserved ex situ. Crop Sci 57(5): 2314–2328. https://doi.org/10.2135/cropsci2017.01.0014

Fuchs C (2008) Convention on international trade in endangered species of wild fauna and flora (CITES)—conservation efforts undermine the legality principle. German Law J 9:1565–1596. https://doi.org/10.1017/S2071832200000584

Garcia SM, Rice J (2020) Assessing progress towards Aichi biodiversity target 6 on sustainable marine fisheries. Technical Series No. 87. Secretariat of the Convention on Biological Diversity, Montreal

Gepts P, Papa R (2003) Possible effects of (trans)gene flow from crops on the genetic diversity from landraces and wild relatives. Environ Biosaf Res 2(2):89–103. https://doi.org/10.1051/ebr:2003009

Ghanimi R, Ouhammou A, Ahouach A, Cherkaoui M (2022) Ethnobotanical study on wild edible plants traditionally used by Messiwa people, Morocco. J Ethnobiol Ethnomed 18(1):16. https://doi.org/10.1186/s13002-022-00500-4

Gnanesh BN, Fetch JM, Zegeye T, McCartney CA, Fetch T (2014) Oat. In: Pratap A, Kumar J (eds) Alien gene transfer in crop plants, vol 2. Springer, New York, pp 51–73

Godfray HCJ, Beddington JR, Crute IR, Haddad L, Lawrence D, Muir JF, Pretty J, Robinson S, Thomas SM, Toulmin C (2010) Food security: the challenge of feeding 9 billion people. Science 327(5967):812–818. https://doi.org/10.1126/science.1185383

Goettsch B, Urquiza-Haas T, Koleff P, Acevedo Gasman F, Aguilar-Meléndez A, Alavez V, Alejandre-Iturbide G, Aragón Cuevas F, Azurdia Pérez C, Carr JA, Castellanos-Morales G, Cerén G, Contreras-Toledo AR, Correa-Cano ME, De la Cruz Larios L, Debouck DG, Delgado-Salinas A, Gómez-Ruiz EP, González-Ledesma M, González-Pérez E, Hernández-Apolinar M, Herrera-Cabrera BE, Jefferson M, Kell S, Lira-Saade R, Lorea-Hernández F, Martínez M, Mastretta-Yanes A, Maxted N, Menjívar J, Ángeles Mérida Guzmán M, Morales Herrera AJ, Oliveros-Galindo O, Orjuela-R. MA, Pollock CM, Quintana-Camargo M, Rodríguez A, Ruiz Corral JA, Sánchez González J d J, Sánchez-de la Vega G, Superina M, Tobón Niedfeldt W, Tognelli MF, Vargas-Ponce O, Vega M, Wegier A, Zamora Tavares P, Jenkins RKB (2021) Extinction risk of Mesoamerican crop wild relatives. Plants People Planet 3(6):775–795. https://doi.org/10.1002/ppp3.10225

Guarrera PM, Savo V (2016) Wild food plants used in traditional vegetable mixtures in Italy. J Ethnopharmacol 185:202–234. https://doi.org/10.1016/j.jep.2016.02.050

Gupta C, Salgotra RK, Mahajan G (2020) Future threats and opportunities facing crop wild relatives and landrace diversity. In: Salgotra RK, Zargar SM (eds) Rediscovery of genetic and genomic resources for future food security. Springer, Singapore, pp 351–364

Hajjar R, Hodgkin T (2007) The use of wild relatives in crop improvement: a survey of developments over the last 20 years. Euphytica 156(1–2):1–13. https://doi.org/10.1007/s10681-007-9363-0

Hanson AA (1952) The origin, variation, immunity, and breeding of cultivated plants: selected writings of N. I. Vavilov, translated from Russian by K. Starr Chester. Chronica Botanica, Vol. 13, No. 1/6 (pp. viii + 366). Waltham, Mass.: The Chronica Botanica Co., New York City: Stechert-Hafner, Inc. $7.50. Agron J 44(2):102–102. https://doi.org/10.2134/agronj1952.00021962004400020016x

Harlan JR, Wet JMJ (1971) Toward a rational classification of cultivated plants. Taxon 20(4): 509–517. https://doi.org/10.2307/1218252

Hay FR, Probert RJ (2013) Advances in seed conservation of wild plant species: a review of recent research. Conserv Physiol 1(1):cot030. https://doi.org/10.1093/conphys/cot030

Heywood V, Casas A, Ford-Lloyd B, Kell S, Maxted N (2007) Conservation and sustainable use of crop wild relatives. Agric Ecosyst Environ 121(3):245–255. https://doi.org/10.1016/j.agee.2006.12.014

Hickey GM, Pouliot M, Smith-Hall C, Wunder S, Nielsen MR (2016) Quantifying the economic contribution of wild food harvests to rural livelihoods: a global-comparative analysis. Food Policy 62:122–132. https://doi.org/10.1016/j.foodpol.2016.06.001

Holness S, Hamer M, Brehm JM, Raimondo D (2019) Priority areas for the in situ conservation of crop wild relatives in South Africa. Plant Genet Resour 17(2):115–127

Hunter D (ed) (2012) Crop wild relatives. Routledge

Idohou R, Assogbadjo AE, Fandohan B, Gouwakinnou GN, Glele Kakai RL, Sinsin B, Maxted N (2013) National inventory and prioritization of crop wild relatives: case study for Benin. Genet Resour Crop Evol 60:1337–1352. https://doi.org/10.1007/s10722-012-9923-6

Iriondo JM, Fielder H, Fitzgerald H, Kell SP, Labokas J, Negri V, Phillips J, Rubio Teso ML, Sensen S, Taylor N, Maxted N (2016) National strategies for the conservation of crop wild relatives. In: Maxted N, Dulloo ME, Ford-Lloyd BV (eds) Enhancing crop genepool use. Capturing wild relative and landrace diversity for crop improvement. CAB International, Wallingford, pp 161–171

IUCN (2020) The IUCN red list of threatened species. Version 2021-3. https://www.iucnredlist.org. Accessed 2020-08-01

Jarvis A, Lane A, Hijmans RJ (2008) The effect of climate change on crop wild relatives. Agric Ecosyst Environ 126(1–2):13–23. https://doi.org/10.1016/j.agee.2008.01.013

Ju Y, Zhuo J, Liu B, Long C (2013) Eating from the wild: diversity of wild edible plants used by Tibetans in Shangri-la region, Yunnan, China. J Ethnobiol Ethnomed 9(1):28. https://doi.org/10.1186/1746-4269-9-28

Kahane R, Hodgkin T, Jaenicke H, Hoogendoorn C, Hermann M, (Dyno) Keatinge JDH, d'Arros Hughes J, Padulosi S, Looney N (2013) Agrobiodiversity for food security, health and income. Agron Sustain Dev 33(4):671–693. https://doi.org/10.1007/s13593-013-0147-8

Khakurel D, Uprety Y, Łuczaj Ł, Rajbhandary S (2021) Foods from the wild: local knowledge, use pattern and distribution in Western Nepal. PLoS One 16(10):e0258905. https://doi.org/10.1371/journal.pone.0258905

Khoury CK, Greene S, Wiersema J, Maxted N, Jarvis A, Struik PC (2013) An inventory of crop wild relatives of the United States. Crop Sci 53(4):1496–1508

Khumalo S, Chirwa PW, Moyo BH, Syampungani S (2012) The status of agrobiodiversity management and conservation in major agroecosystems of Southern Africa. Agric Ecosyst Environ 157:17–23. https://doi.org/10.1016/j.agee.2012.01.028

Kidane L, Kejela A (2021) Food security and environment conservation through sustainable use of wild and semi-wild edible plants: a case study in Berek Natural Forest, Oromia special zone, Ethiopia. Agric Food Secur 10(1):29. https://doi.org/10.1186/s40066-021-00308-7

Kumar A (2021) How can India leverage its botanic gardens for the conservation and sustainable utilization of wild food plant resources through the implementation of a global strategy for plant conservation? J Zool Bot Gard 2(4):586–599. https://doi.org/10.3390/jzbg2040042

Labokas J, Maxted N, Kell S, Brehm JM, Iriondo JM (2018) Development of national crop wild relative conservation strategies in European countries. Genet Resour Crop Evol 65:1385–1403. https://doi.org/10.1007/s10722-018-0621-x

Ladio AH, Lozada M (2004) Patterns of use and knowledge of wild edible plants in distinct ecological environments: a case study of a Mapuche community from northwestern Patagonia. Biodivers Conserv 13:1153–1173. https://doi.org/10.1023/B:BIOC.0000018150.79156.50

Lala S, Amri A, Maxted N (2018) Towards the conservation of crop wild relative diversity in North Africa: checklist, prioritisation and inventory. Genet Resour Crop Evol 65(1):113–124. https://doi.org/10.1007/s10722-017-0513-5

Lauzer D, Laublin G, Vincent G, Cappadocia M (1992) In vitro propagation and cytology of wild yams, *Dioscorea abyssinica* Hoch. and *D. mangenotiana* Miège. Plant Cell Tissue Organ Cult 28:215–223. https://doi.org/10.1007/BF00055520

Lavorgna A, Rutherford C, Vaglica V, Smith MJ, Sajeva M (2018) CITES, wild plants, and opportunities for crime. Eur J Crim Policy Res 24(3):269–288. https://doi.org/10.1007/s10610-017-9354-1

Lawson C, Humphries F, Rourke M (2019) Legislative, administrative and policy approaches to access and benefit sharing ("ABS") genetic resources: digital sequence information ("DSI") in New Zealand and Australian ABS laws. Intellect Prop Forum 118:38–50

Lemmon ZH, Reem NT, Dalrymple J, Soyk S, Swartwood KE, Rodriguez-Leal D, Van Eck J, Lippman ZB (2018) Rapid improvement of domestication traits in an orphan crop by genome editing. Nat Plants 4(10):766–770. https://doi.org/10.1038/s41477-018-0259-x

Luo B, Liu B, Zhang H, Zhang H, Li X, Ma L, Wang Y, Bai Y, Zhang X, Li J, Yang J, Long C (2019) Wild edible plants collected by Hani from terraced rice paddy agroecosystem in Honghe Prefecture, Yunnan, China. J Ethnobiol Ethnomed 15(1):56. https://doi.org/10.1186/s13002-019-0336-x

Mammadov J, Buyyarapu R, Guttikonda SK, Parliament K, Abdurakhmonov IY, Kumpatla SP (2018) Wild relatives of maize, rice, cotton, and soybean: treasure troves for tolerance to biotic and abiotic stresses. Front Plant Sci 9:886. https://doi.org/10.3389/fpls.2018.00886

Maxted N, Ford-Lloyd BV, Jury S, Kell S, Scholten M (2006) Towards a definition of a crop wild relative. Biodivers Conserv 15(8):2673–2685. https://doi.org/10.1007/s10531-005-5409-6

Maxted N, Scholten M, Codd R, Ford-Lloyd B (2007) Creation and use of a national inventory of crop wild relatives. Biol Conserv 140:142–159. https://doi.org/10.1016/j.biocon.2007.08.006

Maxted N, Kell S, Toledo Á, Dulloo E, Heywood V, Hodgkin T, Hunter D, Guarino L, Jarvis A, Ford-Lloyd B (2010) A global approach to crop wild relative conservation: securing the gene pool for food and agriculture. Kew Bull 65(4):561–576. https://doi.org/10.1007/s12225-011-9253-4

Maxted N, Kell S, Ford-Lloyd B, Dulloo E, Toledo Á (2012) Toward the systematic conservation of global crop wild relative diversity. Crop Sci 52(2):774–785. https://doi.org/10.2135/cropsci2011.08.0415

Maxted N, Avagyan A, Frese L, Iriondo JM, Magos Brehm J, Singer A, Kell SP (2015) ECPGR Concept for in situ conservation of crop wild relatives in Europe. Wild Species Conservation in Genetic Reserves Working Group. European Cooperative Programme for Plant Genetic Resources, Rome, Italy

Mba C, Abang M, Diulgheroff S, Hrushka A, Hugo W, Ingelbrecht I, Jankuloski L, Leskien D, Lopez V, Muminjanov H, Mulila Mitti J, Nersisyan A, Noorani A, Piao Y, Sagnia S (2020) FAO supports countries in the implementation of the second global plan of action for plant genetic resources for food and agriculture. Acta Hortic:197–208. https://doi.org/10.17660/ActaHortic.2020.1267.30

Meilleur BA, Hodgkin T (2004) In situ conservation of crop wild relatives: status and trends. Biodivers Conserv 13(4):663–684

Menendez-Baceta G, Pardo-de-Santayana M, Aceituno-Mata L, Tardío J, Reyes-García V (2017) Trends in wild food plants uses in Gorbeialdea (Basque Country). Appetite 112:9–16. https://doi.org/10.1016/j.appet.2017.01.010

Meyer A, Barton N (2019) Botanic gardens are important contributors to crop wild relative preservation. Crop Sci 59(6):2404–2412. https://doi.org/10.2135/cropsci2019.06.0358

Milla R (2020) Crop Origins and Phylo Food: a database and a phylogenetic tree to stimulate comparative analyses on the origins of food crops. Glob Ecol Biogeogr 29(4):606–614. https://doi.org/10.1111/geb.13057

Mishra A, Swamy SL, Thakur TK, Bhat R, Bijalwan A, Kumar A (2021) Use of wild edible plants: can they meet the dietary and nutritional needs of indigenous communities in Central India. Foods 10:1453. https://doi.org/10.3390/foods10071453

Montenegro de Wit M (2016) Are we losing diversity? Navigating ecological, political, and epistemic dimensions of agrobiodiversity conservation. Agric Hum Values 33(3):625–640. https://doi.org/10.1007/s10460-015-9642-7

Montenegro de Wit M (2017) Stealing into the wild: conservation science, plant breeding and the makings of new seed enclosures. J Peasant Stud 44:169–212. https://doi.org/10.1080/03066150.2016.1168405

Mponya NK, Magombo ZLK, Pungulani L, Brehm JM, Maxted N (2020) Development of a prioritised checklist of crop wild relatives for conservation in Malawi. Afr Crop Sci J 28(2):279–311. https://doi.org/10.4314/acsj.v28i2.12

Mutie FM, Rono PC, Kathambi V, Hu G-W, Wang Q-F (2020) Conservation of wild food plants and their potential for combatting food insecurity in Kenya as exemplified by the drylands of Kitui County. Plants 9(8):1017. https://doi.org/10.3390/plants9081017

Nduche M, Magos Brehm J, Abberton M, Omosun G, Maxted N (2021) West African crop wild relative checklist, prioritization and inventory. Genet Res J 2(4):55–65. https://doi.org/10.46265/genresj.EIFL1323

Norton SL, Khoury CK, Sosa CC, Castañeda-Álvarez NP, Achicanoy HA, Sotelo S (2017) Priorities for enhancing the ex situ conservation and use of Australian crop wild relatives. Aust J Bot 65(8):638. https://doi.org/10.1071/BT16236

Nyakoojo C, Tugume P (2020) Traditional use of wild edible plants in the communities adjacent Mabira Central Forest Reserve, Uganda. Ethnobot Res Appl 20:1–14

Oseni OM, Pande V, Nailwal TK (2018) A review on plant tissue culture, a technique for propagation and conservation of endangered plant species. Int J Curr Microbiol App Sci 7(7): 3778–3786. https://doi.org/10.20546/ijcmas.2018.707.438

Padulosi S, Heywood V, Hunter D, Jarvis A (2011) Underutilized species and climate change: current status and outlook. In: Yadav SS, Redden RJ, Hatfield JL, Lotze-Campen H, Hall AE (eds) Crop adaptation to climate change. Wiley-Blackwell, Oxford, pp 507–521

Pandey A, Bhandari DC, Bhatt KC, Pareek SK, Tomer AK, Dhillon BS (2005) Wild relatives of Cflip plants in India: collection and conservation. National Bureau of Plant Genetic Resources, New Delhi

Pardo-de-Santayana M, Macía MJ (2015) The benefits of traditional knowledge. Nature 518(7540): 487–488. https://doi.org/10.1038/518487a

Pascual-Mendoza S, Saynes-Vásquez A, Pérez-Herrera A (2021) Traditional knowledge of edible plants in an indigenous community in the Sierra Norte of Oaxaca, Mexico. Plant Biosyst 156(2): 515–527. https://doi.org/10.1080/11263504.2021.1887956

Pawera L, Khomsan A, Zuhud EAM, Hunter D, Ickowitz A, Polesny Z (2020) Wild food plants and trends in their use: from knowledge and perceptions to drivers of change in West Sumatra, Indonesia. Foods 9(9):1240. https://doi.org/10.3390/foods9091240

Pence VC, Finke LR, Chaiken MF (2017) Tools for the ex situ conservation of the threatened species, *Cycladenia humilis* var. *jonesii*. Conserv Physiol 5:cox053. https://doi.org/10.1093/conphys/cox053

Perrino EV, Perrino P (2020) Crop wild relatives: know how past and present to improve future research, conservation and utilization strategies, especially in Italy: a review. Genet Resour Crop Evol 67(5):1067–1105. https://doi.org/10.1007/s10722-020-00930-7

Perrino EV, Wagensommer RP (2021) Crop Wild Relatives (CWR) priority in Italy: distribution, ecology, in situ and ex situ conservation and expected actions. Sustainability 13(4):1682. https://doi.org/10.3390/su13041682

Phillips J, Magos Brehm J, van Oort B, Asdal Å, Rasmussen M, Maxted N (2017) Climate change and national crop wild relative conservation planning. Ambio 46(6):630–643. https://doi.org/10.1007/s13280-017-0905-y

Pilling D, Bélanger J, Hoffmann I (2020) Declining biodiversity for food and agriculture needs urgent global action. Nat Food 1(3):144–147. https://doi.org/10.1038/s43016-020-0040-y

Pinela J, Carvalho AM, Ferreira ICFR (2017) Wild edible plants: nutritional and toxicological characteristics, retrieval strategies and importance for today's society. Food Chem Toxicol 110: 165–188. https://doi.org/10.1016/j.fct.2017.10.020

Porch T, Beaver J, Debouck D, Jackson S, Kelly J, Dempewolf H (2013) Use of wild relatives and closely related species to adapt common bean to climate change. Agronomy 3:433–461. https://doi.org/10.3390/agronomy3020433

Punchay K, Inta A, Tiansawat P, Balslev H, Wangpakapattanawong P (2020) Traditional knowledge of wild food plants of Thai Karen and Lawa (Thailand). Genet Resour Crop Evol 67(5): 1277–1299. https://doi.org/10.1007/s10722-020-00910-x

Rahman W, Magos Brehm J, Maxted N (2019) Setting conservation priorities for the wild relatives of food crops in Indonesia. Genet Resour Crop Evol 66(4):809–824. https://doi.org/10.1007/s10722-019-00761-1

Rajasekharan PE, Sahijram L (2015) In vitro conservation of plant germplasm. In: Bahadur B, Venkat Rajam M, Sahijram L, Krishnamurthy KV (eds) Plant biology and biotechnology. Springer India, New Delhi, pp 417–443

Ray A, Ray R, Sreevidya EA (2020) How many wild edible plants do we eat—their diversity, use, and implications for sustainable food system: an exploratory analysis in India. Front Sustain Food Syst 4:56. https://doi.org/10.3389/fsufs.2020.00056

Raza A, Razzaq A, Mehmood S, Zou X, Zhang X, Lv Y, Xu J (2019) Impact of climate change on crops adaptation and strategies to tackle its outcome: a review. Plants 8(2):34. https://doi.org/10.3390/plants8020034

Reeve R (2014) Policing international trade in endangered species: the CITES treaty and compliance. Taylor and Francis, Hoboken

Reyes-García V, Guèze M, Luz AC, Paneque-Gálvez J, Macía MJ, Orta-Martínez M, Pino J, Rubio-Campillo X (2013) Evidence of traditional knowledge loss among a contemporary indigenous society. Evol Hum Behav 34(4):249–257. https://doi.org/10.1016/j.evolhumbehav.2013.03.002

Robertson L, Labate J (2006) Genetic resources of tomato (*Lycopersicon esculentum* Mill.) and wild relatives. In: Razdan MK, Mattoo AK (eds) Genetic improvement of Solanaceous crops, vol 2. CRC Press, pp 25–75

Sánchez-Mata M d C, Matallana-González MC, Morales P (2016) The contribution of wild plants to dietary intakes of micronutrients (I): vitamins. In: Sánchez-Mata M d C, Tardío J (eds) Mediterranean wild edible plants. Springer, New York, pp 111–139

Sarkar A, Dasgupta A, Sensarma SR (2019) Climate change and food security in India: adaptation strategies and major challenges. In: Sarkar A, Sensarma SR, vanLoon GW (eds) Sustainable solutions for food security. Springer International Publishing, Cham, pp 497–520

Schanbacher WD (2010) The politics of food: the global conflict between food security and food sovereignty. Praeger Security International, Santa Barbara

Schulp CJE, Thuiller W, Verburg PH (2014) Wild food in Europe: a synthesis of knowledge and data of terrestrial wild food as an ecosystem service. Ecol Econ 105:292–305. https://doi.org/10.1016/j.ecolecon.2014.06.018

Schunko C, Li X, Klappoth B, Lesi F, Porcher V, Porcuna-Ferrer A, Reyes-García V (2022) Local communities' perceptions of wild edible plant and mushroom change: a systematic review. Glob Food Sec 32:100601. https://doi.org/10.1016/j.gfs.2021.100601

Shelef O, Weisberg PJ, Provenza FD (2017) The value of native plants and local production in an era of global agriculture. Front Plant Sci 8:2069. https://doi.org/10.3389/fpls.2017.02069

Siddiqui A, Shukla S (2015) Conservation of plant genetic resources and their utilisation in global perspective. Int J Life Sci 4:46. https://doi.org/10.5958/2319-1198.2015.00007.X

Singh B, Bedi YS (2017) Eating from raw wild plants in Himalaya: traditional knowledge documentary on Sheena tribes along LoC border in Kashmir. Indian J Nat Prod Resour 8(3):269–275

Singh A, Dubey PK, Chaurasia R, Dubey RK, Pandey KK, Singh GS, Abhilash PC (2019) Domesticating the undomesticated for global food and nutritional security: four steps. Agronomy 9(9):491

Smýkal P, Hradilová I, Trněný O, Brus J, Rathore A, Bariotakis M, Das RR, Bhattacharyya D, Richards C, Coyne CJ, Pirintsos S (2017) Genomic diversity and macroecology of the crop wild relatives of domesticated pea. Sci Rep 7(1):17384. https://doi.org/10.1038/s41598-017-17623-4

Tanksley SD, McCouch SR (1997) Seed banks and molecular maps: unlocking genetic potential from the wild. Science 277(5329):1063–1066. https://doi.org/10.1126/science.277.5329.1063

Tardío J, Pardo-De-Santayana M, Morales R (2006) Ethnobotanical review of wild edible plants in Spain. Bot J Linn Soc 152(1):27–71. https://doi.org/10.1111/j.1095-8339.2006.00549.x

Taylor NG, Kell SP, Holubec V, Parra-Quijano M, Chobot K, Maxted N (2017) A systematic conservation strategy for crop wild relatives in the Czech Republic. Divers Distrib 23(4): 448–462. https://doi.org/10.1111/ddi.12539

Tewksbury JJ, Nabhan GP, Norman D, Suzan H, Tuxill J, Donovan J (1999) In situ conservation of wild chiles and their biotic associates. Conserv Biol 13(1):98–107. https://doi.org/10.1046/j.1523-1739.1999.97399.x

Thorn JPR, Thornton TF, Helfgott A, Willis KJ (2020) Indigenous uses of wild and tended plant biodiversity maintain ecosystem services in agricultural landscapes of the Terai Plains of Nepal. J Ethnobiol Ethnomed 16(1):33. https://doi.org/10.1186/s13002-020-00382-4

Thorpe T (2007) History of plant tissue culture. J Mol Microbial Biotechnol 37:169–180

Turner NJ, Łuczaj ŁJ, Migliorini P, Pieroni A, Dreon AL, Sacchetti LE, Paoletti MG (2011) Edible and tended wild plants, traditional ecological knowledge and agroecology. Crit Rev Plant Sci 30(1–2):198–225. https://doi.org/10.1080/07352689.2011.554492

Tyack N, Dempewolf H, Khoury CK (2020) The potential of payment for ecosystem services for crop wild relative conservation. Plants 9(10):1305. https://doi.org/10.3390/plants9101305

Ulian T, Diazgranados M, Pironon S, Padulosi S, Liu U, Davies L, Howes MR, Borrell JS, Ondo I, Pérez-Escobar OA, Sharrock S, Ryan P, Hunter D, Lee MA, Barstow C, Łuczaj Ł, Pieroni A, Cámara-Leret R, Noorani A, Mba C, Nono Womdim R, Muminjanov H, Antonelli A, Pritchard HW, Mattana E (2020) Unlocking plant resources to support food security and promote sustainable agriculture. Plants People Planet 2(5):421–445. https://doi.org/10.1002/ppp3.10145

Upreti BR, Upreti YG (2002) Factors leading to agro-biodiversity loss in developing countries: the case of Nepal. Biodivers Conserv 11:1607–1621. https://doi.org/10.1023/A:1016862200156

Uprety Y, Poudel RC, Shrestha KK, Rajbhandary S, Tiwari NN, Shrestha UB, Asselin H (2012) Diversity of use and local knowledge of wild edible plant resources in Nepal. J Ethnobiol Ethnomed 8(1):16. https://doi.org/10.1186/1746-4269-8-16

van Treuren R, Hoekstra R, Wehrens R, van Hintum T (2020) Effects of climate change on the distribution of crop wild relatives in the Netherlands in relation to conservation status and ecotope variation. Glob Ecol Conserv 23:e01054. https://doi.org/10.1016/j.gecco.2020.e01054

Vavilov NI (1951) The origin, variation, immunity and breeding of cultivated plants. Soil Sci 72: 482. https://doi.org/10.1097/00010694-195112000-00018

Vincent H, Amri A, Castañeda-Álvarez NP, Dempewolf H, Dulloo E, Guarino L, Hole D, Mba C, Toledo A, Maxted N (2019) Modeling of crop wild relative species identifies areas globally for in situ conservation. Commun Biol 2(1):136. https://doi.org/10.1038/s42003-019-0372-z

Vincent H, Hole D, Maxted N (2022) Congruence between global crop wild relative hotspots and biodiversity hotspots. Biol Conserv 265:109432. https://doi.org/10.1016/j.biocon.2021.109432

Viruel J, Kantar MB, Gargiulo R, Hesketh-Prichard P, Leong N, Cockel C, Forest F, Gravendeel B, Pérez-Barrales R, Leitch IJ, Wilkin P (2021) Crop wild phylorelatives (CWPs): phylogenetic distance, cytogenetic compatibility and breeding system data enable estimation of crop wild relative gene pool classification. Bot J Linn Soc 195:1–33. https://doi.org/10.1093/botlinnean/boaa064

von Wettberg E, Davis TM, Smýkal P (2020) Editorial: wild plants as source of new crops. Front Plant Sci 11:591554. https://doi.org/10.3389/fpls.2020.591554

Warschefsky E, Penmetsa RV, Cook DR, von Wettberg EJB (2014) Back to the wilds: tapping evolutionary adaptations for resilient crops through systematic hybridization with crop wild relatives. Am J Bot 101(10):1791–1800. https://doi.org/10.3732/ajb.1400116

Wessels C, Merow C, Trisos CH (2021) Climate change risk to southern African wild food plants. Reg Environ Chang 21(2):29. https://doi.org/10.1007/s10113-021-01755-5

Winstead DJ, Jacobson MG (2022) Food resilience in a dark catastrophe: a new way of looking at tropical wild edible plants. Ambio 51(9):1949–1962. https://doi.org/10.1007/s13280-022-01715-1

Xu Y, Liang D, Wang G-T, Wen J, Wang R-J (2020) Nutritional and functional properties of wild food-medicine plants from the coastal region of South China. J Evid Based Complement Altern Med 25:2515690X2091326. https://doi.org/10.1177/2515690X20913267

Yeşil Y, Çelik M, Yılmaz B (2019) Wild edible plants in Yeşilli (Mardin-Turkey), a multicultural area. J Ethnobiol Ethnomed 15(1):52. https://doi.org/10.1186/s13002-019-0327-y

Yu H, Lin T, Meng X, Du H, Zhang J, Liu G, Chen M, Jing Y, Kou L, Li X, Gao Q, Liang Y, Liu X, Fan Z, Liang Y, Cheng Z, Chen M, Tian Z, Wang Y, Chu C, Zuo J, Wan J, Qian Q, Han B, Zuccolo A, Wing RA, Gao C, Liang C, Li J (2021) A route to de novo domestication of wild allotetraploid rice. Cell 184(5):1156–1170.e14. https://doi.org/10.1016/j.cell.2021.01.013

Yuan Z, Lun F, He L, Cao Z, Min Q, Bai Y, Liu M, Cheng S, Li W, Fuller A (2014) Exploring the state of retention of traditional ecological knowledge (TEK) in a Hani Rice Terrace Village, Southwest China. Sustainability 6(7):4497–4513. https://doi.org/10.3390/su6074497

Zair W, Maxted N, Amri A (2018) Setting conservation priorities for crop wild relatives in the Fertile Crescent. Genet Resour Crop Evol 65(3):855–863. https://doi.org/10.1007/s10722-017-0576-3

Zair W, Maxted N, Brehm JM, Amri A (2021) Ex situ and in situ conservation gap analysis of crop wild relative diversity in the Fertile Crescent of the Middle East. Genet Resour Crop Evol 68(2): 693–709. https://doi.org/10.1007/s10722-020-01017-z

Zamora-Tavares P, Vargas-Ponce O, Sánchez-Martínez J, Cabrera-Toledo D (2015) Diversity and genetic structure of the husk tomato (*Physalis philadelphica* Lam.) in Western Mexico. Genet Resour Crop Evol 62:141–153. https://doi.org/10.1007/s10722-014-0163-9

Zhang H, Mittal N, Leamy LJ, Barazani O, Song B-H (2017) Back into the wild-Apply untapped genetic diversity of wild relatives for crop improvement. Evol Appl 10(1):5–24. https://doi.org/10.1111/eva.12434

Zhu X-G, Zhu J-K (2021) Precision genome editing heralds rapid de novo domestication for new crops. Cell 184(5):1133–1134. https://doi.org/10.1016/j.cell.2021.02.004

Ziska LH (2021) Crop adaptation: weedy and crop wild relatives as an untapped resource to utilize recent increases in atmospheric CO2. Plants 10(1):88. https://doi.org/10.3390/plants10010088

Chapter 13
Databases Relevant to Wild Food Plants

Sushil Kumar, Swarn Singh, Faisal Mushtaq, and Ajay Kumar

Abstract Wild food plants are important for human beings. Their beneficial effects are due to the multiple ingredients of these plants. Besides nutrients, recent evidence also suggests that many non-nutrient bioactive compounds besides promoting optimal health also reduces the risk of chronic diseases. They can also be used as a model plants for the development of resilient crops since they grow in wild conditions. Nowadays databases are very important since they provide the required information at one place. Several databases have been developed over the period of time which provide important information regarding the plants ranging from their taxonomy to phytochemistry to genomics. The present book chapter is an attempt to provide information regarding various databases such as the Food plant international database, eBASIS, Food composition databases, The Chemical Composition of American Food Materials, CMUAP, and TRY-plant trait database that are available and can be used for the study of wild food plants. Besides these, other available databases such as IPNI, The plant list, and efloras can be used for the correct identification and distribution of these wild food plants. There is no dedicated database for wild food plants and therefore, there is a need for developing a comprehensive database for wild food plants so that the information regarding various aspects such as taxonomy, distribution, uses, phytochemistry, etc. can be made available to researchers, planners and other users at one place. This chapter further provides a framework for how a dedicated database can be developed for wild food plants.

S. Kumar (✉) · F. Mushtaq
Department of Botany, Government Degree College, Kishtwar, Jammu and Kashmir, India

S. Singh
Department of Botany, Padma Shri Padma Sachdev Government PG College for Women, Gandhi Nagar, Jammu, Jammu and Kashmir, India

A. Kumar
Department of Plant Science, Central University of Kerala, Kasaragod, Kerala, India

A. Kumar et al. (eds.), *Wild Food Plants for Zero Hunger and Resilient Agriculture*,
Plant Life and Environment Dynamics,
https://doi.org/10.1007/978-981-19-6502-9_13

Keywords Wild food plants · Food plants · Database · Nutrients · Food security · Wild edibles

13.1 Introduction

Food has been an important part of human existence providing energy and nutrition (Ray et al. 2020). It is believed that the world population may reach 9 billion by 2050. Despite the general belief that the present global food system is capable enough to provide food for the world, more than 2 billion people around the world still experience hunger or many of them do not have access to a good quality nutritious diet, especially in the countries of sub-Saharan Africa which are largely import-dependent (Romeo et al. 2015; FAO 2011; FAO 2020). Hunger and nutritional deficiency make the population prone to diseases, and so it can be a major barrier to socio-economic development (FAO 2012). Hence substantial attention to such a situation needs to be given in order to explore new food assets. An important way to enhance the quality of crops and to protect them from diseases and other stresses is genetic manipulation (Colin and Luigi 2010). Due to their diversity and resilience, Wild Food Plants (WFPs) can be of tremendous use in various food development programmes such as to develop new crops as well as they can act as genetic reserves that can be used in various hybridisation experiments (Powell et al. 2015). Wild food plants have immense potential to reshape the current global food system towards more sustainable, nutritious as well as climate-resilient (Powell et al. 2015; Bharucha and Pretty 2010; Hunter and Fanzo 2013).

WFPs are an important group of plants which are collected and used by the tribals and local communities in many parts of the world (Ahmad and Pieroni 2016; Kumar et al. 2021; Pawera et al. 2020; Ghorbani et al. 2012). This practice of consuming wild food plants is not new and is as old as human prehistory (Zohary et al. 2012; Etkin 1994; Sowunmi 1985). They are also called "famine foods" as it has been reported that during the times of disasters whether natural or manmade, affected populations which suffer from acute food shortages become heavily dependent on these wild food plants for their survival (Kalle et al. 2020; Łuczaj et al. 2015; Łuczaj et al. 2012; Guinard and Lemessa 2000; Chopak 2000; Leborgne et al. 2002; Rahmato 1988; World Food Programme 1996). Some WFPs have also been termed as "functional foods" due to the presence of physiologically active ingredients that offers many health benefits beyond their nutritional value (Pinela et al. 2016) while some others may also contain certain compounds which can be potentially toxic to humans (EFSA 2009, 2012; Pinela et al. 2017). WFPs can also play a key role in complementing our staple food crops as they are rich in minerals, vitamins, fatty acids and fibres (Tardío et al. 2006; Scoones et al. 1992; Dansi et al. 2008; Lamien-Meda et al. 2008; Odhav et al. 2007). Some WFPs with favourable nutritional or other properties, can offer new opportunities for the development of mountain agriculture if taken into cultivation for e.g. Rocket salad (*Eruca sativa*), a wild food plant which was earlier consumed locally, is now a widely popular salad (Abbet et al. 2014; Padulosi and Pignone 1997).

Despite being nutritious, WFPs are currently underutilised. This underutilization of wild food plants was promoted by lifestyle changes, the industrial revolution, urbanisation, change in land use and climate change, large-scale cultivation of rice, wheat and maize and also less contact with nature (Borelli et al. 2020; Kalle et al. 2020; Łuczaj et al. 2012). Meanwhile, only a few policy measures are in place, which exclusively target the sustainable use and conservation of wild food plants (Borelli et al. 2020). Therefore, it is very important to record WFPs utilisation knowledge that is being lost at a rapid rate, and also to motivate the local communities to promote the sustainable use of WFPs in their local diet (de Medeiros et al. 2021; FAO 2017; Gold and McBurney 2010; Pieroni et al. 2005; Bortolotto et al. 2017). Wild food plants have also received little attention in valuating, quantifying and mapping studies, due to a lack of data or due to the perceived low importance (Kalle et al. 2020; Schulp et al. 2014). The investigation of WFPs is multidisciplinary and involves ethnobotany, nutrition study, medicine, analytical chemistry, anthropology and phytochemistry (Grivetti and Ogle 2000; Onyechi et al. 1988; Kuhnlein and Turner 1991; Salih et al. 1991; Martin 1995). A database is an information that is set up for easy access, management and updating and they also offer the opportunity to access a wide variety of relevant data in one place. An automatic database must manage the knowledge properly so that users can access it in a simple and quick way (Figueiredo and Pereira 2017). A number of databases are available which can be used for the study of WFPs. In spite of having such a large amount of diverse data, it has not yet been efficiently explored, as many users and researchers are not familiar with all the possible resources to search and analyse the data (Garg and Jaiswal 2016). Further, the increasing number and types of databases makes it more difficult for researchers and prospective users to find out where to go for what information (Rhee and Crosby 2005). Therefore, the present book chapter is an attempt to summarise the available databases (Fig. 13.1) that can be helpful in the study of WFPs.

13.2 Need for Databases

Plant research has undergone various rounds of transformation and in the past few decades, tremendous amounts of data have been produced from plants and are improving with time. For its effective use, it is therefore essential to facilitate and enhance the processing, integration, and interpretation of this large amount of dataset (Rhee and Crosby 2005). A standard way of managing and processing such a huge amount of information has been the creation of databases and this is being done in diverse arenas such as academic disciplines, industries and even in the government sector (Rhee and Crosby 2005). The primary aim of all plant databases has been to acquire, organise, represent and also help to navigate and retrieve the information of interest (Garg and Jaiswal 2016). The retrieval of primary information from various databases can be helpful to gain advanced scientific knowledge (Molina-Venegas et al. 2021a). For example, a recent study drawn on a global database of plant

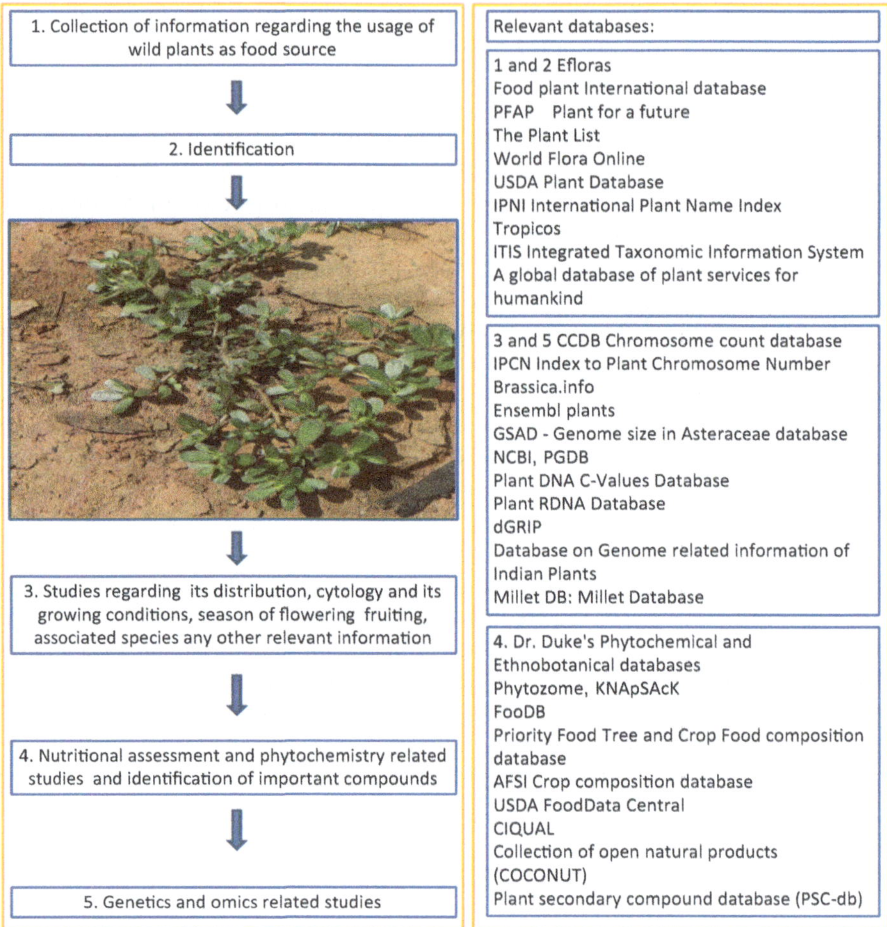

Fig. 13.1 Databases that are relevant to study wild food plants

services for humankind showed that phylogenetic diversity can capture plant services efficiently (Molina-Venegas et al. 2021b), supporting a promising macroevolutionary perspective on global biodiversity conservation (Mooers and Tucker 2021). This is just one case of the use of databases. Besides this, there are several ways in which these databases can expand our current scientific knowledge and will also play an important role to face various biodiversity challenges in the coming decades (Díaz et al. 2020). The databases can also be helpful for the identification and sustainable use of wild food plants. Recently Intergovernmental Science-Policy Platform on Biodiversity and Ecosystem Services (IPBES) has also recognised that there is a need to analyse the utilisation of food plant species, to promote their sustainable utilisation (IPBES 2018).

13.3 Databases for Wild Food Plants

A number of databases can be used to study wild food plants. The relevant databases are hereby categorised into three categories:

13.3.1 Distribution, Taxonomy, and Traditional Uses Related

Documentation of traditional use, correct identification and distribution are important components for the study of wild food plants. There is no single database dedicated to these aspects of wild food plants but the available databases can be used directly or indirectly to gather information on wild food plants. For example, various online floras can be used for the identification of plants and databases such as The Plant List, World Flora Online and IPNI provide information about the correct names of the plants. Similarly, the Food Plant International Database provides information such as scientific names, common names, pictures, nutritional values, etc. The Plants For A Future (PFAP) database contains information on over 8000 edible and useful plants. The list of other databases is given in Table 13.1. Figure 13.1 provides details about the databases relevant to wild food plants. Figure 13.2 outlines the important steps that are required for the development of the databases related to wild food plants.

13.3.2 Chromosome Number and Genomics-Related Databases

Chromosome numbers of a species are useful to study and describe its karyotype and any changes in the number of chromosomes indicate genomic changes such as dysploidy and polyploidy (Mayrose and Lysak 2021). Various databases are available that provide information on the chromosome number of a species. The chromosome count database (CCDB) is one such database that provides information on the chromosome number of a species. It includes 385,707 entries and is an open-access database. Similarly, Index to Plant Chromosome Number (IPCN) also provides information on the chromosome number of a species.

Besides chromosome number, the genome size (c-value) of an organism is also applicable to various scientific fields, including ecology (Kocjan et al. 2022). The Plant DNA C-value database provides information on c-value of about 12,273 species. Genome analysis is one of the important components of plant study. The main aim of genomic analysis is to understand the biology of an organism both in terms of its function and evolution (Koonin and Galperin 2003). Large amounts of genome-related information are stored in public databases and are freely available. Many nucleotide sequence databases are available which provide information on

Table 13.1 Distribution, taxonomy and traditional uses related databases

Sl. no.	Name of database	Information contains	URL
1.	Food Plant International Database	It has information on scientific names, common names, genera, synonyms, description of plants, nutritional value, pictures, etc.	https://foodplantsinternational.com/plants/
2.	PFAP Plant For A Future (PFAP)	It contains information on over 8000 edible and useful plants	https://pfaf.org/
3.	The Plant List	It provides information regarding the accepted Latin name, with links to all synonyms	http://www.theplantlist.org/
4.	World Flora Online	It is an open-access compendium of world's plant species	https://about.worldfloraonline.org/
5.	USDA Plant Database	It provides standardised information about the flora of the USA and its territories	https://plants.sc.egov.usda.gov/home
6.	IPNI International Plant Name Index (IPNI)	It is an open-access database that provides nomenclatural information for the scientific names of vascular plants	https://www.ipni.org/
7.	Tropicos	It provides information on scientific names, classification, images and herbarium specimens of various plants. It links over 1.33M scientific names with over 4.87M and 685K specimens and digital images respectively	https://tropicos.org/home
8.	Euro-Med plant database	Open access database which provides information of Europe flora	http://ww2.bgbm.org/EuroPlusMed/query.asp
9.	ITIS Integrated Taxonomic Information System	Open access database which provides taxonomic information on plants, fungi, etc.	https://www.itis.gov/
10.	A global database of plant services for humankind	A dataset that provides information of plant-use records of 13,489 genera	https://doi.org/10.6084/m9.figshare.13625546.v3

data sequences to the general public and scientific community. One such database is NCBI which is a leading database for analysing molecular and genomic data and is open accessible for the scientific and medical communities, as well as to the general public (Smith 2013). Similarly, plantGDB database is a database of molecular sequence data (Dong et al. 2004). The PlantGDB database contains sequence data of about 70,000 species including the complete sequence of important species such as rice, maize, *Arabidopsis* and *Medicago truncatula* (Lai et al. 2012). Similarly, Phytozome is an important database to access, visualise and analyse various plant genomes (Goodstein et al. 2012). Besides these general databases, there are also species focussed databases such as Graingene is a genetic database for triticaceae, oats and sugarcane. Similarly MilletDB is a genetic and phenotypic database for millets. The other relevant databases are given in Table 13.2.

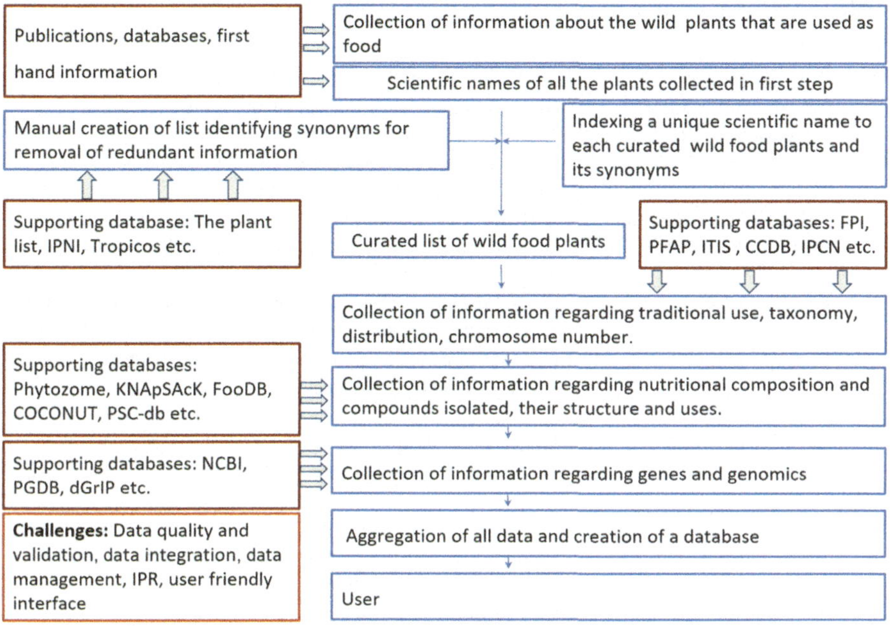

Fig. 13.2 Steps to create a database for WFPs (Adapted from Ningthoujam et al. 2012 and Mohanraj et al. 2018)

13.3.3 Phytochemistry-Related Databases

Plants contain a number of compounds that play an important role in their growth and development. Many of these compounds are useful for human beings as well. Over the past three decades, considerable information on the biological properties as well as the chemistry of phytochemicals has been accumulated. Some of the data on the phytochemical composition of plants have been collected as well as stored in various databases so that they can be automatically updated and retrieved (Scalbert et al. 2011). Phytochemical databases have multiple uses and can provide various information on the phytochemicals such as their source, intake range as well as intercorrelations (Ziegler 2001) PubChem is an open-source database that provides information about various small and large molecules on their physical and chemical properties, chemical structures, identifiers, biological activities, safety, toxicity data, etc. KNApSAcK and the Dictionary of Food Compounds provide information on various compounds and also offer strong taxonomic components. But the USDA Food Composition database and Dr. Duke's Phytochemical databases are strictly nutrient composition databases. Phenol-Explorer is an open-access database that provides information on polyphenol contents of food. The CIQUAL database contains nutritional information on various food plants used mainly in France. The other important phytochemical databases are given in Table 13.3.

Table 13.2 Chromosome number and genomics related databases

Sl. no.	Name of database	Information contains	URL
1.	Chromosome Count Database (CCDB)	It is a community resource for plant chromosome numbers. It includes 385,707 entries, hierarchically ordered by taxonomic conventions	http://ccdb.tau.ac.il/home/
2.	Index to Plant Chromosome Number (IPCN)	It provides information on the chromosome number of various plant species	http://legacy.tropicos.org/Project/IPCN
3.	Brassica.info	This site provides information on Brassica genomics and genetics	https://brassica.info/home/about_site.html
4.	Ensembl plants	Ensembl Plants is an open-source system for handling genomes	https://plants.ensembl.org/index.html; Birney et al. (2004)
5.	GSAD—Genome size in Asteraceae database	The GSAD is an open-access database of genome size data for the family Asteraceae	https://www.asteraceaegenomesize.com/
6.	NCBI	NCBI is a house of many databases which provides access to biomedical and genomic information and also bioinformatics tools	https://www.ncbi.nlm.nih.gov/
7.	PGDB	It is an open-access database that contain genome-related information	http://www.plantgdb.org/; Duvick et al. (2008)
8.	Phytozome	Phytozome is an important database to access, visualise and analyse various plant genomes	https://phytozome-next.jgi.doe.gov/
9.	Plant DNA C-Values Database	Contain information regarding the C-value of DNA. It contains genome size data of about 12,000 species	https://cvalues.science.kew.org/
10.	Plant rDNA Database	It provides information on the numbers and positions of rDNA signals and their structures for a number of plant species	https://www.plantrdnadatabase.com/
11.	PTGBase	It is useful to study tandem duplicated genes of plants	http://ocri-genomics.org/PTGBase/
12.	Database on Genome related information of Indian Plants (dGRIP)	This database shows chromosome numbers and genome sizes of various Indian plant species	http://sbtju.in/Dgrip/index.html
13.	Millet Database (Millet DB)	This database records the genotype-phenotype information of millets	https://db.cngb.org/datamart/plant/DATApla6/
14.	Genome Structural Database (GSDB)	It contains more than 50,000 3D structures of chromosomes and genomes	http://sysbio.rnet.missouri.edu/3dgenome/GSDB/index.php
15.	GrainGene	It is a database for molecular and phenotypic information of wheat, rye, barley, oat and other related species	https://wheat.pw.usda.gov/GG3/

Table 13.3 Phytochemical analysis-related databases

Sl. no.	Name of database	Information	URL
1.	Dr. Duke's Phytochemical and Ethnobotanical Database	It provides information on chemicals, ethnobotany and bioactivity of various plants	https://phytochem.nal.usda.gov/phytochem/search; U.-S. Department of Agriculture, Agricultural Research Service, 1992–2016; Duke (1992)
2.	KNApSAcK	It is a comprehensive species-metabolite database	http://www.knapsackfamily.com/KNApSAcK/
3.	FooDB	It provides information on food constituents, their chemistry and biology	https://foodb.ca/
4.	Priority Food Tree and Crop Food composition database	It provides information on the food crops composition	http://apps.worldagroforestry.org/products/nutrition/index.php/food/
5.	AFSI Crop Composition Database	It contains composition data of various conventional crops	https://www.cropcomposition.org/
6.	USDA FoodData Central	It contains nutrient profile data of various food items	https://fdc.nal.usda.gov/
7.	CIQUAL	It contains nutrient profile data of various foodstuffs mostly used in France	https://ciqual.anses.fr/#
8.	EuroFIR, European Food Information Resource	This database contains information on food composition	https://www.eurofir.org/
9.	Frida Food Data	It contains nutrient profile data of various foodstuffs mostly used in Denmark	https://frida.fooddata.dk/
10.	Plant Secondary Compound Database (PSC-db)	This database contains information on the 3D-structures of secondary metabolites along with their physico-chemical and pharmaceutical properties	http://pscdb.appsbio.utalca.cl/viewIndex/index.php
11.	Collection of open natural products (COCONUT)	It is an open-source portal for natural products cheminformatics	https://coconut.naturalproducts.net/
12.	Naturally occurring plant-based anti-cancer compound activity target database (NPCAT)	It contains information on compounds having anticancer property	http://crdd.osdd.net/raghava/npact/
13.	PubChem	It is an open-access database of various small and large molecules	https://pubchem.ncbi.nlm.nih.gov/
14.	Dictionary of Food Compounds	It provides information on over 30,000 food compounds	https://dfc.chemnetbase.com/faces/chemical/ChemicalSearch.xhtml
15.	Phenol Explorer	It contains polyphenol composition of over 400 foods	http://phenol-explorer.eu/

13.4 Conclusions

Nowadays, databases are very important for managing the huge amount of data that is being generated on a daily basis. WFPs have been used as a food source for centuries. Besides acting as a source of food, these WFPs have tremendous potential for the development of new food crops. They are also used as 'famine food' in many parts of the world. But nowadays, their use is restricted due to various reasons. Their use is limited to a few regions and a few groups such as tribal people only. One of the many reasons for this is the lack of research on wild food plants and their restricted distribution. Databases on wild food plants can be a catalyst in promoting research on wild food plants. But there is no database specifically for wild food plants and various associated databases can be used as a base for generating a dedicated database for wild food plants. Therefore there is a need for forming a database like the recently developed database IMPPAT (Indian medicinal plants phytochemistry and therapeutics) for Indian medicinal plants (Fig. 13.2) to facilitate data management and analysis making coherent information available to planners, researchers and other users.

References

Abbet C, Mayor R, Roguet D, Spichiger R, Hamburger M, Potterat O (2014) Ethnobotanical survey on wild alpine food plants in Lower and Central Valais (Switzerland). J Ethnopharmacol 151(1): 624–634. https://doi.org/10.1016/j.jep.2013.11.022. Epub 2013 Nov 20. PMID: 24269246

Ahmad K, Pieroni A (2016) Folk Knowledge of wild food plants among the tribal communities of Thakht-e-Sulaiman Hills, North-West Pakistan. J Ethnobiol Ethnomed 12:17

Bharucha Z, Pretty J (2010) The roles and values of wild foods in agricultural systems. Philos Trans R Soc B 365:2913–2926. https://doi.org/10.1098/rstb.2010.0123

Birney E, Andrews TD, Bevan P, Caccamo M, Chen Y, Clarke L, Coates G, Cuff J, Curwen V, Cutts T, Down T, Eyras E et al (2004) An overview of Ensembl. Genome Res 14:925–928. https://doi.org/10.1101/gr.1860604

Borelli T, Hunter D, Powell B, Ulian T, Mattana E, Termote C, Pawera L, Beltrame D, Penafiel D, Tan A, Taylor M, Engels J (2020) Born to eat wild: an integrated conservation approach to secure wild food plants for food security and nutrition. Plants 9:1299. https://doi.org/10.3390/plants9101299

Bortolotto IM, Hiane PA, Ishii IH, de Souza PR, Campos RP, Gomes RJB, da Silva Farias C et al (2017) A knowledge network to promote the use and valorization of wild food plants in the Pantanal and Cerrado, Brazil. Reg Environ Chang 17:1329–1341. https://doi.org/10.1007/s10113-016-1088-y

Chopak C (2000) Early warning primer. An overview of monitoring and reporting. United States Agency for International Development Famine Early Warning System Project

Colin K, Luigi G (2010) Back to the roots: wild genes for food security. Samara: The International Newsletter of the Partners of the Millennium Seed Bank Partnership 18(19):1–8

Dansi A, Adjatin A, Adoukonou-Sagbadja H, Faladé V, Yedomonhan H, Odou D, Dossou B (2008) Traditional leafy vegetables and their use in the Benin Republic. Genet Resour Crop Evol 55: 1239–1256 de Medeiros PM, dos Santos GMC, Barbosa DM, César L, Gomes A, de Costa Santos EM, Vasconcelos da Silva RR (2021) Local knowledge as a tool for prospecting wild food plants: experiences in northeastern Brazil. Sci Rep 11:594. https://doi.org/10.1038/s41598-020-79835-5

Díaz S, Zafra-Calvo N, Purvis A, Verburg PH, Obura D, Leadley P et al (2020) Set ambitious goals for biodiversity and sustainability. Science 370:411–413. PMID: 33093100

Dong Q, Schlueter SD, Brendel V (2004) PlantGDB, plant genome database and analysis tools. Nucleic Acids Res 32(Database issue):D354–D359. https://doi.org/10.1093/nar/gkh046

Duke JA (1992) Database of biologically active phytochemicals and their activity. CRC, Boca Raton. ISBN: 9780849336713, 183 pp

Duvick J, Fu A, Muppirala U, Sabharwal M, Wilkerson MD, Lawrence CJ, Lushbough C, Brendel V (2008) PlantGDB: a resource for comparative plant genomics. Nucleic Acids Res 36(Database issue):D959–D965. https://doi.org/10.1093/nar/gkm1041. Epub 2007 Dec 6. PMID: 18063570; PMCID: PMC2238959

EFSA (2009) EFSA Compendium of botanicals that have been reported to contain toxic, addictive, psychotropic or other substances of concern. EFSA J 7:281

EFSA (2012) Compendium of botanicals reported to contain naturally occurring substances of possible concern for human health when used in food and food supplements. EFSA J 10:2663

Etkin NL (1994) Eating on the wild side. The pharmacologic, ecologic and social implications of using non cultigens. University of Arizona Press, Tucson. 08-165-13694

FAO (2017) Voluntary guidelines for the conservation and sustainable use of crop wild relatives and wild food plants

Figueiredo MSN, Pereira AM (2017) Managing knowledge—the importance of Database in scientific production. Proc Manuf 12:166–173

Food and Agriculture Organisation of the United Nations (2011) The state of food insecurity in the world: how does international price volatility affect domestic economies and food security? FAO, Rome. http://www.fao.org/docrep/014/i2330e/i2330e00.htm

Food and Agriculture Organisation of the United Nations (2012) The state of food insecurity in the world: economic growth is necessary but not sufficient to accelerate reduction of hunger and malnutrition. FAO, Rome

Food and Agriculture Organization of the United Nations (FAO) (2020) State of Food Security and Nutrition in the World 2020 (SOFI). Food and Agriculture Organization of the United Nations, Rome

Garg P, Jaiswal P (2016) Databases and bioinformatics tools for rice research. Curr Plant Biol 7-8: 39–52. https://doi.org/10.1016/J.CPB.2016.12.006

Ghorbani A, Langenberger G, Sauerborn J (2012) A comparison of the wild food plant use knowledge of ethnic minorities in Naban River Watershed National Nature Reserve, Yunnan, SW China. J Ethnobiol Ethnomed 8:17. https://doi.org/10.1186/1746-4269-8-17

Gold K, McBurney RPH (2010) Conservation of plant diversity for sustainable diets. In: Burlingame B, Dernini S (eds) Sustainable diets and biodiversity: directions and solutions for policy, research and action. FAO Headquarters, pp 30–36

Goodstein DM, Shu S, Howson R, Neupane R, Hayes RD, Fazo J, Mitros T, Dirks W, Hellsten U, Putnam N, Rokhsar DS (2012) Phytozome: a comparative platform for green plant genomics. Nucleic Acids Res 40(1):D1178–D1186. https://doi.org/10.1093/nar/gkr944

Grivetti LE, Ogle BM (2000) Value of traditional foods in meeting macro- and micronutrient needs: the wild plant connection. Nutr Res Rev 13:31–46

Guinard Y, Lemessa D (2000) Wild-food plants in Southern Ethiopia. Reflections on the role of 'Famine Foods' at a time of drought. United Nations Development Programme—Emergencies Unit for Ethiopia

Hunter D, Fanzo J (2013) Agricultural biodiversity, diverse diets and improving nutrition. In: Fanzo J, Hunter D et al (eds) Diversifying food and diets: using agricultural biodiversity to improve nutrition and health. Issues in Agricultural Biodiversity, Earthscan, pp 1–13

IPBES (2018) Information on scoping for the thematic assessment of sustainable use of wild species. Intergovernmental Science-Policy Platform on Biodiversity and Ecosystem Services, Bonn

Kalle R, Belichenko O, Kuznetsova N, Kolosova V et al (2020) Gaining momentum: popularization of *Epilobium angustifolium* as food and recreational tea on the Eastern edge of Europe. Appetite 150:104638

Kocjan D, Dolenc Koce J, Etl F, Dermastia M (2022) Genome size of life forms of *Araceae*—a new piece in the C-value puzzle. Plants 11(3):334. https://doi.org/10.3390/plants11030334

Koonin EV, Galperin MY (2003) Chapter 3: Information sources for genomics. In: Sequence—evolution—function: computational approaches in comparative genomics. Kluwer Academic, Boston. https://www.ncbi.nlm.nih.gov/books/NBK20256/

Kuhnlein HV, Turner NJ (1991) Traditional plant foods of Canadian indigenous peoples: nutrition, botany and use. Food and nutrition in history and anthropology. Gordon and Breach Science Publishers, London

Kumar A, Anju T, Kumar S, Chhapekar SS, Sreedharan S, Singh S, Choi SR, Ramchiary N, Lim YP (2021) Integrating omics and gene editing tools for rapid improvement of traditional food plants for diversified and sustainable food security. Int J Mol Sci 22(15):8093

Lai K, Lorenc MT, Edwards D (2012) Genomic databases for crop improvement. Agronomy 2(1): 62–73. https://doi.org/10.3390/agronomy2010062

Lamien-Meda A, Lamien CE, Compaoré MM, Meda RN, Kiendrebeogo M, Zeba B, Millogo JF, Nacoulma OG (2008) Polyphenol content and antioxidant activity of fourteen wild edible fruits from Burkina Faso. Molecules 13:581–594

Leborgne P, Wilkinson C, Montembaut S, Tesse-Ververs M (2002) Scurvy outbreak in Afghanistan. An investigation by Action Contre la Faim (ACF) and the World Health Organization (WHO). Field Exchange No. 17, November 28–29

Łuczaj Ł, Pieroni A, Tardío J, Pardo-de-Santayana M, Sõukand R, Svanberg I, Kalle R (2012) Wild food plant use in 21st century Europe: the disappearance of old traditions and the search for new cuisines involving wild edibles. Acta Soc Bot Pol 81:359–370

Łuczaj Ł, Stawarczyk K, Kosiek T, Pietras M, Kujawa A (2015) Wild food plants and fungi used by Ukrainians in the western part of the Maramureş region în România. Acta Soc Bot Pol 84(3): 339–346. https://doi.org/10.5586/asbp.2015.029

Martin G (1995) Ethnobotany. A methods manual. Chapman & Hall, London

Mayrose I, Lysak MA (2021) The evolution of chromosome numbers: mechanistic models and experimental approaches. Genome Biol Evol 13:2. https://doi.org/10.1093/gbe/evaa220

Mohanraj K, Karthikeyan BS, Vivek-Ananth RP et al (2018) IMPPAT: a curated database of Indian medicinal plants, phytochemistry and therapeutics. Sci Rep 8:4329. https://doi.org/10.1038/s41598-018-22631-z

Molina-Venegas R, Rodríguez MÁ, Pardo-de-Santayana M, Mabberley DJ (2021a) A global database of plant services for humankind. PLoS One 16(6):e0253069. https://doi.org/10.1371/journal.pone.0253069

Molina-Venegas R, Rodríguez MA, Pardo-de-Santayana M et al (2021b) Maximum levels of global phylogenetic diversity efficiently capture plant services for humankind. Nat Ecol Evol 5:583–588. https://doi.org/10.1038/s41559-021-01414-2

Mooers A, Tucker CM (2021) Useful plants have deep evolutionary roots. Nat Ecol Evol 5:558–559. https://doi.org/10.1038/s41559-021-01438-8

Ningthoujam SS, Talukdar AD, Potsangbam KS, Choudhury MD (2012) Challenges in developing medicinal plant databases for sharing ethnopharmacological knowledge. J Ethnopharmacol 141:9–32. https://doi.org/10.1016/j.jep.2012.02.042

Odhav B, Beekrum S, Akula US, Baijnath H (2007) Preliminary assessment of nutritional value of traditional leafy vegetables in KwaZulu-Natal, South Africa. J Food Compos Anal 20:430–435

Onyechi UA, Judd PA, Ellis PR (1988) African plant foods rich in non-starch polysaccharides reduce postprandial blood glucose and insulin concentrations in healthy human subjects. Br J Nutr 30(5):419–428

Padulosi S, Pignone D (eds) (1997) Rocket: a Mediterranean crop for the world. Report of a workshop 13-14 December 1996, Legnaro (Padova), Italy. International Plant Genetic Resources Institute, Rome

Pawera L, Khomsan A, Zuhud EAM, Hunter D, Ickowitz A, Polesny Z (2020) Wild Food plants and trends in their use: from knowledge and perceptions to drivers of change in West Sumatra, Indonesia. Foods 9:1240

Pieroni A, Nebel S, Santoro RF, Heinrich M (2005) Food for two seasons: culinary uses of non-cultivated local vegetables and mushrooms in a south Italian village. Int J Food Sci Nutr 56(4):245–272

Pinela J, Carocho M, Dias MI, Caleja C, Barros L, Ferreira ICFR (2016) Wild plant-based functional foods, drugs, and nutraceuticals. In: Ferreira ICFR, Morales P, Barros L (eds) Wild plants, mushrooms and nuts. Wiley, Chichester, pp 315–351

Pinela J, Carvalho AM, Ferreira ICFR (2017) Wild edible plants: nutritional and toxicological characteristics, retrieval strategies and importance for today's society. Food Chem Toxicol 110: 165–188. https://doi.org/10.1016/j.fct.2017.10.020. Epub 2017 Oct 14. PMID: 29042290

Powell B, Thilsted SH, Ickowitz A et al (2015) Improving diets with wild and cultivated biodiversity from across the landscape. Food Secur 7:535–554. https://doi.org/10.1007/s12571-015-0466-5

Rahmato D (1988) Peasant survival strategies in Ethiopia. Disasters 12:326–344

Ray A, Ray R, Sreevidya EA (2020) How many wild edible plants do we eat—their diversity, use, and implications for sustainable food system: an exploratory analysis in India. Front Sustain Food Syst 4:56. https://doi.org/10.3389/fsufs.2020.00056

Rhee SY, Crosby WL (2005) Biological databases for plant research. Plant Physiol 138:1–3

Romeo R, Vita A, Testolin R, Hofer T (2015) Mapping the vulnerability of mountain peoples to food insecurity. FAO, Rome

Salih OM, Nour AM, Harper DB (1991) Chemical and nutritional composition of 2 famine food sources used in Sudan, Mukheit (*Boscia senegalensis*) and Maikah (*Dobera roxburghii*). J Sci Food Agric 57:367–377

Scalbert A, Andres-Lacueva C, Arita M, Kroon P, Manach C, Urpi-Sarda M, Wishart D (2011) Databases on food phytochemicals and their health-promoting effects. J Agric Food Chem 59(9):4331–4348. https://doi.org/10.1021/jf200591d. Epub 2011 Apr 12. PMID: 21438636

Schulp CJE, Thuiller W, Verburg PH (2014) Wild food in Europe: a synthesis of knowledge and data of terrestrial wild food as an ecosystem service. Ecol Econ 105:292–305

Scoones I, Melnyk M, Pretty NJ (1992) The hidden harvest: wild foods and agricultural systems. A literature review and annotated bibliography. IIED, London

Smith K (2013) A brief history of NCBI's formation and growth. In: The NCBI handbook [Internet], 2nd edn. National Center for Biotechnology Information (US), Bethesda. https://www.ncbi.nlm.nih.gov/books/NBK148949/

Sowunmi MA (1985) The beginnings of agriculture in West Africa: botanical evidence. Curr Anthropol 26:127–129

Tardío J, Pardo-de-Santayana MA, Morales R (2006) Ethnobotanical review of wild edible plants in Spain. Bot J Linnean Soc 152:27–71

U.S. Department of Agriculture, Agricultural Research Service. (1992–2016) Dr. Duke's phytochemical and ethnobotanical databases. Home Page: http://phytochem.nal.usda.gov/; https://doi.org/10.15482/USDA.ADC/1239279

World Food Programme (1996) Vulnerability analysis and mapping. World Food Programme of the United Nations, Rome

Ziegler RG (2001) The future of phytochemical databases. Am J Clin Nutr 74(1):4–5. https://doi.org/10.1093/ajcn/74.1.4

Zohary D, Hopf M, Weiss E (2012) Domestication of plants in the old world. The origin and spread of domesticated plants in South-West Asia, Europe, and the Mediterranean Basin, 4th edn. Oxford University Press, New York

Chapter 14
A Comprehensive Update on Traditional Agricultural Knowledge of Farmers in India

Avinash Sharma, Chowlani Manpoong, Himanshu Pandey, Chandan Kumar Gupta, Yani Baja, Mayanglambam Sanjit Singh, and Chau Chiktiya Mounglang

Abstract Traditional agricultural knowledge is an important information created by the forefathers in past civilizations. The forefathers practised traditional agriculture during Harappa, Vedic and Iron Age civilizations. The present small and marginal farmers utilize traditional information in crop production and management, crop protection, farm machinery, tools, soil and water management, medicinal and aromatic plants for diseases diagnosis, animal husbandry, stored grain pest management, weed management and value-added value products. The traditional information in agriculture practices is collected from the different geographical states of India. The information from traditional practices used in specific activities by the farmers. The farmer utilizes compositions of natural resources in the geographical states for crop husbandry and farm-linked activities. The Southern and North-Eastern states of India have more traditional agricultural practices. The farmer applies specific information in crop production and management, crop protection, farm machinery and tools, soil and water management, medicinal and aromatic plants for disease diagnosis, animal husbandry, stored grain pest management, weed management, and value-added food products. The farmer preserves and transfers the information to the rural community. The farmer transmits information to the present generation to create mobilisation. Traditional agriculture information transforms agriculture resources, maintains biodiversity ethics and enlightens historical and practical approaches to the present generations. This chapter outlines and

A. Sharma (✉) · C. Manpoong · M. S. Singh · C. C. Mounglang
Faculty of Agricultural Sciences, Arunachal University of Studies, Namsai, India

H. Pandey · C. K. Gupta (✉)
Division of Plant Physiology and Biochemistry, Indian Institute of Sugarcane Research, Lucknow, India
e-mail: chandan.gupta@icar.gov.in

Y. Baja
Department of Agricultural Entomology, Bidhan Chandra Krishi Viswavidyalaya, Mohanpur, West Bengal, India

A. Kumar et al. (eds.), *Wild Food Plants for Zero Hunger and Resilient Agriculture*, Plant Life and Environment Dynamics,
https://doi.org/10.1007/978-981-19-6502-9_14

provides comprehensive information on the traditional agricultural knowledge of the farmers.

Keywords Indigenous practices · Traditional agricultural knowledge · Indian agriculture · Farmers

14.1 Introduction

The collection and utilisation of agricultural information from crop production, plant protection, animal husbandry, farm machine and tools, stored grains pest management, rituals in agriculture, and value-added product preparations for agriculture activities in the present period is called traditional knowledge of agriculture. Land preparations, cropping systems, cropping patterns, and preservation of grains have emerged in the Harappa civilizations. The rotation of crops, processing of crops, and preservation were initiated in the Vedic and Iron Age civilizations. Jhum and Kumari cultivation originated in the Aryavarta period in the eleventh century AD. The south Indian farmers cultivated rice in jhum cultivation and kumari cultivation. The local paddy variety known as Sali was cultivated in the Aristobalus period in 320 BC. The transplantation of seedlings in highlands and lowlands around the volume of water was initiated in the Aristobalus period. Huge traditional knowledge emerged and was utilized in ancient India. Traditional knowledge is still useful for economic development and policy formulation.

The large and small-scale industry applies traditional agriculture knowledge for producing commodities without Intellectual Property Rights (IPRs) policy. The government of India amended the IPR policy to use traditional knowledge of agriculture (Sofia and Deepa 2020). The traditional knowledge enlightens and transfers the agriculture education and research that advanced the agriculture technology in India, transforming sustainable agriculture into modern smart agriculture. The small and marginal farmers involve more in agriculture occupation, countering the high cost of agriculture technology in smart agriculture. The Indian farmers are determined to identify the modern agriculture technology in Digital India. However, the small and marginal farmers are unable to purchase modern technology for agricultural activity because of the low land area, less per capita income, and less support from the government schemes. Some farmers still follow traditional knowledge of agriculture for crop production, farm-linked activities, and management. The farmers utilize conventional knowledge in crop production, soil and water management, pest and disease management, animal husbandry, and farm machine and tools (Anupam et al. 2020) (Fig. 14.1). The traditional knowledge in agriculture has originated from the various cultures and communities of the farmers. The farmer utilizes ancient agriculture information to reduce the cost of cultivation and maintain naturally selected agriculture (Ajay 2018). The traditional knowledge of agriculture is available in the form of myths, lyrics, quotes, dances, cultural age and taxonomy, cultural materials, and species. The goals of traditional knowledge are to restrict the overexploitation of natural resources and restore long-term natural

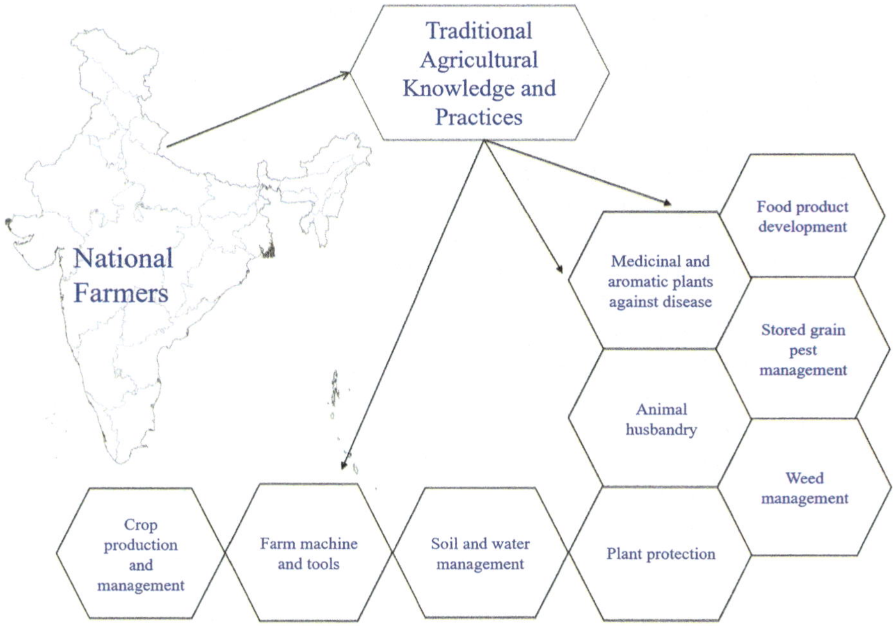

Fig. 14.1 Overview of farmers' traditional agriculture knowledge (TAK) in India

resources (Kareemulla et al. 2020). The traditional knowledge is affordable and non-detrimental to the environment. The conventional knowledge of crop production and management, animal husbandry, disease diagnosis, stored grain pests management, weed management, and food products was compiled from respective geographical states in India and are presently utilised by regional farmers for crop production and farm-linked activities. This chapter discusses indigenous knowledge practices of farmers in (1) crop production and management, (2) plant protection, (3) farm machine and tools, (4) soil and water management, (5) usage pf medicinal and aromatic plants against diseases, (6) animal husbandry, (7) stored grain pest management, (8) weed management, and (9) food product development.

14.2 Various Traditional Practices of Farmers

14.2.1 Crop Production and Management

The farmers of Himachal Pradesh utilized a grafted mango pit, dug out 3 × 3 × 3 ft, 25 ft apart on either side. The pit is dried in the presence of sunlight for 3 months. The community of weeds, pests, and insects declined in the pit. The burned leaves and twigs are buried near the mango tree in the early morning and late evening, promoting flowering and mitigating hopper populations (Kranthi et al. 2016). The

Tamil Nadu and Northeastern farmers practice sunflower cultivation between mango trees that attract honey bees and raise pollination and fruit production (Das et al. 2019). On the day of the solar eclipse, 3–5 ft of the mango tree's bark is excised from the ground. In Andhra Pradesh, farmers cover mango fruit with paddy straw in a closed room for a week for uniform ripening of the mango fruit (Mangal 2009).

In Manipur, farmers use Pit Nursery Method for crop seedlings in the pit. The pit maintains water requirements and forbids evapotranspiration loss (Ansaria et al. 2021). The Jharkhand farmers use *Paira* cropping system in rice lowlands and broadcasted *Lathyrus* seeds in the main field of rice for water management (Dey and Sarkar 2011). The farmers of the Western Ghat zone of Maharashtra cultivate indigenous rice Tulshi tall, and the farmer of the Konkan region of Maharashtra cultivate indigenous rice Vikram which has several medicinal benefits against diabetes, obesity, and cardiovascular diseases. The farmers of West Bengal grow Tulaipanji's traditional aromatic rice variety for aroma seed quality. In the Kullu valley of Himachal Pradesh, the farmers grow the traditional aromatic rice cultivar Jatu rice for aroma and taste. Matali and Lal Dhan local rice cultivars are grown in Himachal Pradesh and these rice cultivars possess medicinal properties against fever and blood pressure. In the hills of Himachal Pradesh and Uttar Pradesh, farmers grow the traditional red rice variety Kafalya for curing leucorrhoea and complicated abortion. The Tamil Nadu farmers cultivate traditional rice cultivars Kari Kagga and Atikaya to regulate heat in the human body and drug preparation. The Tamil Nadu farmer grows the ancient rice cultivar Neelam Samba to regulate milk in the mother. The Tamil Nadu farmers also grow local rice Maappillai Samba to increase fertility in the mother. The Assam farmers cultivate traditional cultivar black rice for mitigating cancer disease and rice bran is used to decrease inflammation, allergies, and asthma. In Kerala, farmers grow local rice Karinjan and Karimalakaran which posses antidiabetic properties. In Kerala, farmers grow traditional rice Mundakan to increase the stamina of humans, Vella chennellu and Chuvanna chennellu to reduce, menopause, and hormone problems. In Bihar and Chhattisgarh, farmers cultivate Jonga and Maharaji traditional rice to increase lactation in the mother. The Assam farmers grow local rice Bora for curing jaundice. In Chhattisgarh and Jharkhand, farmers cultivate local rice Karhani to formulate a drug for epilepsy treatment (Ann et al. 2019; Krishnankutty et al. 2021).

In Kerala, farmers burn dry leaves, twigs, and trash on the ground for the fresh planting of the banana sucker. A 500-g groundnut cake is applied to the planted banana sucker for better crop growth and yield (Alexander et al. 2009). Onion, tomato, cowpea, and okra are intercropped with the banana crop in the early phase for 4 months. The leaves of *Basella alba* (Basale) are kept in banana gunny bags/basket. The gunny bags are covered airtight with a cover lid. A 10-ft pit is dug out before planting the grape. In Meghalaya, farmers utilize green leaf manures such as kolangi (*Tephrosia purpurea*), Agave spp., and Ekka (*Calotrpis* spp.), which are poured into the 10-ft pit. The 10-ft pit is covered with a layer of soil for the decomposition process for 3 months. After 3 months, the grape plant is grown in the 10-ft pit. The powdered groundnut cake is dissolved in the water. The powdered groundnut cake containing water is stored overnight. The next day, the solution is poured into each pit of the planted grape tree for better fruit quality and yield. The tip of the grape bunch is thinned to promote better fruit size. The tip of the pineapple is nipped to obtain more fruit weight and size (Zizira 2015). The farmers of Tripura hills follow Mai kaim/jhum cultivation. The farmers harvest parboiled paddy in the aluminium container and the parboiled rice is dried in the sunlight. The farmers collect the forest part and the betel vine batch to prepare the broom (Boroj). The east Siang farmers cultivate local cucurbit cultivars such as pumpkin (tupa), ash gourd (pani lao), cucumber (makung), ash gourd (pao), bottle gourd (pani lao), smell melon (pakum barey), snap melon (mare/makungmari), sponge gourd (bul), bitter gourd (karela), cho-cho marrow (tupop), ridge gourd (jhika), snake gourd (dunduli), sweet gourd (bhat karela), pointed gourd (patal), ivy gourd/little gourd (kunduli), and watermelon (kumarah) from the last five decades (Pandey et al. 2021). The farmers in Punjab collect the harvested plant, and stubbles are burned in the field to manage termite populations. The farmers believe that tobacco residue in the soil control termite populations (Jaskarn and Simerjeet 2021). Farmers from the North-East, Bihar, Jammu, and Maharashtra apply ash to the onion nursery bed and field to improve bulb quality (Bhowmick et al. 2010). Table 14.1 presents the indigenous knowledge of farmers in crop production and management.

14.2.2 Plant Protection

The farmers of Punjab practice fire in the field after wheat crop harvesting to control pests (Bhuvaneshwari et al. 2018). The Uttarakhand farmers collect the dry leaves of pine and burn them in the field to control white grubs. The farmers also apply table salt (NaCl) to challah ash to control white grub. The farmers do deep ploughing up to 30 cm depth with indigenous/country ploughs for damaging insect pests pathogens harboured in the soil (Surya et al. 2021). In Kerala and Jharkhand, farmers prepare a mixture of wood ash/kitchen ash + farm yard manure to arrest the chewing and biting mouth part insect (Manoj 2016; Das et al. 2003). In North-eastern region, Uttarakhand, and Madhya Pradesh, farmers use crops such as mustard, cauliflower, and cabbage that were severely infested with an aphid infestation and were buried in the soil for aphid control (Ajay 2018). The Bihar and Uttarakhand farmers earthen to control potato tuber greening and check sunlight exposure (Manish et al. 2011). The Odisha and Kerala farmers prepares cow dung + cow urine compound for controlling pesticide, wilt symptoms, and onion blight. In Uttarakhand, Odisha, and Gujarat, farmers promote Madira or Barnyard millet (*Echinochloa* sp.) and Konri millet crop, mustard intercropped with paddy crop for controlling insect-pests infestation (Rajendra et al. 2018). In Chhattisgarh, farmers prepare a solution with Bicchu booti (*Urtica dioica*) mixed with 8–10 L of cow urine for 24 h and the prepared solution is dispersed in the vegetable crops such as tomato, capsicum, onion, radish, and cucurbits to control fungal diseases. In Tamil Nadu, farmers collect healthy seeds, performs constant smoking with edible and non-edible oil for drying the seeds, mixes ashes of firewood, mixes with neem powder, and stores them in the

Table 14.1 Indigenous knowledge of farmers in crop production and management

Sl. No.	Indigenous Traditional Knowledge (ITK)	Use	State	References
1	Digging and drying of grafted Mango pit	Eradication of weeds, pests, insects	Himachal Pradesh	Kranthi et al. (2016)
2	Sunflower cultivation in between mango trees	Attract honey bees and raises pollination and fruit production	Tamil Nadu and North East	Das et al. (2019)
3	Storing of mango fruit on the paddy straw	Uniform ripening of the mango fruit	Andhra Pradesh	Mangal (2009)
4	Burning of dry leaves and twigs	Fresh planting of the banana sucker	Kerala	Alexander et al. (2009)
5	Green leaf manures such as kolangi (*Tephrosia purpurea*), Agave spp., and Ekka (*Calotrpis* spp.)	Cultivation of grape plant after 3 months	Meghalaya	Zizira (2015)
6	Cultivation of local cucurbit cultivars such as pumpkin (tupa), ash gourd (pani lao), cucumber (makung), ash gourd (pao), bottle gourd (pani lao), smell melon (pakum barey), snap melon (mare/makungmari), sponge gourd (bul), bitter gourd (karela), cho-cho marrow (tupop), ridge gourd (jhika), snake gourd (dunduli), sweet gourd (bhat karela), pointed gourd (patal), ivy gourd/little gourd (kunduli), and water melon (kumarah) since five decades	Local cultivars conservation	East siang, Arunachal Pradesh	Pandey et al. (2021)
7	Burning of harvested plant and stubbles	Termite population control	Punjab	Jaskarn and Simerjeet (2021)
8	Ash in the onion nursery bed and field	Progressing bulb quality	North-East, Bihar, Jammu, and Mahrashtra	Bhowmick et al. (2010)
9	Pit nursery method	Water and evapo-transpiration loss	Manipur	Ansaria et al. (2021)
10	*Paira* cropping system in rice lowlands with broad-casted Lathyrus seeds in main field	Water management	Jharkhand	Dey and Sarkar (2011)
11	Cultivation of indigenous rice Tulshi tall	Medicinal properties against type II diabetes, obesity, and cardiovas-cular diseases	Western Ghat zone of Maharashtra	Ann et al. (2019); Krishnankutty et al. (2021)

(continued)

Table 14.1 (continued)

Sl. No.	Indigenous Traditional Knowledge (ITK)	Use	State	References
12	Cultivation of indigenous rice Vikram	Mitigation of type II, diabetes, obesity, and cardiovascular diseases	Konkan region of Maharashtra	Ann et al. (2019); Krishnankutty et al. (2021)
13	Cultivation of Tulaipanji traditional aromatic rice variety	Aroma seed quality	Northern district of West Bengal	Ann et al. (2019); Krishnankutty et al. (2021)
14	Cultivation of traditional aromatic rice cultivar Jatu rice	Aroma and taste	Kullu valley of Himachal Pradesh	Ann et al. (2019); Krishnankutty et al. (2021)
15	Cultivation of Matali and Lal Dhan local rice cultivars	Fever and reducing blood pressure	Himachal Pradesh	Ann et al. (2019); Krishnankutty et al. (2021)
16	Cultivation of traditional red rice variety Kafalya	Curing leucorrhoea and complicated abortion	Himachal Pradesh and Uttar Pradesh	Ann et al. (2019); Krishnankutty et al. (2021)
17	Cultivation of traditional rice cultivar Kari Kagga and Atikaya	Human body heat regulation and drugs preparation	Tamil Nadu	Ann et al. (2019); Krishnankutty et al. (2021)
18	Cultivation of ancient rice cultivar Neelam Samba	Milk regulation in mother	Tamil Nadu	Ann et al. (2019); Krishnankutty et al. (2021)
19	Cultivation of local rice Maappillai Samba	Increase mother fertility	Tamil Nadu	Ann et al. (2019); Krishnankutty et al. (2021)
20	Cultivation of traditional cultivar black rice	Anticancer properties	Assam	Ann et al. (2019); Krishnankutty et al. (2021)
21	Cultivation of local rice Karinjan and Karimalakaran	Diabetes control	Kerala	Ann et al. (2019); Krishnankutty et al. (2021)
22	Cultivation of traditional rice Mundakan	Increase human stamina	Kerala	Ann et al. (2019); Krishnankutty et al. (2021)
23	Cultivation of traditional rice Vella chennellu and Chuvanna chennellu	Reduction of menopause, and hormone problems	Kerala	Ann et al. (2019); Krishnankutty et al. (2021)
24	Cultivation of Jonga and Maharaji traditional rice	Increasing mother lactation	Bihar and Chhattisgarh	Ann et al. (2019); Krishnankutty et al. (2021)
25	Cultivation of local rice Bora	Curing jaundice	Assamese	Ann et al. (2019); Krishnankutty et al. (2021)
26	Cultivation of local rice Karhani	Epilepsy treatment	Chhattisgarh and Jharkhand	Ann et al. (2019); Krishnankutty et al. (2021)

container for controlling pests and diseases in the leguminous seeds, vegetable seeds, and maize seeds. In the North Eastern region and Tamil Nadu, farmers use citronella grass and lemon grass to control the maize crop's weevil and grain attack. In Tripura and Uttarakhand, farmers use *Zanthoxylum acanthopodium* leaves in the paddy grains for pest and pathogen control (Santosh and Chhetry 2012). In Tripura, farmers recommend wood ash of the leaves of vegetable crops to control the aphid pod borer. The farmers use tobacco dust powder to control aphid pests. The Tripura farmers apply Hookah water, which is very efficient for restricting major and minor pests and diseases such as rice blasts, pod borers, and sucking bugs in vegetable crops. In Punjab, farmers manage insect pests of the paddy crop with the leaves of *Artemisia vulgaris, Croton caudatus, Munromia wallichii,* and *Adhatoda vesica* (Ahuja et al. 2015). The farmer uses Pomace (wine residue, made from millet) for controlling leaf folders and rice blasts. In the North-Eastern region, farmers apply oak tree bark through irrigation for controlling insect pests in rice. The farmers use 5-month old paddy husks to control blast disease and improve soil fertility (Firake et al. 2012). The farmers recommend *Chrysanthemum coronarium* and *Tagetes erecta* in the border of the crop field for controlling nematode of turmeric, tomato, chilli, and ginger, enhancing the soil property and nutrient enrichment. In Tamil Nadu, farmers inhibit stem borer in paddy with a pegging branch of *Cymbopogon khasianum* and *Saccharum spontaneum* (Gopal and Lassaad 2013). In Assam, farmers distribute banana sucker, a black colocasia, wild turmeric, and bamboo perch corner of the field during rice transplanting to control pests (Sarodee et al. 2020). In Madhya Pradesh, farmers formulate onion or garlic juice to control grasshopper and leaf insects in the maize crop (Shakrawar et al. 2018). The pouring of liquid lime into the Mandarin trunk to control Gummosis disease and bark-eating caterpillar trunk borer (Gohain et al. 2019). The burning of plant debris in the vegetable crop and paddy crop field to kill harbour insects. In Tamil Nadu, Kerala, and North-Eastern region, farmers practice zero tillage practice (dibbling method) to encourage mycorrhizal root growth and nodulated Frankia (Ngachan 2019). In Meghalaya, Jaipur, Punjab, and Himachal Pradesh, farmers utilise decompose mulch to inhibit the pathogens in the soil (Rana 2011). In Nagaland, farmers manage pests of paddy, airborne pathogens, and microclimate augmentation with mixed cropping of rice + maize, rice + legume crops, rice + job's tear (*Coix lacryma-jobi* L.), and rice + sorghum (Rakesh et al. 2017). The Indian farmers also practise deep ploughing after crop harvesting to control insect pests, arthropods, and nematodes from the soil. In Tripura, Meghalaya, and Assam, farmers apply the burnt paddy husk and dry chili plants to control rodents in the jhum field (Satyapriya et al. 2021). In Kerala, Tamil Nadu, and Uttar Bengal, farmers apply dried peel of mandarin in the transplanted rice for controlling stem borer. The termite population is controlled with fermented products of *Agave sisalana* (Agave), *Piper nigrum, Vernonia amygdalina,* and *Nicotiana tabacum*. The Kerala and Odisha farmers use Datura extract mixed with cow urine to control the ant population in the soil (Patnaik 2011). In Punjab, farmers apply neem leaves in the grains storage to control weevil and grain moths (Subash 2017). In Punjab, Haryana, and Telangana, people use turmeric powder mixed with water to control weevil and grain moths (Satyagopal et al. 2014).

In Sikkim farmers prevent the outbreak of aphids and white flies in the tomato and chili crop with the extract of titey pati, banjara, and *Lantana camara* (Gopi et al. 2016). The Karnataka and North-Eastern farmers dissolve chili seedlings in a 1:3 solution of cow urine water. The cow urine water assists in protecting seedlings against damping off (Rakesh et al. 2013).

In Kerala, farmers dissolve 400-mL neem oil in 400-L water to produce a liquid solution mixed with 500 g detergent soap and spray it on the mango tree to control hoppers (Alexander et al. 2009). The farmers from Jorhat, Assam and Southern states use 1-kg cow dung and 5-g detergent soap dissolved in 10 L of water to control sooty mould disease of mango. In Tamil Nadu, Kerala, and Himachal Pradesh, farmers formulate 1-L neem oil dissolved in 100 L of water and mixed with 500 g detergent powder. The banana sucker is immersed in hot water for 30 min and controls rhizome rot. The Tamil Nadu farmers release the dried outer banana bark once in 4 months to prevent root primordia growth, lodging, and side sucker emergence. The dried dropping leaves are excised once in 3 months to avoid shade effect, mitigate wind damage, and prevent lodging. 150-g neem cake powder is applied in the hole of the oozed gum portion. The South Indian and North-Eastern farmers remove wilted banana crops from the pits and burn and bury them in the ground, applying 1–2 kg lime in the individual pits. The Tamil Nadu, Kerala, and Himachal Pradesh farmers prepare a liquid formulation with 1 kg neem powder and 1 kg tobacco powder dissolved separately in 5 L of water. The next day, the neem powder solution and tobacco powder solution are filtered into a fine solution. The neem powder solution and tobacco powder solution are mixed to produce a compound solution. The banana sucker is dissolved in the compound solution to prevent nematode attack. In Tamil Nadu and Kerala, farmers use 2 kg *Calotropis* spp. +3 kg Neem cakes soaked in 20 L of water and stored for 4 days, later, the solid parts of *Calotropis* spp. and neem cake are extracted in 200 L of water. 500 g detergent powder is dissolved in 200 L of extracted water. Two hundred litres of the extracted water is recommended in the 1 acre of guava orchard for controlling white flies (Rohini 2010).

In Tamil Nadu and North-Eastern region, farmers mix cotton seeds with ash and cow dung slurry for protecting against damping off. This method is also explained in Kautilya's Arthashastra (Manickam et al. 2013). The farmers from Assam and South Indian states use clay layer, cow dung ball is poured into the cut stalk of the banana to prevent spoilage and ripening of the banana. The Karnataka, Kerala, and Tamil Nadu farmers prepare a liquid formulation with 4 kg powdered neem seed dissolved in 100 L of water for managing leafminer pests and other diseases in citruss (Vanaja et al. 2009). The farmers in Southern states produce liquid formulation with dried neem fruits crushed to prepare fine power to manage nematode attacks. The farmers of Karnataka, Mysore, and Tamil Nadu are pouring leaves of Kasarka (*Strychnos nux-vomica*) with cow dung to the citrus tree to manage grub insects (Ravi 2021). The dried forest leaves are mulched in mandarin orange (*Citrus reticulata*) to maintain soil moisture and temperature. The farmers of Assam and Maharashtra apply lime wash/lime-soaked cotton in the holes of mandarin orange for controlling stem borer (Ahuja and Chattopadhyay 2015). The South Indian farmers cut green

Aloe vera plants and apply them to mandarin orange trees during the flowering phase to control powdery mildew disease. The Tamil and Odisha farmers cultivate wild sugarcane with paddy to control leaf folder disease (Mayabini and Dangar 2020). The Tamil Nadu and Assam farmers recommend Parasi (*Cleistanthus collinus*) and Sali (*Boswellia serrata*) for controlling caseworms (*Nymphula depunctalis*) in rice (Kudada et al. 2020).

The Jharkhand farmers apply 10 kg Parso/Persu leaves in 100 m^2 of the paddy field for managing gall fly (Sinha and Singh 2020). The farmers apply 50–200 kg of fresh Karada leaves to control Gundhi bugs (Chard) from the paddy field (Richa et al. 2017). The Tamil Nadu farmers use extract *Cynodon dactylon* leaf extract for managing fruit borer, wilt, leaf curl, and early blight in tomatoes. The Odisha farmers use compost of cow dung +10 kg Kochila (*Strychnos nux-vomica*) seed powder +25 kg kochila leaf compost + compost for managing fruit and shoot borer in the Brinjal crop (Das et al. 2004a, b). The Jharkhand farmers formulate cow urine mixed with tobacco-soaked powder for managing leaves and fruit diseases of cucurbits, cowpea, and lady finger (Devendra 2020). The farmers prepare a liquid solution with rice starch + cow urine for controlling insects and pests in vegetable crops such as lady's finger and tomato (Khudada et al. 2020). The Bihar farmers control shoot and fruit borer in brinjal crop with tobacco-soaked water (Choubey 2020). The Tamil Nadu farmers apply cow dung slurry to control rhinoceros beetles (Koodalingam 2020).

In Madhya Pradesh, farmers recommend dry mahua flower for controlling *Scolopendra* spp. (Gaygwalan) in the soybean crop. The Tamil Nadu and Madhya Pradesh farmers mix 100–150 g Asafoetida with 1 L boiled water for 10–15 min; the boiled solution is poured in 40–50 L water, and used in the field for controlling the larvae of *Heliothis* sp. and other small insects. The solution of dung waste, crop waste, and cow urine is used in controlling pest populations (Ranjay et al. 2014). The Odisha farmers dissolve dry tobacco leaves in boiled water against the larvae of *Heliothis armigera*. The liquid formulation of 1000-1200 fresh leaves + buds of Ipomea bushes is mixed with 30-35 liters for controlling *Heliothis armigera*, spotted bollworm and army worm. In the North-Eastern region, Odisha, Tamil Nadu, and Bihar, farmers use fresh neem leaves boiled in 10 L of water for controlling specific pests (Sandhya 2022). Table 14.2 summarises the traditional practices of farmers in crop protection.

14.2.3 Farm Machine and Tools

The farmers have utilize farm machines and tools since the onset of agriculture. The land preparation implements like la-nogala (small plough) and si-ra (heavy plough), la-n.gala, si-ra (rod), varatra (rope), pha-la (share), yoke (yuga) are traction/animal-drawn are described in the Yajurveda. The farmer employs oxen, sheep, and camels for land preparation.

The tools like corn-cutting tools, a sort of sickle in the shape of a cooking knife, a sickle, and a reaping hook were applied in the Rigvedic period. The farmers utilized the tools for crop harvesting. In the Rigveda period, the corn grain cleaning was completed with a sieve and winnowing fan. The farmers transported harvested grains with the following carts ie., ana-sa (carts) and sfakat.a (wagon). The cart was constructed with Acacia trees, Dalbergia trees, bamboo poles and metal tyres (pavi). The wooden cart is transported by ox, stallion, ram, and dog. The Bodo farmer utilizes Ruwa (Axe), Kodhal (Digging hoe), and Danger (roke) for the eradication of unwanted plant material from the crop field. The Assam farmers apply Nangal (plough), Jugal, Mwi, Khodal (digging hoe), Kontha (spua), Gandhi or Dangan (leveller) for the field preparation (Sibisan 2019). The farmers from Assam use Lauthi (digging stick), khopri, Mukha/Kho (mask) for the seed sowing. The farmers use Phalla (Weighing tool), Nareal Koltha (Coconut cover), Kurai Kowrai, Kurai Guhai for measuring harvested agricultural produce. The farmers use Moscow Giri (Bullock cart) for transporting agricultural commodities The Assam farmers recommend Kashi (sickle), Sika (Knife), Sika-gobla (cleaver) for crop harvesting; applies Baukha, Hukhen (grain separator), Royna, Sandanga (Sieve), Songri (winnower), Khada (Basket made of Bamboo), Duli (grains store), Dingkhi (Grinder), Sundri (small kind of sieve), Khasa (rice store basket), Gan/Gaihen (milling tools), Val/Ural (Milling tools), and Don (Bamboo pan) for post harvesting processings. The Chhattisgarh farmers use chilli (water lifter) for field irrigation (Nijra and Luke 2017). The Tamil Nadu farmers utilises the country plow (Kalappai) for land preparation. The Tamil Nadu farmers uses sickle (karukkarival), knife (kambar Kathi), Tamarind harvester (Puli kokki), and lemon harvesting tool (Ezhumichai karandi) for crop harvesting. The farmers use weeder (aruguvetti), dry land weeder (cycle gundu), and spade (Mammootty) for inter-cultural operations. The farmers use a grain separator (kodun Kol), wooden thresher (thattuppalagai), stone roller (uruttu kal), bamboo grinder (chekku), and milling tool (ulakkai) for post-harvesting of the grains. The farmers use Pukka, Marakaal, and Naali for measuring the agricultural produce. The farmers apply floor cleaner (Sakkai piratti) and bamboo pan (moonghil thattu) for cleaning of the grains in the southern region (Karthikeyan et al. 2009). The farmers of the eastern region apply plough (lungal), spader (plough), khurpa (khurpi), weeder rack, spader (kodal/phaura), guity, sickle (kaste/daw), Daw (katruri), long handle dauli, axe, sabal, hand stone mill, silpata, paddy spader, bamboo sieve, winnower, silo, bamboo basket, nanda, bankua, megara (Gila), pola, khalui, and panki (boti) in crop production and management (Bikash et al. 2015). The farmers of Odisha, Uttar Pradesh, and Gujarat apply bullock-drawn dhanti for effective control of weed populations in the crop. The indigenous guddeli tools require less power in the operation, preventing the loss of ginger harvesting (Swain et al. 2020; Shamkuwar et al. 2019). Table 14.3 provides a summary of the traditional knowledge of the farmers in agricultural machines and tools.

Table 14.2 Indigenous knowledge of farmers related to plant protection

Sl. No.	Indigenous Traditional Knowledge (ITK)	Cause	State	References
1	Fire in the wheat crop field after harvest	Pest control	Punjab	Bhuvaneshwari et al. (2018)
2	Burning of dry leaves of pine	White grubs control	Uttarakhand	Surya et al. (2021)
3	Preparation of a mixture of wood ash/kitchen ash + farm yard manure	Chewing and bitting mouth part insect	Kerala and Jharkhand	Manoj (2016); Das et al. (2003)
4	Decomposition of aphid infestation crops such as mustard, cauliflower, cabbage	Aphid control	North-eastern, Uttarakhand and Madhya Pradesh	Ajay (2018)
5	Earthen up of potato	Greening of potato tubers and check exposure in sunlight	Bihar and Uttarakhand	Manish et al. (2011)
6	Preparation of cow dung + cow urine compound	Wilt symptoms and Onion blight control	Odisha and Kerala	Rajendra et al. (2018)
7	Intercropping of Madira or Barnyard millet (*Echinochloa* sp.) and Konri millet crop, mustard with Paddy crop	Insect-pests infestation control	Uttarakhand, Odisha and Gujarat	Rajendra et al. (2018)
8	Liquid solution of Bicchu booti (*Utrica dioca*) mixes with 8–10 L cow urine	Anthracnose in chili, Tomato late blight and fruit rot and Cucurbit Alternaria blight	Chhattisgarh	Santosh and Chhetry (2012)
9	Collection of healthy seeds and constant smoking with edible and non-edible oil	Drying of the seeds	Tamil Nadu	Santosh and Chhetry (2012)
10	Aromatic plants: citronella grass, lemon grass	Maize weevil and grain storage, pests of pomelo	North East and Tamil Nadu	Santosh and Chhetry (2012)
11	Application of *Zanthoxylum acanthopodium* leaves in the paddy grains	Pest and Pathogen control	Tripura and Uttarakhand	Santosh and Chhetry (2012)
12	Application of wood ashes in the leaves of vegetable crops	Aphid pod borer control	Tripura	Ahuja et al. (2015)
13	Application of Hookah water in vegetable crops	Major and Minor pest and diseases such as rice blast, pod borers, sucking bugs controls	Tripura	Ahuja et al. (2015)
14	Application of oak tree bark	Insect-pests control in rice	North-Eastern states of India	Firake et al. (2012)
15	Bordering of *Chrysanthemum coronarium*, *Tagetes erecta* in the crop field	Turmeric, Tomato, Chlilli and Ginger Nematode Control	Jammu and Kashmir and Southern states	Gopal and Lassaad (2013)

(continued)

Table 14.2 (continued)

Sl. No.	Indigenous Traditional Knowledge (ITK)	Cause	State	References
16	*Cymbopogon khasianum*, and *Sacharrum spantaneum* pegging branches	Paddy stem borer control	North-Eastern states and Tamil Nadu	Gopal and Lassaad (2013)
17	Burning of plant debris	Killing harbour of insects in paddy fields and vegetable fields	Punjab and Haryana	Ngachan (2019)
18	Decomposed mulch	Inhibition of the pathogen in the soil	Meghalaya, Jaipur, Punjab, and Himachal Pradesh	Rana (2011)
19	Mixed cropping: rice + maize, rice + legume crops, rice + job's tear, rice + sorghum	Pest control of paddy, airborne pathogen control and augmenting microclimate	Nagaland	Rakesh et al. (2017)
20	Burning of paddy husk and dry chilli plant in the jhum field	Rodent controls	Tripura, Meghalaya, and Assam	Satyapriya et al. (2021)
21	Application of dried peel of mandarin in the transplanted rice	Stem borer controls	Kerala, Tamil Nadu, and West Bengal	Patnaik (2011)
22	Datura extracts mixes with cow urine	Ant controls	Kerala and Odisha	Patnaik (2011)
23	Application of neem leaves in the grains storage	Weevil and grain moth controls	Punjab	Subash (2017)
24	Irrigation of turmeric powder with water	Weevil and grain moth controls	Punjab, Haryana and Telangana	Satyagopal et al. (2014)
25	Fermented extract of titey pati, banmara and *Lantana camara*	Aphids and white flies controls in the tomato and Chilli crop	Sikkim	Gopi et al. (2016)
26	Treatment of 1:3 solution cow urine water of seedlings of chilli	Damping off control	Karnataka and North-East	Rakesh et al. (2013)
27	Formulation of liquid solution with 400 mL neem +400-L water +500 gm detergent soap	Hopper control in Mango	Kerala	Alexander et al. (2009)
28	Formulation of liquid solution with 1 kg cow dung +10-L water +5 gm detergent	Sooty mold control in Mango	Jorhat and South	Rohini (2010)
29	Immersion of Banana sucker in hot water for 30 min	Rhizome rot control	Tamil Nadu, Kerala, and Himachal Pradesh	Rohini (2010)
30	Burn and bury of banana crop in the ground	Wilt control	South Indian and North-Eastern	Rohini (2010)

(continued)

Table 14.2 (continued)

Sl. No.	Indigenous Traditional Knowledge (ITK)	Cause	State	References
31	Dissolving of banana crop in the formulated solution with 1 kg neem powder +1 kg tobacco powder +5 L water	Nematode control	Tamil Nadu, Kerala, and Himachal Pradesh	Rohini (2010)
32	Formulation of liquid solution with 2 kg *Calotrpis* spp. + 3 kg Neem +20 L water, 500 g detergent powder +200 L water	White fly control in Guava	Tamil Nadu and Kerala	Rohini (2010)
33	Mixing of cotton seed with ash + cow dung slurry	Damping off control	Tamil Nadu and North-Eastern states	Manickam et al. (2013)
34	Pouring of clay layer/cow dung ball in the cut stalk of banana	Prevent spoilage and ripening of Banana	Assam and Southern states	Vanaja et al. (2009)
35	Application of dried neem fruits fine powder	Nematode control	Southern states	Ravi (2021)
36	Application of 5 L cow urine +50 L water	Sooty mold control in citrus tree	Southern states	Ravi (2021)
37	Application of leaves of Kasarka (*Stychnos nux-vomica*) + cow dung	Grub insect control in citrus tree	Karantaka, Karnataka, and Tamil Nadu	Ravi (2021)
38	Application of lime wash/ lime-soaked cotton in the holes of mandarin orange	Stem borer control	Assam, Maharashtra	Kudada et al. (2020)
39	Application of cut green aloe vera plant in mandarin orange tree	Powdery mildew disease control	South Indian	Kudada et al. (2020)
40	Cultivation of wild sugarcane with paddy	Leaf folder disease control	Tamil and Odisha	Kudada et al. (2020)
41	10 kg Parso/Persu leaves in the paddy field	Gall fly control	Jharkhand	Sinha and Singh (2020)
42	50–200 kg fresh Karada leaves in the paddy field	Gundhi bug control	Tamil and Odisha	Mayabini and Dangar (2020)
43	Compost Preparation: cow dung +10 kg Kochila seed powder +25 kg kochila leaf	Fruit and shoot borer control in the brinjal crop	Odisha	Das et al. (2020)
44	Blending of cow urine and tobacco powder	Diseases control of cucurbits, cowpea, and lady finger	Jharkhand	Devendra (2020)
45	Application of cow dung slurry	Rhinoceros beetle control	Tamil Nadu	Koodalingam (2020)
46	Application of dry mahua flower in the Soyabean crop	*Scalopendra* spp. (Gaygwalan) control	Madhya Pradesh	Ranjay et al. (2014)

(continued)

Table 14.2 (continued)

Sl. No.	Indigenous Traditional Knowledge (ITK)	Cause	State	References
47	100–150 g Asafoetida +1-L boiled water for 10–15 min, the boiled solution is poured in 40–50-L water	Control of *Heliothis armigera* larvae and other small insects	Tamil Nadu and Madhya Pradesh	Ranjay et al. (2014)
48	Dry tobacco leaves mixed with 5–6 L boiled water	Larvae of *Heliothis armigera* control	Odisha	Ranjay et al. (2014)
49	Fresh leaves + buds Ipomea bushes is mixed 30–35 L boiled water	*Heliothis armigera*, spotted bollworm, and army worm control	Odisha	Ranjay et al. (2014)
50	Fresh neem leaves are boiled in 10 L water	Girdle beetle, Bihar hairy caterpillar and other pests control	North-Eastern states, Odisha, Tamil Nadu, and Bihar	Ranjay et al. (2014)
51	Distribution of banana sucker, a black colocasia, wild turmeric, and bamboo perch corner of the main field	Rice pests control	Assam	Sarodee et al. (2020)
52	Formulation of onion or garlic juice	Grasshoppers and leaf insects control	Madhya Pradesh	Shakrawar et al. (2018)
53	Pouring of liquid lime in the Mandarin trunk	Gummosis disease and bark-eating caterpillar, trunk borer control	North-Eastern states	Gohain et al. (2019)

14.2.4 Soil and Water Management

The farmers constructed a water reservoir, dam, pond, chauka system, and haveli system for harvesting the water in the Sindhu Valley Civilization. The farmers prepare farm ponds, check dams, and shallow wells for harvesting rainwater. They use the bamboo drip method in irrigation for controlling water-borne diseases in terrace farming. The farmers of Tamil Nadu apply traditional micro-depression methods for managing water in the Neem, teak and Mango tree for controlling soil erosion, improving soil properties, and progressing growth and development of the tree (Hiswaran 2020). The farmers of Andhra Pradesh utilize the rolu method for collecting the rainwater (Maruthi and Subba 2020). In Himachal Pradesh, chaal (small water storage ponds) for drinking and irrigation purposes in hill areas are used (Pradeep et al. 2020). The farmers recommend cow dung slurry for improving soil property and water retention capacity. In Kerala, Madhya Pradesh, Punjab, and Uttar Pradesh, farmers improve the property of the soil and moisture of the soil with a mixture of wood ash, rice husk and cow dung cake. The wood ash is rich in phosphorus. The farmers recommend paddy straw mulching for preserving water in the soil and sandbags for controlling soil degradation. The farmers apply a mixture

Table 14.3 Indigenous knowledge practices of farmers in farm machine and tools

Sl. No.	Indigenous Traditional Knowledge (ITK)	Use	State	References
1	Field preparation—Nangal (plough), Jugal, Mwi, Khodal (digging hoe), Kontha (spua), Gandri or Dangan (leveller); seed sowing—Lauthi (digging stick), khopri, Mukha/Kho (mask); measuring harvested agricultural produce—Phalla (Weighing tool), Nareal Koltha (Coconut cover), Kurai Kowrai, Kurai Guhai; transporting agriculture commodities—Mosow giri (Bullock cart)	Field preparation, seed sowing, measuring harvested agricultural produce, transporting agriculture commodities	Assam	Nijra and Luke (2017)
2	Field preparation—Nangal (plough), Jugal, Mwi, Khodal (digging hoe), Kontha (spua), Gandri or Dangan (leveller)	Field preparation	Assam	Sibisan (2019)
3	Post-harvesting processings—Kashi (sickle), Sika (Knife), Sika-gobla (cleaver) for the crop harvesting; applies Baukha, Hukhen (grain separator), Royna, Sandanga (Sieve), Songri (winnower), Khada (Basket made of Bamboo), Duli (grains store), Dingkhi (Grinder), Sundri (small kind of sieve), Khasa (rice store basket), Gan/Gaihen (milling tools), Val/Ural (Milling tools), Don (Bamboo pan)	Post harvesting operations	Assam	Nijra and Luke (2017)
4	Chili (water lifter)	Field irrigation	Chhattisgarh	Nijra and Luke (2017)
5	Land preparation—country plough (Kalappai); crop harvesting—sickle (karukkarival), knife (kambar kathi), Tamarind harvester (Puli kokki), lemon harvesting tool (Ezhumichai karandi); inter-cultural operations—weeder (aruguvetti), dry land weeder (cycle gundu), spade (mammutty); post harvesting of the grains—grain separater (kodun kol), wooden thresher (thattuppalagai), stone roller (uruttu kal), bamboo grinder (chekku), milling tool (ulakkkai); measuring the agricultural produce—Pukka, Marakaal, Naali; cleaning of the grains—(Sakkai piratti), bamboo pan (moonghil thattu)	Land preparation, crop harvesting, inter-cultural operations, post harvesting of the grains, measuring the agricultural produce, cleaning of the grains	Tamil Nadu	Karthikeyan et al. (2009)
6	Application of plough (lungal), spader (plough), khurpa (khurpi), weeder rack (hashkini), spader (kodal/phaura), guity, sickle (kaste/daw), Daw (katruri), long handle dauli, Axe, sabal, hand stone mill, silpata, paddy spader, bamboo sieve, winnower, silo, bamboo basket, nanda, bhaungi, mugara (gila),pola, khalui, panki (boti)	Crop production and management	Eastern region	Bikash et al. (2015)
7	Application of bullock drawn dhanti	Control of weed populations	Odisha, Uttar Pradesh, and Gujarat	Swain et al. (2020); Shamkuwar et al. (2019)

of salt ash and coco pit in the field before transplanting. The cocopeat contains potassium that improves the physical and chemical properties of the soil (Yadav et al. 2013; Balasubramanian et al. 2009). The Uttarakhand farmers construct water catchment reservoirs like Tals, Khals, Chals, and Rou for collecting water for domestic and agricultural purposes. In Assam, farmers construct Bari system for harnessing water. In Rajasthan, farmers build Saza Kuva open well for domestic and irrigation uses (Anwesha and Pardeep 2020). The farmers from Southern states grow Aduthininapalai (*Aristolochia bracteolate*) for conserving soil water. The mixed cropping and intercropping of leguminous plants facilitate soil improvement. 200 tonnes of tank silt is applied in the field for land measures. The sheep/cattle penning improves soil fertility during the summer season. In Tamil Nadu, farmers restrict soil erosion and moisture with the cultivation of Kolingi (*Tephrosia purpurea*) between fruit trees in sloppy land. The Nuna tree (*Morinda tinctoria*) improves moisture retention in the soil. The deep plowing encourages moisture content in the soil. In Chhattisgarh, Kerala, Tamil Nadu, and Uttar Pradesh, farmers prepare a liquid solution with ingredients of 10 kg Neem +10 L Cow urine + ½ kg asafetida waste, stored overnight. The extracted liquid manure is applied to the 1-acre land for soil productivity. The mixtures of Neem oil, fine sand, and cow dung are stored in the moist area for 3 days. The formulated mixtures are dissolved in 150-L water and are recommended in the soil for soil amelioration and sucking pest control (Ravisankar et al. 2017). In Karnataka, Andhra Pradesh, and Kerala, farmers cultivate Vetiver (Khus grass) for managing land degradation and soil conservation. The perennial vegetation is cultivated in the field for controlling soil erosion. The farmer apply farm yard manure (FYM) for improving soil properties (Prakasa et al. 2015; Mishra et al. 2011). The Sikkim farmers construct a terrace in a field for promoting terrace farming and land reformation (Prabuddh et al. 2020). In Maharashtra, Kerala, and Assam, farmers cultivate Diancha (*Sesbania* sp.) and Sun hemp (*Crotalaria juncea*) to improve water holding capacity and soil property of the alkali soil (Shobha et al. 2020). In Kerala and Tamil Nadu, farmers cultivate Poorvarasu (*Thespesia populnea*) which mitigates water loss from the soil (Somasundaram et al. 2021). In Tamil Nadu, Kerala, and Karnataka, farmers apply bagasse of sugarcane, leaves and branches of Indian gooseberry (*Phyllanthus distichus*) for improving saline soil. In Kerala and Hyderabad, Telangana, farmers cultivate *Tea quadrifolia* and *Cyanodan dactylon* weed which encourages high yield from the soil (Binoo et al. 2016). In Tamil Nadu, Punjab, Haryana, Himachal Pradesh, and Maharashtra, farmers grow the population of Pirandai (*Cissus quadrangularis*) for improving alkaline soil properties. The farmers from Southern states and Uttar Pradesh cultivate Diancha and Nutgrass (*Cyperus rotundus*) to ameliorate the property of alkali soil (Somasundaram et al. 2020).

The Indian farmers apply cow dung, pig dung, sheep dung, and goat dung for progressing soil property. The application of cattle manure in garden soil and wetland and leaf manure in wetlands enhance the property of the soil (Modupe et al. 2020). The farmers of Tamil Nadu, Karnataka, Gujarat, and Bihar use mixture of cow dung, *Calotropis gigantea* leaves, and neem cake powder for improving soil properties (Krishan 2005; Krishna et al. 2019). The Bihar and North-Eastern farmers apply water hyacinth as compost or burnt ash for soil and water improvement and provides potassium nutrients in the soil. The farmers recommend goat manure for improving soil properties (Ganesh et al. 2011). The farmers from Rajasthan cultivate green leaf manure such as *Tephrosia purpurea*, *Calotropis gigantea*, *Morinda tinctoria*, *Pongamia pinnata*, *Azadirachta indica*, *Thespesia populnea*, and *Adhatoda vasica* facilitating crop growth and soil improvement. The leguminous plant red gram acts as a green manure crop and improves soil properties (Dagar and Tewari 2016). Table 14.4 presents traditional knowledge of farmers related to soil and water management.

14.2.5 Animal Husbandry

The fish farmers of Dimapur, Assam, Kanyakumari, Tamil Nadu and Goa use pseudostem banana for fish. The pseudostem banana increases pH and oxygen in the pond water and raises fish production (Bhalerao et al. 2015). The farmers of Maharashtra prepare a mixure of 500 g maida +500 g behada powder in 2.5-L water after boiling and use for controlling foot and mouth disease (FMD; Choubey 2005). The farmers of West Bengal, Rajasthan, Uttar Pradesh, Tamil Nadu, Jharkhand, Himachal Pradesh, Uttaranchal, and Odisha, farmers prepare liquid medicine with peach leaf extract and milk and apply to control cattle lesion of the mouth and hooves (Das et al. 2004a, b). In Uttar Pradesh, Maharashtra, and Odisha, farmers control foot and mouth disease in the cattle with a mixture of babool bark and jamun bark paste (Rajesh and Bharathi 2012; Sarita et al. 2003). The farmers control foot and mouth disease of cattle with camphor and coconut oil in the cattle. The Uttar Pradesh farmers recommend the paste of Bantulsi (*Ocimum gratissimum*) leaf along with water for controlling Khurha (FMD) disease in cattle and buffalo (Swarup and Pradhan 2020). The Uttarakhand farmers prepare liquid with stone apple (bael) and water for controlling diarrhoea (Mahesh 2020). The Uttar Pradesh and Uttarakhand farmers prepare pigeon waste mixed with jaggery, and is applied to the heifers for inducing the oestrus cycle (Swarup et al. 2020). The Uttarakhand farmers use juice of gurhal (urhul) flower and administer orally in the goat for controlling diarrhoea (Dakshinkar and Vihan 2020). The Jharkhand farmers apply a crushed paste of Pojo (*Listea authapoly*) for treating diarrhoea and dehydration (Haque et al. 2020). The farmers recommend oral administration of flower juice of Takala (*Cassia tora*) in goats for controlling diarrhoea. In Himachal Pradesh, farmers grind the leaves of ridge gourd or ekdandi to extract the juice and the extracted juice is smeared on the wound of the animals (Varshney 2020). In Maharashtra, farmers crushed 200–250 g stem and leaf of Bhangariya (*Eclipta alba*) to produce a paste and 50–60 mL mustard oil paste is fried and applied to the cattle, buffaloes, and goats for controlling the blain (Jangde and Dhanan 2020). The farmers of Maharashtra and Uttar Pradesh prepare mixture of 30 g geru, 50 g snail shell/sippi and boiled with castor oil and 20 g Alua, 50 g kudru/sahjam gum are mixed with the mixture to produce a paste. The prepared paste is used for the bullocks or bulls for controlling swelling (Swarup and

Table 14.4 Indigenous knowledge practices of farmers related to soil and water management

Sl. No.	Indigenous Traditional Knowledge (ITK)	Use	State	References
1	Traditional micro-depression method in the Neem tree, teak tree and Mango	Managing water, soil erosion control, improving soil properties	Tamil Nadu	Hiswaran (2020)
2	Rolu method	Determining the rain water and collecting the rainwater	Andhra Pradesh	Maruthi and Subba (2020)
3	Chaal (small water storage ponds)	Drinking and irrigation purposes	Himachal Pradesh	Pradeep et al. (2020)
4	Compound mixture of wood ash, rice husk, and cow dung cake	Improving the property of the soil and moisture of the soil	Kerala, Madhya Pradesh, Punjab, Uttar Pradesh	Yadav et al. (2013), Balasubramanian et al. (2009)
5	Grow of Aduthininapalai (*Aristolochia bracteolacia*)	Evaluating soil water	Southern states of India	Ravisankar et al. (2017)
6	Fruit trees-Kolingi (*Tephrosia purpurea*) cultivation	Prevent soil erosion and moisture	Tamil Nadu	Ravisankar et al. (2017)
7	Liquid solution with ingredients of 10 kg Neem +10-L cow urine + ½ kg asafoetida waste	Improving the soil productivity	Chhattisgarh, Kerala, Tamil Nadu, and Uttar Pradesh	Ravisankar et al. (2017)
8	Cultivation of Vetiver (Khus grass)	Managing land degradation and soil conservation	Karnataka, Andhra Pradesh and Kerala	Prakasa et al. (2015); Mishra et al. (2011)
9	Cultivation and ploughs of Diancha (*Sesbania* sp.) snd Sun hemp (*Crotolaria juncea*)	Improves water holding capacity and soil property	Mahrashtra, Kerala, and Assam	Shobha et al. (2020)
10	Cultivation of Poorvarasu (*Thespesia populnea*)	Water loss from the soil	Kerala and Tamil Nadu	Binoo et al. (2016)
11	Application of bagasse of sugarcane, leaves and branches of Indian gooseberry (*Phyllanthus distichus*)	Improving saline soil	Tamil Nadu, Kerala, and Karnataka	Binoo et al. (2016)
12	Cultivation of *Tea quadrifolia* and *Cyanodan dactylon*	Encourages better yield on the soil	Kerala and Hyderabad	Binoo et al. (2016)
13	Cultivation of population of Pirandai (*Cissus quandrangularis*)	Improving alkali soil	Tamil Nadu, Punjab, Haryana, Himachal Pradesh, and Maharashtra	Somasundaram et al. (2020)

(continued)

Table 14.4 (continued)

Sl. No.	Indigenous Traditional Knowledge (ITK)	Use	State	References
14	Cultivation of Diancha and Nut grass	Improving alkali soil	South Indian and Uttar Pradesh	Somasundaram et al. (2020)
15	Decomposed manure of cowdung, *Calotropis gigantea* leaves, neem cake powder	Improving soil property	Tamil Nadu, Karnataka, Gujarat, and Bihar	Krishan (2005); Krishna et al. (2019)
16	Water hyacinth as compost or burnt ash	Progressing soil, water improvement, provides Potassium (K) nutrient in the soil	Bihar and North-East	Ganesh et al. (2011)
17	Cultivation of green leaf manure such as *Tephrosia purpurea*, *Calotropis gigantea*, *Morinda tinctoria*, *Pongamia pinnata*, *Azadirachta indica*, *Thespesia populnea* and *Adathoda vasica*	Crop growth and soil improvement	Rajasthan	Dagar and Tewari (2016)
18	Water catchment reservoirs—Tals, Khals, Chals and Rou	Water collection	Uttarakhand	Anwesha and Pardeep (2020)
19	Bari system	Water collection	Assam	Anwesha and Pardeep (2020)
20	Saza Kuva open well	Water collection	Rajasthan	Anwesha and Pardeep (2020)
21	Terrace construction	Terrace farming and land reformation	Sikkim	Prabuddh et al. (2020)

Dhakate 2020). In Maharashtra, farmers prepare the paste of Kala jeera which is applied to the animal for managing haemorrhagic septicaemia (Vihan 2020). The dairy farmers cultivate high salt plants, and avoids shade trees for controlling ticks. The farmers of Jharkhand and Himachal Pradesh develop Hajore paste for recovering bone fractures in animals (Roy and Varshney 2020).

The fish farmers of West Bengal apply ghani, chero/kero, chokhia, and atal for trapping fish. Aran bata/aran pata is used for creating barriers for fish. The farmer uses circular-shaped earthen rings/earthen pots for encouraging catfish breeding in waterlogged fields/paddy fields. *Channa gachua* (Changmachh) is a local fish of Majuli island of Assam for curing Asthma and body pain. The oil of *Mystus vittatus* is used for healing burn injuries, fever, and bacterial dysentery. The chela fish (*Salmophasia bacaila*) is used for promoting lactation in women. The mortality rate of prawn seed is forbidden with *Cinnamomum tamala*. The leaves have Vitamin A, Vitamin C, and antimicrobial activity. The traditional tool Ankar/Anksi is used for catching mud crabs (*Scylla serrata*). The unripe gaoh (*Diospyros embryteris*) is used for strengthening fishing nets in Sunderbans (Aparna et al. 2020). The fish farmers of Tripura produce Lau Macha local fish through fish cum vegetable (bottle gourd). The farmer produces snakehead murrel (*Channa striata*), climbing perch (*Anabas testodeneous*), catfish (*Clarias batrachus*), and bloch (*Heteroneutes fossilis*) in paddy fields. The fish controls weed population and soil loosening. The fish cum duckery method grows fish along with duck (Ratan and Dilip 2013).

The West Bengal fishing community manages bloat disease with formulation of 10 g bark aswatha (Banyon, ficuspa) + 10 g ada (ginger) + 10 g salt. The disease recovers in 7 days. 50 mL liquid common guava leaves are applied for managing diarrhoea. The West Bengal farmers prepare extracts of ganda (Marigold) leaves for curing wounds in animals. The halud (turmeric) is ground and applied to animal wounds (Amitendu et al. 2004). Saltation and sun drying is a prominent process in fish preservation. The mustard oil and salt and turmeric powder are applied to cut fish for controlling fish spoilage. The paste is prepared with roots of the bonson tree and 21 pieces of black pepper, fed into the dog-bit portion. The Odisha and Gujarat farmers grind the stems and leaves of anantamul for releasing juice, which is mixed with honey for managing animal dysentery. The Gujarat farmers formulate a liquid solution with 100 g tulsi leaves and 100 g basak boiled with water. The extracted juice is mixed with 1 teaspoon of honey and fed to the animal for controlling cold and cough (Bikram et al. 2012; Patel et al. 2016). The farmers of Hyderabad use tamarind bark to prepare glue for strengthening the nets. The farmers of Uttar Pradesh, Gujarat, and Rajasthan pour root of *Acacia arabica* with mustard in 1:3 proportions for managing arthritis. The oestrous cycle of animal treats for 2 days with a combination of *Musa paradisiaca* along with sugar (Ram et al. 2013).

The farmers of Uttar Pradesh use vinegar for tympany medication, castor oil for deworming, mustard oil for body heat regulation, turmeric lime paste for sprain heal, and black pepper butter oil mixture for pneumonia fever control in animals (Gyan et al. 2016). In Assam, Nagaland, Madhya Pradesh, and Haryana, farmers prepare drugs with *Glyricidia* and roasted soaked tamarind seeds and feed the cows for increasing lactation. The seeds of subabul cater to animals for improving milk secretion. In Assam, Nagaland, Madhya Pradesh, and Haryana, farmers formulate liquid products with bottle gourd, fenugreek, coconut, black gram, and palm jaggery mixed with water, and feed to animals for 3 days to increase milk. In Assam, Nagaland, Madhya Pradesh, and Haryana, farmers feed dried flowers of *Madhuca latifolia* to bullock for improving work efficiency. The Assam, Nagaland, Madhya Pradesh, and Haryana farmers prepare a powdered formulation with pepper, jaggery, and betel leaf, which is fed to animals for increasing digestion rate. The farmers of Jammu and Kashmir serve grinded *Iris kashmiriana* and jaggery to enhance milk yield (Shubeena et al. 2018; Deepandita et al. 2021).

The farmers of Hyderabad and Karnataka use cow dung slurry for managing Euglena bloom. The dry fish is prepared with the intervention of cow dung slurry. 200 g termite mound soil is boiled with water, the prepared solution is applied to the animal for controlling mastitis, the poisonous bite of insects, and mechanical injury (Swamy et al. 2015). The ray fish oil is applied for vanishing boats and controlling leakage. The farmers of Bihar and Hyderabad preserves boat and net with cashew shell oil, coal tar, and sardine oil. In Hyderabad, farmers store and transport fish by mixing sawdust and rice. The fish net is strengthened with boiled tamarind seed powder and kalasha bark. The farmers treat bloat disease in animals with mango pickle, spices and neem leaves. The farmers of Bihar, Hyderabad, and Odisha control cattle constipation with *Gardenia resinifera* leaves and *Dendrophthoe falcata* seeds (Sumit and Shivani 2021). A small quantity of curd/buttermilk is stored overnight to receive blue-green colour, the solution is used for deworming in young calves (Shenoy 2021). The whey milk, onion, and custard apple leaves are applied to the animal for managing excess grazing (Dipika et al. 2017). Table 14.5 presents farmers knowledge related to animal husbandry.

14.2.6 Medicinal and Aromatic Plants against Diseases

The farmers recommend a diversity of medicinal plants for trating diseases. The people of Northern part of India utilise several plants for treatment of many diseases, For example, root of *Acacia catechu* (khair) for asthma, and bronchitis remedy, and *root of Aconitum ferox* Wall. (Vatsnabh) for treating rheumatism, root of *Aconitum heterophyllum* Wall. (Atees) for treating fever, cough, piles, and stomach, fruit and bark of *Aegle marmelos* (L.) correa (Bell) for curing dysentery, diarrhoea, and fever, bulb of *Alpinia galalnga* (L.) Wild. (Kulanjan) as health tonic, and *Andrographis paniculata* (Burm. F.) Wall to control malaria, liver, and as blood purifier, whole plant *Aquillaria malaccensis* Lamk. (Agaru) for removing fish spine from the throat, whole plant *Artemisia maritima* L. (Kunja) as tonic, blood purifier, fever, root and stem of *Berberis aristata* DC. (Kingora) for curing eye disease, root of *Cassia augustifoila* Vahl (Senna) for curing rheumatism, tuber of *Chlorophytum tuberosum* Bak. (Safed musli) for curing leucorrhea, sexual tonic, root of *Coleus barbatus* Benth. (Patharchur) as tonic and blood pressure, resin and bark of *Cammiphora wightii* (Arn.) Bhandari for treating asthma, typhoid, root of *Curculigo orchioides* Haerten (Kali Musli) for curing asthma, dysentery, and as tonic, rhizome of *Curcuma zedoaria* to treat jaundice, blood pressure, seed and fruit of *Embelia ribes* to treat skin problem, leprosy, fruit of *Garcinia indica* Choisy (Kokam) for curing skin disease, rhizome of *Gloriosa superb* L. (Kalibari) for treating snake bite, leprosy, root and leaf of *Gymnema sylvestre* (Retz.) (Gudmar) for curing gastric diseases, eye disease, root of *Hemidesmus indicus* (L.) Br. for curing cough, hypertension, dysentery, rhizome of *Myrica esculenta* Ham. Exdon (Kaphal) for curing bronchitis, blood purifier, hysteria, fruit and seed of *Nelumbo nucifera* barten (kamal phool) for curing cholera, diarrhoea, leaf and seed of *Ocimum sanctum* L. for treating fever, vomiting, liver and blood purifier, seed and leaf of *Phyllanthus emblica* L. (amla) for curing fever, vomiting, liver, blood purifier, root of *Picrorhiza kurroa* Benth. (Katuki) for curing headache, fever, dysentery, fruit of *Pistacacia chinenesis* Bunge (Kakadshingi) for curing cholera, fever, cough, root of *Piper longum* L. to cure indigestion, childbirth, dysentery, root of *Pistacacia chinensis* Bunge (Sarapagandha) for treating malaria fever, snake bite, wood of *Santallum*

Table 14.5 Indigenous knowledge practices of farmers in animal husbandry

Sl. No.	Indigenous Traditional Knowledge (ITK)	Disease/Use	State	References
1	Pseudostem banana	Production of pond fish	Assam, Tamil Nadu and Goa	Bhalerao et al. (2015)
2	500 g maida +500 g behada powder + water after boiling	Foot and Mouth Disease control	Maharashtra	Choubey (2005)
3	Extract of peach leaves + fresh milk	Lesion of mouth and hooves control	West Bengal, Rajasthan, Uttar Pradesh, Tamil Nadu, Jharkhand, Himachal Pradesh, Uttaranchal, and Odisha	Das et al. (2004a, b)
4	Paste of babool bark and Jamun bark	Foot and Mouth Disease control	Uttar Pradesh, Maharashtra, and Odisha	Rajesh and Bharathi (2012); Sarita et al. (2003)
5	Paste of Bantulsi (*Ocimum gratissimum*) leaf + water	Khurha (FMD) disease control	Uttar Pradesh	Swarup and Pradhan (2020)
6	liquid medicine with stone apple (bael) + water	Diarrhoea control	Uttaranchal	Mahesh (2020)
7	Mixing of pegion waste + jaggery	Inducing oestrus cycle	Uttar Pradesh and Uttaranchal	Swarup and Mahesh (2020)
8	Extracted juice of gurhal (urhul) flower treated orally in the goat	Diarrhoea control	Uttaranchal	Dakshinkar and Vihan (2020)
9	Paste of Pojo (*Litsaea authapoly*)	Diarrhoea and dehydration control	Jharkhand	Haque and Vihan (2020)
10	Leaves of ridge gourd or ekdandi	Wound of the animals	Himachal Pradesh	Varshney (2020)
11	Paste of 200–250 g stem and leaf of Bhangariya (*Ecliptaalba*)+50–60 mL mustard oil	Cattle, buffaloes, and goats for cure blain control	Maharashtra	Jangde and Dhanan (2020)
12	Paste of 30 g geru +50 g snail shell/sippi are boiled with castor oil +20 g Alua +50 g kudru/ sahjam gum	Bullocks or bulls for swelling control	Maharashtra and Uttar Pradesh	Swarup and Dhakate (2020)
13	Paste of kalajeera	*Haemrrhagic septicaemia* control	Maharashtra	Vihan (2020)
14	Hajore paste	Recovering bone fracture	Jharkhand and Himachal Pradesh	Roy and Varshney (2020)
15	Fish trapping—Ghuni, chero/kero, chokhia and atal; fish barrier-Aran	Fish trapping, fish barrier, catfish breeding	West Bengal	Aparna et al. (2020)

(continued)

Table 14.5 (continued)

Sl. No.	Indigenous Traditional Knowledge (ITK)	Disease/Use	State	References
	bata/Aran pata; catfish breeding—earthen rings/ earthen pots			
16	Channa gachua (Changmachh) local fish	Curing asthama and Body pain	Assam	Aparna et al. (2020)
17	Oil of *Mystus vittatus*			
18	Lau Macha local fish cum vegetable (bottle gourd); murrels (*Channa* spp.), climbing perch (*Anabas testodeneous*) and cat fish (*Clarias batrachus*) and Bloch (*Heteroneutes fossilis*) in paddy field	Weed population and soil loosening control	Tripura	Ratan and Dilip (2013)
19	10 g Bark Aswatha (Banyon, ficuspa) + 10 g Ada (Ginger) + 10 g salt	Managing bloat disease	West Bengal	Amitendu et al. (2004)
20	Stem, leaves of Anantamul + honey	Managing animal dysentery	Odisha and Gujarat	Bikram et al. (2012); Patel et al. (2016)
21	100 g tulsi leaves +100 g basak are boiled + water + honey	Cold and cough control	Gujarat	Bikram et al. (2012); Patel et al. (2016)
22	Preparation of glue with tamarind	Strengthening of the nets	Hyderabad, Telangana	Ram et al. (2013)
23	Root of Babul (*Acacia arabica*) + mustard	Arthritis control	Uttar Pradesh, Gujarat and Rajasthan	Ram et al. (2013)
24	Kala/*Musa paradisiaca* L. + sugar	Oestrous cycle control	Uttar Pradesh, Gujarat and Rajasthan	Ram et al. (2013)
25	Cow dung slurry	Managing Euglena bloom	Hyderabad, Telangana and Karnataka	Swamy et al. (2015)
26	200 g termite mound soil + water	Mastitits, poisonous bite of insects and mechanical injury control	Hyderabad, Telangana and Karnataka	Swamy et al. (2015)
27	Application of cashew shell oil, coal tar, and sardine oil	Boats and nets preservation	Bihar and Hyderabad, Telangana	Sumit and Shivani (2021)
28	Application of saw dust and rice	Fish preservation and transport	Hyderabad, Telangana	Sumit and Shivani (2021)
29	Spices of mango pickles and neem leaves	Bloat disease control	Bihar and Hyderabad, Telangana	Sumit and Shivani (2021)

(continued)

Table 14.5 (continued)

Sl. No.	Indigenous Traditional Knowledge (ITK)	Disease/Use	State	References
30	*Gardenia resinifera* Leaves and *Dendrophthoe falcata* seeds	Constipation control	Bihar, Hyderabad and Odisha	Sumit and Shivani (2021)
31	Whey milk, onion and custard apple leaves	Excess grazing control	Bihar, Hyderabad and Maharashtra	Dipika et al. (2017)
32	Vinegar	Tympany medication	Uttar Pradesh	Gyan et al. (2016)
33	Castor oil	Deworming	Uttar Pradesh	Gyan et al. (2016)
34	Mustard oil	Body heat regulation	Uttar Pradesh	Gyan et al. (2016)
35	Turmeric lime paste	Sprain heal	Uttar Pradesh	Gyan et al. (2016)
36	Black pepper butter oil mixture	Pneumonia fever control	Uttar Pradesh farmer	Gyan et al. (2016)
37	*Glyricidia* and roasted soaked tamarind seeds	Lactation improvement	Assam, Nagaland, Madhya Pradesh, and Haryana	Deepandita et al. (2021)
38	Seeds of subabul	Milk secretion growth	Assam, Nagaland, Madhya Pradesh, and Haryana	Deepandita et al. (2021)
39	Liquid formulation product with Bottle gourd, fenugreek, coconut, black gram, palm jaggery, water	Increase milk	Assam, Nagaland, Madhya Pradesh, and Haryana	Deepandita et al. (2021)
40	Dried flowers of *Madhuca latifolia*	Increase bullock work efficiency	Assam, Nagaland, Madhya Pradesh, and Haryana	Deepandita et al. (2021)
41	Powdered formulation with Pepper, jaggery, and betel leaf	Increase digestion rate	Assam, Nagaland, Madhya Pradesh, and Haryana	Deepandita et al. (2021)
42	Grinded *Iris kashmiriana* and jaggery	Increase milk	Jamm and Kashmir	Shubeena et al. (2018)

album (Chandan) for curing dysentery and skin disease, bark and leaf of *Saraca asoca* (Ashok) for treating heart diseases, root of *Saussurea costus* (Falc.) Lipsch. (Kut) for treating dysentery, asthma, and ulcer, *Smilax* sp. (Chopchini) for treating menstrual complain and smallpox, whole plant *Solanum nigrum* (Giloe) for curing jaundice, bone fracture, root and leaf of *Valeriana jatamansi* (Tagar) for treating epilepsy, urinary complain, root and leaf of *Withania somnifera* (Ashwagandha) for treating eye, asthma, cough, bark and latex of *Wrightia tinctoria* (Indra java) for treating toothache, piles, dysentery (Chandra et al. 2006).

The Tripura people use diverse medicinal plant species in curing mild and acute diseases. The Tripura people utilise leaves of traditional plant *Andrographis panicular* for curing dog bite, root of *Phylogacanthus thyrsiflorus* used for curing cold, cough, asthama, root of *Achyranthes aspera* is used for curing epilepsy, bark and root of *Mangifera indica* L. utilised for treating toothache, whole plant *Centella asiatica* L. used for treating toothache, latex and shoot of *Alstonia scholaris* L. used for enhancing mother milk, leaves of *Holarrhena antidysentria* used for treating dysentery, diarrhoea, anthelmintic, leaves and latex of *Homalonema aromatic* utilized for curing snake bite, leaves of *Ageratum conyzoides* utilized for curing wounds, leaves of *Enydra fluctuans* used for treating bleeding, whole plant of *Spilanthes paniculata* utilized for treating gastric, stomach problem, throat, diabetes, leaves of *Kalanchoe pinnata* used for curing dysentery, leaves of *Coccinia grandis* used for curing diabetes, leaves and fruits of *Momordica cacharantia* used for curing hand pimples, foot pimples, leaves of *Ricinus communis* used for treating swelling, rheumatism, *Acacia concinna* used for treating diabetes and body pain, leaves of *Cajanus cajan* used for treating jaundice, fruits of *Cassia fistula* for curing laxative, leaves of *Cassia accidentalis* for treating skin disease, leaves and roots of *Mimosa pudica* for curing ring worm, piles, fruits of *Parkia javanica* utilized for curing gastric problem, leaves and flowers of *Lecuas aspera* used for curing pain, gastric problem, swelling, leaves and bark of *Ocimum basilicum* uses for curing gastric problem, stomach problems, leaves of *Ocimum sanctum* L. used for treating cough, cold, leaves of *Premna* sp. uses for treating ant bite, bark and leaves of *Litsea glutinosa* uses for curing muscle pain, bone fracture, roots, leaves and buds of *Hibiscus rosa-sinensis* uses for treating irregular menstruation, leaves of *Sterculli aviliosa* utilized for treating menstruation pain, fruits and leaves of *Moringa oleifera* used for the cooling effect, leaves of *Psidium guajava* is used for treating diarrhoea, dysentery, piles, and vomiting, leaves of *Nyctanthes arbor-tristis* used for curing asthma, and stomach disorders, Leaves of *Aporosa octandra* used for curing injury, fruits and leaves of *Phyllanthus acidus* utilises for treating chicken pox, leaves of *Scoparia daclis* utilized for treating body pain, whole plant *Cynodon dactylon* used for treating toothache, rhizome of *Drynaria quercifolia* used for treating swelling, fruits and leaves of *Ageles marmelos* used for curing high fever, and malaria, roots of *Murraya paniculata* utilizes for curing toothache, fruits of *Flacourita jangomas* is used for curing dysentery, diarrhoea, rhizome of *Aloe barbadensis* is utilised for curing cold, rhizome of *Curcuma zeodaria* is used for curing stomach, urinary disorder (Maria et al. 2017).

The people in Uttarakhand recommend a diversity of medicinal plants for curing human diseases. In Uttarakhand, people use medicinal plants such as - *Aconitum balfourii* (meetha/Bhngwa) for diaphoretic, diuretic, analgesic, anti-inflammatory, anit-pyretic, vermifuge. *Aconitum heterophyllum* (Atees) is used for treating anti-inflammatory, anti-pyretic, anti-bacterial, anthelmintic. *Ajuga parviflora* (Neel Kanthi) is used for curing hypertension, malaria, pneumonia, edema, anti-fungal, hypoglycemic, and anti-microbial agents. *Allium cepa* (Pyaj) is anti-tumour, anti-diabetic, anti-allergic and anti-molluscicidal. The whole plant of *Allium sativum* (Lehsum) is used in burns and cuts; the leaves of *Allium wallichii* is used in treating gastric; the leaves of *Angelica glauca* Edgew (choru) uses for treating gastric disorderseaves; the leaves of *Artemisia nilagirica* (kunja) uses for treating cuts and wounds; the roots of *Asparagus filicinus* (Jhirna) uses for treating weakness; the root of *Berberis aristata* (kingod) is used for curing eye ailments; and the root of *Bergenia stracheyi* (Pashanbhed) is used for curing stone problems. The leaves of *Centella asiatica* (Brahmi) is used for treating coolant diseases The leaves and bark of *Cinnamomum tamla* (tejpat) is used for curing blood pressure. The seeds of *Cirisium wallichii* (kanjelu) are used for treating fever. The seeds of *Cucumis sativus* (kakdi) is used for curing diuretic diseases. The root of *Cucurma longa* (Haldu) is used for treating cut wounds. The tuber of *Dioscorea bulbifera* (Tairu) is used for treating coolant disease. The leaves of *Eupatorium adenophorum* (Basya) is used for treating cuts and wounds The root of *Girardinia diversifolia* (kandali) is used for curing fever. The fruit of *Hippophae salicifolia* (Amesh) is used for treating coolant. The fruit peel of *Juglans regia* (Akhrot) is used for curing skin diseases. The root of *Jurinea macrocephala* (Biskhanada) is used for curing fever. The root of *Macrotyloma uniflorum* (gahat) is used for curing stone diseases. The root of *Megacarpaea polynadra* (Barmolu) is used for treating gastric problems. The leaves of *Mentha pipertia* (Pudina) are used for curing coolant disease. The leaves of *Mirabilis jalapa* is used for curing cuts and wounds. The leaves of *Nardostachys jatamansi* (Maasi) is used for treating jaundice. The aerial part of *Ocimum corniculata* (Almodu) is used for treating boils. The leaves of *Paeoni emodi* (chandra) is used for treating fever. The leaves of *Picrorhiza kurrooa* (Kadwi) is used for treating fever. The rhizome of *Polygonatum verticillatum* (Mahamaida/ salampanja) is used for curing fever. The fruits of *Potentilla lineata* (Bajradanti) is used for treating anaemia. The roots of *Rheum moorcroftianum* (Dolu) is used for curing injuries, cuts, and wounds. The leaves of *Rhododendron campanulatum* (Syamru) is used for curing skin diseases. The root of *Rumex nepalensis* (khuldya) is used for curing pneumonia, cuts, and wounds. The roots and leaves of *Saussurea costus* (kuth) is used for treating skin diseases. The aerial part of *Saussurea obvallata* (kaunl) is used for treating the immune system. The root of *Selinum vaginatum* (bhutkesh) is used for curing coolant. The aerial part of *Swertia chiraytia* (chiraitu) is used for curing fever and stomach ache. The leaves of *Tagetes erecta* (gainda) is used for for curing ear aches. The bark of *Taxus wallichiana* (thuner) is used for treating high blood pressure. The aerial part of *Tinospora sinesis* (giloe) is used for curing fever, and stomach ache. *Utrica dioca* (kundali) is used for treating anaemia. The seed, stem and aerial parts of *Zanthoxylum armatum* (Timru) are used for curing toothache (Ankit et al. 2019).

The Habb-e-Asgand is a unani formulation for controlling Wajalal mafasil (rheumatoid arthritis) (gaathia) (Verma et al. 2021). The powdered forms of root of *Anacyclus pyrethrum*, *Withania somnifera*, *Chlorophytum borivilianum*, *Asparagus racemosus* and tuber of *Pueraria tuberosa* are used for stimulating male sex hormone (Kumar et al. 2021). The liquid of pseudostem of *Ensete glaucum* (Roxb.) Cheesman contains amino acids, cardiac glycosides, flavonoids, polyphenol, alkaloids, reducing sugars, starch, saponins, tannins, terpenoids, oil and fats used against diarrhoea (Jogaa et al. 2021). The people of Solan, Himachal Pradesh recommend *Cryptolepis buchananii*, *Eucalyptus citriodora*, *Ligustrum japonicum*, *Pinus roxburghii*, *Rosa alba*, *Ziziphus nummularia* and *Sonchus oleraceus* for treating skin infections, *Rhododendron arboreum*, *Zanthoxylum armatum*, *Viola canescens*, *Quercus leucotrichophora*, *Rubus ellipticus*, *Punica granatum*, *Ocimum sanctum*, *Morus nigra*, *Mentha arvensis*, *Justicia adhatoda*, *Ficus benghalensis*, *Eriobotrya japonica*, *Debregeasia longifolia*, *Cissampelos pareira*, *Datura innoxia*, *Eucalyptus citriodora*, *Cynodon dactylon*, *Colebrookea oppositifolia* and *Cannabis sativa* for treating diarrhoea, diabetes, dysentery, cough, cold, and fever (Kumar et al. 2021).

The Adi community of Arunachal Pradesh treats asthma, bronchitis, cough, sinusitis, diabetes, malaria, typhoid, and jaundice with Frangipani, periwinkle, turkey berry, Nightshade, Indian trumpet flower, and Giloy (Ranjay et al. 2021). In the Himalayan cold desert region of Ladakh, people treat ulcer and liver diseases with *Juniperus polycarpus* C. Koch (Himalayan pencil cedar) (Dorjeya and Mauryab 2021). The Uttarakhand people recover skin disease problems with hairy beggaricks with Deodar. Constipation and liver disorder are treated with *Eclipta alba*, *Mallotus philippensis*, *Boehmeria rugulosa*, *Celtis australis*. Fever, inflammation, headaches, bleeding, and malaria are treated with *Artemisia annua;* insect bites, infertility problem with *Parthenium hysterophorus*, and the human stone problem is treated with *Chenopodium album* and *Berginia ciliata*. The human tooth problem is treated with *Xanthium stramonium*. The blood dysentery is cured with *Boerhavia diffusa* and *Sterculia villosa*. The human muscular pain and swelling are cured with *Helicteres isora*, epilepsy problems are cured with *Artemisia japonica*. The cut and wound are treated with *Ageratum conyzoides, Brassica campestris, Betula utilis*, *Achyranthus aspera*, *Colebrookia oppositifolia*, *Rumex hastus* and *Bergenia ciliata*. The urinary disorder, headache and menstrual disorder are treated with *Fagopyrum esculatum*. The children worms are restricted with *Amaranthus paniculatus*. The human stomach problem is cured with *Artimisia maritime, Cyanodon dactylon* and *Syzgium cumini*. *Bombax ceiba* is used against piles disease. *Treminalia chebula* is used in indigestion problem. The fractured bone is treated with *Litsea chinenesis*. The bite of *scorpion* is cured with *Amaranthus spinosus*. The human memory is stimulated with *Centella asiatica* (Akasha and Bhandari 2021). The garo tribes utilise 36 trees, 5 shrubs, and 2 creeper/climbers for medicinal purposes (Singh and Mathew 2021). The community of Dongria Kandha tribes use the traditional medicinal plants such as *Discorea bulbifera* L. for curing cancer, HIV, anti-inflammatory, anti-microbial, cardioprotective and anti-hyperthyroid activities (Parida and Sarangi 2021). The seed of *Manikara zapota*, *Caatinga biome*, *Moringa oleifera*, *Carica papaya*, *Myracrodruon urundeuva* are used in controlling *Ades agegypti* mosquito. The seed protein of *Jatropha curcas* and leaves of *Solanum villosum* are recommended for restricting *Culex quiinquefasciatus* and *Ades aegypti* (Manisha and Neelam 2021). In Uttar Pradesh, people control Bovine herpes virus type I, foot and mouth disease virus and Newcastle disease virus in animals with sacred plant *Ocimum tenuiflorum* and *Ocimum sanctum* (holy basil/tulsi) (Goel and Bhatia 2021). The Jammu and Kashmir, people use non-timber forest products *Aconitum heterophyllum* wall. (Patis, Aconite, Dhar buti, Attees or Bis Mohra) for intestinal worms, diarrhoea, dysentery, high fever and anti-rheumatic. The disease of fever, cold, cough, hypertension, muscle spasms, parasitic worms and malaria root are treated with the rhizomes of *Viola odorata* (Bnafsaha, wild violet, sweet violet). The root of *Valeriana jatamansi* (mush khala, jatamansi, balchhari, mansi, nihani) is used for treating eye, blood, liver problem, hysteria, and nervous and urinal stress. The root and rhizomes of *Picrorhiza kurroa* (Kaud, kaur, kutki) are used in fever, cold cough, hypertension, muscle spasms, parasitic worms, and malaria. The root of *Bergenia ligulata* (patharchoor, pashanbeda) is used for healing of wounds (Bagala et al. 2021). The powdered medicine is recommended in the management of longevity, anti-viral, analgesics, ascites, hypoglycemic, anti-arthritis and is anti-ageing (Manosi et al. 2021). Table 14.6 presents traditional' knowledge related to the usage of plants for the treatment of diseases.

14.2.7 Stored Grain Pests' Management

The storage of the grains has been an important post harvest agricultural activity. The Uttar Pradesh and Tamil Nadu farmers constructed godowns with straw, leaves, and godowns mounted with cow dung. The grains are stored in various structures (Vishal et al. 2021). The cleft of the godowns mounted with the rod, cow dung led mud. The construction and storage of food grains were described in the Vishwakarma vastu sastra. The stored grains are prone to pest infestaion. The Tamil Nadu farmers mix 200 g of common salt in red gram/arhar for controlling stored grain pests. The Tamil Nadu farmers mix and treat sorghum seed at a 1:4 ratio in jute gunny bags for 6 months of storage of seed and controlling pest problems. They control storage pests and insect pests with the liquid solution of neem oil + coconut oil/castor oil. The application of 5-L groundnut oil and ¼ kg tamarind in the container and the container is covered with cotton cloth tight for groundnut oil storage. The farmer applies 100 g coriander seeds, a litre of oil, and a spoon of salt in the container. The coriander seed releases odour in the oil that prevents oil spillage and oil spoilage (Karthikeyan et al. 2009). To control flat grain borer, lesser grain borer and saw-toothed beetles, blending ragi grains into neem leaves, thumbai, and any strong odour leaves (Kaddi patta, tulsi, lemon grass. etc) is done. Some farmers expose stored pulse grain in the open sunlight at 20 °C for controlling *Callosobruchus chinensis* eggs and grubs (Adesina et al. 2019). The Karnataka farmers prepare custard apple seed powder and use it for pulse grains to control bruchid adult and eggs (Prakash et al. 2016).

The Tamil Nadu, Kerala, Telangana, and Karnataka farmers stored pigeon pea seed with horse gram seed dust in an air-tight container. The horse gram dust assimilates excess moisture and encourages long-term storage (Shaila and Nafeesa 2021).

The Tamil Nadu, Uttar Pradesh, and Maharashtra farmers construct godowns/ granary rooms with brick and wooden boards for controlling rice moths and restricting the moisture of the grains (Parimala et al. 2013). The Manipur and

Table 14.6 Indigenous knowledge related to medicinal and aromatic plants for the treatment of diseases

Sl. No.	Traditional medicinal plants and part used	Diseases	State/Part of India	References
1	Root part of *Acacia catechu* (Khair)	Asthama, and bronchitis	Northern India	Chandra et al. (2006)
2	Root part of *Aconitum ferox* Wall. (Vatsnabh)	Rheumatism	Northern India	Chandra et al. (2006)
3	Root part of *Aconitum heterophyllum* Wall. (Atees)	Fever, cough, piles, and stomach control	Northern India	Chandra et al. (2006)
4	Fruit and bark part of *Aegle marmelos* (L.) Correa (Bell)	Dysentery, diarrhoea, fever, and health tonic	Northern India	Chandra et al. (2006)
5	Bulb part of *Alpinia galalnga* (L.) Wild. (Kulanjan)	Health tonic	Northern India	Chandra et al. (2006)
6	Bulb part of *Andrographis paniculata* (Burm. F.) wall.	Malaria, liver and blood purifier	Northern India	Chandra et al. (2006)
7	Whole part of *Aquillaria malaccensis* Lamk. (Agaru)	Removing fish spine from throat	Northern India	Chandra et al. (2006)
8	Whole part of *Artemisia maritima* L. (Kunja)	Tonic, blood purifier, and fever	Northern India	Chandra et al. (2006)
9	Root and stem of *Berberis aristata* DC. (Kingora)	Eye diseases	Northern India	Chandra et al. (2006)
10	Root of *Cassia augustifoila* Vahl (Senna)	Rheumatism control	Northern India	Chandra et al. (2006)
11	Tuber of *Cholorphytum tuberosum* Bak. (Safed musli)	Leucorrhea and sexual tonic	Northern India	Chandra et al. (2006)
12	Root of *Coleus barbatus* Benth. (Patharchur)	Tonic and blood pressure control	Northern India	Chandra et al. (2006)
13	Resin and bark of *Cammiphora wightii* (Arn.) Bhandari	Asthma, and typhoid control	Northern India	Chandra et al. (2006)
14	Root of *Curculigo orchioides* haerten (Kali musli)	Asthma, dysentery, and tonic	Northern India	Chandra et al. (2006)
15	Rhizome of *Curcuma zedoaria* (Christm.) Rosc.	Control of jaundice, and blood pressure	Northern India	Chandra et al. (2006)
16	Seed and fruit of *Embelia ribes* Burm. f. (Jheum)	Control of skin diseases, and leprosy	Northern India	Chandra et al. (2006)
17	Fruit of *Garcinia indica* choisy (Kokam)	Skin disease control	Northern India	Chandra et al. (2006)
18	Rhizome of *Gloriosa superb* L. (Kalibari)	Snake bite, and leprosy control	Northern India	Chandra et al. (2006)

(continued)

Table 14.6 (continued)

Sl. No.	Traditional medicinal plants and part used	Diseases	State/Part of India	References
19	Root and leaf of *Gymnema sylvestre* (Retz.) (Gudmar)	Gastric disorder, and eye disease	Northern India	Chandra et al. (2006)
20	Root of *Hemidesmus indicus* (L.) Br.	Curing cough, hypertension, and dysentery	Northern India	Chandra et al. (2006)
21	Rhizome of *Myrica esculenta* Buch.-Ham. ex D.Don (Kaphal)	Curing bronchitis, blood purifier, and hysteria	Northern India	Chandra et al. (2006)
22	Fruit and seed of *Nelumbo nucifera* Gaertn. (Kamal phool)	Curing chlorea, diarrhoea	Northern India	Chandra et al. (2006)
23	Leaf and seed of *Ocimum sanctum* L.	Treating fever, vomiting, liver and blood purifier	Northern India	Chandra et al. (2006)
24	Leaf and seed of *Phyllanthus emblica* L. (Amla)	Curing fever, vomiting, liver, blood purifier	Northern India	Chandra et al. (2006)
25	Root of *Picrorhiza kurrooa* Benth. (Katuki)	Curing Headache, fever, dysentery	Northern India	Chandra et al. (2006)
26	Fruit of *Pistacacia chinenesis* Bunge (Kakadshingi)	Curing cholera, fever, and cough	Northern India	Chandra et al. (2006)
27	Root of *Piper longum* L.	Curing indigestion, child birth, and dysentery	Northern India	Chandra et al. (2006)
28	Root of *Pistacia chinensis* Bunge (Sarapagandha)	Curing malaria fever, and snake bite	Northern India	Chandra et al. (2006)
29	Wood of *Santallum album* (Chandan)	Curing dysentery and skin disease	Northern India	Chandra et al. (2006)
30	Bark and leaf of *Saraca asoca* (Roxb.) Willd. (Ashok)	Heart disorder	Northern India	Chandra et al. (2006)
31	Root of *Saussurea costus* (Falc.) Lipsch. (Kut)	Dysentery, asthama, and ulcer	Northern India	Chandra et al. (2006)
32	Whole part of *Smilex* sp. (Chopchini)	Menstrual complaints and small pox	Northern India	Chandra et al. (2006)
33	Whole plant of *Solanum nigrum* Acerbi ex Dunal (Giloe)	Curing jaundice, and bone fracture	Northern India	Chandra et al. (2006)
34	Root and leaf of *Valeriana jatamansi* D.Don (Tagar)	Curing epilepsy, and urinary complain	Northern India	Chandra et al. (2006)
35	Root and leaf of *Withana somnifera* L. (Ashwagandha)	Curing eye, asthama, and cough	Northern India	Chandra et al. (2006)
36	Bark and latex of *Wrightia tinctoria* R.Br. (Indra java)	Curing toothache, piles, and dysentery	Northern India	Chandra et al. (2006)
37	Leaves of *Andrographis paniculata* (Burm.f.) Nees	Curing dog bite	Tripura	Maria et al. (2017)

(continued)

Table 14.6 (continued)

Sl. No.	Traditional medicinal plants and part used	Diseases	State/Part of India	References
38	Root of *Phylogacanthus thyrsiflorus* Nees	Curing cold, cough, and asthama	Tripura	Maria et al. (2017)
39	Bark and root of *Achyranthes aspera* L.	Toothache	Tripura	Maria et al. (2017)
40	Whole plant of *Centella asiatica* L.	Tooth problem	Tripura	Maria et al. (2017)
41	Latex and shoot of *Alstonia scholaris* L.	Curing mother milk	Tripura	Maria et al. (2017)
42	Leaves of *Holarrhena antidysentrica* (L.) Wall.	Curing dysentery, diar-rhoea, and anthelmintic	Tripura	Maria et al. (2017)
43	Leaves and latex of *Homalonema aromatica* Schott	Curing snake bite	Tripura	Maria et al. (2017)
44	Leaves of *Ageratum conyzoides* Sieber ex Steud.	Curing wounds, and cut	Tripura	Maria et al. (2017)
45	Leaves of *Enydra fluctuans* Lour.	Treating bleeding	Tripura	Maria et al. (2017)
46	Whole plant of *Spilanthes paniculata* Wall. ex DC.	Treating gastric, stomach problem, throat, and diabetes	Tripura	Maria et al. (2017)
47	Leaves of *Kalanchoe pinnata* (Lam.) Pers.	Curing dysentery	Tripura	Maria et al. (2017)
48	Leaves of *Coccinia grandis* (L.) Voigt	Curing diabetes	Tripura	Maria et al. (2017)
49	Leaves and fruits of *Momordica cacharantia* L.	Curing hand and foot pimples	Tripura	Maria et al. (2017)
50	Leaves of *Acacia concinna* (Willd.) DC	Treating diabetes and body pain	Tripura	Maria et al. (2017)
51	Leaves of *Cajanus cajan* (L.) Millsp.	Treating jaundice	Tripura	Maria et al. (2017)
52	Fruits of *Cassia fistula* L.	Curing laxative	Tripura	Maria et al. (2017)
53	Leaves of *Cassia accidentalis* L.	Treating skin disease	Tripura	Maria et al. (2017)
54	Leaves and root of *Mimosa pudica* L.	Curing ring worm, piles	Tripura	Maria et al. (2017)
55	Fruits of *Parkia javanica* (Lam.) Merr.	Curing gastric problem	Tripura	Maria et al. (2017)
56	Leaves and flower of *Lecuas aspera* (Willd.) Link.	Curing pain, gastric prob-lem, and swelling	Tripura	Maria et al. (2017)
57	Leaves and bark of *Ocimum basilicum* L.	Curing gastric problem, and stomach problem	Tripura	Maria et al. (2017)

(continued)

Table 14.6 (continued)

Sl. No.	Traditional medicinal plants and part used	Diseases	State/Part of India	References
58	Leaves of *Ocimum sanctum* L.	Treating cough, cold	Tripura	Maria et al. (2017)
59	Leaves of *Premna* sp.	Treating ant bite	Tripura	Maria et al. (2017)
60	Bark and root of *Litsea glutinosa* (Lour.) C.B.Rob.	Curing muscle pain, and bone fracture	Tripura	Maria et al. (2017)
61	Root, leaves and bud of *Hibiscus rosa-sinensis* L.	Treating irregular menstruation	Tripura	Maria et al. (2017)
62	Leaves of *Sterculia avillosa* Roxb.	Treating menstruation pain	Tripura	Maria et al. (2017)
63	Leaves and fruits of *Moringa oleifera* Lam.	Treating cooling effect	Tripura	Maria et al. (2017)
64	Leaves of *Psidium guajava* L.	For treating diarrhoea, dysentery, piles, and vomiting	Tripura	Maria et al. (2017)
65	Leaves of *Nyctanthes arbor-tristis* L.	Curing asthama, stomach disorder	Tripura	Maria et al. (2017)
66	Fruits and leaves of *Phyllanthus acidus* (L.) Skeels	Treating chicken pox	Tripura	Maria et al. (2017)
67	Leaves of *Scoparia daclis* L.	Treating body pain	Tripura	Maria et al. (2017)
68	Whole plant of *Cyanodon dactylon* (L.) Pers.	Treating toothache	Tripura	Maria et al. (2017)
69	Rhizome of *Drynaria quercifolia* (L.) J.Sm.	Treating swelling	Tripura	Maria et al. (2017)
70	Leaves and fruits of *Aegele marmelos* (L.) Corrêa	Curing high fever, and malaria	Tripura	Maria et al. (2017)
71	Root of *Murraya paniculata* Kaneh.	Curing toothache	Tripura	Maria et al. (2017)
72	Fruits of *Flacourtia jangomas* Steud.	Curing dysentery, and diarrhoea	Tripura	Maria et al. (2017)
73	Rhizome of *Aloe barbadensis* Mill.	Curing cold, cand ough	Tripura	Maria et al. (2017)
74	Rhizome of *Curcuma zeodaria* Roxb.	Curing stomach, and urinary disorders	Tripura	Maria et al. (2017)
75	*Aconitum balfourii* Stapf (Meetha/Bhngwa)	Curing diaphoretic, diuretic, analgesic, anti-inflammatory, anti-pyretic, and vermifuge	Uttarakhand	Ankit et al. (2019)
76	*Aconitum heterophyllum* Wall. (Atees)	Treating anti-inflammatory, anti-pyretic, anti-bacterial, and anthelminthic	Uttarakhand	Ankit et al. (2019)
77	*Ajuga parviflora* Benth. (Neel Kanthi)	Curing hypertension, malaria, pneumonia,	Uttarakhand	Ankit et al. (2019)

(continued)

Table 14.6 (continued)

Sl. No.	Traditional medicinal plants and part used	Diseases	State/Part of India	References
		edema, anit-fungal, hypo-glycemic, and anti-micro-bial agents		
78	*Allium cepa* L. (Pyaj)	Anti-tumour, anti-diabeteic, anti-allergic and anti-mollusicidal	Uttarakhand	Ankit et al. (2019)
79	*Allium sativum* L. (Lehsun)	Burns and cuts	Uttarakhand	Ankit et al. (2019)
80	*Allium wallichii* Kunth	Treating gastric disorders	Uttarakhand	Ankit et al. (2019)
81	*Angelica glauca* Edgew (Choru)	Treating gastric disorders	Uttarakhand	Ankit et al. (2019)
82	*Artemisia nilagirica* (C.B. Clarke) Pamp. (Kunja)	Cut and wounds	Uttarakhand	Ankit et al. (2019)
83	*Asparagus filicinus* Buch.-Ham. ex D.Don (Jhirna)	Treating weakness	Uttarakhand	Ankit et al. (2019)
84	*Berberis aristata* Sims (Kingod)	Curing eye ailments	Uttarakhand	Ankit et al. (2019)
85	*Bergenia stracheyi* (Hook. f. & Thomson) Engl. (Pashanbhed)	CUring stone problem	Uttarakhand	Ankit et al. (2019)
86	*Centella asiatica* (L.) Urb. (Brahmi)	Treating coolant disease	Uttarakhand	Ankit et al. (2019)
87	*Cinnamomum tamla* (Buch.-Ham.) T.Nees & C. H.Eberm. (Tejpat)	Curing blood pressure	Uttarakhand	Ankit et al. (2019)
88	*Cirsium wallichii* DC. (Kanjelu)	Treating fever	Uttarakhand	Ankit et al. (2019)
89	*Cucumis sativus* L. (Kakdi)	Curing diuretic disease	Uttarakhand	Ankit et al. (2019)
90	*Cucurma longa* L. (Haldu)	Treating cuts, wounds	Uttarakhand	Ankit et al. (2019)
91	*Dioscorea bulbifera* L. (Tairu)	For cooling effect	Uttarakhand	Ankit et al. (2019)
92	*Eupatorium adenophorum* Spreng. (Basya)	Treating cuts and wounds	Uttarakhand	Ankit et al. (2019)
93	*Girardinia diversifolia* (Link) Friis (Kandali)	Curing fever	Uttarakhand	Ankit et al. (2019)
94	*Hippophae salicifolia* D. Don (Amesh)	For cooling effect	Uttarakhand	Ankit et al. (2019)
95	*Juglans regia* L. (Akhrot)	Curing skin disease	Uttarakhand	Ankit et al. (2019)
96	*Jurinea macrocephala* Benth. ex Hook. f. (Biskhanada)	Curing fever	Uttarakhand	Ankit et al. (2019)

(continued)

Table 14.6 (continued)

Sl. No.	Traditional medicinal plants and part used	Diseases	State/Part of India	References
97	*Macrotyloma uniflorum* (Lam.) Verdc. (Gahat)	Curing stone disease	Uttarakhand	Ankit et al. (2019)
98	*Megacarpaea polyandra* Benth. (Barmolu)	Treating gastric problem	Uttarakhand	Ankit et al. (2019)
99	*Mentha pipertia* L. (Pudina)	For cooling effect	Uttarakhand	Ankit et al. (2019)
100	*Mirabilis jalapa* L.	Curing cut and wound	Uttarakhand	Ankit et al. (2019)
101	*Nardostachys jatamansi* (D.Don) DC. (Maasi)	For treating jaundice	Uttarakhand	Ankit et al. (2019)
102	*Oxalis corniculata* L. (Almodu)	Treating boils	Uttarakhand	Ankit et al. (2019)
103	*Paeoni emodi* Royle (Chandra)	Treating fever	Uttarakhand	Ankit et al. (2019)
104	*Picrorhiza kurrooa* Royle (Kadwi)	Treating fever	Uttarakhand	Ankit et al. (2019)
105	*Polygonatum verticillatum* (L.) All. (Mahamaida/ salampanja)	Curing fever	Uttarakhand	Ankit et al. (2019)
106	*Potentilla lineata* Trevir. (Bajradanti)	Treating Anaemia	Uttarakhand	Ankit et al. (2019)
107	*Rheum moorcroftianum* Royle (Dolu)	Curing injury, cut and wound	Uttarakhand	Ankit et al. (2019)
108	*Rhododendron campanulatum* D.Don (Syamru)	Curing skin disease	Uttarakhand	Ankit et al. (2019)
109	*Rumex nepalensis* Spreng. (Khuldya)	Curing pneumonia, cut, wound	Uttarakhand	Ankit et al. (2019)
110	*Saussurea costus* (Falc.) Lipsch. (Kuth)	Treating skin disease	Uttarakhand	Ankit et al. (2019)
111	*Selinum vaginatum* C.B. Clarke (Bhutkesh)	Curing coolant disease	Uttarakhand	Ankit et al. (2019)
112	*Swertia chirayita* (Roxb.) Buch.-Ham. ex C.B. Clarke (Chiraitu)	Curing fever, stomach, ache	Uttarakhand	Ankit et al. (2019)
113	*Tagetes erecta* L. (Gainda)	Curing ear ache	Uttarakhand	Ankit et al. (2019)
114	*Taxus wallichiana* Zucc. (Thuner)	Treating high blood pressure	Uttarakhand	Ankit et al. (2019)
115	*Tinospora sinensis* (Lour.) Merr. (Giloe)	Curing fever, stomach, ache	Uttarakhand	Ankit et al. (2019)
116	*Utrica dioica* L. (Kundali)	Treating anaemia, weakness	Uttarakhand	Ankit et al. (2019)

(continued)

Table 14.6 (continued)

Sl. No.	Traditional medicinal plants and part used	Diseases	State/Part of India	References
117	*Zanthoxylum armatum* DC. (Timru)	Curing teeth, toothache	Uttarakhand	Ankit et al. (2019)
118	Habb-e-Asgand (*Trachyspermum ammi, Withania somnifera, Argyreia speciosa, Piper longum, Ficus religiosa, Zingiber officianale, Asparagus racemosus, Curculigo orchioides* and Jaggery) A Unani medicine	Wajalal mafasil (Rheumatoid arthritis) (gaathia)	Uttar Pradesh	Verma et al. (2021)
119	Powdered form of root of *Anacyclus pyrethrum* (L.) Link, *Withania somnifera* (L.) Dunal, *Chlorophytum borivilianum* Santapau & R.R.Fern., *Asparagus racemosus* Willd. and tuber of *Pueraria tuberosa* (Roxb. ex Willd.) DC.	Stimulates sexual hormone in male gender	Uttar Pradesh	Kumar et al. (2021)
120	Liquid of pseudostem of *Ensete glaucum* (Roxb.) Cheesman	Diagnosing diarrhoea	Meghalaya	Jogaa et al. (2021)
121	*Frangipani, periwinkle, turkey berry, Night shade, Indian trumpet flower* and *Giloy*	Asthma, bronchitis, cough, sinusitis, diabetes, malaria, typhoid and jaundice controls	Adi community, Arunachal Pradesh	Ranjay et al. (2021)
122	*Juniperus polycarpus* C. Koch (Himalayan pencil cedar)	Monastery constructions, increases preparations, fuelwood and fodder crops	Himalayan cold desert region of Ladakh	Dorjeya and Mauryab (2021)
123	*Bidens pilosa* L., *Cedrus deodara* (Roxb. ex D.Don) G.Don	Skin disease treatment	Uttarakhand	Akasha and Bhandari (2021)
124	*Eclipta alba* L. ex B.D. Jacks., *Mallotus philippensis* (Lam.) Müll. Arg., *Boehmeria rugulosa* Wedd., *Celtis australis* L.	Constipation disorder, lier disorder	Uttarakhand	Akasha and Bhandari (2021)

(continued)

Table 14.6 (continued)

Sl. No.	Traditional medicinal plants and part used	Diseases	State/Part of India	References
125	*Aretmisia annua* Pall.	Cosmetic product development	Uttarakhand	Akasha and Bhandari (2021)
126	*Parthenium hysterophorus* Adans.	Insect bites, infertility problem	Uttarakhand	Akasha and Bhandari (2021)
127	*Chenopodium album* Bosc ex Moq., *Berginia ciliata* (Haw.) Sternb.	Curing stone problem	Uttarakhand	Akasha and Bhandari (2021)
128	*Xanthium stramonium*	Curing tooth problem	Uttarakhand	Akasha and Bhandari (2021)
129	*Boerhavia diffusa* L., *Sterculia villosa* Roxb.	Curing blood dysentery	Uttarakhand	Akasha and Bhandari (2021)
130	*Helicteres isora* L., *Artemisia japonica* Lauener	Curing epilepsy	Uttarakhand	Akasha and Bhandari (2021)
131	*Betula utilis* D.Don, *Achyranthes aspera* L.	Curing muscular pain and swelling	Uttarakhand	Akasha and Bhandari (2021)
132	*Bergenia ciliata* (Royle) A. Braun ex Engl., *Colebrookia oppositifolia* Donn, *Rumex hastatus* D.Don, *Ageratum conyzoides* Sieber ex Steud., *Brassica campestris* L.	Cut and wound treatment	Uttarakhand	Akasha and Bhandari (2021)
133	*Fagopyrum esculentum* Moench	Curing urinary disorder, headache and menstrual disorder	Uttarakhand	Akasha and Bhandari (2021)
134	*Amaranthus paniculatus* L.	Destroying worms in children	Uttarakhand	Akasha and Bhandari (2021)
135	*Cynodon dactylon* (L.) Pers., *Syzygium cumini* (L.) Skeels, *Artemisia maritima* Kitag.	Curing stomach problem	Uttarakhand	Akasha and Bhandari (2021)
136	*Bombax ceiba* Burm.f.	Curing piles disease	Uttarakhand	Akasha and Bhandari (2021)
137	*Treminalia chebula* Retz.	Curing indigestion problem	Uttarakhand	Akasha and Bhandari (2021)

(continued)

Table 14.6 (continued)

Sl. No.	Traditional medicinal plants and part used	Diseases	State/Part of India	References
138	*Litsea chinensis* Lour.	Curing fractured bone	Uttarakhand	Akasha and Bhandari (2021)
139	*Amaranthus spinosus* L.	Curing scorpion bite	Uttarakhand	Akasha and Bhandari (2021)
140	*Centella asiatica* (L.) Urb.	Memory enrichment	Uttarakhand	Akasha and Bhandari (2021)
141	*Discorea bulbifera* L.	Curing cancer, HIV, anti-inflammatory, anti-microbial, cardioprotective and anti-hyperthyroid activities	Dongria Kandha tribes, Odisha	Parida and Sarangi (2021)
142	*Ocimum tenuiflorum* Burm.f., *Ocimum sanctum* L. (Holy basil/tulsi)	Herpes virus, foot and mouth disease virus and New castle disease virus control	Uttar Pradesh	Goel and Bhatia (2021)
143	*Aconitum heterophyllum* Wall. (Patis, Aconite, Dhar buti, Attees or Bis Mohra)	Intestinal worms, diarrhoea, dysentery, high fever and anti-rheumatic	Jammu and Kashmir	Bagala et al. (2021)
144	*Viola odorata* L. (Bnafsaha, wild violet, sweet violet)	Fever, cold, cough, hypertension, muscle spasms, parasitic worms, malaria controls	Jammu and Kashmir	Bagala et al. (2021)
145	*Valeriana jatamansi* D.Don (Mush khala, jatamansi, balchhari, mansi, nihani)	Treating eye, blood, liver problem, hysteria, nervous andurinal stress	Jammu and Kashmir	Bagala et al. (2021)
146	*Picrorhiza kurroa* Royle ex Benth. (Kaud, kaur, kutki)	Fever, cold cough, hypertension, muscle spasms, parasitic worms and malaria treatments	Jammu and Kashmir	Bagala et al. (2021)
147	*Bergenia ligulata* (Wall.) Engl. (Patharchoor, pashanbeda)	Longevity, anti-viral, analgesiscs, ascites, hypoglycemic, anti-arthritic and anti-ageing	Jammu and Kashmir	Manosi et al. (2021)
148	*Cryptolepis buchananii* Roem. & Schult., *Eucalyptus citriodora* Hook., *Ligustrum japonicum* Thunb., *Pinus roxburghii* Sarg., *Rosa alba* G.Gaertn., B.Mey. & Scherb., *Ziziphus numm ularia* (Burm.f.) Wight & Arn. and *Sonchus oleraceus* L.	Skin infection treatment	Solan district, Himachal Pradesh	Kumar et al. (2021)

Tamil Nadu farmers maintain short-term grain storage with 1 g camphor per 5 kg grains in jute bags. The farmers prepare plate-like round-shaped structure (Varati) with the help of fresh cow dung, the seed were enclosed in the Varati for 2–3 days under sunlight. The enclosed seed is stored into wooden boxes for up to 1 year for seed storage (Adesina et al. 2019). The seed materials of the crops are poured into ¾th height earthen pot; the pot covers with a rough cloth containing neem leaves, pungam leaves, and notchi leaves. The quantity of sand covers the mouth of the container. The pulses and food grains are immersed in a 10% salt solution and dried for controlling pest attacks. The application of Neem leaves/Pungam leaves manages the storage pest of cereal crops. The Karnataka, Assam, and Kerala farmers use 10 g lime per kg grains in jute gunny bags for storing 1-year grains storage. The Karnataka and Tamil Nadu farmers mix gingelly seed with 100 g paddy in a container for the 3 months for controlling the Indian meal moth (*Plodia interpunctella*) (Bhavani and Ningdalli 2015).

The long-term storage by blending 1 kg pulse seed in 20 mL neem oil and controls weevils, long-headed flour beetle, red flour beetles and fig moth during storage. The salt treatment conducts breakage of seed dormancy and increases drought stress tolerance (Marziyeh et al. 2017). The farmers control Angoumois grain moth and rice weevils by using Pungam leaves in the paddy gunny bags and managing long-term storage (Sahu et al. 2022). The North-East, Tamil Nadu, Punjab, and Haryana farmers place paddy husk up to 5 cm in the top portion of the earthen pot for seed damage and pest control. The Tamil Nadu, Kerala, Odisha, and North-East farmer practice 2 kg paddy seed +1 kg salt +10-L water placed in the sunlight for an hour. The chaffy seed is separated from the hard seed. The hard seed is dried in the shade. The addition of salt increases the density, and separates light seed and chaffy seed. It also increases seed germination (Bordoloi et al. 2017; Singh 2018).

The farmers pour paddy seed in the water for 12 h. The dried paddy seed is placed in the pit containing tree sawdust and sheep manure. The pit is covered tightly with plastics/cotton cloth. The seeds are excised after 2 days of drying for spawning. The proportion of 1:10 salt and water solution is poured into 10 kg of paddy seed. The dried seed is recommended for sowing after 72 h. The North-Eastern, Tamil Nadu, Madhya Pradesh, and Karnataka farmers dissolve 1 kg sorghum seed dissolves in 100 g dry cow dung powder +250 mL cow urine for an hour before sowing and it improves seed germination. The dissolved sorghum seed in 1 kg lime +10-L water is kept for 10 days. The seeds dried in shade before sowing. The lime prevents attack seed-borne diseases such as smut and bunt. The healthy ear head led with an awn of sorghum is kept with dried paddy grass heap (banave) for controlling seed damage and improving seed longevity. The North-Eastern, Telangana and Karnataka farmers store pigeon pea seed with dry powder bitter gourd and drumstick seed for 3—6 months for controlling insect pests. The North-Eastern and Karnataka farmers mix 10 kg green gram seed with 250 g chilli powder +1 kg ragi/finger millet flour. The prepared mixtures are stored in the bamboo pot along with paddy husk. The chilli powder and flour prevent the attack of storage pests. This practice is mentioned and explained in Varabamihira's Brihat Jataka (Rakesh et al. 2013; Ambika et al. 2014).

The North-Eastern, Uttar Pradesh, Himachal Pradesh, and Kerala farmers dissolve dry cow dung with ghee and honey for seed treatment in the Kautilya period (SCERT 2016). The Uttar Pradesh, Karnataka, and Odisha farmers treat pigeon pea seed with dry pongamia leaf for controlling storage pests. The farmers store pigeon pea seeds with dry Guntur chilli powder and neem leaf powder for controlling insect pests and seed senescence. The control insect pests in seed is achieved either by mint leaf powder or sweet flag root powder. Farmers apply dried fruit of sponge gourd after removing seed in the sunflower seed store. The dried sponge gourd fruit containing sunflower seed is kept in an air-tight container. Farmers use a protective capsule of sponge gourd to protect against storage pests of Sunflower seed. All crops, dried seeds, and grains are restricted from pest attack on new moon day (Usharani et al. 2019; Jyoti et al. 2020). 5 kg pigeon pea/chickpea seed is mixed with pearl millet/finger millet. The mixed seed is placed in the earthen pot and sealed with cow dung smear. The millet assimilates moisture content in the pot for pulse seed storage. The green gram seed is stored in the layer of ash in the earthen pot. The earthen pot is smeared with cow dung. The insect population dies of ash suffocation, the seed is stored for a longer period. The farmers manage insect pests and microbes of chickpeas such as *Alternaria* sp. Or *Fusarium* sp. by blending 100 kg chickpea seed with citronella leaf oil/cottonseed oil/soyabean oil/castor seed oil (Ruparao and Gadi 2018). The farmers store dried leaves of neem in the grains warehouse, to stop the attack of stored grains pests (Yallappa et al. 2012). The dried leaves of notchi (*Vitex negundo*) helps in stopping the attack of stored pests (Shivankar et al. 2020). Pouring 1 kg Vasambu (*Acorus calamus*) in 50 kg grains prevents the invasion of stored pests and enhances 1 year storage period (Kathirvelu et al. 2019). Table 14.7 presents farmers' knowledge related to stored grains pest management.

14.2.8 Weed Management

The Uttarakhand and Karnataka farmers cultivate jethi rice, finger millet, black soybean, and horse gram for weed control and moisture conservation (Nautiyal et al. 2017; Reddy et al. 2008). The volunteer plant emerges in the rice crop field, eradicated with the help of Danala implements for weed control. The Jammu and Kashmir, Haryana, Himachal Pradesh, Uttar Pradesh, parts of Sikkim, West Bengal, and Arunachal Pradesh farmers disperse dry leaves of pine into the field in the middle of June, fire in the field for weed control. The dried biomass of weed population available in the dryland field then conserves soil moisture. The Meghalaya and Maharashtra farmers apply common salt (NaCl) for eradicating *Ageratum conyzoides* and *Crassocephalum creidioides* weeds (Patel et al. 2015).

The North-Eastern farmers maintain the weed population of *Cyanodon dactylon* in the soil field for 3 years and it conserves the soil moisture (Gulab et al. 2018). The North-Eastern, Odisha, West Bengal, Tamil Nadu, and Karnataka farmers inhibit the weed population with the production of green leaf manure such as *Sesbania* sp. and *Tephrosia purpurea*. The farmer prepares 1 kg salt + 100 g sarvodaya solution for

Table 14.7 Indigenous knowledge practices of farmers in stored grain pest management

Sl. No.	Indigenous Traditional Knowledge (ITK)	Cause	State	References
1	Construction of godowns with straw, leaves and the godowns mounted with cow dung	Grains storage	Uttar Pradesh and Tamil Nadu	Vishal et al. (2021)
2	Mixing of 200 g of common salt + red gram/ Arhar	Controlling stored grains pests	Tamil Nadu	Karthikeyan et al. (2009)
3	Neem oil blends with coconut oil/castor oil (1: 1)	Storage and pests control	Tamil Nadu	Karthikeyan et al. (2009)
4	5 L groundnut oil + ¼ kg tamarind	Prevents oil spillage	Tamil Nadu	Karthikeyan et al. (2009)
5	Neem leaves, thumbai and any strong odour leaves (Kaddi patta, tulsi, lemon grass etc) + ragi grains	Controlling lesser grain borer, saw toothed beetle and flat grain borer	Tamil Nadu, Kerala and Karnataka	Shaila and Nafeesa (2021)
6	Mixing of pegion pea seed with horse gram seed dust	Assimilates excess moisture and encourages long term storage	Tamil Nadu, Kerala, Telangana and Karnataka	Shaila and Nafeesa (2021)
7	Construction of godowns/ granary room with brick and wooden boards	Controlling rice moth and restrict moisture of the grains	Tamil Nadu, Uttar Pradesh and Mahrashtra	Parimala et al. (2013)
8	Jute bags: 1 g camphor/ 5 kg grains	Short term grains storage	Manipur and Tamil Nadu	Adesina et al. (2019)
9	10 g lime per kg grains in jute gunny bags	1 year grains storage	Karnataka, Assam and Kerala	Bhavani and Ningdalli (2015)
10	Mixing of gingelly seed with 100 g paddy	3 months gingelly seed storage, controlling Indian meal moth (*Plodia interpunctella*)	Karnataka and Tamil Nadu	Bhavani and Ningdalli (2015)
11	Mixing of 1 kg pulse seed in 20 mL of neem oil	Long term storage and weevils, red flour beetles, long headed flour beetle and fig moth controls	Tamil Nadu	Marziyeh et al. (2017)
12	Application of Pungam leaves in paddy storage in the gunny bags	Angoumois grain moth, rice weevils controls and long term storage	Gujarat and Odisha	Sahu et al. (2022)
13	Paddy husk upto 5 cm in top portion of the earthen pot	Seed damage control and pest control	North-Eastern India, Tamil Nadu, Punjab and Haryana	Bordoloi et al. (2017); Singh (2018)
14	2 kg paddy seed +1 kg salt +10 L water	Releasing chaffy seed	Tamil Nadu, Kerala, Odisha and North-East	Bordoloi et al. (2017); Singh (2018)

(continued)

Table 14.7 (continued)

Sl. No.	Indigenous Traditional Knowledge (ITK)	Cause	State	References
15	Pouring of paddy seed in the water overnight	Seed germination	North-Eastern India and Karnataka	Rakesh et al. (2013); Ambika et al. (2014)
16	1 kg sorghum seed dissolves in 100 g dry cow dung powder +250 mL cow urine	Improving seed germination	North-Eastern, Tamil Nadu, Madhya Pradesh and Karnataka	Rakesh et al. (2013); Ambika et al. (2014)
17	Pegion pea seed with dry powder bitter gourd and drum stick seed for 3–6 months	Controlling insect-pests	North-Eastern India, Telangana and Karnataka	Rakesh et al. (2013); Ambika et al. (2014)
18	10 kg green gram seed with 250 g chilli powder +1 kg ragi/finger millet flour + paddy husk	Prevents attack of storage pests	North-Eastern India and Karnataka	Rakesh et al. (2013); Ambika et al. (2014)
19	Dry cow dung with ghee + honey	Seed treatment	North-Eastern India, Uttar Pradesh, Himachal Pradesh and Kerala	SCERT (2016)
20	Treatment pegion pea seed with dry pongamia leaf	Controlling storage pests	Uttar Pradesh, Karnataka and Odisha	Usharani et al. (2019); Jyoti et al. (2020)
21	Pegion pea seed with dry guntur chilli powder and neem leaf powder	Controlling insect-pests and seed senescence	North-Eastern India, Kerala, Andhra Pradesh and Karnataka	Usharani et al. (2019); Jyoti et al. (2020)
22	Either mint leaves powder or sweet flag root powder	Inhibiting insect-pests	Kerala, Andhra Pradesh and Karnataka	Usharani et al. (2019); Jyoti et al. (2020)
23	Chilli seed in the gunny bag and kept in a hot water for a day	Improves seed viability and vigour	North-Eastern India, Kerala and Karnataka	Usharani et al. (2019); Jyoti et al. (2020)
24	Blending of citronella leaf oil/cotton seed oil/soyabean oil/castor seed oil in 100 kg chickpea seed	Controlling insect pest and microbes such as *Alternaria* sp. or *Fusarium* sp.	Odisha, Telangana and Maharashtra farmer	Ruparao and Gadi (2018)
25	Dried leaves of neem in the grains warehouse	Stop attack of stored grains pest	Punjab, Haryana, Rajasthan, Uttar Pradesh and Karnataka	Yallappa et al. (2012)

(continued)

Table 14.7 (continued)

Sl. No.	Indigenous Traditional Knowledge (ITK)	Cause	State	References
26	Dried leaves of notchi (*Vitex negundo*)	Stops the attack of stored pests	Tamil Nadu, Madhya Pradesh, Assam, Uttar Pradesh and Uttarakhand	Shivankar et al. (2020)
27	1 kg Vasambu (*Acorus calamus*) in 50 kg grains	Forbids invasion of stored pests and enhances 1 year storage period	Tamil Nadu	Kathirvelu et al. (2019)
28	Exposure stored pulse grain in open sunlight at 20 °C	*Callosobruchus chinensis* eggs and grubs control	Tamil Nadu	Adesina et al. (2019)
29	Custard apple seed powder in pulse grains	Bruchid adult and eggs control	Karnataka	Prakash et al. (2016)

restricting the growth of nut grass weed plant. The cultivation of *Calotropis gigantea* restricts Aarai (Mars/*Tea quadrifolia*) weed population (Ramyajit and Saumi 2019). The farmer discharges a volume of water into the field for managing volunteer seeds and plants. The Tamil Nadu farmera apply 200 g salt dissolved in a 1-L water solution in controlling congress grass weed (*Parthenium hysterophorus*). Some farmers apply 50-kg neem cake in the field for controlling nut grass (Surinder et al. 2018). Table 14.8 presents farmers' knowledge related to weed management.

14.2.9 Food Products

The farmers from North-Eastern part of India prepare indigenous pickle and chatni with ingredients of wild-type Mesta and Roselle (*Hibiscus subdoriffa*). The biochemical contents such as Citric acid, ß carotene, malic acid, Vitamin C, allo-hydroxycitric acid protein, total sugar, and tartaric acid are found in Roselle. The North-East tribal community consumes fruit and leaves of wild brinjal in the home. The tribal people of Arunachal Pradesh prepare fermented food such as gundruk, sinki, anishi, bhatooru, marchu and chilra, kienma, tungrymbai, mesu, soibum, ngari, hentak, kadi, churpa/churpi and nadu, ghanti, jann/jaan and daru (Nazish 2013).

The Nagaland community cooks meat using plants such as *Amaranthus* sp., Bamboo shoot, *Brassaiopsis* sp., *Chenopodium album*, *Colocasia esculenta*, *Curcuma angustifolia*, *Fagopyrum esculentum*, *Hibiscus sabdariffa*, *Oenanthe stolonifera*, *Persicaria chinensis*, *Polygonum molle*, *Zanthoxylum armatum* and *Zanthoxylum rhetsa*. The Nagaland community prepares galho rice either with wild leaves; or mixtures of salt, garlic, potatoes, tomatoes, dry fish and fermented

Table 14.8 Indigenous knowledge practices of farmers in weed management

Sl. no.	Indigenous Traditional Knowledge (ITK)	Use	State	References
1	Cultivation of jethi rice, finger millet, black soybean, horse gram	Weed control and moisture conservation	Uttarakhand and Karnataka	Nautiyal et al. (2017); Reddy et al. (2008)
2	Dispersion and burning of dry leaves of pine into the field	Weed control	Jammu and Kashmir, Haryana, Himachal Pradesh, Uttar Pradesh, parts of Sikkim, West Bengal and Arunachal Pradesh	Patel et al. (2015)
3	Application of common salt (NaCl)	*A. conyzoides* and *Crassocephalum creidioides* control	Meghalaya and Maharashtra	Patel et al. (2015)
4	Cultivation of green leaf manure such as Diancha (*Sesbania* sp.), Kolingi (*Tephrosia purpurea*)	Weed control	North-Eastern, Odisha, West Bengal, Tamil Nadu and Karnataka	Ramyajit and Saumi (2019)
5	200 g Salt +1 L water	Controlling congress weed (*Parthenium hysterophorus*)	Tamil Nadu	Surinder et al. (2018)
6	50 kg Neem cake in the field	Controlling nut grass	Tamil Nadu, Kerala, Himachal Pradesh, Assam, Meghalaya and Kerala	Surinder et al. (2018)

soybean, and *Perilla frustescens* seeds for increasing taste. The Nagaland community also prepares tathu chutney with chilli paste, leaves, and dry meat or fermented fish. The Nagaland tribe prepare ghabe food by boiling leaves with the addition of spices, chilli, fermented Soyabean or dry fish. They also prepare galkemeluo food by boiling wild leaves with bamboo shoots, garlic, tomato, potato, dry or smoked meat, dry fish, fermented soybean, *Zanthoxylum rhetsa* and *Zanthoxylum armatum* (Singh and Teron 2017). The Mizoram community prepares fermented pig fat with chopped pieces of the inner abdominal portion of the pig. They extract oil with fermented Seasame for cooking the food. They performs sun-drying pf leaves of *Hibiscus sabdariffa* and prepare either with seasonal vegetables and fish, chicken, beef, or pork for eating source. The people of Mizoram use fermented crabs with sesame oil, smoked meat from wild animals such as barking deer, sambar deer, wild boar, macaque, birds, squirrels, and rodents with thick pointed Bamboo sticks (Lalthanpuii et al. 2016). The Manipur tribe prepare tunateinzi food with ingredients of rice flour and sugar, and lengchiphon food with ingredients of rice flour and liquid sugar. The Manipur and Nagaland tribes prepare ganang tamdui food with fermented mustard and banana leaves. The Manipur, Mizoram, Sikkim, and Darjeeling tribes prepare gundruk food with dried mustard leaves. The Manipur tribals prepare bi-kang food with boiling and drying of *Colocasia*. The Manipur, Mizoram, Sikkim, and Darjeeling community prepare fermented Soyabean for use in curry and chutney. The Manipur community prepares Gankhiang-khui food with alkaline-fermented seeds of *Hibiscus canabinus*. The Manipur people also prepare food with *Auricularia auriculari*, *Schizophyllum commune*, and *Lentinula edodes* (wild mushroom) (Thangjam et al. 2018).

The Manipur people produce several dishes from edible Bamboo species such as *Bambusa Cephalostachyum*, *Chimono Bambusa*, *Dendrocalamus* sp., and *Melocanna* sp. For example, bamboo shoot curry (Usoi Ooti), bamboo shoot salad (Usoi Kangsu), bamboo shoot chutney (Soibum), fermented shoot curry (Soibum Thonga), fried bamboo shoot (Laiwa Kanghou), boiled bamboo shoot (Usoi Champhut), bamboo shoot pickles (Usoi aachar) (Premlata et al. 2020). The Bhotia community of Uttarakhand use wild edible fruit species such as *Cotoneaster* sp., *Fragaria* sp., *Malus* sp., *Prunus* sp., *Rosa* sp., *Sorbaria* sp. and *Sorbus* sp. for preparing local beverages (Aygar), tobacco pickles, chutney oil, furniture, and agriculture tools and implements (Badal et al. 2022) (Table 14.9). Table 14.9 presents traditional knowledge related to ethnic food products/preparations.

14.3 Conclusions and Future Directions

The traditional knowledge of agriculture is followed by Indian farmers for crop production and farm-linked activities. The farmer complies with rituals in agriculture for production and other activities in rural areas. The agriculture aspirants will receive the scope and imperative of indigenous agriculture. The traditional agricultural knowledge pertains to various activities and not only to the crop production. India is a diverse country with diverse agroclimatic zones and crops. The farmers from various parts of the country hold diverse traditional agricultural knowledge. The traditional agricultural knowledge in India has become a part of Indian culture. Various activities related to traditional agriculture are celebrated as a part of family festivals. For example, beginning of the preparation of the agriculture and harvesting of the crops is considered major parts of the culture in various parts of India. Other intermediate activities are also celebrated with great fervor. The inclusion of traditional agriculture in the course the curriculum will enlighten the students of agriculture. It is crucial to preserve diverse knowledge that is associated with agri-activities in India. The research in the future should focus on the identification of those traditional practices that are still relevant and can be patented. Beneficial and traditional knowledge which is still relevant should be widely popularised among the farmers and scientific rsarch also should be carried out to validate the traditional knowledge.

Table 14.9 Indigenous knowledge related to the culinary products/dishes

Sl. No.	Indigenous Traditional Knowledge (ITK)	Preparation/Use	State/Part of India	References
1	Fermented food—Gundruk, Sinki, Anishi, Bhatooru, Marchu and Chilra, Kienma, Tungrymbai, Mesu, Soibum, Ngari, Hentak, Kadi, Churpa/Churpi and Nadu ghanti, Jann/Jaan and Daru	Fermented vegetable food, fermented pulse food, fermented bamboo food, fermented fish food, fermented milk food, fermented alcoholic beverage	Arunachal Pradesh	Nazish (2013)
2	Edible bamboo species—*Bambusa Cephalostachyum*, *Chimono Bambusa*, *Dendrocalamus* sp. and *Melocanna* sp. for culinary uses-Bamboo shoot curry—(Usoi Ooti), Bamboo shoot salad—(Usoi Kangsu), Bamboo shoot chutney—(Soibum), Fermeneted shoot curry—(Soibum Thonga), Fried Bamboo shoot—(Laiwa Kanghou), Boiled Bamboo shoot—(Usoi Chamfat), Bamboo shoot pickles—(Usoi aachar)	Culinary and product uses, bamboo shoot curry, Bamboo shoot salad, Bamboo shoot chutney, Fermeneted shoot curry, Boiled Bamboo shoot, Bamboo shoot pickles	Manipur	Premlata et al. (2020)
3	Wild edible fruits such as *Cotoneaster* sp., *Fragaria* sp., *Malus* sp., *Prunus* sp., *Rosa* sp., *Sorbaria* sp. and *Sorbus* sp.	Preparation of local beverage (Aygar), tobacco pickles, chutney oil, furniture and agriculture tools and implements	Bhotia community of Uttarakhand	Badal et al. (2022)
4	Galho rice either with wild leaves; mixtures of Salt, garlic, potatoes, tomatoes, dry fish and fermented soyabean and *Perilla frustescens* Seeds	Diet	Nagaland community	Singh and Teron (2017)
5	Preparation of Tathu chutney with chilli paste, leaves and dry meat or fermented fish	Diet	Nagaland community	Singh and Teron (2017)
6	Preparation of Modi with a piece of mithun, beef or pork, ginger, garlic, onion, chilli, and salts	Diet	Nagaland community	Singh and Teron (2017)

(continued)

Table 14.9 (continued)

Sl. No.	Indigenous Traditional Knowledge (ITK)	Preparation/Use	State/Part of India	References
7	Preparation of Ghabe food with boiling of leaves with addition of spices, chilli, fermented Soyabean or dry fish	Diet	Nagaland community	Singh and Teron (2017)
8	Preparation of Galkemeluo food with boiling of wild leaves with bamboo shoot, garlic, tomato, potato, dry or smoked meat, dry fish, fermented soyabean, *Zanthoxylum rhetsa* and *Zanthoxylum armatum*	Diet	Nagaland community	Singh and Teron (2017)
9	Preparation of fermented pig fat with chopped pieces of inner abdominal portion of pig	Diet	Mizoram community	Lalthanpuii et al. (2016)
10	Oil extraction with fermented Seasame	Diet	Mizoram community	Lalthanpuii et al. (2016)
11	Preparation of Sun drying leaves of *Hibiscus sabdariffa* either with seasonal vegetables and fish, chicken, beef, pork	Diet	Mizoram community	Lalthanpuii et al. (2016)
12	Preparation of smoked meat from wild animals such as barking deer, sambar deer wild boar, macaque, birds, squirrels and rodents with thick pointed bamboo sticks	Diet	Mizoram community	Lalthanpuii et al. (2016)
13	Preparation of Tunateinzi food with ingredient of rice flour and sugar	Diet	Manipur tribe	Thangjam et al. (2018)
14	Preparation of Lengchiphon food with ingredient of rice flour and liquid sugar	Diet	Manipur tribe	Thangjam et al. (2018)
15	Preparation of Ganang Tamdui food with fermented mustard leaves and banana leaves	Diet	Manipur and Nagaland Tribe	Thangjam et al. (2018)
16	Preparation of Gundruk food with dried mustard leaves	Diet	Manipur, Mizoram, Sikkim and Darjeeling Tribe	Thangjam et al. (2018)

(continued)

Table 14.9 (continued)

Sl. No.	Indigenous Traditional Knowledge (ITK)	Preparation/Use	State/Part of India	References
17	Preparation of Bi-kang food with boiling and drying of *Colocasia*	Diet	Manipur tribe	Thangjam et al. (2018)
18	Preparation of fermented Soyabean	Diet	Manipur, Mizoram, Sikkim and Darjeeling community	Thangjam et al. (2018)
19	Preparation of Gankhiang-khui food with alkaline fermented seeds of *Hibiscus canabinus*	Diet	Manipur community	Thangjam et al. (2018)
20	Preparation of food with *Auricularia auriculari*, *Schizophyllum commune* and *Lentinula edodes* wild mushrooms	Diet	Manipur community	Thangjam et al. (2018)

References

Adesina JM, Nameirakpam B, Dinabandhu S, Yallappa R (2019) Traditional methods of food grains preservation and storage in Nigeria and India. Ann Agric Sci 64:196–205. https://doi.org/10.1016/j.aoas.2019.12.003

Ahuja DB, Chattopadhyay C (2015) Pests of fruit trees (citrus, banana, mango, pomegranate and sapota) E-pest surveillance and pest management advisory. ICAR-National Research Centre for Integrated Pest Management New Delhi and State Department of Horticulture Commissionerate of Agriculture Pune (Maharashtra) 124:1–134. https://vdocument.in/pest-of-fruit-integrated-pest-ncipm-foreword-the-project-on-e-pest-surveillance.html

Ahuja SC, Siddharth A, Uma A (2015) Nirgundi (*Vitex negundo*)-nature's gift to mankind. Asian Agri-History 19(1):1–28. https://www.asianagrihistory.org/pdf/volume19/nirgundi-natures-gift-to-mankind.pdf

Ajay SR (2018) Booklet on Indigenous Technical knowledge (ITKS) Crop wise with reference to promotion of organic farming. Thirty Days Certificate Course on Organic Farming, pp 1–40. https://www.vetextension.com/indian-itk-traditional-knowledge-for-organic-agriculture/

Akasha N, Bhandari BS (2021) Ethnobotanical plants used in health care and traditional practices by local inhabitants (Gujjars) of Rajaji Tiger Reserve, Uttarakhand, India. Indian J Trad Knowl 20:91–105. https://doi.org/10.56042/ijtk.v20i1.27596

Alexander D, Rajan S, Rajamony L, Ushakumari K, Sajan K (2009) The adhoc package of practices recommendations for organic farming. Directorate of Research Vellanikkara Thrissur Kerala India, pp 1–209. https://keralaagriculture.gov.in/wp-content/uploads/2018/12/package_2015.pdf

Ambika S, Balakrishnan K, Sujatha K (2014) Enhancing the seed germination and vigour in coarse cereals by bovine urines. J Agroecol Nat Resour Manag 1:40–43

Amitendu D, Arya HPS, Babulal T, Goswami A (2004) Indigenous technical knowledge in animal husbandry. Livest Res Rural Dev 16(8):1–9. http://www.lrrd.org/lrrd16/8/arun16059.htm

Ankit S, Robbie H, Sudeep C, Nautiyal MC, Alexander KS (2019) Traditional herbal knowledge among the inhabitants: a case study in Urgam Valley of Chamoli Garhwal, Uttarakhand, India. Evid Based Complement Alternat Med 14:1–22. https://doi.org/10.1155/2019/5656925

Ann RLEN, Kavitha R, Usha A (2019) The impact of the Green Revolution on indigenous crops of India. J Ethnic Foods 6:1–10. https://doi.org/10.1186/s42779-019-0011-9

Ansaria MA, Sharma SK, Roya SS, Ramakrishna Y, Shiv D, Arati N, Ajitkumar NS, Solei L, Prakasha N (2021) Documenting the agriculture based indigenous traditional knowledge in Manipur State of North Eastern India. Indian J Tradit Knowl 20(4):1065–1074. http://op.niscair.res.in/index.php/IJTK/article/view/32549/0

Anupam M, Singh SRK, Raut AA (2020). Traditional knowledge in agriculture. Division of Agricultural Extension ICAR New Delhi, p 39. http://zpd7icar.nic.in/download/IITKA_DATA/IITKA_Book_(Traditional%20Knowledge%20in%20Agriculture-English%20).pdf

Anwesha B, Pardeep S (2020). Indigenous knowledge systems in sustainable water conservation and management. Chapter, Water Conservation and Wastewater Treatment in BRICS Nations, pp 321–328. https://www.elsevier.com/books/water-conservation-and-wastewater-treatment-in-brics-nations/singh/978-0-12-818339-7

Aparna R, Archana S, Ranjan KM, Aftabuddin MD, Sanjay KD (2020) Traditional knowledge of the fishermen community of Indian Sundarbans: an assessment of rationality and effectiveness. Indian J Fish 67(2):94–101. http://epubs.icar.org.in/ejournal/index.php/IJF/article/view/86752

Badal S, Krishna MR, Kailash CB, Rekha C, Saurabh KD, Sudhir PA (2022) Ethnobotanical notes on some potential wild edible fruits used by the Bhotia community of Niti Valley, Uttarakhand, India. Indian J Tradit Knowl 21(1):135–144. http://op.niscair.res.in/index.php/IJTK/article/view/35647

Bagala YS, Nandaa R, Sharmaa LK, Raina NS (2021) Traditional use of non-timber forest products: boon for livelihood security of people in Shivalik range of Jammu region. Indian J Tradit Knowl 21(1):180–185. http://op.niscair.res.in/index.php/IJTK/article/view/28562

Balasubramanian AV, Nirmala DTD, Merlin FM (2009). Use of animal products in traditional agriculture, A Pilot Project in Southern India. Centre for Indian Knowledge Systems Chennai, pp 1–156. https://www.yumpu.com/en/document/view/36475739/use-of-animal-products-in-traditional-agriculture-compas

Bhalerao AK, Kumar B, Singha AK, Jat PC, Bordoloi R, Deka BC (2015). Dimapur district inventory of Agriculture. ICAR-Agricultural Technology Application Research Institute Umiam Meghalaya India, pp 1–101. http://www.icarzcu3.gov.in/district_agri_inventory/Dimapur.pdf

Bhavani K, Ningdalli M (2015) Documentation of traditional and indigenous seed and foodgrain storage practices of Bidar district. Int J Process Post Harv Technol 6:118–124. http://researchjournal.co.in/ONLINE/IJPPHT/IJPPHT%206(1)/6_118-124_A.pdf

Bhowmick BC, Dutta SK, Puzari NN, Goswami J, Saud RK, Ray BK (2010) Package of practices for horticultural crops, fisheries and home science. Agricultural Information Wing Department of Agriculture Assam Khanapara Guwahati. pp 1–301. http://kvkjorhat.nic.in/docs/pop/horti_fish_home.pdf

Bhuvaneshwari S, Hiroshan H, Meegoda JN (2018) Crop residue burning in India: policy challenges and potential solutions. Int J Environ Res Public Health 16(832):1–19. https://doi.org/10.3390/ijerph16050832

Bikash S, Prem KS, Amitava D, Ujjwal K, Kamal S, Bhatt BP (2015) Traditional agricultural tools used by tribal farmers in Eastern India. Res J Agric Sci 6(1):215–219. 2106-1209-2014-054

Bikram KM, Tribhuban P, Rabindra NP (2012) Ethnoveterinary practices of aborigine tribes in Odisha, India. Asian Pac J Trop Biomed 16:S1520–S1525. https://doi.org/10.3389/fvets.2022.815294

Binoo P, Bonny S, Helen AS (2016) Package of practices recommendations crops 2016. In: Estelitta S (ed), Directorate of Extension, Kerala Agricultural University, Thrissur, pp 1–401. https://kau.in/document/11029

Bordoloi R, Bidyut CD, Singha AK, Bagish K, Jat PC, Sarma CK, Borgohain R (2017). Technology Inventory for Northeast India. ICAR—Agricultural Technology Application Research Institute Umiam Meghalaya, pp 1–327. http://icarzcu3.gov.in/book_publications/Technology_inventory/Technology%20Inventory.pdf

Chandra PK, Pitamber PD, Bikram SS (2006) Developing the medicinal plants sector in northern India: challenges and opportunities. J Ethnobiol Ethnomed 2(32):1–15. https://doi.org/10.1186/1746-4269-2-32

Choubey TK (2005) Working plan for the forests of bramhapuri forest division north chandrapur circle. Deputy conservator of forests, working plan division Chandrapur, pp 1–261. https://mahaforest.gov.in/writereaddata/managementpdf/1290165780Bramhapuri%20%20Introduction.pdf

Choubey S (2020) Control of shoot and fruit borer through use of tobacco (*Nicotiana tobaccum*) soaked water in brinjal (*Solanum melongena*). Traditional Knowledge in Agriculture Division of Agricultural Extension Indian Council of Agricultural Research New Delhi, pp 1–52. https://www.academia.edu/39598441

Dagar JC, Tewari JC (2016) Agroforestry research developments. Nova Science Publishers, New York, pp 1–599. https://www.researchgate.net/publication/315728440

Dakshinkar NK, Vihan VS (2020) Traditional knowledge in agriculture. Division of Agricultural Extension Indian Council of Agricultural Research New Delhi, pp 1–52. http://zpd7icar.nic.in/download

Das P, Das SK, Mishra A, Arya HPS, Bujarbarua KM, Singh RP, Verma LR, Subba RG, Geetha, RM, Gupta HS, Kavia ZD, Ray DP (2003) Inventory of indigenous technical knowledge in agriculture. Directorate of Information and Publications of Agriculture (DIPA) Indian Council of Agricultural Research Krishi Anusandhan Bhavan Pusa New Delhi, pp 1–192. https://icar.org.in/sites/default/files/Inventory

Das P. Arya PS, Das SK, Rani MG, Mishra A, Verma LR, Singh RPR, Ray DP, Subba G (2004a) Cross-sectoral validation of indigenous technical knowledge in agriculture. Indian council of agricultural research New Delhi, pp 1–225. https://atarijabalpur.icar.gov.in/upload/publication

Das P, Subba GR, Das SK, Geetha R, Mishra A, Gupta HS, Verma LR, Ray DP, Singh RP, Kavia ZD, Arya HPS (2004b) Validation of indigenous technical knowledge in agriculture. Mission mode Project on Collection, Documentation and Validation of Indigenous Technical Knowledge. Indian Council of Agricultural Research, New Delhi, pp 1–492. https://icar.org.in/sites/default/files

Das SC, Datta M, Ray P, Singh SK, Jena RK, Das B, Ray SK (2019) Mango (*Mangifera indica*) cultivation in North-Eastern Region of India. Adv Agric Res Technol 3(1):1–13. https://www.semanticscholar.org/paper

Das AK, Nayak N, Mandal SK, Naskar SK (2020) Use of kochila (*Strychnos nux-vomica*) mixed cowdung compost in brinjal for controlling fruit and shoot borer. In: Traditional knowledge in agriculture. Division of Agricultural Extension, Indian Council of Agricultural Research, New Delhi, pp 1–52. https://icar.org.in/ITK

Deepandita B, Arunoday D, Maneesh KA, Anup KS, Vikram R (2021) Practice of traditional herbal medicine in animal husbandry of rural India. Ann Phytomed 10(2):507–514. https://doi.org/10.21276/ap.2021.10.2.67

Devendra P (2020) Control of insect-pests in cucurbits, cowpea and lady's finger by spraying cow urine mixed with tobacco-soaked water. Traditional Knowledge in Agriculture Division of Agricultural Extension Indian Council of Agricultural Research New Delhi, pp 1–52. http://zpd7icar.nic.in/download/IITKA

Dey P, Sarkar AK (2011) Revisiting indigenous farming knowledge of Jharkhand (India) for conservation of natural resources and combating climate change. Indian J Tradit Knowl 10(1):71–79. http://nopr.niscair.res.in/handle/123456789/11067

Dipika D, Khodke KU, Bidwe RR, Shelke, Kahate PA (2017) Preparation of whey based custard apple (*Annona squamosa* L.) pulp beverage. Res J Anim Husband Dairy Sci 8:1–5. https://doi.org/10.15740/HAS/RJAHDS/8.2/94-98

Dorjeya K, Mauryab AK (2021) Ethnobotany of *Juniperus polycarpos* C. Koch (Cupressaceae) in the Himalayan cold desert of Union Territory of Ladakh, India. Indian J Trad Knowl 20:83–90. https://doi.org/10.56042/ijtk.v20i1.27057

Firake DM, Behere GT, Uttam SN (2012) Common traditional pest management practices of north east India. ICAR Research Complex for NEH Region Umroi Road Umiam, pp 1–11. https://citeseerx.ist.psu.edu/viewdoc/download

Ganesh CD, Roshan WS, Ajay SK (2011) Agitated pile composting of water hyacinth. In: 2nd International Conference on Environmental Science and Development, vol 4. pp 79–83. http://www.ipcbee.com/vol4/17-ICESD2011D039.pdf

Goel A, Bhatia AK (2021) *Ocimum sanctum*: in vitro antiviral potential against animal viruses. Indian J Tradit Knowl 21(1):120–125. http://nopr.niscair.res.in/handle/123456789/59155

Gohain S, Neog M, Bhattacharyya HC (2019) Innovative traditional pest management practices in horticultural crops. Asian Agri-History 23(1):61–64. https://www.asianagrihistory.org/pdf/home/january-march.pdf

Gopal KNC, Lassaad B (2013) Indigenous pest and disease management practices in traditional farming systems in North East India: a review. Afr J Bot 2(2):070–073. https://www.internationalscholarsjournals.com/articles

Gopi R, Avasthe RK, Kalita H, Chandan K, Ashish Y, Subhash B, Das SK (2016) Traditional pest and disease management practices in Sikkim Himalayan Region. Int J Bio-resour Stress Manag 7(3):471–476. https://krishi.icar.gov.in/jspui/handle/123456789/51910

Gulab SY, Anup D, Rattan L, Subhash B, Ram SM, Somanagouda BP, Poulami S, Mrinmoy D (2018) Conservation tillage and mulching effects on the adaptive capacity of direct-seeded upland rice (*Oryza sativa* L.) to alleviate weed and moisture stresses in the North Eastern Himalayan Region of India. Arch Agron Soil Sci 31:1–16

Gyan PM, Rajnish K, Yogesh (2016) Revival of ITK for sustainable agriculture under Eastern Uttar Pradesh (India). Int J Theor Appl Sci 8(2):40–44. https://researchtrend.net/ijtas/ijtas.php

Haque S, Shekhar P, Vihan VS (2020) Traditional knowledge in agriculture. Division of Agricultural Extension Indian Council of Agricultural Research New Delhi, pp 1–52. http://zpd7icar.nic.in/download/IITKA

Hiswaran SVK (2020) Rain-water management for teak (*Tectona grandis* Linn, f.), mango (*Mangifera indica* Linn.) and neem (*Azadirachta indica* A. Juss) in arid and semi-arid regions. Traditional Knowledge in Agriculture Division of Agricultural Extension Indian Council of Agricultural Research New Delhi, pp 1–52. https://doi.org/10.1080/03650340.2018.1423555

Jangde CR, Dhanan G (2020) Traditional knowledge in agriculture. Division of Agricultural Extension Indian Council of Agricultural Research New Delhi, pp 1–52. http://zpd7icar.nic.in/download/IITKA

Jaskarn SM, Simerjeet K (2021) Package of practices for crops of Punjab. Punjab Agricultural University PAU Printing Press Ludhiana, pp 1–476. https://www.pau.edu/content/ccil

Jogaa RJ, Sangmaa E, Karmakara B, Lyngdoha V, Aochenb C (2021) Phytochemical investigations on the therapeutic properties of *Ensete glaucum* (Roxb.) Cheesman. Indian J Tradit Knowl 20(1):68–73. http://op.niscair.res.in/index.php/IJTK/article/view/26278

Jyoti K, Tripathy MK, Harekrushna D (2020) Diversity of insect and non insect pest infesting Karanja, *Pongamia pinnata* (L.) Pierre at Bhubaneswar, Odisha, India and their natural enemies. Int J Curr Microbiol App Sci 9(7):1577–1596. https://doi.org/10.20546/ijcmas.2020.907.183

Kareemulla K, Ravichandran S, Suvangi R (2020) COVID-19 pandemic and its impact on Indian and global economies. Econ Aff 65:451–458. https://doi.org/10.46852/0424-2513.3.2020.20

Karthikeyan K, Veeraragavathatham D, Karpagam D, Firdouse AS (2009) Traditional tools in agricultural practices. Indian J Tradit Knowl 8(2):212–217. https://www.cabdirect.org/cabdirect/abstract/20093162475

Kathirvelu C, Mangayarkarasi S, Kulothungan K (2019) Evaluation of phyto tablet formulation against key insect pests of stored produce under laboratory conditions. J Pharmacogn Phytochem SP2:387–390. https://doi.org/10.22271/phyto

Koodalingam K (2020) Control of rhinocerous beetle in coconut by using cowdung slurry. Traditional Knowledge in Agriculture Division of Agricultural Extension Indian Council of Agricultural Research New Delhi, pp 1–52. http://zpd7icar.nic.in/download/IITKA

Kranthi TK, Babu BR, Sankara CD, Gangadhara PR, Venkata MR (2016) Application of indigenous technical knowledge in organic fruit production. Progress Res 11(4761–4765):1–6. http://zpd7icar.nic.in/download/IITKA

Krishan C (2005) Organic manures. Production and Quality Control of Organic Inputs Kottayam Kerala, pp 1–46. https://rvskvv.net/images/Organic-Manures

Krishna KR, Suresh VB, Ravindra SG (2019) Sustainable livelihoods and adaptation to climate change, training manual. Centre for Natural Resource Management National Institute of Rural Development and Panchayati Raj Ministry of Rural Development Government of India Rajendranagar Hyderabad, pp 1–258. http://nirdpr.org.in/nird_docs/tps/tps070521.pdf

Krishnankutty J, Blakeney M, Raju RK, Siddique KHM (2021) Sustainability of traditional rice cultivation in Kerala, India—a socio-economic analysis. Sustainability 13:1–16. https://doi.org/10.3390/su13020980

Kudada N, Rabindra P, Sinha KK, Ratan RPS (2020) Control of caseworm (*Nymphula depunctatis*) in rice by leaves of parasi and sali. In: Traditional knowledge in agriculture. Division of Agricultural Extension, Indian Council of Agricultural Research, New Delhi, pp 1–52. https://icar.org.in/ITK

Kumar M, Radha DH, Prakash S, Rathore S, Thakur M, Puri S, Pundir A, Bangar SP, Changan S (2021) Ethnomedicinal plants used in the health care system: survey of the mid hills of Solan District, Himachal Pradesh, India. Plants 10:1–24. https://doi.org/10.3390/plants10091842

Lalthanpuii PB, Lalruatfela B, Zoramdinthara, Lalthanzara H (2016) Traditional food processing techniques of the Mizo people of Northeast India. Sci Vis 15(1):39–45. https://doi.org/10.1186/s42779-021-00104-5

Mahesh K (2020) Use of Stone Apple (Bael) to check diarrhoea in animals. In: Traditional knowledge in agriculture. Division of Agricultural Extension, Indian Council of Agricultural Research, New Delhi, pp 1–52. https://icar.org.in/ITK

Mangal R (2009) Research update. Indian Council of Agricultural Research Krishi Bhawan New Delhi India, pp 1–12. https://icar.org.in/dare-icar-annual-reports

Manickam S, Sankaranarayanan K, Prakash AH (2013) Relevance and techniques of organic cotton production. Central Institute for Cotton Research Regional Station Lawley Road Coimbatore Tamil Nadu, pp 1–211. https://www.cicr.org.in/pdf/Lecture_Notes_Organic_Cotton.pdf

Manish C, Surya R, Kumar B (2011) Indigenous pest management practices prevalent among hill farmers of Uttarakhnad. Indian J Tradit Knowl 10(2):311–315. http://nopr.niscair.res.in/handle/123456789/11510

Manisha K, Neelam S (2021) Plant proteins as natural, biodegradable, low cost larvicides against mosquitoes. Indian J Tradit Knowl 21(1):89–96. http://op.niscair.res.in/index.php/IJTK/article/view/28350

Manoj KV (2016) Agriculture crop health management. Vocational Higher Secondary Education, Second Year, State Council of Educational Research and Training Poojappura Thiruvananthapuram Kerala, pp 1–192. https://scert.kerala.gov.in

Manosi D, Ravindra S, Rajesh B, Avijit B, Souvik R, Jayram H (2021) Development of quality standards of Triphala Kwatha churna with its ingredients through HPTLC and mass spectroscopy. Indian J Tradit Knowl 21(1):113–119. http://op.niscair.res.in/index.php/IJTK/article/view/41049

Maria D, Nazir AP, Munesh K, Rainer WB (2017) Traditional knowledge of medicinal plants in tribes of Tripura in northeast, India. Afr J Tradit Complement Altern Med 14(4):156–168. https://doi.org/10.21010/ajtcam.v14i4.19

Maruthi V, Subba GR (2020) Traditional knowledge in agriculture. Division of Agricultural Extension Indian Council of Agricultural Research New Delhi, pp 1–52. http://zpd7icar.nic.in/download/IITKA

Marziyeh C, Dzolkhifli O, Mahiran B, Norhayu A (2017) Preparation and characterization of neem oil nano emulsion formulations against *Sitophilus oryzae* and *Tribolium castaneum* adults. J Pestic Sci 42:158–165. https://doi.org/10.1584/jpestics.D17-032

Mayabini J, Dangar TK (2020) Planting of wild *Saccharum spontaneum* in paddy field for controlling leaf-folder. In: Traditional knowledge in agriculture. Division of Agricultural Extension, Indian Council of Agricultural Research, New Delhi, pp 1–52. https://icar.org.in/ITK

Mishra PK, Osman M, Satendra, Venkateswarlu B (2011) Techniques of water conservation & rainwater harvesting for drought management. SAARC Training Program 18–29 July, 2011 Central Research Institute for Dryland Agriculture Hyderabad India, pp 1–720. https://krishi.icar.gov.in/jspui/bitstream

Modupe SA, Oluwaseyi SO, Olubukola OB, Olu O (2020) Waste management through composting: challenges and potentials. Sustainability 12:1–23. https://doi.org/10.3390/su12114456

Nautiyal P, Pokhriya P, Panwar P, Arunachalam K, Arunachalam A (2017) Cropping systems and agro-management practices in relation to soil carbon in the temperate regime of Uttarakhand. Int J Environ Rehabil Conserv 8:60–72. https://essence-journal.com/2017/09/24

Nazish N (2013) Knowledge of traditional fermented food products harbored by the tribal folks of the Indian Himalayan Belt. Int J Agric Food Sci Technol 4:401–414. https://www.ripublication.com/ijafst_spl/ijafstv4n5spl_03.pdf

Ngachan SV (2019) Hill agriculture in North East India: status and option for sustainable development. ICAR Research Complex for NEH Umiam Meghalaya, pp 1–194. http://www.kiran.nic.in/pdf/IJHF/Special_Issue_2020/3.pdf

Nijra B, Luke D (2017) The traditional agricultural tools and technology used by the Bodos. J Hum Soc Sci 22(5):65–72. https://www.iosrjournals.org/iosr-jhss/papers

Pandey DK, Singh PK, Adhiguruc P, Kumar P (2021) Vocal for local: diversity of local cucurbits cultivars in East Siang, Arunachal Pradesh. Indian J Tradit Knowl 20(1):1–6. http://op.niscair.res.in/index.php/IJTK/article/view/28502

Parida S, Sarangi M (2021) Medicinal uses of few edible tuber crops by "Dongria Kandha" tribes of Kandhamal district of Odisha, India. Indian J Tradit Knowl 20(1):122–131. http://nopr.niscair.res.in/bitstream/123456789

Parimala K, Subramanian K, Mahalinga KS, Vijayalakshmi K (2013) Seed storage techniques—a primer. Centre for Indian Knowledge Systems Chennai Revitalising Rainfed Agriculture Network, pp 1–25. http://ciks.org/wp-content/

Patel DP, Anup D, Munda GC (2015) Physiological efficiency of some weeds species under hill farming systems of subtropical Meghalaya. International Grassland Congress on Sustainable use of Grassland Resources for Forage Production, Biodiversity and Environmental Protection Range Management Society of India New Delhi India, pp 1–4. https://uknowledge.uky.edu/cgi/viewcontent

Patel SJ, Patel AS, Patel JH, Patel NR, Parmar VN (2016) Indigenous technical knowledge regarding animal husbandry practices. Adv Life Sci 5(5):1610–1617. http://advancesinlifesciencesjournal.com

Patnaik HP (2011) Indigenous technical knowledge (itk) in pestmanagement for sustainable agriculture. ENVIS Centre of Odisha's State of Environment, Ministry of Environment Forests & Climate Change Govt of India, pp 1–8. http://orienvis.nic.in

Prabuddh KM, Aman R, Suresh CR (2020) Indigenous knowledge of terrace management for soil and water conservation in the Sikkim Himalaya, India. Indian J Tradit Knowl 19(3):475–485. http://op.niscair.res.in/index.php/IJTK/article/view/41446

Pradeep KS, Kapur OC, Masand SS (2020) Traditional knowledge in agriculture. Division of Agricultural Extension Indian Council of Agricultural Research New Delhi, pp 1–52. http://zpd7icar.nic.in/download/IITKA

Prakasa REVS, Srinivas A, Gopinath CT, Ravindra NS, Aparna H, Nagendra P (2015) Sustainable agriculture reviews. Springer International Publishing Switzerland Lichtfouse (ed.), pp 1–19. https://link.springer.com/book

Prakash BG, Raghavendra KV, Gowthami R, Shashank R (2016) Indigenous practices for eco-friendly storage of food grains and seeds. Adv Plants Agric Res 3(4):101–107. https://doi.org/10.15406/apar.2016.03.00101

Premlata T, Vivek S, Madho SB, Chongtham N (2020) Edible bamboo resources of Manipur: consumption pattern of young shoots, processing techniques and their commercial status in the local market. Indian J Tradit Knowl 19(1):73–82. http://op.niscair.res.in/index.php/IJTK/article/view/30843

Rajendra RC, Prabhakar G, Shyam P, Das IK, Vilas AT (2018) Improved millets production technologies and their impact. ICAR-All India Coordinated Research Project on Small Millets ICAR-Indian Institute of Millets Research Rajendranagar Hyderabad, pp 1–88. https://www.millets.res.in/annual_report/ar18-19.pdf

Rajesh K, Bharathi KA (2012) Folk veterinary medicine in Sitapur distrct of Uttar Pradesh, India. Indian J Nat Prod Resour 3(2):267–277. http://nopr.niscair.res.in/bitstream/123456789

Rakesh CM, Vasudevan SN, Naveen CM, Patil SB (2013) Traditional seed treatment and storage methods of Northeastern Region of Karnataka. Asian Agri-History 17:233–239. https://www.academia.edu/7905371

Rakesh K, Khesi Y, Rajesha G, Bidyut CD (2017) Performance of jobs tears lines (*Coix lacryma-jobi*) under food hill condition of Nagaland. Environ Ecol 35(1B):440–444. https://krishi.icar.gov.in/jspui/bitstream

Ram N, Dinker S, Yadav SM, Balai LP (2013) Traditional wisdom for diseases treatment in animal husbandry. Popular Kheti 1(2):1–9. http://www.popularkheti.com/documents/2013-2/PK-1-2-7-30-38.pdf

Ramyajit M, Saumi G (2019) Sesbania—an important aspect of INM for sustainable productivity of rice: a review. Int J Chem Stud 7(6):2832–2837. https://www.chemijournal.com/archives

Rana SS (2011) Organic farming. Department of Agronomy College of Agriculture CSK Himachal Pradesh Krishi Vishvavidyalaya Palampur, pp 1–90. http://www.hillagric.ac.in

Ranjay KS, Dwivedi BS, Anshuman S, Sarvesh T (2014) Farmers' knowledge and creativity in eco-friendly pest management: lessons in sustainable agriculture. Indian J Tradit Knowl 13(3): 574–581. http://nopr.niscpr.res.in/handle/123456789/29130

Ranjay KS, Legoc YJ, Surejaa AK, Srivastavae RC, Hazarikaa BN (2021) People and plant: learning with Adi community on ethnomedicinal practices and conservation in Arunachal Pradesh, India. Indian J Tradit Knowl 20(1):74–82. http://nopr.niscair.res.in/bitstream/123456789

Ratan KS, Dilip N (2013) Indigenous Technical Knowledge (ITK) of fish farmers at Dhalai district of Tripura, NE India. Indian J Tradit Knowl 12(1):80–84. http://nopr.niscair.res.in/bitstream/123456789

Ravi VRK (2021) Agro India. Keltech India Limited, Infantry Road Bangalore, pp 1–41. https://www.keltechenergies.com

Ravisankar N, Panwar AS, Kamta P, Vipin K, Bhaskar S (2017) Organic farming crop production guide. Network Project on Organic Farming Indian Institute of Farming Systems Research Modipuram Meerut Uttar Pradesh India, pp 1–602. https://ncof.dacnet.nic.in

Reddy MR, Veena S, Savalgi VP (2008) Nutritional profile of millet tempeh. Asian J Bio Sci 3: 355–360. http://www.researchjournal.co.in/upload/assignments/3_355-360.pdf

Richa V, Jalali SK, Ramanujam B, Chandish RB (2017) Proceedings of the XXVI Biocontrol Workers' Group Meeting and Technical Programme. Yashwant Singh Parmar University of Horticulture and Forestry Nauni Solan Himachal Pradesh, pp 1–93. https://www.nbair.res.in

Rohini R (2010) Traditional practices in agriculture. SARRA, pp 1–151. https://www.researchgate.net/publication/

Roy BK, Varshney AC (2020) Traditional knowledge in agriculture. Division of Agricultural Extension Indian Council of Agricultural Research New Delhi, pp 1–52. http://zpd7icar.nic.in/download/IITKA

Ruparao TG, Gadi VPR (2018) Management of insect pests in the production and storage of minor pulses. Ann Entomol Soc Am 111(4):172–183. https://agris.fao.org/agris-search

Sahu KC, Patro B, Kar AK (2022) Quality of paddy seeds stored by farmers of Orissa. Environ Ecol 26:1709–1712. https://www.cabi.org/isc/abstract/20093002105

Sandhya NS (2022) The Indigenous Technical Knowledge (ITK) & its application for sustainability in agriculture. XSM Division NAARM, pp 1–3. https://morungexpress.com

Santosh T, Chhetry GKN (2012) Agro-biodiversity management related ITKs in North-Eastern India. J Biol Agric Healthcare 2(6):1–11. https://doi.org/10.7176/JBAH

Sarita D, Dash SK, Padhy SN (2003) Ethno-medicinal Informations from Orissa State, India. A review. J Hum Ecol 14(3):165–227. https://doi.org/10.1080/09709274.2003.11905616

Sarodee B, Sanjoy B, Neog M (2020) Indigenous technological knowledge in pest and disease management of agricultural crops—a review. Int J Curr Microbiol App Sci 9(9):2867–2876. https://doi.org/10.20546/ijcmas.2020.909.354

Satyagopal K, Sushil SN, Jeyakumar P, Shankar G, Sharma OP, Sain SK, Boina DR, Srinivasa NR, Sunanda BS, Ram A, Kapoor KS, Arya S, Kumar S, Patni CS, Chattopadhyay C, Jacob TK, Jadav RG, Shukla A, Bhale U, Singh SP, Khan ML, Sharma KC, Dohroo NP, Suseela BK, Santosh JE, Hanumanthaswamy BC, Srinivas KR, Thakare AY, Halepyati AS, Patil MB, Sreenivas AG (2014) AESA based IPM package for turmeric. National Institute of Plant Health Management Rajendranagar Hyderabad, p 41. https://link.springer.com/book

Satyapriya S, Biswajit D, Anup D, Sujan M, Hidangmayum LD, Ranjeet SG, Alok KS, Manas RS (2021) Indigenous plant protection practices of Tripura, India. J Ethnobiol Ethnomed 1–24. https://doi.org/10.1007/s40974-020-00158-2

SCERT (2016) Agriculture science and processing technology, Reference Book. Department of Education Government of Kerala, pp 1–189. https://scert.kerala.gov.in

Shaila M, Nafeesa B (2021) Ancient farming methods of seed storage and pest management practices in India—a review. Plant Arch 21:499–509. https://www.semanticscholar.org

Shakrawar M, Naberia S, Pande AK (2018) Indigenous technical knowledge for pest, disease and weed management in agriculture. Int J Chem Stud 6(4):497–498. https://doi.org/10.22271/chemi

Shamkuwar SV, Baral SS, Budhe V, Gupta KP, Swarnkar R (2019) A critical study on weed control techniques. Int J Adv Agric Sci Technol 6(12):1–22. http://ijaast.com/publications/vol6issue12/V6I1201.pdf

Shenoy NS (2021). Indigenous technical knowledge and its relevance for sustainability. ICAR-National Academy of Agricultural Research Management Rajendranagar Hyderabad, pp 1–18. https://naarm.org.in

Shobha RT, Kumar TS, Nagabhushanam U, Jaganmohan RP (2020) Rainfed evaluation of genotypes for seed yield and yield components in sunhemp (*Crotalaria juncea* L.) and dhaincha (*Sesbania aculeata* L.). Int J Curr Microbiol App Sci 9:1022–1029. https://doi.org/10.20546/ijcmas.2020.904.121

Shivankar VJ, Rao CN, Shyam S (2020) Bio-efficacy and selectivity of bio-pesticides against citrus insect pests and their natural enemies. J Eco-friendly Agric 1(1):29–31. https://doi.org/10.21608/eajbsf.2015.17243

Shubeena S, Hai A, Hamdani SA, Akand AH (2018) Indigenous Technical Knowledge (ITKs) used by farmers of central Kashmir to increase production and reproduction in livestock. Int J Livestock Res 8(8):294–302. https://doi.org/10.5455/ijlr.20171004030110

Sibisan N (2019) Traditional agricultural tools of Boros: used in rice cultivation. Rev Res 8(6):1–5. http://oldror.lbp.world/UploadedData/7745.pdf

Singh MS (2018) Influence of rice husk ash on controlling insect pests on storage of maize (*Zea maize* L.) seeds under Manipur condition. Int J Sci Res 9:1466–1467. https://www.ijsr.net/archive/v9i8/SR20825162356.pdf

Singh KD, Mathew B (2021) Ethnobotanical study on wild edible fruits, spices and aquatic plants traditionally used by the Garo tribe of Meghalaya. Indian J Tradit Knowl 20(1):1–5. http://op.niscair.res.in/index.php/IJTK/article/view/28366

Singh AB, Teron R (2017) Ethnic food habits of the Angami Nagas of Nagaland state, India. Int Food Res J 24(3):1061–1066. http://www.ifrj.upm.edu.my/.../(22).pdf

Sinha KK, Singh RPR (2020) Control of gall fly (*Pachydiplosis oryzae*) in rice. In: Traditional knowledge in agriculture. Division of Agricultural Extension, Indian Council of Agricultural Research, New Delhi, pp 1–52. https://icar.org.in/ITK

Sofia K, Deepa RS (2020) Traditional knowledge and its efficacy in economic growth. Int J Law Manag Human 3:1–14

Somasundaram J, Sinha NK, Ram CD, Rattan L, Mohanty M, Naorem AK, Hati KM, Chaudhary RS, Biswas AK, Patra AK, Chaudhari SK (2020) No-till farming and conservation agriculture in South Asia—issues, challenges, prospects and benefits. Crit Rev Plant Sci 39(3):236–279. https://doi.org/10.1080/07352689.2020.1782069

Somasundaram E, Nandhini DU, Meyyappan M (2021) Principles of organic farming. In: Earth sciences environment & agriculture, 1st edn. CRC Press, London, pp 1–421. https://doi.org/10.1201/9781003260844

Subash S (2017) Natural plant products—As protectant during grain storage: a review. J Entomol Zool Stud 5(3):1873–1885. https://doi.org/10.22271/j.ento

Sumit S, Shivani R (2021) Indigenous Technical Knowledge (ITK) for sustainable agriculture in India. Agric Food Newslett 3:31–35. https://agris.fao.org

Surinder SR, Suresh K, Man CR (2018) Practical manual on weed management. Department of agronomy CSK Himachal Pradesh Krishi Vishvavidyalaya Palampur, pp 1–79. http://hillagric.ac.in/edu/coa/agronomy

Surya R, Manish C, Rupan R, Manmeet K, Kundan VS (2021) Indigenous pest management practices of Indian hill farmers: introspecting their rationale and communication pattern for secure ecosystems. Sustainability 13:1–17. http://hillagric.ac.in/edu/coa/agronomy

Swain SK, Dash AK, Mohapatra AK, Das DM, Behera D, Nayak BR, Mohapatra M (2020) Effect of mechanization on cost-economics of maize cultivation by small farmers of Gajapati District, Odisha. Int J Chem Stud 8(4):3103–3107. https://doi.org/10.22271/chemi.2020.v8.i4al.10126

Swamy SGS, Sondur SN, Venkatesh KN, Savitha G, Prajanya GP (2015) Student project programme, biofuel projects. Karnataka state council for science and technology Indian Institute of Science Campus Bangalore, pp 1–203. http://www.kscst.iisc.ernet.in

Swarup D, Dhakate MS (2020) Treatment of swelling of shoulders in bullocks or bulls. In: Traditional knowledge in agriculture. Division of Agricultural Extension, Indian Council of Agricultural Research, New Delhi, pp 1–52. http://zpd7icar.nic.in/download/IITKA

Swarup D, Mahesh K (2020) Determination of efficacy of pigeon waste in showing oestrus symptoms in heifers. In: Traditional knowledge in agriculture. Division of Agricultural Extension, Indian Council of Agricultural Research, New Delhi, pp 1–52. http://zpd7icar.nic.in/download/IITKA

Swarup D, Pradhan NR (2020) Use of Babool (*Acacia nilotica*) and Jamun (*Syzygium cumini*) bark extract to cure foot. In: Traditional knowledge in agriculture. Division of Agricultural Extension, Indian Council of Agricultural Research, New Delhi, pp 1–52. http://zpd7icar.nic.in/download/IITKA

Swarup D, Mahesh MS, Satish K, Mahesh C, Mahesh K, Prasad JK (2020) Determination of efficacy of pigeon waste in showing oestrus symptoms in heifers. In: Traditional knowledge in agriculture. Division of Agricultural Extension, Indian Council of Agricultural Research, New Delhi, pp 1–52. https://icar.org.in/sites/default/files/IITKA_Book_%28Traditional%20Knowledge%20in%20Agriculture-English%20%29_0.pdf

Thangjam AS, Prakash KS, Joykumar S (2018) Traditional process foods of the ethnic tribes of western hills of Manipur, India. Int J Curr Microbiol App Sci 7(10):1100–1110. https://doi.org/10.20546/ijcmas.2018.710.121

Usharani KV, Dhananjay N, Manjunatha RL (2019) *Pongamia pinnata* (L.): composition and advantages in agriculture: a review. J Pharmacogn Phytochem 8(3):2181–2187

Vanaja R, Srikanthamurthy HS, Ningappa K, Shivakumar BN, Ningaraju SD, Vijay AR, Shivanna M, Obanna N, Pandu AC, Satish R, Madhu S, Pacchapur V, Suma S (2009) A handbook on "sustainable agricultural practice". Green Foundation, Bangalore, pp 1–101. https://greenfoundation.in/publications_eng/

Varshney AC (2020) Traditional knowledge in agriculture. Division of Agricultural Extension, Indian Council of Agricultural Research New Delhi, pp 1–52. http://zpd7icar.nic.in/download/IITKA

Verma RS, Khan P, Ayub S, Afza S, Akhtar J, Ahmad S, Khan AA (2021) Efficacy and safety of a Unani compound drug—Habb-e-Asgand in Wajaal-Mafasil (Rheumatoid Arthritis) cases—a preliminary study. Indian J Trad Knowl 20:1–7. https://doi.org/10.56042/ijtk.v20i1.27232

Vihan VS (2020) Traditional knowledge in agriculture. Division of Agricultural Extension Indian Council of Agricultural Research New Delhi, pp 1–52. http://zpd7icar.nic.in/download/IITKA

Vishal S, Deepak KV, Prem PS (2021) Food grain storage structures: introduction and applications. Engineering Interventions in Foods and Plants, pp 1–39. https://www.phytojournal.com/archives

Yadav SK, Babu S, Yadav MK, Singh K, Yadav GS, Suresh P (2013) A review of organic farming for sustainable agriculture in Northern India. Int J Agron 18:1–9. https://doi.org/10.1155/2013/718145

Yallappa R, Nandagopal B, Thimmappa S (2012) Botanicals as grain protectants. Psyche: J Entomol 2012:646740. https://doi.org/10.1155/2012/646740

Zizira (2015) Organic pineapples provides livelihood for a Meghalaya farmer. Exotic Fruits, Meghalaya, India. https://www.zizira.com/blogs/people-and-process/tagged/exotic-fruits; https://www.zizira.com

The manufacturer's authorised representative in the EU is Springer Nature Customer Service Centre GmbH, Europaplatz 3, 69115 Heidelberg, Germany. If you have any concerns regarding our products, please contact ProductSafety@springernature.com

Printed and bound by CPI Group (UK) Ltd, Croydon, CR0 4YY
30/09/2026
02225314-0012